Flora of Mount Rainier National Park

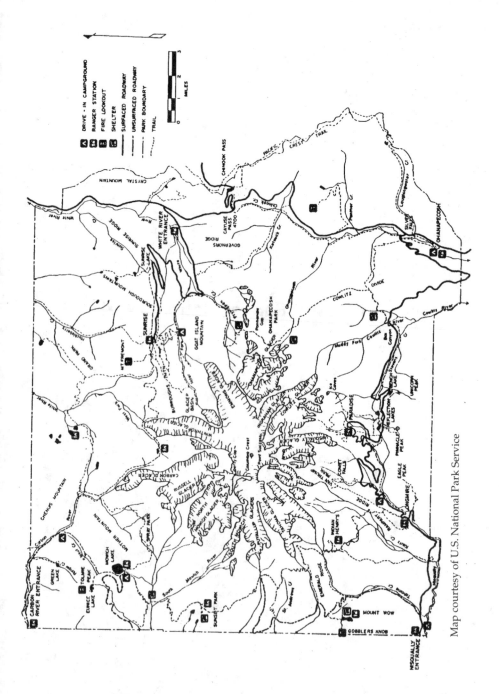

Map courtesy of U.S. National Park Service

The home page for *The Flora of Mount Rainier* is at
www.wolfe.net/~biek

Flora of Mount Rainier National Park

by

David Biek

Oregon State University Press
Corvallis

Substantial gifts from the following donors
helped make publication of this book possible.
The Oregon State University Press
is grateful for their support.

JiJi Foundation

Northwestern Chapter, North American Rock Garden Society

*Dedicated to the memory of James C. Hickman (1941–1993),
editor of The Jepson Manual. He wrote to me after I had moved
to Washington and said that, during the period he taught at
Washington State University, he had contemplated a revision of
Jones's flora. I hope Jim would have enjoyed this book.*

The line drawings are reprinted by permission of the University of
Washington Press, from *Vascular Plants of the Pacific Northwest,* by C.L.
Hitchcock, copyright © 1959-1967 by the University of Washington Press.

The paper in this book meets the guidelines for permanence and durability
of the Committee on Production Guidelines for Book Longevity of the
Council on Library Resources and the minimum requirements of the
American National Standard for Permanence of Paper for Printed Library
Materials Z39.48-1984.

Library of Congress Cataloging-in-Publication Data
Biek, David.
    Flora of Mt. Rainier National Park/by David Biek—1st ed.
      p. cm.
    Includes bibliographical references (p.)
    ISBN 0-87071-470-8 (alk. paper)
    1. Botany—Washington (State)—Mount Rainier National Park. 2. Plants—
Identification. 3. Mount Rainier National Park (Wash.). I. Title.
QK192.B54 1999
581.979'782—dc21   99-057564

Oregon State University Press
101 Waldo Hall, Corvallis OR 97331-6407
541-737-3166 •fax 541-737-3170
http://osu.orst.edu/dept/press

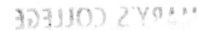

Contents

Acknowledgments

Regina Rochefort, Park Botanist, provided access to the herbarium at Mount Rainier National Park, and answered a great number of questions I had. She also provided a number of significant leads to other researchers. Ted Stout, collections manager for the Park, was often on hand to let me into the herbarium.

Access to herbarium collections was made possible through arrangements made by Sarah Gage at the University of Washington, Larry Hufford and Linda Cook at Washington State University, John Main at Pacific Lutheran University, and Dennis Paulson and Kathy Ann Miller at the University of Puget Sound.

Peter Dunwiddie lent me copies of his Mount Rainier field books and shared other valuable information from his research in the Park.

A number of correspondents offered advice and information on special topics, including Bruce Barnes, Bill Burley, Jerry Davison, Mark Egger, John Kartesz, Betsy Kirkpatrick, Aaron Liston, Richard Olmstead, and W. H. Wagner. Ed Alverson reviewed the section on the ferns.

Mary Fries has hiked the Park and studied its flora for decades, and was most generous in sharing her knowledge, including her views on common names. Cathy Maxwell, who has researched the life and work of J. B. Flett, answered several questions.

Gary Reese, of the Special Collections department of the Tacoma Public Library, guided me through the collections of historical information related to the Mountain and answered many questions about place names.

Nancy Hori, head of the National Park Service Library in Seattle, was very helpful, and the interlibrary loan staff at the Tacoma Public Library provided many of the books and articles cited in the bibliography.

The marvelous line drawings of Jeanne R. Janish appear courtesy of the University of Washington Press.

Introduction

"Every one of these parks, great and small, is a garden filled knee-deep with fresh, lovely flowers of every hue, the most luxuriant and the most extravagantly beautiful of all the alpine gardens I ever beheld in all my mountain-top wanderings."

Thus wrote John Muir, describing the subalpine meadows of Mount Rainier. Words of high praise from a man who spent most of his life trekking across North America, from the northwoods of Wisconsin, to the Appalachians, west to the "Range of Light," California's Sierra Nevada, and beyond it to the "sweet bee-garden" of the Great Valley. In 1888, at the age of fifty, with much of his roaming behind him, Muir made an ascent of Mount Rainier, in the company of Philemon Van Trump and seven others. Leaving the deep forests at the Mountain's feet, Muir and his party departed the Inn at Longmire, climbed the ridge bordering the Nisqually River, and passed through the flower fields of the Paradise valley, the greatest of the Mountain's "parks." The party bivouacked high above the Paradise meadows, on the south side of the Mountain at 10,062 feet, a place now known as Camp Muir.

Muir's voice was one of those calling for the establishment of a national park to preserve the beauty and grandeur of Mount Rainier, and on March 2, 1899, by an act of Congress, Mount Rainier National Park was founded.

In 1893, land in the area of the present national park had been set aside as the Pacific Forest Reserve (called on some maps the Mount Rainier Forest Reserve). Additions of adjacent land were made through 1931, but by 1945 it had become clear that aggressive logging to the north, south, and west of the parcel precluded the incorporation of any more land within the National Park.

Although at the present time some of the Park is surrounded by National Forests and Wilderness Areas, much neighboring land is owned by private timber companies, and in places their clearcuts reach to the Park boundary. The Park, to an unhappy extent, is an island in the midst of a landscape that has been greatly modified by human activities over the past one hundred years, a biogeographical island where plant communities are preserved that outside the Park are becoming more rare.

The most recent comprehensive flora, *The Flowering Plants and Ferns of Mount Rainier*, was published in 1938 by George Neville Jones, of the University of Washington. Since then, a number of books have been written, designed to acquaint visitors and students of natural history with the flora of the Park. Park Naturalist C. Frank Brockman published a popular flora in 1947, and beautifully illustrated flower books continue to appear, books such as *Wildflowers of Mount Rainier and the Cascades*, by Mary Fries, with photographs by Bob and Ira Spring; *Mountain Plants of the Pacific Northwest*, by Ronald Taylor and George Douglas; and *Wayside Wildflowers of the Pacific Northwest*, by Dee Strickler.

1

Since the publication of Jones's 1938 flora, many additional plants have been discovered in the Park and numerous changes in nomenclature have been made. Meanwhile, decades of geological, ecological, and botanical research have greatly increased our understanding of the plants and plant communities of Mount Rainier. A new flora of the Park, therefore, is needed. It is my hope that the present work will serve as well and as long as Professor Jones's work did.

The flora of Mount Rainier National Park comprises 871 species, subspecies, and varieties. The 723 that are native plants constitute about 30% of the total native flora of Washington state. The land area of the Park is just under 0.6% of the state's 66,511 square miles, making Mount Rainier National Park an excellent place to see a significant portion of the flora of the state.

Throughout this book you will find Mount Rainier referred to as "the Mountain." The use of the noun as a proper name goes back a long way and is meant to convey awe and familiarity, respect and pride.

How to Use This Book
Scope and Intent

My aim in this treatment of the flora of Mount Rainier National Park is to provide a complete list of the native and introduced plants found in the Park, with keys and descriptions sufficient to distinguish each species from the others. Some species get more descriptive detail than others, for the keys serve also to provide descriptive data. The descriptions do not, however, constitute complete data on each plant. Such information is widely available in a number of works cited in the bibliography.

The keys are designed to work for the plants known to occur in the Park and should not be relied on to provide trustworthy results in regions away from the Park. The plants listed have been verified by examination of herbarium specimens, field observations, and reliable reports. No plant is listed solely because it is "expected to occur" in the Park, although a small number of such plants are mentioned at appropriate places in the text.

Much research has been done in the Park since 1938, when George Neville Jones published *The Flowering Plants and Ferns of Mount Rainier*. Many additional plants have been discovered, among them an unfortunate number of non-native plants (sometimes called "weeds") that have become established in the Park. Seventy-five plants not previously recorded as growing in the Park were noted in the course of the preparation of the book. I also show that 34 species attributed to the Park's flora, by Jones and by other writers, are not actually to be found on Mount Rainier. The new and excluded species are discussed in an appendix.

This new flora, then, is both a commemoration of the centennial of the founding of Mount Rainier National Park and, I hope, a contribution toward a deeper understanding of the natural systems preserved within the Park.

Introduction

Information about Each Plant

Each entry for a plant shows the name of the plant in boldface italic. The name of the genus and species is followed by an abbreviation of the name of the author of that scientific name. Common names are included, and largely follow Hitchcock and Cronquist (1976), supplemented by names used by Nelsa Buckingham (Buckingham et al., 1995).

The descriptions are written with a minimum number of technical terms. The glossary (page 477) should be consulted for unfamiliar terms, and from professional botanists I ask indulgence where I use common English terms instead of precise scientific language. Each species is described, and when there is more than one species in a genus the essential characteristics of that genus are described as well. The same is true when there is more than one genus in a family.

A general statement describes habitat and ecological information about the plant. This is followed by a list of places where the plant is known to occur, with an emphasis on those places within easy reach from a road. Location information came from herbarium collections, reliable reports, and my own observations; the locations given are not meant to completely represent the occurrence of the plant, but should serve to help predict other places in which the plant might be found.

A common name is given for each species. Some of these names enjoy wide recognition and use, such as mountain hemlock, red alder, and thimbleberry. Others are less well-known but still genuinely common, such as white-vein shinleaf and clustered rose. A third group of common names are those that are essentially translations of the scientific names, such as Fendler's waterleaf and creeping sibbaldia. In some cases, the common name is a compound hyphenated name, such as dog-fennel (*Anthemis*) or marsh-marigold (*Caltha*). This convention is followed to show that these plants may resemble but are not, respectively, a true fennel (*Foeniculum*) or a true marigold (*Tagetes*).

Measurements of plants are expressed in metric units. Measurements of distance and elevation are in the common English units found on maps and odometers. A metric scale is printed on the back cover for easy reference.

The following abbreviations are used:

cm	centimeter	Cr	Creek	fl	flower
ft	foot	Hwy	Highway	fls	flowers
mm	millimeter	Lk	Lake	lf	leaf
m	meter	Lks	Lakes	lvs	leaves
ssp.	subspecies	Mtn	Mountain		
var.	variety	R	River		
		Rd	Road		
		V	Valley		

Nonvascular plants—mosses, liverworts, and lichens — are prominent members of the Park's flora but are not included in this book. The best available guidebooks for their study are *Mosses, Lichens and Ferns of Northwest North America* by Dale H. Vitt (1993) and *Macrolichens of the Pacific Northwest* by Bruce McCune and Linda Geiser (1997).

Using the Keys

Each key consists of one or more pairs of contrasting statements, each pair called a couplet. Observe the specimen in light of the characteristics described in the statements and choose the statement that best describes the specimen. At this point, you may have the name of a species, or you may have a number. The number refers to another couplet, and so it proceeds until you've "keyed out" the plant. Keying out isn't the same thing as identification, however, and the final step in the process always is to read the full description of the plant, compare it to the illustration, and be sure of the match.

You may find that the description doesn't seem to fit; in that case, it is necessary to work your way back through the key. Consider this example from the key to the huckleberries:

7(6) Leaf margins markedly toothed from base to tip; tips of the leaves long-pointed .... *V. membranaceum*

7 Leaf margins nearly toothless to weakly toothed, generally below the midpoint; tips of the leaves short-pointed to rounded 8

Suppose you've arrived at couplet 7 and selected the first statement, but discovered after reading the description of *Vaccinium membranaceum* that the plant fails in some way to match the description. You would go on to try couplet 8, but it might happen that no species you find further in the key fits either. You know you will have to backtrack, but to what point? The "6" in parentheses indicates that you reached couplet 7 from a reference in couplet 6. Refer back to couplet 6 and reconsider the choice you made there. In a short key, this back reference number may seem unnecessary, but in a long key it is easy to lose track. Even in a short key, the reference will be helpful for those new to the use of identification keys.

Within each major division (except for the ferns), families are listed alphabetically. This is also true for genera within families and species within genera.

Plant Names

Many of the names used by Jones in 1938 are no longer accepted as correct. The plants haven't changed, but botanists' concepts of how to define species and how species are related to each other have, and changes of this sort are often accompanied by changes in names. Even the most recently published flora for the region, Hitchcock and Cronquist's *Flora of the Pacific Northwest* (1976), is out of date with respect to advances in botany. (Indeed, this book, too, will be out of date the instant it is printed. To keep abreast of changes of this sort, consult the sources listed below.)

The authority for plant names for this book, at the level of genus, species, and subspecies or variety, is *A Synonymized Checklist of the Vascular Flora of the United States, Canada, and Greenland* by John T. Kartesz (second edition, Timber Press, 1994). This work is out of print but is widely available through libraries. Throughout the text, unless otherwise noted, a reference to "Kartesz" is a reference to the 1994 edition. A revision, with changes incorporated by Kartesz through the summer of 1998, is not available in print, but will be found on the World Wide Web at: <http://www.csdl.tamu.edu/FLORA/newgate/cr1famzz.htm>.

The PLANTS database established by agencies of the U.S. Department of Agriculture (USDA) as the standard for the federal government uses Kartesz's 1998 online work as its name authority. The PLANTS database provides information on the geographic occurrence of plants, wetland status, other basic information, and photographs. It, too, does not exist in print, but may be consulted at: <http://plants.usda.gov/plants>.

Botanical nomenclature, the scientific names applied to plants, has been subject to change, and the literature on the flora of the Park is replete with names that are no longer accepted for use. Many of these are synonyms, while some are valid names that have been used incorrectly (or "misapplied") for plants at Mount Rainier. The index to this book includes entries for synonyms and misapplied names.

Family names and definitions follow the system proposed by James L. Reveal. A number of systems for the classification of higher groups of plants have been proposed to replace the system used in Hitchcock and Cronquist's flora for the Northwest. Those interested in matters of classification above the level of the genus may want to examine the USDA's APHIS concordance of family names at: <http://www.inform.umd.edu/PBIO/usda/usdaindex.html>.

At a few particular points, I have chosen not to follow these authorities. In each such case, I present my reasoning, which is usually based on advice from experts on a species or group of species, especially those who are writing treatments for the *Flora of North America*. This massive project has published three volumes to date (for the ferns, conifers, and two subclasses of the flowering plants). Published text may be found on the Web at: <http://www.fna.org/Libraries/plib/WWW/online.html>.

The full name of each species always includes the name of the author, the scientist who first published the name and description of that species. For example, *Draba aureola* S. Watson was first described by Sereno Watson. Authors' names are standardized as well, and are taken from "W3 Tropicos" <http://mobot.mobot.org/Pick/Search/pick.html>, which follows and updates Brummit and Powell (1992).

About the Park
Location and Topography

Mount Rainier National Park is located in Pierce and Lewis Counties, Washington, about 55 miles southeast of downtown Seattle. Established by an Act of Congress as the nation's fifth national park on March 2, 1899, the original boundaries formed a square, 18 miles on each side. Earlier, in 1893, Congress had established the Pacific Forest Reserve, a much larger tract of land whose western border fell just east of Mowich Lake and whose eastern border was about halfway between the Cascade Divide and Yakima. Following the establishment of the Park, the southern border was expanded to include the Ohanapecosh area and the eastern border was drawn along the Cascade Divide. Today, the Park covers 235,612 acres, or 368 square miles.

The summit of Mount Rainier, the highest of the Cascade volcanoes, is located at 46°51" north latitude and 121°46" west longitude. The height of the Mountain has been disputed: the U.S. Geological Survey gives the height as 14,410 feet, whereas the Washington Centennial Commission, using satellite instrumentation, set the figure at 14,411 feet. The Mountain itself occupies about one-third of the area of the Park, and one-quarter of the land in the Park is bare alpine rock, ice, and permanent snowfields. Although dwarfed by the mass of Rainier, Little Tahoma Peak, on the east side of the Mountain, is the third highest peak in Washington State, at 11,138 feet.

The major entrances to the Park lie between 2 and 2.5 miles below the summit: The Ohanapecosh and Carbon River entrances are both at 1,760 feet, the Nisqually entrance is at 2,003 feet, and the White River entrance is at 2,725 feet.

Much of the Park is surrounded by National Forest land: the Mount Baker–Snoqualmie National Forest on the north and east sides, the Gifford Pinchot National Forest on the south. Four wilderness areas also abut the Park: the Glacier View Wilderness at the southwest corner, the Clearwater Wilderness on the northwest, the Naches Peak Wilderness on the northeast, and the William O. Douglas Wilderness along much of the eastern border of the Park. A good portion of the western border, in Pierce County, is privately owned and much has been clearcut. Indeed, the view of this side presented by satellite photography shows a stark border between the deep green forests of the Park and the lighter green of cut-over and second-growth woodlands.

The Park itself lies wholly to the west of the Cascade Divide. Mount Rainier is located a short distance west of the center of the Park, and its snow and ice give rise to each of the major rivers of the Park. These are, beginning on the north side and moving counterclockwise around the Mountain, the White, Carbon, Mowich, the north and south forks of the Puyallup, Nisqually, Paradise, Cowlitz, and Ohanapecosh Rivers. The latter two rivers join and meet the Columbia River at Longview, Washington, downstream from Portland, while the others all reach Puget Sound.

The valleys of these rivers are narrow and steep-sided on the flanks of the Mountain, wider and broader as the lowlands are reached. They can lie 1,000 to 3,000 feet below the rocky ridges that divide them. Rivers flow swiftly and there are relatively few lakes in the Park. Those lakes that are found tend to be small and located on the periphery of the Park. The largest is Mowich Lake, at almost 123 acres.

Ninety-seven percent of the Park is designated wilderness; areas reachable by road are few, and were made even fewer in the 1990s by recurring damage to the road approaching the Carbon River entrance and to Westside Road. The Park's annual count of about two million visitors is funneled, then, into a relatively small amount of space, some of the consequences of which are described in subsequent sections of the Introduction.

Climate

The location of the Park on the west side of the Cascade Divide is significant and well illustrated by the dramatic change in vegetation one sees along the road east from Chinook Pass. Although Mount Rainier is located at a latitude roughly equal to that of Michigan's Upper Peninsula, the area enjoys a moderate climate, for the influence of the nearby Pacific Ocean brings both cool temperatures and heavy rain and snow.

About 25% of the annual precipitation falls between June and September, although only July and August can be considered reliably warm and dry (barring the occasional thundershower). The winter's first snows typically arrive late in September. Snow depths can range from an average midwinter accumulation at Longmire of some 4 feet to more than 50 feet at Paradise. A record for North America was set in the winter of 1971–1972, when 1,122 inches (more than 93 feet) of snow fell at Paradise. Snow turns to rain sometime between March and April, when the blooming season begins at the Park's lower elevations. In some years, flowing in the meadows at Paradise doesn't begin until well into July.

The most complete climate records are kept at the Paradise Ranger Station. The figures in the table on the following page are averages, compiled for the period 1965–1994.

It has been said by many that Mount Rainier creates its own weather. The Mountain is so massive that it produces its own rain shadow, as winter storms arriving from the southwest drop their moisture on the south and west sides. This leaves Sunrise, on the northeast side, comparatively dry: winter snow accumulation there is typically half that at Paradise and the growing season is longer. Special microclimates around the Park result from unique interactions of landforms and weather patterns. For example, snow accumulations at Cayuse Pass, due east of the Mountain, can exceed those at Paradise. Snow depth at some alpine locations is reduced by winds that sweep away the snow, and on ridges around Sunrise the winter ground can be free of snow. Some studies there suggest that weather at Sunrise is

Flora of Mount Rainier National Park

	Jan	Feb	Mar	Apr	May	Jun	Jul	Aug	Sep	Oct	Nov	Dec	Year
Temperature (°F)													
Max.	31.8	34.0	36.4	40.4	47.8	53.6	60.5	61.7	56.0	47.4	35.5	32.0	44.8
Min.	20.2	21.6	22.7	25.1	30.8	36.6	42.2	43.6	39.2	32.7	24.0	20.4	29.9
Mean	26.0	27.8	29.6	32.8	39.4	45.1	51.4	52.7	47.6	40.1	29.7	26.2	37.4
Precipitation (inches)													
	18.2	13.2	11.6	8.7	5.1	3.9	1.9	2.4	4.8	8.7	17.4	18.4	114.2
Snowfall (inches)													
	131.6	101.3	98.7	72.1	25.2	6.2	0.6	0.0	3.4	28.8	97.2	121	686.1

Note: No day was recorded during this period when the temperature exceeded 90°F. Indeed, summer days reaching 80°F are rare. A temperature below 0°F was recorded just twice. Freezing temperatures and snow can occur any day of the year at Paradise.

Source: Weather America, edited by Alfred N. Garwood. Toucan Valley Publications, 1996.

more unstable earlier in the summer but more moderate later. At lower elevations, of course, much more precipitation falls in the form of rain.

The heaviest amount of rain (180–210 inches per year) falls in the northwest corner of the Park, in the Carbon River valley. Here, the abundant moisture and low elevation produce a true temperate rain forest, of the type found on the west side of the Olympic Peninsula.

While rain is sparse in the summer months, it sometimes happens that the higher ridges are enveloped in fog and clouds. At such times, enough moisture can condense on the needles of conifers that a small amount of "rain" actually falls at the base of a tree.

Forest and Plant Communities

At the time Jones wrote his flora, in 1938, plant distribution was analyzed in terms of "life-zones," a concept most famously associated with the work of C. Hart Merriam. Jones elaborated on this concept in the Pacific Northwest in his earlier book on the flora of the Olympic Peninsula, *A Botanical Survey of the Olympic Peninsula, Washington* (1936). In Merriam's view, temperature was the controlling factor in the distribution of plants, and it had been observed that as one ascended a mountain, the average temperature fell at a steady rate, in the same fashion that average temperatures fell as one moved north from the equator. Thus, the life-zones defined by Merriam were mapped against the North American continent, and named Transition, Canadian, Hudsonian, and Arctic-Alpine. (A fifth zone, the Austral, lies far to the south of the Pacific Northwest.)

When Jones indicated that *Oxyria digyna*, for example, was a plant of the Arctic-Alpine zone, he was making a statement about both its geographic and its elevational range within the Park. The difficulty in making practical use of the life-zone scheme is that, especially in mountainous areas at elevations near timberline, the effects of exposure and topography can result in adjacent zones becoming hopelessly intertwined. On warm ridges facing south or west, for example, trees can advance to an altitude higher than the average timberline on that part of the mountain. On the other hand, in the valley next to this ridge, the effects of shading and cold air drainage may push the timberline lower. Avalanche, fire, and soil development add complexity to the picture. There is, then, only a rough correspondence between altitude and Merriam's life-zones at Mount Rainier.

What remains valid of the life-zone concept, especially as elaborated by Jones, is the idea that predictable associations of plants and animals are to be found in, and indeed characterize, each life-zone. These associations he called "phytosociological units," and they relate to the life-zones (or "biogeographical units,") as shown in the following table from Jones's Mount Rainier flora. The table below includes Jones's range of altitudes for these zones in the Park.

Published in 1973, *Natural Vegetation of Oregon and Washington* by Jerry F. Franklin and C. T. Dyrness provided a new framework for descriptive analysis of the forest and plant communities of the Northwest. Of the four major vegetation areas recognized by Franklin and Dyrness, two are

Biogeographical units	Phytosociological units	Elevation (ft)
Transition Zone	Tsuga-Thuja Climax Forest Pseudotsuga Subclimax Forest	2,000–3,000
Canadian Zone	Tsuga-Abies-Pinus Forest	3,000–4,500
Hudsonian Zone	Abies-Tsuga-Chamaecyparis Climax Forest	4,500–6,000
Arctic-Alpine Zone	Treeless	6,000+

relevant to Mount Rainier National Park: forests and timberline/alpine regions. The forest area is subdivided into a series of zones that are characterized and demarcated by "areas in which a single tree species is the major climax dominant" (Franklin and Dyrness, 1973). For the Park, which lies wholly west of the Cascade Divide, these are the *Tsuga heterophylla* Zone, the *Abies amabilis* Zone, and the *Tsuga mertensiana* Zone, known also by the common names Western Hemlock Zone, Silver Fir Zone, and Mountain Hemlock Zone. At higher elevations a number of meadow communities are found which will be discussed later in this section.

The boundaries of a zone are set largely by elevation, through a direct correlation between elevation and temperature. Within a zone, systematic variations in the plant cover at a site can be ascribed chiefly to moisture and to the history of disturbance.

The Forests of Mount Rainier

The zones themselves show regular patterns that allow them to be more finely divided into units called "associations," so named because certain shrubs and herbs predictably associate with the dominant, climax tree species. The nature of the understory vegetation is largely determined by the amount of moisture available on a site. For example, the western hemlock/devil's club association is found on low ground, often along streams, on very moist soil, whereas the western hemlock/salal association is typical of relatively drier sites. Another kind of grouping is called the "plant community type," which may be conveniently distinguished from an association because the dominant tree species is a seral rather than a climax species; that is, the vegetation of the site is transitory between initial revegetation following a disturbance and the final, old-growth tree cover. In the Park, seral stands are typically less than about 200 years old. *Pseudotsuga menziesii*, or Douglas fir, is often the significant tree in seral plant community types. "Phases" are sometimes used to distinguish geographic varieties of a plant association, often based on a west-to-east gradient, or to highlight climax-seral varieties. The term "associes" is used to indicate a grouping of species, resembling an association, as near to but not yet at its climax form. (In a letter to the National Park Service written in 1983, Jerry Franklin estimated that 65% of the forested acreage in the Park was old-growth forest.)

The importance of this analytic structure is that each unit of the classification "characterizes a particular kind of environment and can therefore be used as a basis for distinguishing land areas of differing environments and biological potential" (Franklin, 1988).

The chart on page 11, simplified from one presented in *The Forest Communities of Mount Rainier National Park*, by Jerry Franklin and collaborators (1989), presents a classification of the forests of the Park. In an appendix, Franklin also includes a key and a convenient synopsis of the forest types.

Forest and Plant Communities

Association or plant community type	Forest zone
1. *Tsuga heterophylla/Achlys triphylla* association	*Tsuga heterophylla*
2. *Tsuga heterophylla/Polystichum munitum* association	
a. *Tsuga heterophylla* phase	*Tsuga heterophylla*
b. *Abies amabilis* phase	*Abies amabilis*
3. *Tsuga heterophylla/Oplopanax horridus* association	*Tsuga heterophylla*
4. *Alnus rubra/Rubus spectabilis* community type	*Tsuga heterophylla* and *Abies amabilis*
5. *Abies amabilis/Oplopanax horridus* association	
a. Valley phase	*Abies amabilis*
b. Slope phase	*Abies amabilis*
6. *Abies amabilis/Tiarella unifoliata* association	
a. Climax phase	*Abies amabilis*
b. Seral phase	*Abies amabilis*
7. *Abies amabilis/Vaccinium alaskense* association	
a. *Vaccinium alaskense* phase	*Abies amabilis*
b. *Berberis nervosa* phase	*Abies amabilis*
c. *Rubus pedatus* phase	*Abies amabilis*
d. *Chamaecyparis nootkatensis* phase	*Abies amabilis*
8. *Tsuga heterophylla/Gaultheria shallon* association	*Tsuga heterophylla*
9. *Pseudotsuga menziesii/Ceanothus velutinus* community type	*Tsuga heterophylla* and *Abies amabilis*
10. *Pseudotsuga menziesii/Xerophyllum tenax* community type	*Abies amabilis*
11. *Pseudotsuga menziesii/Viola sempervirens* community type	*Abies amabilis*
12. *Abies amabilis/Gaultheria shallon* association	*Abies amabilis*
13. *Abies amabilis/Berberis nervosa* association	*Abies amabilis*
14. *Abies amabilis/Xerophyllum tenax* association	
a. *Tsuga heterophylla* phase	*Abies amabilis*
b. *Tsuga mertensiana* phase	*Tsuga mertensiana*
c. Seral phase	*Abies amabilis*
15. *Abies amabilis/Rubus lasiococcus* association	
a. *Rubus lasiococcus* phase	*Abies amabilis* and *Tsuga mertensiana*
b. *Erythronium montanum* phase	*Tsuga mertensiana*
16. *Abies amabilis/Valeriana sitchensis* association	*Tsuga mertensiana*
17. *Abies lasiocarpa/Rhododendron albiflorum* association	*Tsuga mertensiana*
18. *Chamaecyparis nootkatensis/Vaccinium ovalifolium* association	*Tsuga mertensiana*
19. *Abies amabilis/Menziesia ferruginea* association	
a. Climax phase	*Abies amabilis* and *Tsuga mertensiana*
b. Seral phase	*Abies amabilis* and *Tsuga mertensiana*

Tsuga heterophylla Zone (Western Hemlock Zone)

The *Tsuga heterophylla* Zone is found at elevations below about 3,000 feet and is best developed in the valleys of the major rivers. On the northwest side of the Park, in this zone in the Carbon River valley is found one of the few examples of an inland temperate rain forest in the United States. While *Tsuga heterophylla*, western hemlock, is the dominant tree in mature forests, *Pseudotsuga menziesii*, Douglas fir, is frequently found and can grow to be as large in size as the hemlock. Douglas fir, in fact, is often the more prevalent tree within this zone on ground that has suffered disturbance from fire, flood, avalanche, or wind, for its seeds germinate rapidly on open soils. The western hemlock, however, will come to dominate in the long run (barring further disturbances), for its seeds can germinate and grow in the shade of the Douglas fir, eventually replacing it. Western hemlock often gets its best start on the fallen logs of Douglas fir: the hemlock seeds germinate easily within the crevices of the bark of the rotting logs (called "nurse logs"), the roots eventually reaching down to the soil. Except at its upper elevational limit, Jones's "Humid Transition Zone" corresponds closely with the *Tsuga heterophylla* Zone.

Another significant conifer in this zone is *Thuja plicata*, the western red-cedar. *Abies grandis*, grand fir, and *Pinus monticola*, western white pine, may also be present. In valley bottoms and on nearby slopes, several deciduous trees often grow to massive size, including *Acer macrophyllum*, big-leaf maple, *Alnus rubra*, red alder, and *Populus balsamifera* ssp. *trichocarpa*, black cottonwood. *Acer circinatum*, vine maple, is a smaller broadleaf tree of this zone.

Understory plants may nearly cover the forest floor on rich, moist sites, and a hiker's line of sight may be severely limited by the lush growth of shrubs. Ferns, perennial herbs, and shrubs common in the understory include the following:

Ferns
Athyrium filix-femina, lady-fern; *Blechnum spicant*, deer-fern; *Gymnocarpium dryopteris*, Pacific oak-fern; *Polystichum munitum*, sword fern; *Pteridium aquilinum* var. *pubescens*, bracken

Herbs
Achlys triphylla, vanilla-leaf; *Chimaphila umbellata*, prince's pine; *Circaea alpina*, enchanter's nightshade; *Cornus unalaschkensis*, bunchberry; *Linnaea borealis* ssp. *longiflora*, twinflower; *Tiarella trifoliata* var. *trifoliata*, foam-flower; *Viola glabella*, stream violet; *Viola sempervirens*, redwood violet; *Xerophyllum tenax*, beargrass

Shrubs
Berberis nervosa, dwarf barberry; *Gaultheria shallon*, salal; *Oplopanax horridus*, devil's club; *Rubus spectabilis* var. *spectabilis*, salmonberry; *Rubus ursinus* ssp. *macropetalus*, Pacific blackberry; *Vaccinium alaskense*, Alaska blueberry; *Vaccinium membranaceum*, thin-leaf huckleberry; *Vaccinium parvifolium*, red huckleberry

Abies amabilis Zone (Silver Fir Zone)

The Park's signature forest zone, found between 2,500 and 4,700 feet on level ground to steep slopes, the *Abies amabilis* Zone is dominated by silver fir. *Pseudotsuga menziesii*, Douglas fir, may be a codominant, especially in younger forests, and *Tsuga heterophylla*, western hemlock, is often found as well. Soils may be moist to dryish and are well-drained. On moist soils *Chamaecyparis nootkatensis*, Alaska yellow-cedar, and *Abies procera*, noble fir, are also found. At its upper elevational limit, this zone intergrades with the subalpine parklands. In the upper White River valley, on moraines and other places with young glacial soils, *Picea engelmannii*, Engelmann spruce, and *Pinus contorta* var. *murrayana*, lodgepole pine, can be found in large numbers. Several associations and plant community types are found at the upper edge of this zone and pass into the *Tsuga mertensiana* zone. The *Abies amabilis* zone includes most of the forests considered by Jones, in his day, as the "Canadian Zone."

Forests in this zone have a more open appearance: the understory shrubs are fewer and shorter, and seldom is the forest floor more than half-covered by herbaceous plants. The canopy is more or less closed, but not as densely as in the *Tsuga heterophylla* Zone. Ferns, perennial herbs, and shrubs common in the understory include the following:

Ferns
Athyrium filix-femina, lady-fern; *Blechnum spicant*, deer-fern

Herbs
Clintonia uniflora, queen's cup; *Cornus unalaschkensis*, bunchberry; *Linnaea borealis* ssp. *longiflora*, twinflower; *Rubus lasiococcus*, dwarf bramble; *Rubus pedatus*, five-leaved bramble; *Tiarella trifoliata* var. *unifoliata*, foam-flower; *Xerophyllum tenax*, beargrass

Shrubs
Berberis nervosa, dwarf barberry; *Gaultheria shallon*, salal; *Menziesia ferruginea*, fool's huckleberry; *Vaccinium membranaceum*, thin-leaf huckleberry; *Vaccinium ovalifolium*, early blueberry; *Vaccinium parvifolium*, red huckleberry

Tsuga mertensiana Zone (Mountain Hemlock Zone)

Between the closed-canopy forests of the *Abies amabilis* Zone and the open subalpine meadows lies the *Tsuga mertensiana* Zone. *Tsuga mertensiana*, mountain hemlock, is present, of course, but this tree often claims only the rocky ridges in the zone, leaving the nearly flat to moderate or steep ground and better soils to other conifers, including *Abies lasiocarpa*, subalpine fir, *Abies amabilis*, silver fir, and *Chamaecyparis nootkatensis*, Alaska yellow-cedar. This zone is found above 4,000 feet, with the tree cover becoming progressively thinner as higher elevations are reached, reaching timberline between about 5,500 and 6,000 feet. *Pinus albicaulis* is an early colonizer of newly opened or disturbed ground near timberline. Jones's "Hudsonian

Zone" corresponds, except at its lower edge, with the *Tsuga mertensiana* Zone.

Brush fields are common in the *Tsuga mertensiana* Zone, and *Rhododendron albiflorum*, the cascade azalea, may form dense thickets. Herb cover is usually low, and several species of the subalpine meadows, including the avalanche lily and sitka valerian, can reach down into these forests.

At its upper limits there occurs a fascinating and continually shifting interplay of forest and meadow, and it is this "parkland" pattern—trees in clumps and stringers interspersed with meadows—that lends the timberline reaches so much aesthetic appeal. Park Botanist Regina Rochefort has monitored the advance of forest trees in the Paradise meadows (Rochefort and Peterson, 1996) and has correlated changes in the forest/ meadow ecotone with climatic changes. Historically fire has been infrequent at this elevation; nevertheless, fire, along with the advance and retreat of glaciers, has been important in influencing the overall distribution of trees. In the shorter run, however, the duration of the snowpack into the spring, and its depth through the winter, have been more significant factors.

Trees, therefore, in the parkland zone, tend to be found in clumps on slopes that melt free early and along ridgelines where snow accumulations are less. Where trees invade meadows, it is often because climate change has resulted in less winter snowfall.

A clump may result from the establishment of a single tree, often *Pinus albicaulis*. *Abies lasiocarpa* seedlings may then become established under its protection. The subalpine fir, in turn, tends to spread by layering, the rooting of lower branches at their tips and the subsequent development of new trunks and a repetition of the process. The expanding clump of trees can become home to a number of shrubs, including *Rhododendron albiflorum* and several *Vaccinium* and *Rubus* species. Sunrise, on the northeast side of the Mountain, is a good place to observe this phenomenon.

On rocky ridgelines, where winter winds sweep away much of the snow, trees typically form a "krummholz," German for "crooked wood." The trees, including *Chamaecyparis nootkatensis*, *Picea engelmannii*, *Pinus contorta*, and *Tsuga mertensiana*, are shrublike even when many decades old, with twisted branches and no leading trunk. Between the parklands and the ridges occupied by krummholz, trees often take on a "flagged" appearance, bent by the prevailing wind and mostly stripped of branches along the windward side.

Herbs
Erythronium montanum, avalanche lily; *Orthilia secunda*, sidebells pyrola; *Rubus lasiococcus*, dwarf bramble; *Rubus pedatus*, five-leaved bramble; *Valeriana sitchensis*, sitka valerian

Shrubs
Menziesia ferruginea, fool's huckleberry; *Rhododendron albiflorum*, cascade azalea; *Vaccinium membranaceum*, thin-leaf huckleberry; *Vaccinium ovalifolium*, early huckleberry

The story of the evolution of the forests of Mount Rainier, including the complex interplay of the forces of fire, ice, avalanche, and volcanism, lies outside the scope of this book. A fine account for nonspecialists is William Moir's *Forests of Mount Rainier*. Those interested should also consult the bibliography for articles and books by Martha Cushman, LeRoy Detling, Peter Dunwiddie, Jerry Franklin, Miles Hemstrom, and Regina Rochefort.

The Subalpine Meadows of Mount Rainier

Surveys show that for most of the two million annual visitors, Mount Rainier National Park is a place not of dense forests but of open, flowery meadows, by far the most visited areas. These are the famous "parks," borrowing a term from English landscaping style. John Muir was greatly impressed when he visited and climbed the Mountain in 1888: "Every one of these parks, great and small, is a garden filled knee-deep with fresh, lovely flowers of every hue, the most luxuriant and the most extravagantly beautiful of all the alpine gardens I ever beheld in all my mountain-top wanderings."

Summertime traffic at the Paradise meadows testifies to the popularity of Mount Rainier's flower fields. In fact, so much traffic crossed the Paradise meadows, and the meadows at Sunrise and around Tipsoo Lake, that large tracts were trampled and nearly ruined. The 1980 dissertation by Ola Edwards, "The Alpine Vegetation of Mount Rainier National Park: Structure, Constraints, and Development," presented a detailed study of the meadows in the Park and laid the groundwork for painstaking meadow restoration efforts that began in 1986. This work is described in articles by Regina Rochefort, Park Botanist.

Jan Henderson studied the subalpine meadows of Mount Rainier, with an emphasis on succession and the invasion of meadows by forests. He distinguished five major community type groups (this presentation of his research is summarized from Franklin and Dyrness's *Natural Vegetation of Oregon and Washington* [1988]):

1. *Phyllodoce–Cassiope–Vaccinium* group, characterized by low-growing heather and huckleberry shrubs.
2. *Valeriana sitchensis–Carex spectabilis* group, featuring an abundance of lush, herbaceous perennials.
3. *Carex nigricans* group, dominated by dwarf sedges.
4. "Rawmark" and low herbaceous group, a group of communities that colonize newly opened ground.
5. *Festuca viridula* group, dominated by bunchgrasses.

1. *Phyllodoce–Cassiope–Vaccinium* Communities

This group of attractive communities of low shrubs is well-displayed on the slopes of Paradise Park and at Indian Henrys Hunting Ground, and other places on the south and west sides of the Mountain. The plants form a dense cover and their proportion of occurrence will be found to vary with elevation. Soils are moderately well-drained and moist through the growing season, which averages three to four months. Important species include *Phyllodoce empetriformis* (and *P. glanduliflora* at higher elevations), *Cassiope mertensiana*, *Vaccinium deliciosum*, and *Lupinus arcticus* ssp. *subalpinus*, along with *Antennaria lanata*, *Castilleja parviflora*, *Luetkea pectinata*, *Pedicularis ornithorhyncha*, *Polygonum bistortoides*, and *Vahlodea (Deschampsia) atropurpurea*. Fruitful stands of a huckleberry, *Vaccinium deliciosum*, attract a number of people to these places.

2. Valeriana sitchensis–Carex spectabilis Communities

Found around the Mountain, on moderate to steep, well-drained slopes that are watered in the summer by melting snow on higher slopes. The best sites can support tall, dense, stands of a variety of perennial wildflowers. The growth of shrubs, including the heathers and huckleberries, and trees is thought to be suppressed by avalanches. The lushest stands are found on the south and west sides of the Park, including the meadows at Paradise, while on the east and the north these communities are less diverse and productive, and may grade into the *Festuca viridula* group of communities. Dominant species include *Carex spectabilis*, *Lupinus arcticus* ssp. *subalpinus*, *Polygonum bistortoides*, *Valeriana sitchensis*, and *Veratrum viride*. Other common members are *Anemone occidentalis*, *Castilleja parviflora*, *Erigeron peregrinus*, *Heracleum maximum*, *Ligusticum grayi*, *Potentilla flabellifolia*, and both species of *Erythronium*. A special group of plants is found on low, wet ground along streams, and includes *Carex nigricans*, *Mimulus lewisii*, *Mimulus tilingii*, *Parnassia fimbriata*, *Petasites frigidus*, and both subspecies of *Caltha*.

3. Carex nigricans Communities

These communities face severe challenges: snow cover persists late into the spring and the growing season is short, typically less than three months. Sometimes seen at the edges of meltwater ponds, the soils are cold and wet even as the summer progresses. *Carex nigricans* is the dominant species, growing in dense mats. *Carex spectabilis* and *Potentilla flabellifolia* are found on somewhat more favorable patches of ground, and *Aster alpigenus*, *Epilobium anagallidifolium*, *Luetkea pectinata*, and *Vahlodea (Deschampsia) atropurpurea* may also be present.

4. "Rawmark" and Low Herbaceous Communities

Much of the Mount Rainier landscape is new, or relatively new, and at subalpine elevations this sort of ground gives rise to distinctive communities. These, in turn, may be succeeded by other community types. Soils are unstable, fine to stony, often of pumice, and may be moist to dry, depending on the availability of meltwater seepage. The growing season is short. Mosses are often among the dominant plants. *Saxifraga tolmiei*, a very attractive plant, does well where moisture is available. *Antennaria lanata*, *Carex nigricans*, *Hieracium gracile*, *Luetkea pectinata*, *Vahlodea (Deschampsia) atropurpurea*, and rather dwarfed *Valeriana sitchensis* are the most frequently seen species. Patches of ground may be bare. On the east side of the Park, especially where the "newness" is caused by disturbance due to human activities, *Eriogonum pyrolifolium* var. *coryphaeum* and *Cistanthe umbellata* var. *caudicifera* are common.

5. *Festuca viridula* Communities

The timberline zone at Sunrise, in the rain shadow on the east side of the Mountain, presents a very different picture from similar elevations at Paradise: grassy meadows, punctuated by small groups of trees, extend for great distances, on flats and on gentle to moderate slopes. Soils are dry and loose, and the prevailing wind has, over the centuries, favored this side with pumice and ash from eruptions of the volcano. In other parts of the Park, these communities may develop where wind removes snow as it falls, leaving locally drier conditions. *Festuca viridula*, a bright green-leaved bunchgrass, is the dominant species, and distinct communities within this type are characterized by the presence of other, codominant species, including *Lupinus arcticus* var. *subalpinus* and *Potentilla flabellifolia* in moister places, *Aster ledophyllus* in drier places. Other significant species include *Anemone occidentalis, Carex spectabilis, Claytonia lanceolata, Ligusticum grayi, Polygonum bistortoides, Ranunculus suksdorfii*, and *Veronica cusickii*.

Marcia Hamann surveyed 150 subalpine and alpine vegetation stands on the north side of Mount Rainier and, in her 1972 thesis, "Vegetation of Alpine and Subalpine Meadows of Mount Rainier National Park, Washington," recognized a larger number of distinct plant associations in the subalpine and alpine zones. She points out that complex factors make it impossible to fix a boundary between the subalpine and alpine zones. She also notes that the north side of the Mountain, which she studied, enjoys milder and drier weather than the south side and that soil development has been influenced by prevailing winds that have deposited pumice from eruptions in the recent era to the north and east sides, yielding a loose and porous soil. She found that significant factors in the distribution of these associations were depth and duration of the snow cover, nature of the soil parent material, and exposure.

Alpine Plants

Above the last outposts of trees of the *Tsuga mertensiana* Zone is perhaps the most floristically interesting zone of plant life on Mount Rainier. Conditions here, between about 6,000 and 7,500 feet, are harsh — the growing season is just two or three months long, and plants must adapt to constant winds, strong sunlight, periods of dryness, and generally infertile and poorly developed soils. Plants grow as cushions or mats, leaves are often insulated and protected by hairs, and roots dig deeply. The best growth is often seen on "fellfields," flats or shallow slopes littered with small rocks, which provide protected niches.

Plants here are all perennials, accompanied by a few low or matted shrubs. Grasses, sedges, and rushes make up about one-third of the flora at these highest elevations. Characteristic wildflowers include *Cistanthe umbellata* var. *caudicifera, Draba aureola, Erigeron aureus* var. *aureus, Eriogonum pyrolifolium* var. *coryphaeum, Lupinus lepidus* var. *lobbii, Polemonium elegans, Saxifraga tolmiei,* and both species of *Smelowskia.* The Burroughs Mountain trail provides the best opportunity to observe this habitat.

Floyd Schmoe tells the story, in *A Year in Paradise,* of his discovery of a patch of bare ground melted out from the middle of the Paradise Glacier at 7,500 feet. He speculated that this was "newborn land, exposed to air and light for the first time perhaps in more than a hundred thousand years." He watched closely to see what would happen. Within a month, three plants had appeared: a moss, alpine willow-weed (possibly *Epilobium anagallidifolium*), and a "tiny grasslike sedge" (perhaps *Carex engelmannii, C. nardina,* or *C. pyrenaica*). A number of insects were also present. The next summer, the bare area had expanded and 25 species of plants and animals were present.

At the Extreme

A few species ascend the Mountain to great heights, following rocky ridges bordered by permanent snowfields. Far too few in number to constitute any type of plant community, they nevertheless deserve mention here, for, with the exception of a few lichens and mosses, they represent the final outpost of plant life on Mount Rainier. Two species reach 11,000 feet: *Poa lettermanii* and *Smelowskia ovalis* var. *ovalis* (which probably holds the elevation record on the Mountain). Eight others reach 10,000 feet: *Agrostis humilis, Carex nardina* var. *hepburnii, Collomia larsenii, Poa suksdorfii, Polemonium elegans, Draba aureola, Draba oligosperma,* and *Smelowskia calycina.*

Plant Geography and Distribution

For all its 14,411 feet of height, Mount Rainier may be thought of as an island. More than 100 years ago, geologist Bailey Willis described the Mountain as "an arctic island in a temperate sea."

Arthur R. Kruckeberg, emeritus professor of botany at the University of Washington, has taken up this idea and has described the Cascade Mountains as "an archipelago of high 'islands' extending from northern California to British Columbia" (Bilderback, 1987). He has elaborated on the concept of a mountain as an island, pointing out that there are three interrelated factors that serve to isolate a mountain, much as the sea isolates a conventional island. These "geo-edaphic" factors, to use his phrase, are topography, substrate, and time.

High elevation imposes constraints on plants, forcing them to adapt to short growing seasons, intense light, seasonally dry conditions, shallow soils, and other challenges. "Substrate" refers to the underlying rocks and the soils that develop from those rocks. As far back as 1906, Charles Piper noted that the volcanic rocks of Mount Rainier and peaks to the south serve to limit the southern expansion of a number of species found in the North Cascades, a region where the substrates are chiefly metamorphic and sedimentary rocks. Time is an interesting factor with respect to Mount Rainier: The Mountain's eruptive and glacial history, fire and ice, plus frequent avalanches, have repeatedly forced plants away from the area now encompassed by the Park boundaries. Much of the land surface in the Park is very young, and more land is being "created" in many valleys as glaciers recede.

One result of these geo-edaphic factors is that the count of species for the Park, disregarding the introduced plants, is relatively low, considering the size of the Park and the elevational and ecological possibilities, from lowland rain forests available to alpine crevices. The total count of species for the Park is 871 (species for which two varieties or subspecies are present are counted twice). Nonnatives comprise 148 of these, leaving a total of 723 native taxa. Latitude may be a factor in constraining species richness: Lassen Volcanic National Park in northeastern California, with only about half the area of Mount Rainier National Park, is home to 745 native taxa. The geo-edaphic features significant at Mount Rainier were even more pronounced at Mount Saint Helens, and its pre-eruption flora was even more limited, with about 315 species known from the mountain (Bilderback, 1987).

Another result is that a number of species reach the northern or southern limits of their ranges in or near the Park. In some cases of widespread species, these are limits in western North America or in the Cascade Mountains. In terms of speciation — the evolution of new species from parent stocks — populations at the limits of a plant's range are interesting, for they may be subject to selection pressures beyond those affecting more central parts of the range, and new variations may be favored by such pressures. Also, the likelihood is increased that a population at the edge of the range may be cut off from the parent stock, allowing selection and random genetic processes to perhaps produce a new species. The obliteration of the summit of Mount Saint Helens in 1980 removed a significant amount of the subalpine and alpine habitat lying between Mount Rainier and Mount Hood and may well have increased the isolation of some of the plants at the northern limit of their ranges on Mount Rainier. Perhaps a greater factor, at the close of the twentieth century, is the artificial isolation of species in Mount Rainier National Park as the cutting of forests reduces the possibilities of genetic flow among populations along the Cascade Range.

Marcia Hamann, who surveyed subalpine and alpine vegetation stands on the north side of Mount Rainier for her 1972 thesis, discussed an unusual plant association, dominated by *Ivesia tweedyi* and *Astragalus alpinus*, found at 6,300 feet on stony soil on a ridge top on Brown Peak. She writes, "It is not too unusual to find eastern Washington species at Mount Rainier, considering the proximity of the peak to the crest of the Cascade Mountains.

It is possible that these species . . . are relics of a more xeric flora [plants adapted to dry conditions] similar to those found on lithosols [shallow, well-drained, stony soils] of peaks west of the Oregon Cascade Mountains."

She cites a pioneering plant ecology study by LeRoy Detling of species typically found in eastern Oregon that are found also west of the Cascades on isolated peaks in Lane and Douglas Counties. Detling attributed the presence of these "outlier species" to special environmental conditions on those peaks. Shallow and rapidly drained volcanic soils allow only limited competition from trees and shrubs, and early snowmelt lengthens the growing season, producing conditions similar to regions east of the Cascades.

On Mount Rainier, the same prevailing southwest winds that leave the northeast side of the Mountain in a rain (and snow) shadow have, through the centuries, deposited an extra measure of pumice on the ground. Not only is there less precipitation, but the soils are sharply drained and can become quite dry late in the summer.

Following Detling, it may be concluded that a number of species found in the northeast quarter of the Park represent relics from a time when drier conditions were found on the west side of the Cascades, plants that were left behind as conditions became cooler and moister. Detling cites studies of pollen recovered from bogs to show that there was a period of warm, dry conditions culminating 6,000 to 8,000 years ago and postulates that "the earlier immigrants of the ponderosa pine flora reached the Puget Area from the south, when the upper limits of this flora reached the level of the Cascade crest under the influence of extreme dryness, [and] there probably was an influx of additional species from the east."

Among the plants that may have crossed the Cascades and entered the Park during this warm, dry interval are two conifers that are rare in the Park, *Picea engelmannii* and *Pinus ponderosa*. (A third conifer, western larch, *Larix occidentalis*, may well have made the journey, but if did, it failed to persist to the present day.) Other plants that are frequently found associated with ponderosa pine east of the Cascades but that are also found at Mount Rainier include *Arabis holboellii* var. *retrofracta*, *Arnica cordifolia*, *Arnica parryi*, *Arnica rydbergii*, *Dicentra uniflora*, *Lewisia triphylla*, *Ligusticum canbyi*, *Penstemon confertus*, *Penstemon fruticosus* var. *fruticosus*, *Ribes triste*, *Ribes viscosissimum*, *Sanicula graveolens*, *Sedum rupicola*, *Senecio streptanthifolius* var. *streptanthifolius*, *Tonestus lyallii*, and *Woodsia scopulina* ssp. *scopulina*.

A second location for plants not found more widely in the Park is Mount Wow, in the southwest corner of the Park. Here the rocks are much older than those in most of the rest of the Park. Mount Wow is also high ground that, to a significant extent, served as an ice-free refuge during the last glacial period. *Oxytropis monticola* is found only here in the Park. Other plants found on Mount Wow that are more typically found east of the Cascade Divide include *Allium cernuum*, *Cryptantha affinis*, *Saxifraga bronchialis* ssp. *vespertina*, and *Senecio integerrimus* var. *ochroleucus*. (Mount Wow, and nearby ridges, are also the location for a number of species that enter the Park from the lower elevations to the west, including *Antennaria howellii* ssp. *neodioica*, *Erigeron subtrinervis* var. *conspicuous*, *Fritillaria affinis*, and *Satureja douglasii*.)

No plant species is endemic, or entirely restricted in its occurrence, to Mount Rainier National Park, although several come close to that distinction. Each of these plants—*Castilleja cryptantha*, *Pedicularis rainierensis*, and *Tauschia stricklandii*—was once thought to occur only within the boundaries of the Park but has subsequently been found to range, slightly or significantly, beyond that arbitrary demarcation.

Tauschia stricklandii, a plant in the parsley family, has a widely disjunct distribution: it occurs in the Park in meadows between 5,000 and 6,500 feet and also in a small area on the south side of the Columbia River Gorge above Bonneville Dam on Moffett Creek. *Tauschia* is a small genus growing across the Great Basin and the Southwest. The other species in Washington, *Tauschia hooveri*, grows on sagebrush scablands in Yakima County. The likely hypothesis is that *T. stricklandii* once grew more widely along the east side of the Cascades, reaching downstream along the Columbia into the Gorge, but that changing conditions since the end of the last glacial epoch have eliminated suitable habitat in all but these two places.

The other two plants, *Castilleja cryptantha* and *Pedicularis rainierensis*, both have a number of relatives through the Cascades and in eastern Washington. The fact that their ranges are limited to the Park and adjacent mountains suggests that they evolved in place, perhaps from parent stocks isolated from more wide-ranging species. Among the paintbrushes, *Castilleja cryptantha*'s closest relatives appear to be *C. chrysantha*, of the Wallowa and Blue Mountains, and *C. pulchella*, from the northern Rockies. *Pedicularis rainierensis* may have evolved from *P. capitata* or *P. oederi*, both plants of northeast Asia and Alaska that extend along the mountains into British Columbia.

Weeds and Rare Plants in the Park
Weeds

This flora includes 148 species that are not native to the Park, called variously "weeds," "invasives," and "exotics." A few of these are descended from garden plants introduced in the early years of settlement in the area. More arrived in the days before the automobile, when horses and their feed were present. Most, however, are casual introductions, having migrated from regions at lower elevations: Many have seeds that are easily carried in on shoes and tires and by the wind.

For the most part, these species are restricted in their distributions, growing in places where disturbance of the soil is routine, often following roadsides and trails. A few, including English ivy, *Hedera helix*, and wall lettuce, *Mycelis muralis*, are able to spread into forests. The species of greatest concern are those that have the ability to aggressively colonize and push aside native vegetation. Such species include the following: *Cytisus scoparius*, Scot's broom; *Geranium robertianum*, herb robert; *Hieracium atratum*, polar hawkweed; *Hypericum perforatum*, Klamath weed; *Hypochaeris radicata*, rough cat's-ear; *Leucanthemum vulgare*, oxeye daisy; *Phalaris arundinacea*, reed canary-grass; *Polygonum cuspidatum*, Japanese knotweed; *Senecio jacobaea*, tansy ragwort; *Tanacetum vulgare*, common tansy.

In 1965, Robert Wakefield did a baseline study of exotic plants in the Park, and the Park Service continues to monitor Wakefield's plots and conduct systematic eradication efforts.

Only one plant in the Park is designated as a Federal Noxious Weed, *Heracleum mantegazzianum*, or giant hogweed. Young plants have been found along the road in the vicinity of Kautz Creek.

The Noxious Weed Control Board of the state of Washington maintains a list of species of concern and rates them according to the threat posed to the natural environment and the potential for control. Degree of presence in the state, not severity of the threat, is the key factor in the ratings.

A rating of "A" means that the plant is not significantly present in Washington and that control still remains feasible. Prohibitions on commerce in the plant are in place and property owners may be directed to remove plants on their property. Only one class A species occurs in the Park: *Heracleum mantegazzianum*, giant hogweed.

Class "B" is used for plants that are present but not widespread in Washington. Decisions regarding control measures are the responsibility of counties in areas where the plants occur and subject to state laws where the plants are not yet established. The following class B species occur in the Park: *Centaurea biebersteinii*, spotted knapweed; *Centaurea diffusa*, diffuse knapweed; *Centaurea nigra*, black knapweed; *Cytisus scoparius*, Scot's broom; *Hieracium aurantiacum*, orange hawkweed; *Hypochaeris radicata*, rough cat's-ear; *Leucanthemum vulgare*, oxeye daisy; *Senecio jacobaea*, tansy ragwort.

Class C species are widespread and the subject of local measures only. The following are found in the Park: *Cirsium arvense*, Canada thistle; *Cirsium vulgare*, bull thistle; *Hypericum perforatum*, Klamath weed; *Linaria vulgaris*, yellow toad-flax; *Phalaris arundinacea*, reed canary-grass; *Polygonum cuspidatum*, Japanese knotweed; *Silene latifolia* ssp. *alba*, white cockle; *Tanacetum vulgare*, common tansy; *Verbascum thapsus*, common mullein.

A number of species, while native plants, can behave in a "weedy" manner. Often annuals, they are able to invade and maintain themselves on ground that is subjected to disturbances, especially at roadsides and along trails. Examples include *Equisetum arvense* (field horsetail), *Prunella vulgaris* ssp. *lanceolata* (self-heal), and *Rorippa curvisiliqua* (western yellowcress).

Rare Plants

Federal and Washington state law designate certain plants as rare on the basis of several factors, including number of individuals of a species in existence, the scope of the geographic area in which the species occurs, and threats to the continued existence of the species. However, a plant formally designated as "rare" may actually be frequently seen in a particular place. Thus, Rainier lousewort, *Pedicularis rainierensis*, is fairly common in subalpine meadows in the Park.

Federal designation of rare plants is made by the U.S. Fish and Wildlife Service (USFWS). The terms "endangered" and "threatened" are used. No species found in the Park has either status. One plant, *Castilleja cryptantha*, found only in the Park and nearby in the William O. Douglas Wilderness, is considered by the USFWS to be a "species of concern," not a formal status but nevertheless an indication that conservation of the species is an important matter (Washington Natural Heritage Program, 1997).

The Washington Natural Heritage Program (WNHP), of the Washington State Department of Natural Resources, maintains its own list of "endangered, threatened, and sensitive" plant species. Although the Endangered Species Act of the state of Washington does not cover plants, the WNHP list is important to local, state, and federal agencies, including the National Park Service.

No plants from the Park are listed by the WNHP as "endangered" or "threatened." Several, however, are considered to be "sensitive," a status defined as "vulnerable or declining and [which] could become endangered or threatened in the state without active management or removal of threats." These "sensitive" plants are listed below and the WNHP assessment of the plant's status is given for each.

Botrychium lanceolatum, Botrychium pinnatum, and *Saxifraga rivularis*: plants of worldwide distribution, but rare and of limited occurrence in Washington.

Castilleja cryptantha: range limited to a small area of Washington and imperiled because of rarity.

Luzula arcuata: a plant of worldwide distribution but critically imperiled in the Washington because of extreme rarity.

Microseris borealis: of limited occurrence in North America and imperiled in Washington because of rarity.

Pedicularis rainierensis: range limited to a small area in Washington and imperiled because of rarity.

In addition to these listed species, the state program maintains a list of "review" species. Plants for which more information is needed in order to determine rarity and threats are on the "review 1" list, and in the Park only *Whipplea modesta* has this status. "Review 2" list plants are those for which taxonomic questions exist; in other words, the Washington plants may perhaps constitute a separate species or variety and therefore be of interest and concern, or, alternatively, the actual definition of the species may not be well settled. *Botrychium minganense* is included here.

Finally, there is a group of plants on a "watch list," plants that were once listed by the WNHP but which have proven to be more abundant or less threatened than was previously thought. A number of plants in the Park are in this group: *Arnica rydbergii, Asplenium trichomanes-ramosum*,

Cephalanthera austiniae, Diphasiastrum (Lycopodium) alpinum, Draba aureola, Draba ventosa var. *ruaxes, Hemitomes congestum, Hulsea nana, Ivesia tweedyi, Lloydia serotina, Physaria alpestris, Pleuricospora fimbriolata, Poa suksdorfii, Rainiera (Luina) stricta, Saxifraga lyallii,* and *Thelypteris nevadensis.*

Explorations and Studies
Early Botanical Explorations

Botanists were relatively late in reaching the territory presently located within the Park boundaries. In the 1790s, Archibald Menzies, a member of George Vancouver's expedition, collected plants in the Puget lowlands. John Scouler visited in 1825, and the eminent David Douglas passed through on his west coast travels in the 1820s and early 1830s. None ventured far from the Puget Sound into the Cascade Range and none reached the present area of the Park.

William F. Tolmie, a surgeon of the Hudson's Bay Company with a strong interest in the medicinal uses of plants, was the first botanist, as well as the first person of European descent, to visit the area of the Park. Tolmie Peak, in the northwest corner of the Park, was named in his honor, although research suggests that Tolmie actually reached Hessong Rock, about 2.5 miles southeast of Tolmie Peak and closer to the Mountain. Here, on September 2, 1833, he collected "a vasculum [small case] of plants."

Charles Pickering and W. D. Brackenridge, botanists attached to the Wilkes Expedition, left Fort Nisqually, on Puget Sound, in 1841 and, following the White River, crossed the Cascades at Naches Pass. They passed just north of the present boundary of the Park, but collected extensively and compiled the first catalog of plants of the region.

Charles Vancouver Piper explored the area in 1888, 1889, and 1895 and was the first resident botanist to study the flora. Published in 1901 and 1902 in articles in *The Mazama* entitled "The Flora of Mount Rainier," Piper's account of the flora listed 295 species. A revision of this work was published in 1916 as a chapter in E. S. Meany's *Mount Rainier, a Record of Exploration.* The count of species in the latter work had reached 315.

Other collectors of note in the late 1890s included Ernest C. Smith, who visited the south slopes of the mountain, E. L. Greene, who collected on the north side, and M. W. Gorman. Oscar D. Allen, professor of botany at Yale University, settled in the upper Nisqually Valley in 1895 and collected widely on the south and west sides of the Mountain through 1905.

Park Ranger J. B. Flett made the most extensive collections in the Park, working between 1913 and 1921; he had earlier visited the Park in 1895 and 1896. He was the author of a popular booklet entitled *Features of the Flora of Mount Rainier National Park,* published in 1922. Flett stated that 532 species were known in the Park. Park Naturalist Floyd Schmoe brought national attention to the Park and its wildflowers with his 1925 book, *Our Greatest Mountain.*

In 1929, systematic collecting was undertaken by the National Park Service, directed at creating a Park herbarium. On the basis of this work, Park Naturalist C. Frank Brockman published a special issue of *Mount Rainier Nature Notes* in 1938 that listed 677 species for the Park. In 1949, Brockman used this list as the basis for the first nontechnical treatment of the plants of the Park, *Flora of Mount Rainier National Park*.

Fred A. Warren, a seasonal Park Ranger from 1926 to 1933, collected widely in the Park and coauthored a 1937 paper with Harold St. John that appeared in the *American Midland Naturalist* entitled "The Plants of Mount Rainier National Park." St. John and Warren listed 695 species.

George N. Jones, of the University of Washington, began studies at Mount Rainier in 1925 and published in 1938 the first comprehensive flora of the Park with identification keys, *The Flowering Plants and Ferns of Mount Rainier*. This flora included 729 species.

Other collectors whose material may be found in Washington herbaria include Winona Bailey, J. W. Thompson, Harold W. Smith, T. C. Frye, and George B. Rigg. H. E. Bailey, working on a forestry type map, collected for the National Park Service in 1934 and 1935. J. R. Slater, of the University of Puget Sound, worked on both plants and reptiles in the Park in the 1950s and 1960s, while many collections made by Irene Creso in the 1970s can be found at Pacific Lutheran University (she had a special affection for grasses and sedges).

Many species were first scientifically described from specimens collected within the area of the Park, including several first found by William Tolmie on his journey to Mount Rainier in 1833. Such plant specimens are called "type" specimens, and they are important because the description of the type specimen serves to define the concept of the species.

Types were collected in the area of the Park for the following: *Castilleja cryptantha*, found in Yakima Park in 1937 by F. W. Pennell and E. Y. Danner; *Castilleja rupicola*, found by Charles V. Piper in 1895 at Sluiskin Falls; *Gentiana calycosa, Aster alpigenus, Pedicularis contorta, Pedicularis ornithorhyncha, Penstemon procerus* var. *tolmiei, Potentilla flabellifolia*, and *Saxifraga tolmiei* discovered by William Tolmie in 1833; *Lysichiton americanus*, found by O. D. Allen in 1895 in the "upper valley of the Nisqually"; *Pedicularis rainierensis*, collected around 1927 by Fred A. Warren; *Penstemon rupicola*, discovered by Charles V. Piper in 1899, above Paradise; *Petasites frigidus* var. *nivalis* and *Rainiera stricta* first collected by E. L. Greene in 1889; *Tauschia stricklandii*, discovered in 1896 in Paradise Park by O. D. Allen and a "Mr. Strickland"; and *Vaccinium deliciosum*, described by C. V. Piper from a collection made by O. D. Allen in 1896.

Recent Studies of the Park Flora and Ecology
In an appendix to his 1983 dissertation, "Holocene Forest Dynamics on Mount Rainier, Washington," Peter W. Dunwiddie compiled what was to that date the most complete list of the flora of the Park, including 804 species

and varieties. Jerry Franklin's work analyzing forest communities has been the underpinning of contemporary work in the field.

Several researchers, including Ola M. Edwards, Regina Rochefort (the present Park Botanist), and Joseph C. Van Horn, have studied human impact on fragile alpine and subalpine meadows in the Park, chiefly in the Paradise and Sunrise areas. Rochefort has written extensively of revegetation work in these meadows. She has also supervised rare plant studies, maintained weed control efforts, and developed an inventory of species.

Marcia J. Hamman, in 1972, and Jan A. Henderson, in 1974, studied subalpine meadows in the Park. In 1995, Betsy Kirkpatrick, of the University of Puget Sound, investigated factors governing the species composition of subalpine meadows, and included meadows in Grand Park and Spray Park along with her North Cascades study sites.

Disruptions to which the plant cover of the Park is susceptible have been intensively studied: Miles A. Hemstrom has investigated the role of fire in shaping the forests of the Park; Hans K. Frehner and Peter M. Frenzen have studied soil development and plant succession on the Kautz Creek mudflow; Robert S. Sigafoos examined botanical evidence of recent glacial activity in the Park; and Martha J. Cushman studied the influence of recurrent avalanches in the Butter Creek Research Natural Area, on the south slope of the Tatoosh Range, in her 1981 dissertation.

References to these writings, and more, will be found in the bibliography.

Key to the Major Groups of Vascular Plants at Mount Rainier National Park

1 Plants reproducing by spores borne in sporangia; fls and true seeds not present .... Ferns and Fern Allies (below)
1 Plants reproducing by seeds, produced through the union of pollen and ovules in fls or in cones 2

2(1) Seeds borne on the surfaces of the scales of cones or in berrylike structures; ovary absent; fls absent (Conifers) .... (Gymnosperms, p. 55)
2 Seeds borne within a fruit that develops from an ovary; true fls present (Angiosperms) 3

3(2) Parts of the fl in 4s or 5s (rarely 3s); lvs net-veined; cotyledons 2 .... Dicots (p. 67)
3 Parts of the fl in 3s; lvs parallel-veined; cotyledons 1 ... Monocots (p. 368)

Ferns and Fern Allies

Grouped here are the nonflowering vascular plants. Like the other major groups of vascular plants, and unlike the nonvascular mosses and lichens, these plants have leaves and a vascular system, the roots and stems through which water is conducted in specialized tubes. The ferns and the so-called fern allies are distinguished from other vascular plants by reproducing by spores instead of true seeds.

The most conspicuous members of the group are the "true" ferns, comprising seven families, which were formerly treated as just one family, the Polypodiaceae. Somewhat similar in appearance are the Ophioglossaceae, or the grapeferns, none of which are common or abundant in the Park. Horsetails, quillworts, and the clubmosses are minor elements of the Park flora, although they may be dominant features in some habitats.

An exception to the alphabetical arrangement of the families is made for the ferns and fern allies. An alphabetical arrangement would result in the families of true ferns being dispersed in the text and the separation of the seven families would make it difficult to comprehend the features that unite and the features that separate the members of the group.

The arrangement adopted is the one used in the treatment of the ferns in the *Flora of North America*. In a linear order this is Lycopodiaceae, Selaginellaceae, Isoetaceae, Equisetaceae, Ophioglossaceae, Pteridaceae, Dennstaedtiaceae, Thelypteridaceae, Blechnaceae, Aspleniaceae, Dryopteridaceae, and Polypodiaceae. Since the arrangement is not alphabetical, page numbers are given in the key below.

1	Stems hollow, jointed; the joints sheathed with greatly reduced, scalelike whorled lvs (horsetails) .... Equisetaceae (p. 35)
1	Stems solid, not jointed; lvs well-developed (if scalelike, then not whorled at the joints) 2

2(1)	Lvs attached directly to the stem, undivided, scalelike to grasslike 3
2	Lvs stalked, often lobed or divided (true ferns) 5

3(2)	Lvs grasslike, tufted on a short, fleshy stem (quillworts) .... Isoetaceae (p. 34)
3	Lvs scalelike or needlelike, clothing a usually branched stem (clubmosses) 4

4(3)	Lvs lacking ligules and mostly longer than 3 mm; the stems and lvs appearing "round" in cross section .... Lycopodiaceae (p. 29)
4	Lvs with ligules and mostly less than 3.5 mm long; the stems and lvs appearing "square" in cross section .... Selaginellaceae (p. 33)

5(2)	Plants with mostly 1 frond; sporangia large, borne in clusters on simple or branched spikes (grapeferns) .... Ophioglossaceae (p. 38)
5	Plants with 2 to many fronds; sporangia minute, grouped into sori, borne on the underside or at the lower margin of the frond (true ferns) 6

6(5)	Fronds of 2 different types; the leaflets of the fertile fronds linear or greatly narrowed 7
6	Fronds all alike, the leaflets of the fertile fronds not narrowed 8

7(6)	Fronds once-pinnate .... Blechnaceae (p. 44)
7	Fronds 2 or more times pinnate .... Pteridaceae (p. 40)

8(6)	Sori borne on the underside of the leaflets, more or less covered by the turned-under margin of the leaflets 9
8	Sori borne on the underside of the leaflets, away from the margin, naked or covered by an indusium 10

9(8)	Fronds usually longer than 100 cm, arising individually from the rootstock .... Dennstaedtiaceae (p. 43)
9	Fronds seldom longer than 80 cm, in tufted clumps .... Pteridaceae (p. 40)

10(8)	Sporangia not clustered into distinct sori; indusium absent .... Pteridaceae (p. 40)
10	Sporangia borne in distinct sori; indusium present or absent 11

11(10)	Indusium present, although it may be obscure or disappear as the sorus matures 12
11	Indusium absent at all stages of development 15

12(11)	Lf stalk with 2 strands of veins (visible in cross section); midvein of lf blade short-hairy .... Thelypteridaceae (p. 43)
12	Lf stalk with 3 or more strands of veins; midvein of lf blade may be scaly or chaffy, but not short-hairy 13

13(12)   Sori and indusium round, kidney-shaped, or otherwise, but not elongated .... Dryopteridaceae (p. 46)

13   Sori and indusium elongate 14

14(13)   Fronds once-pinnate .... Aspleniaceae (p. 45)

14   Fronds 2 or more times pinnate .... Dryopteridaceae (p. 46)

15(11)   Fronds deeply pinnately lobed to once-pinnate ....
             .... Polypodiaceae (p. 53)

15   Fronds 2 or more times pinnate .... Dryopteridaceae (p. 46)

Lycopodiaceae P. Beauv. ex Mirb.
CLUBMOSS FAMILY

Superficially, these plants are similar to the true mosses, but structurally and evolutionarily they differ. The clubmosses are vascular plants with small, evergreen leaves that are, in general, spirally arranged about the stems. Spores are produced in sporangia which are found in the axils of modified leaves called "sporophylls"; the sporophylls are usually lighter in color and of different size than the normal leaves. The sporophylls may be grouped into cones, which may be either stalkless at the branch tips or on specialized stalks. Early in the year, before the sporangia are produced, the clubmosses are difficult to distinguish. The key below requires mature plants.

Kartesz merges *Diphasiastrum* with *Lycopodium*, but Warren H. Wagner Jr. and Joseph M. Beitel, who wrote the treatment of the Lycopodiaceae for the *Flora of North America*, recognize *Diphasiastrum* as a separate genus on the basis of the characteristics cited in the family key, below, as well as the shape of the gametophyte. (The gametophyte represents the sexual generation of the clubmoss life cycle; this is a tiny body, usually hidden under debris on the forest floor, which bears eggs and sperm. The sporophyte generation is the familiar clubmoss plant, which produces spores.) A second means of reproduction is found in *Huperzia*: very short, budlike branches called "gemmae" form on the stems, break off, and can form new plants.

Should you find a "clubmoss" that does not key satisfactorily in this family, try the Selaginellaceae.

1   Sporangia borne in the axils of green, leaflike sporophylls, not in cones .... *Huperzia*

1   Sporangia borne in cones formed of modified, light green sporophylls; cones at the ends of the stems 2

2(1)   Cones usually solitary and not stalked on the fruiting stems; if more than 1 cone then the stalks of each are of equal lengths; lvs 6–12-ranked .... *Lycopodium*

2   Cones 2 to several on branches of unequal lengths; lvs 4–5 ranked ....
            .... *Diphasiastrum*

Diphasiastrum Holub—CLUBMOSS

The most obvious distinguishing feature of *Diphasiastrum* is that the fertile stems bear two or more cones on stalks of unequal length. Close examination of the vegetative stems will reveal that these three species are also distinguished from the other clubmosses in having leaves that are arranged on the stems in 4 or 5 ranks in a pattern that is fairly easy to discern. In species of *Lycopodium* the leaves are in 6–10 ranks, and lose, in effect, the appearance of being "ranked" at all. In addition, the leaves in *Diphasiastrum* are rarely longer than 3.5 mm.

1   Branches flattened; the lvs appressed, with free tips to 2 mm ....
.... *D. complanatum*
1   Branches appearing round; the free tips of the lvs 2–3.5 mm 2

2(1)   Lvs 4-ranked, every other pair winged at their bases .... *D. alpinum*
1   Lvs more loosely 4- or 5-ranked, none winged at their bases . *D. sitchense*

Diphasiastrum alpinum (L.) Holub—ALPINE CLUBMOSS

Very different from all the other clubmosses: the lvs are 4-ranked, to quote C. L. Hitchcock (1973), "decussately opposite, every other pair decurrent on the stem as a pair of flanges, the flange continuous with one margin of the lf, so that the stem appears somewhat flattened or wing-margined with the lateral rows of lvs twisted." The creeping stems can reach 50 cm in length, with the upright fertile branches densely tufted and reaching about 7 cm tall. The cones are 10–20 mm long.

Listed as "watch" in *The Endangered, Threatened, and Sensitive Vascular Plants of Washington* (Washington Natural Heritage Program, 1997).

Uncommon, growing on the ground on rocky slopes in open forests, above 5000 ft. Reported from Berkeley Park and Paradise.

A collection made by J. B. Flett was determined by J. M. Beitel in 1980 to be a hybrid: *Diphasiastrum alpinum* x *sitchense*. It was collected in 1904 at "Mt. Rainier," presumably in the south or west quarter of the Park.

Diphasiastrum complanatum (L.) Holub
GROUND-CEDAR

With stems reaching impressive lengths, to 3 m, this rare species differs in having upright branches that are markedly flattened, with relatively short lvs, to 2 mm long, which are 4-ranked and flattened against the stem (hence the reference to "cedar" in the common name). The upright branches can reach 30 cm, with three cones per branch; each cone as much as 10 cm long.

Lycopodiaceae

A rare clubmoss of forests at middle elevations. Known from along the trail to Indian Henrys Hunting Ground (elevation not recorded). Also collected by J. B. Flett on Mount Wow, and in Stevens Canyon.

Diphasiastrum sitchense (Rupr.) Holub
ALASKA CLUBMOSS

The horizontal stems are about 40 cm long and the cone-bearing stems, sparsely arranged in tufts, are 8–12 cm tall. The slender but firm, scalelike lvs are generally 4-ranked, but not infrequently 5-ranked as well, and to 3.5 cm long; the horizontal stems are only sparsely leafy. The cones are 1.5–2.5 cm long, in 2s or 3s on the fertile branches.

The common clubmoss of open subalpine meadows. Collected along Ipsut Cr; at Indian Henrys Hunting Ground; in the meadow above Bench Lk; and in Paradise Park; also collected at an unrecorded elevation in Box Canyon and reported from Green and Eunice Lks. Occasionally found in forests, as at about 3,200 ft on the Wonderland Trail up Rampart Ridge.

Huperzia Bernh.—FIR-MOSS

Differing from the two other clubmoss genera in the absence of cones: instead, the sporangia are borne singly in the axils of leaves along the stems, and consequently they are easy to overlook. The sporophylls (the leaves that bear the sporangia) are usually found to be arranged in "zones" about the stem, alternating with zones of regular leaves. Plants in this genus occurring in the Park were formerly all considered to be *Lycopodium selago*, a species now known not to occur in Washington. None are common. In each, a prostrate stem is nearly or completely absent. Instead, the plants grow as tufts of upright stems 10–30 cm tall. The leaves are 5–20 mm long, spreading to ascending, and sharp-tipped. Gemmae are borne along the stems in the axils of the regular leaves: these are budlike branches off the main stems that can fall free and grow into new plants.

1 Largest lvs widest above the midpoint; plants typically growing on the ground or on rotted wood in forests .... *H. occidentalis*
1 Largest lvs widest at the base; typically growing on mossy rocks, in marshes, or in meadows 2

2(1) Stems 8–11 cm tall; spore-bearing structures borne at the tips of each year's new growth .... *H. chinensis*
2 Stems 12–18 cm tall; spore-bearing structures borne along the length of the mature shoots .... *H. haleakalae*

31

Huperzia chinensis (Christ) Czern.—PACIFIC FIR-MOSS

Stems clustered and erect, 12–18 cm tall. Lvs uniform on the stems, without annual constrictions. The yellowish green lvs of the juvenile stems are narrowly lanceolate and 4.5–7 mm long, and the juvenile stems themselves tend to be curled downward at the top. The lvs of the mature stems are triangular and 3.5–5.5 mm long. The gemmae are produced in 2 or 3 bands at the top of the mature stems.

Known in the Park from one specimen collected by E. C. Smith in 1890, the precise location of which is not given. In other parts of the Northwest, it grows on ledges of cliffs and on mossy talus.

Huperzia haleakalae (Brack.) Holub
ALPINE FIR-MOSS

Stems erect, 8–11 cm tall, somewhat bent over at the base. Lvs uniform on the stems, without annual constrictions. The lvs of the juvenile stems are lanceolate and 4.5–6 mm long; those of the mature stems are ovate and 3–4 mm long. The gemmae are produced all along the mature stem.

Uncommon, on rocky ridges and alpine slopes: on Westside Rd; at St. Andrews Cr; on Pinnacle Peak; and at Little Tipsoo Lk.

Huperzia occidentalis (Clute) Kartesz & Gandhi
WESTERN FIR-MOSS

Stems erect, 12–20 cm tall and leaning over on the ground at their bases. The lvs are produced in distinct annual "zones," with constrictions in between. The light green lvs are oblanceolate, 6–10 mm long, and reflexed to spreading. The smaller lvs are narrowly triangular. Gemmae are produced in a single zone at the top of the mature stem.

Uncommon, in low-elevation forests, and often growing on decaying logs. Known from near the North Fork of the Puyallup R and collected in 1902 by J. B. Flett in the "foothills of Mt. Rainier."

Lycopodium L.—CLUBMOSS, GROUND-PINE

Species of *Lycopodium*, as the genus is presently defined, have leaves that can reach 6–8 mm arranged in 6–10 ranks on the stems. As noted above, this can be hard to see, and perhaps the best that can be said is that the genus does not have leaves in clear and discrete ranks. Varieties of the two species below are recognized by Kartesz; in the *Flora of North America* they are not.

1   Cones 2 to several on a 2-branched stalk that differs in appearance from the rest of the stem; lvs tapering to a thin, hairlike tip .... *L. clavatum*
1   Cone solitary on the stem, lacking a distinct stalk; lvs with a sharp but not hairlike tip .... *L. annotinum*

Lycopodium annotinum L.
STIFF CLUBMOSS

Horizontal stems reaching about 2 m in length give rise to erect stems that are usually branched and 10–25 cm tall. The lvs are about 6–7 mm long, spreading, and pointed but lacking a hairlike thin tip. The cones, up to 4 cm long, appear at the ends of the erect branches; no structure resembling a separate stalk is present.

Rare, evidently favoring middle-elevation forests. Known from below Narada Falls on the Paradise R; in Stevens Canyon on old burned ground, as at Bench Lk; and on the roadside at the Chenuis Falls trailhead. Reported from Green Lk. According to a note file at the University of Puget Sound, a collection was made by J. R. Slater on Westside Rd, 2.4 mi north of the main Park road, evidently in the 1960s; the specimen has been lost.

Lycopodium clavatum L.
RUNNING GROUND-PINE

"Running" refers to the great length to which the prostrate stems can reach: 3–4 m is not uncommon. The erect stems are usually branched and about 15 cm tall. The lvs are 6–8 mm long, pliable, slender, and tipped with a long, hairlike bristle; the horizontal and vertical stems are about equally leafy. The cones, which reach about 15 cm long, are borne in pairs on stalks of approximately equal length at the top of a slender stem.

Common in forests up to about 5000 ft. Collected at Ipsut Cr; on the trail to Indian Henrys Hunting Ground; at 4600 ft in Stevens Canyon; on the Olallie Cr trail at 3000 ft; on the trail from White R to Owyhigh Lks at 4400 ft; and on the trail to Carter Falls at an unspecified elevation. Also seen in the forest between Kautz Cr and Longmire; on Westside Rd at 2700 ft; on Mazama Ridge above Reflection Lks; and at Green Lk.

Selaginellaceae Willk.
LESSER-CLUBMOSS FAMILY

Selaginella Beauv.—SPIKE-MOSS

Somewhat similar to the Lycopodiaceae, which are also known as clubmosses. The points that distinguish the two families are rather technical and not easy to see without magnification. The two selaginellas that occur in the Park tend to grow as small to largish mats with slender stems. The cones are borne on short, upright stems; the leaves of the cones are densely arranged in 4 ranks and consequently the cones are square in cross section. The cones are generally 10–20 mm long. (In the Lycopodiaceae, the cones are round in cross section.)

1   Stems loosely branched and spreading .... *S. wallacei*
1   Stems compactly branched, tufted .... *S. densa* var. *scopulorum*

Selaginella densa Rydb. var. *scopulorum* (Maxon) R. M. Tryon
COMPACT SPIKE-MOSS

Similar to the next species, but tending to grow in smaller, denser mats, or even as small cushions; the whole plant reaches about 10 cm across. In addition, the sterile branches are more flattened. The lvs are ovate-lanceolate, and about 3 mm long.

Evidently rare. Known from two collections made by Irene Creso in the 1970s in the western portion of the Park: on rocks at Klapatche Park and on Gobblers Knob.

Selaginella wallacei Hieron.—WALLACE'S SPIKE-MOSS

Matlike and reaching perhaps 20 cm across, the prostrate sterile stems are densely clothed with closely overlapping, lanceolate-linear lvs 3–4 mm long.

Uncommon, typically found on dry rocks and soil on exposed ridges, from about 2500 ft to about 5500 ft. Known from collections made at "Longmire" (but more likely on nearby Rampart Ridge); on lower slopes of Paradise Park; and in the Carbon R V below the glacier.

Isoetaceae Rchb.
QUILLWORT FAMILY

Isoetes L.—QUILLWORT

Quillworts grow as small tufts of perennial, rushlike lvs that rise from small, bulblike stems on ground that are submerged at least part of the growing season. Two types of spores are produced, megaspores and microspores, both borne in separate, saclike sporangia located on the inner surface at the base of the leaves and enclosed by a velum, or flap, formed by the infolded leaf margins.

Distinguishing the species by characteristics of the leaves is far from satisfactory, for *I. bolanderi* especially is subject to variation in this feature. Examination of the megaspores is important, but since plants are destroyed in the process of this examination, it's best left undone for plants in the National Park.

Equisetaceae

1   Lvs soft, flexible, gradually tapered to a slender tip .... *I. echinospora*
1   Lvs rigid, abruptly tapered to the tip ....*I. bolanderi* var. *bolanderi*

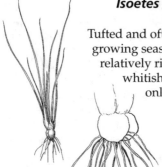

Isoetes bolanderi Engelm. var. *bolanderi*
BOLANDER'S QUILLWORT

Tufted and often found out of standing water late in the growing season, with lvs 6–25 cm long which are relatively rigid and unbending. The megaspores are whitish to almost bluish, 0.3–0.5 mm wide with only scattered tubercules.

Easily overlooked among grasses and sedges at the margins of lakes and bogs, from about 4,000 ft up to nearly timberline.

Isoetes echinospora Durieu
BRISTLE-LIKE QUILLWORT

Almost always found to be submerged throughout the growing season, with lvs 3–20 cm long. The tapered lf tips are slender and grasslike. The megaspores are white, 0.4–0.6 mm wide, and covered with short spines.

In shallow standing water of ponds and lakes below about 5000 ft, as at Reflection and Mirror Lks. Also found in Dewey Lk.

Equisetaceae DC.
HORSETAIL FAMILY

Equisetum L.—HORSETAIL

Named for the tail-like appearance of some of the branched species. These are perennial herbaceous plants, with extensive horizontal rootstocks. The hollow stems are jointed (and therefore easily pulled apart) and may be annual and deciduous, or perennial and persisting for more than one season. The stems are of two types, sterile and fertile, and in some species the two are dimorphic and differ greatly. The fertile stems and the sterile stems do not usually appear at the same time. They are prominently grooved lengthwise and rough to the touch, from tiny crystals of the mineral silica. The stems may be branched, with the branches arising in whorls from joints in the stems. These branches might be mistaken for leaves, but in fact the plant's lvs are represented by the ring of dark teeth that sheathe the stem at each joint. Spores are produced in conelike structures at the tops of the fertile stems.

1   Stems with whorled branches, of 2 types or not; annual species 2
1   Stems not branched, all alike; perennial species 5

2(1)   Stems of 2 types, the sterile green and branched, the fertile brown and simple 3

2   Fertile and sterile stems alike 4

3(2)   Sterile stems 10–60 cm tall; teeth of the sheaths 6–14 .... *E. arvense*

3   Sterile stems 30–100 cm tall; teeth of the sheaths 14–28 ....
.... *E. telmateia* var. *braunii*

4(2)   Teeth of the sheaths 5–10 .... *E. palustre*

4   Teeth of the sheaths 9–25 .... *E. fluviatile*

5(1)   Stem slender, 2–4 mm in diameter, with 5–10 lengthwise grooves ....
.... *E. variegatum* var. *variegatum*

5   Stem stouter, 4–6 mm in diameter, with 15–50 lengthwise grooves .... *E. hyemale* var. *affine*

Equisetum arvense L.—FIELD HORSETAIL

Along with *E. telmateia* ssp. *braunii*, the most common of the Park's horsetails. It differs in its markedly smaller size overall. The sterile stems reach about 60 cm in height and are bright green and much-branched, with 10–14 grooves. The sheaths are 4–5 mm long. The fertile stems appear early in the spring and are somewhat shorter, thicker, unbranched, and brownish to nearly whitish in color, with a cone 2–4 cm long.

Common and occasionally "weedy," on moist, shaded ground, on stream banks, in thickets, and at roadsides, up to 6000 ft. At Mowich Lk; around the Nisqually Entrance; in the Longmire swamp; at the head of the Paradise R; at Ohanapecosh; and on the Owyhigh Lks trail at 5000 ft.

Equisetum fluviatile L.
SWAMP HORSETAIL

Similar to but less branched than *E. arvense*, this species is easily distinguished by the fact that the fertile and sterile stems are alike. The stem is green and grows to 50–100 cm tall. It has 10–30 grooves and 9–25 sheath teeth. The cone is 1–2 cm long.

Uncommon; typically found below 4000 ft on wet soils and in swamps. Collected at the Longmire meadow and near the foot of the Nisqually Glacier in a bog; also reported from Green Lk.

Equisetum hyemale L. var. *affine* (Engelm.) A. A. Eaton
SCOURING-RUSH

An unbranched horsetail with perennial stems reaching 100–200 cm. The stems are green, all alike, with 15–50 grooves; it has the roughest feel of the Park's horsetails and magnification will show rows of silica-bearing bumps on the ridges separating the grooves. The teeth of the sheath are dark brown and a blackish band is found at the base of the sheath. The 1–2.5 cm cone is distinctive, with a small, sharp point at the top.

Scattered through the Park, on wet soils below 3000 ft. Collected at Ohanapecosh.

Equisetum palustre L.
MARSH HORSETAIL

A smaller and more slender version of E. *fluviatile*, with stems less than 50 cm tall. The sterile and fertile stems are alike, branched, with 5–10 grooves on the stem and 5–10 teeth in the sheath. The cones are 1–2.5 cm long.

In swamps; uncommon. Known from Longmire, Box Canyon, and Ohanapecosh.

Equisetum telmateia Ehrh. var.
braunii (J. Milde) Milde
GIANT HORSETAIL

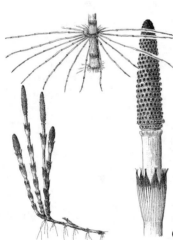

A striking horsetail with robust stems. The sterile stems reach about 100 cm, but are sometimes less than 50 cm, with 20–40 grooves and numerous branches. The fertile stems are stout, to about 60 cm tall, and have 15–30 teeth in the sheath; they are yellow-brown and appear early in the spring. The cones are 4–8 cm long.

Very common below about 3000 ft, in swamps and on wet hillsides. At Ohanapecosh, Longmire, and in the Carbon R area.

Equisetum variegatum Schleich. ex F. Weber & D. Mohr var. *variegatum*
NORTHERN SCOURING-RUSH

Very much like *E. hyemale*, with perennial, unbranched stems. It is smaller overall, reaching about 50 cm in height, with just 5–10 lengthwise grooves. The cones are 5–10 mm long.

Not previously reported for the Park and known from one specimen collected at Mountain Meadows in 1970 by J. R. Slater.

Ophioglossaceae (R. Br.) C. Agardh
ADDER'S-TONGUE FAMILY

Botrychium Sw.—GRAPEFERN, MOONWORT

These are not true ferns at all, although they are somewhat similar in appearance. In *Botrychium*, each plant consists of a single leaf, comprised of a sterile "fernlike" blade and a specialized structure called the sporophyll that rises from the leaf and bears a panicle of sporangia crowded on the branches. These clusters of sporangia account for the "grape" in the common name. The grapeferns are quite appealing plants, but none is common and they are easily overlooked.

Two little grapeferns have been found just outside the Park boundary on the upper slopes of the Crystal Mtn ski area. *Botrychium ascendens* W. H. Wagner was found at 4575 ft, 0.5 mi south of the base lodge; it has once-pinnate lvs and is unusual for having a few sporangia on the lower leaflets of the "sterile" blade. *Botrychium simplex* E. Hitchc. was found between 5600 and 5800 ft near Lk Elizabeth; it has leaflets that are rounded across the apex and squared off at the base. Both are no more than 5 cm tall. They very well may turn up in the Park.

1   Fertile portion of the blade longer than wide; plants seldom more than 10 cm tall overall 2
1   Fertile portion of the blade wider than long, or about as wide as long; plants usually more than 10 cm tall overall 3

2(1) Leaflets shiny green, in 7 or fewer pairs .... *B. pinnatum*
2   Leaflets dull green, in 7–10 pairs .... *B. minganense*

3(1) Sterile lf blade stalked, arising near the ground; leaflets leathery and persisting through the winter .... *B. multifidum*
3   Sterile lf blade attached directly about halfway up the stem; leaflets thin, deciduous .... *B. lanceolatum* var. *lanceolatum*

Botrychium lanceolatum (S. G. Gmel.) Angstrom var. *lanceolatum*—LANCE-LEAVED MOONWORT

Usually less than 15 cm tall but rarely reaching 20 cm. The sterile blade is attached directly to the stalk. The sterile blade is roughly triangular, to 6 cm long and 8 cm wide, once- to twice-pinnate with the leaflets deeply toothed. The fertile part of the lf is 1–4 times divided, with the sporangia rather loosely clustered.

Listed as "sensitive" in *The Endangered, Threatened, and Sensitive Vascular Plants of Washington* (Washington Natural Heritage Program, 1997).

Rare, collected in the Longmire meadow in 1899 by John Allen, and near Panorama Point along the Pebble Cr trail.

Botrychium minganense Vict. MINGAN MOONWORT

Growing 10–12 cm tall, with the sterile lf blade on a short stalk. The sterile blade is narrowly triangular to oblong, 5–8 cm long, and once- or twice-pinnate with the leaflets fan- or wedge-shaped. The fertile portion is long-stalked, with the sporangia crowded on numerous short branches.

Listed as "review – group 2" in *The Endangered, Threatened, and Sensitive Vascular Plants of Washington* (Washington Natural Heritage Program, 1997).

Uncommon in the Park; usually seen in dense forests at lower elevations, but also in more open places up to about 6500 ft.

Botrychium multifidum (S. G. Gmel.) Rupr. LEATHERY GRAPEFERN

The largest and most robust of the grapeferns in the Park, the plants reach 35 cm tall. The sterile blade is borne on a short, heavy stalk and is broadly triangular, up to 30 cm wide and long, and 2- or 3-times pinnate. It persists well into the next growing season. The toothed leaflets are ovate to diamond-shaped. The fertile portion of the blade is on a stalk, and may reach to about 40 cm long, with 4–6 major branches.

The most frequently seen grapefern in the Park, perhaps because it is the largest, found in moist meadows and wet thickets below 4000 ft. Collected at Longmire and Ohanapecosh. According to a note file at the University of Puget Sound, a collection was made by J. R. Slater at Tipsoo Lk, in 1962; the specimen has been lost—the identification or location information may be incorrect, given the elevation. Also in Stevens Canyon.

Botrychium pinnatum H. St. John
NORTHWESTERN MOONWORT

The smallest species in the Park, reaching just 8–10 cm tall. The sterile blade is stalkless or very short-stalked, oblong, to about 4 cm long, and once- to twice-pinnate, with mostly ovate, lobed or unlobed leaflets. The fertile blade is on a short to long stalk, twice-divided, with narrow, ascending branches.

Listed as "sensitive" in *The Endangered, Threatened, and Sensitive Vascular Plants of Washington* (Washington Natural Heritage Program, 1997).

Rare. Collected at Mount Rainier in 1888 by the Reverend E. C. Smith, at an unspecified location. According to a note file at the University of Puget Sound, a collection was made by J. R. Slater in the Park, evidently in the mid 1960s, also at an unspecified location; the specimen has been lost. It has also been found in the Crystal Mtn ski area.

Pteridaceae E. D. M. Kirchn. (formerly in Polypodiaceae)—MAIDENHAIR FERN FAMILY

These are small ferns and, except for maidenhair, mostly of drier, rocky places where conditions are too harsh for the more luxurious forest ferns to thrive. A proper indusium is not present, although in some species the sori are protected by the inrolled margin of the leaflet.

1   Fronds of 2 types; leaflets of the fertile fronds long and narrow ....
    .... *Cryptogramma*
1   Fertile and sterile fronds alike 2

2(1) Sporangia not clustered into distinct sori; sporangia not covered
    by turned-back leaflet margins .... *Pentagramma*
2   Sporangia borne in distinct sori that are covered by turned-back
    leaflet margins 3
3(2) Fronds twice-pinnate, the blade lanceolate .... *Cheilanthes*
3   Fronds as wide as long, palmate-pinnate (fan-like in outline) .... *Adiantum*

Adiantum L.—MAIDENHAIR

Adiantum aleuticum (Rupr.) C.A. Paris—NORTHERN MAIDENHAIR

Unique in appearance among the ferns of the Park: the glossy black stalk branches at its summit into 2 divisions, each of these then divided into 3–7 divisions of approximately equal length, giving the frond a rounded outline. The fronds are 25–100 cm tall, rising singly from creeping rootstocks. The rectangular-oblong to sometimes triangular, bright green leaflets are crowded on each midrib, overlapping each other. The sori are oblong and placed at the margin of the leaflets where they are partially covered by the recurved margin of the leaflets.

Common on wet, shaded banks and cliffs in the western and southern portions of the Park, up to about 5000 ft. At Mowich Lk; along Westside Rd up to about 3000 ft; along Crater Cr and St. Andrews Cr; on Tahoma Cr to about 3000 ft; along Rampart Ridge trail; along the trail to the Grove of the Patriarchs; in the Ohanapecosh area; at 5100 ft above Cayuse Pass; at Ipsut Cr campground; and at the trailhead to Windy Gap.

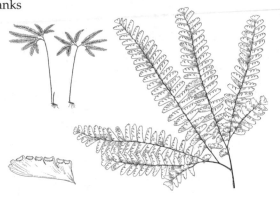

Cheilanthes Sw.—LIP FERN

Cheilanthes gracillima D. C. Eaton
LACE LIP FERN

A graceful fern, with tufted fronds 5–25 cm tall that are twice-pinnate, rather firm, and oblong-lanceolate, rather a dull yellowish green in color. The sori are roundish and covered by a recurved and modified leaflet margin.

A rather common fern (if not often seen), found in crevices and among rocks, 4000-7000 ft. On the summit of Pinnacle Peak; at McClure Rock; at Tipsoo Lk; near Shadow Lk; at Moraine Park; 0.25 mi south of Dewey Lk; and on Burroughs Mtn just above Sunrise.

Cryptogramma R. Br.—ROCK-BRAKE

Small, tufted ferns, with fronds seldom reaching more than 15 cm tall and 1–3 times pinnate. The fronds are dimorphic. The sterile are shorter than the fertile, and have ovate to obovate, flat, toothed leaflets. In the fertile fronds, the leaflets are oblong with a "podlike" appearance; the sori form a continuous line beneath the turned-under margins. Subtle anatomical differences distinguish the Park's two parsley ferns, most easily seen in older plants at the height of the growing season, when the presence or absence of remnant lf-stalk bases is evident. The two plants may grow in rough proximity, as near the Nisqually Glacier.

1	Fronds thin, translucent when pressed and dried; bases of the lf stalks deciduous .... *C. cascadensis*
1	Fronds leathery, not translucent; bases of the lf stalks persistent .... *C. acrostichoides*

Cryptogramma acrostichoides R. Br.
AMERICAN PARSLEY FERN

Overall, a fern of more robust appearance than the next species. The sterile lvs are 6–22 cm tall, and the leaflet blade is dark green and thick enough that it is not translucent when pressed and dried. As the plant ages, a ruff is formed of old, persistent lf-stalk bases. The sterile fronds generally remain green until the next year's growth emerges.

Common, on rocky slopes and in crevices, throughout the Park, 2500-6000 ft. At Mowich Lk; on Westside Rd and lower slopes of Mount Wow; along the trail to Klapatche Park at about 3000 ft; near the snout of the Nisqually Glacier; in Box Canyon; at Tipsoo Lk and along the trail to Dewey Lk; on the Naches Peak loop trail at 5800 ft; and at Eunice Lk.

Cryptogramma cascadensis E. R. Alverson
CASCADE PARSLEY FERN

In this species, the sterile lf blades are 3–20 cm tall, with thin leaflet blades that are light to medium green, translucent when pressed. The bases of the stalks of the lvs fall from the plant as the lvs wither and the sterile fronds turn brown at the end of the season.

Infrequent, on moist sites on subalpine talus slopes, with a marked preference for open ground where the snow is late to melt. Known from the slopes above Nisqually Glacier; on the east fork of Edith Cr at 5900 ft and elsewhere in Paradise V; around Reflection Lks; near Snow Lk; on Wapowety Cleaver; and at Glacier Basin. Perhaps easiest to see on talus slopes on the trail up to Frozen Lk. First described by Ed Alverson (Alverson, 1989).

Pentagramma Yatsk., Windham & E. Wollenw.
GOLD FERN

Pentagramma triangularis (Kaulf.) Yatsk., Windham & E. Wollenw. ssp. *triangularis*—GOLDBACK FERN

A most attractive fern, growing in tufts. The fronds are 15–30 cm tall, with wiry, dark brown stalks about twice as long as the blade. The blade is triangular to pentagonal in outline, dark green above, and decorated with a yellowish, waxy powder beneath. The sori are linear and follow the veins of the leaflets, but tend to merge together as they mature; an indusium is absent.

Infrequent, preferring dry rocky slopes at low elevations, usually in the open. According to a note file at the University of Puget Sound, a collection was made by J. R. Slater at Ohanapecosh in the 1960s; the specimen has been lost. In *Ferns of the Northwest* by T. C. Frye (1934), it was said that this fern had been found at "Ohanapecosh Hot Springs."

Dennstaedtiaceae Lotsy (formerly in Polypodiaceae)—BRACKEN FAMILY

Pteridium Gled. ex Scop.—BRACKEN

Pteridium aquilinum (L.) Kuhn var. *pubescens* (L.) Underw.—NORTHERN BRACKEN

A coarse, tough, almost weedy fern, growing in drier places than most ferns and often found colonizing newly opened ground. The fronds arise singly from a widely spreading rootstock and can reach 1–3 m in height. The blades are roughly triangular and mostly 3 times pinnate. The leaflets are hairy on the underside. The sori are borne in lines at the margins of the leaflets and are protected by the inrolled margin of the leaflet; a true indusium is absent.

Widespread and abundant in open woods and on dry slopes, especially along roads and on burned-over ground, below about 3000 ft in the valleys of the major rivers of the Park.

Thelypteridaceae Pic. Serm. (formerly in Polypodiaceae)—MARSH FERN FAMILY

Thelypteris Schmidel

Thelypteris nevadensis (Baker) Clute ex C.V. Morton NEVADA MARSH FERN, OREGON SHIELD FERN

Much like the several *Dryopteris* ferns that occur in the Park, this plant merits placement in a separate genus in a separate family because of an obscure but important feature of the lf stalk: a cross-section view shows 2 strands of veins (that appear as heavy bundles) rather than 3, which is more typical among ferns. Otherwise, *Thelypteris* is a graceful, medium-sized fern with thin, light green, twice-pinnate fronds up to 50 cm long. The leaflets are shallowly to deeply divided and the lower leaflets may be again pinnately divided. The 2 vascular bundles are given expression as a lengthwise groove in the hairy midvein of the lf. The sori and indusia are horseshoe-shaped.

Listed as "watch" in *The Endangered, Threatened, and Sensitive Vascular Plants of Washington* (Washington Natural Heritage Program, 1997).

Very rare in the Park. J. R. Slater's specimen in the herbarium of the University of Puget Sound describes the location as "Ohanapecosh, 2.7 miles north on trail to Cowlitz Divide." J. B. Flett's collection in the Park herbarium was made "3 miles from Ohanapecosh on Paradise trail, 3000 ft." Presumably the location is on the east-facing slope of Cowlitz Divide, roughly along the course of the present road. The Flett collection, made in 1929, is labeled "new to Washington." The species is more common in California.

Blechnaceae (C. Presl) Copel. (formerly in Polypodiaceae)—CHAIN FERN FAMILY
Blechnum L.—DEER FERN

Blechnum spicant (L.) With.—DEER FERN
A handsome fern, with bright green fronds of two distinct types. The sterile fronds tend to lie flat on the ground, in a rosette. These are 20–75 cm long, simple, but so deeply divided as to appear pinnate; the leaflets are oblong, about 15 mm long and 5 mm wide. The fertile fronds are longer and stiffly erect at the center of the rosette of sterile fronds. Their leaflets are widely spaced, 10–20 mm long, just 2–3 mm wide, and inrolled at the margins. The sori form continuous bands on each side of the midvein; an indusium is present.

Very common below 3000 ft, on moist, well-drained soil in forests and at the borders of swamps, but found as high as 5000 ft at Mowich Lk. Found in all the river valleys in the Park.

Aspleniaceae Newman (formerly in Polypodiaceae)—SPLEENWORT FAMILY

Asplenium L.—SPLEENWORT

Elegant little ferns, with tufted, once-pinnate lvs. The sori are elongated and angled from the midveins of the more or less oval leaflets. An elongated indusium is attached along one side to a vein of the leaflet; it can be hard to see when the sori are mature.

1   Lf stalk reddish brown to dark brown toward base, greenish above ....
      .... *A. trichomanes-ramosum*
1   Lf stalk purple-brown its entire length ....*A. trichomanes* ssp. *trichomanes*

Asplenium trichomanes L. ssp. *trichomanes*
MAIDENHAIR SPLEENWORT

The glossy purple-brown lf stalk of this fern is quite attractive, set against the dark green leaflets. The evergreen fronds can be somewhat longer than the next species, measuring up to 20 cm long, but are usually less than 10. The leaflets are in 15–35 pairs, each 3–7 mm long.

Fairly common, in crevices of rocks and on mossy cliffs. Known from along Westside Rd at about 2200 ft; on Rampart Ridge at 3000 ft; and near Ohanapecosh. Often found with *Cryptogramma acrostichoides* and *Woodsia oregana*.

Asplenium trichomanes-ramosum L.—GREEN SPLEENWORT

The color of the lf stalk is a reliable feature to distinguish the two species. In addition, the leaflets of the green spleenwort are in 12–20 pairs, each measuring 4–10 mm long. The fronds, which are 3–15 cm long, are deciduous.

Listed as "watch" in *The Endangered, Threatened, and Sensitive Vascular Plants of Washington* (Washington Natural Heritage Program, 1997).

Rare in the Park, typically in crevices on moist cliffs above 5000 ft, although sometimes at lower elevations. Collected along the Carbon R 1.5 mi from the glacier snout and above Owyhigh Lks at 5500 ft.

Dryopteridaceae Herter (formerly in Polypodiaceae)—WOOD FERN FAMILY

These might be thought of as "generalist" ferns, with numerous species found in forests and on rocky ridges throughout the Park. The fronds are all alike and have 3 strands of veins in the leaf stalk. The sori are distinct and borne away from the margins of the leaflets. The presence and nature of the indusium is critical in separating the genera. This has also been called the "shield fern family."

1	Indusium cuplike, splitting into radiating fragments, often obscure .... *Woodsia*
1	Indusium covering the sorus from above or from the side, or indusium absent 2
2(1)	Indusium absent 3
2	Indusium present 4
3(2)	Fronds (excluding the stalk) much longer than wide; base of the plant with persistent lf stalk bases .... *Athyrium*
3	Fronds (excluding the stalk) wider than long; base of the plant lacking persistent lf bases .... *Gymnocarpium*
4(2)	Indusium curved, attached from the side .... *Athyrium*
4	Indusium round to kidney-shaped, variously attached 5
5(4)	Indusium round, attached on the underside to a tiny, central stalk .... *Polystichum*
5	Indusium kidney-shaped or hoodlike, attached at a cleft on the side 7
6(5)	Small fern; indusium hoodlike .... *Cystopteris*
6	Medium-sized to large ferns; indusium kidney-shaped .... *Dryopteris*

Athyrium Roth—LADY FERN

Small to large ferns with tufted, deciduous fronds that are 2 or 3 times pinnately compound. The blade of the frond is widest at the middle, and tapered above and below. The lf stalk is chaffy with scales near the base, while the bases of the dead fronds tend to persist and form a ruff.

1	Indusium absent; plant of subalpine slopes .... *A. americanum*
1	Indusium present; plant of low- to mid-elevation forests .... *A. filix-femina* ssp. *cyclosorum*

Athyrium americanum (Butters) Maxon—ALPINE LADY FERN

Although the fronds can reach 90 cm, plants growing in exposed places are often much smaller. The leaflets are ovate-lanceolate and on short stalks. The sorus is round; an indusium is absent, but a tooth on the lf margin sometimes covers the sorus.

A common fern on open rocky slopes and moraines above about 5000 ft. Near Sluiskin Falls at 5400 ft; on talus slopes above Owyhigh Lks; on Seattle Peak at 5000 ft; and at Eunice Lk.

Athyrium filix-femina (L.) Roth ssp. *cyclosorum* (Rupr.) C. Chr.
WESTERN LADY FERN

A tall fern, reaching 1–2 m, growing in large clumps. The leaflets are lanceolate to oblong and not stalked. The sorus is slender and curved (like a crescent moon or the letter "J"), with a curved indusium.

Quite common and often abundant on swampy ground in forests and along streams below 4000 ft. Along the trail to Indian Henrys Hunting Ground at 3000 ft; around Longmire; on Shaw Cr at the White R road at 3600 ft; and at the trailhead to Windy Gap.

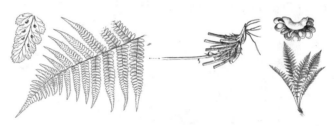

Cystopteris Bernh.
BLADDER FERN

Cystopteris fragilis (L.) Bernh.
BRITTLE FERN

A small fern, with deciduous, once- or twice-pinnately compound fronds 5–40 cm long. Its name *"fragilis"* describes it well: the leaflets are thin and light green and the lf stalks brittle. The sori are round and partially covered by a hoodlike indusium, which is soon obliterated by the growing sporangia.

Uncommon, but widespread throughout the Park up to nearly 7000 ft, typically on moist cliffs. Near Longmire; in Paradise Park; at Owyhigh Lks; on Pinnacle Peak trail; between Cayuse Pass and Tipsoo Lk; at Summerland; at 6800 ft at Frozen Lk; on McNeely Peak; and at 2000 ft 1 mi northwest of Chenuis Falls.

Dryopteris Adans.—WOOD FERN

Mostly large ferns, with tufted fronds reaching about 100 cm. They grow from stout, chaffy creeping rootstocks; the lower leaf stalk is chaffy or scaly as well. The fronds are broadly triangular oblong-lanceolate in outline. The sori are rounded to somewhat kidney-shaped, partly covered by an indusium of the same shape. The members of this genus are sometimes called "shield ferns."

Dryopteris cristata (L.) A. Gray is listed in the National Park Service flora, but the basis for this is not known; see "Doubtful and Excluded Species," p. 469.

| 1 | Fronds 3-times pinnate .... *D. expansa* |
| 1 | Fronds nearly but not fully twice-pinnate .... *D. filix-mas* |

Dryopteris expansa (C. Presl) Fraser-Jenk. & Jermy
SPREADING WOOD FERN

The blade of the frond is lanceolate to deltate and the longest leaflets are found near the base of the blade. The leaflets of the 3-times pinnate fronds are tipped with short, sharp points.

Very common in dense forests up to about 4000 ft, often growing with western lady fern. Around Ohanapecosh; on the trail to Green Lk; and 1 mi northwest of Chenuis Falls at 2000 ft.

Dryopteris filix-mas (L.) Schott
MALE FERN

The blade of the frond is elliptical, with longest leaflets near the middle. The blade is usually not quite completely twice-pinnate. The leaflets are blunt at the tips while the teeth of the leaflets are serrate and not spine-tipped.

An uncommon fern of forests below 4,000 ft, on moist ground. One herbarium specimen was found, collected 0.5 mi below the snout of the Carbon Glacier.

Dryopteridaceae

Gymnocarpium Newm.—OAK FERN

Gymnocarpium disjunctum (Rupr.) Ching
PACIFIC OAK FERN

A medium-sized fern with broadly triangular fronds to about 40 cm tall rising singly from a slender, creeping rootstock, the blade with 3 primary divisions of nearly equal size. The leaflets are oblong and toothed. The sori are round and lack an indusium.

Common on streambanks and in moist forests to about 3500 ft. On Ipsut Cr trail; at the Tahoma Cr picnic area; along the trail to Indian Henrys Hunting Ground at 3000 ft; along the Trail of the Shadows; around Ohanapecosh; and 1 mi northwest of Chenuis Falls at 2000 ft. A collection at the Slater Herbarium at the University of Puget Sound labeled "Chinook Pass" was undoubtedly made well below the Pass. A plant was seen along the trail to the Grove of the Patriarchs that had veins of a deep violet color, like a Japanese painted fern. The name above is taken from the *Flora of North America*.

Polystichum Roth—SWORD FERN

Lower-elevation forests are practically defined by the abundant P. *munitum*, western sword fern, the only common member of the genus. They all grow as tufts of once- or twice-pinnate, evergreen fronds, typically with tough, dark-green, toothed leaflets; the lower leaflets often have an earlike lobe at the base. The lf stalks are scaly. The sori are largish and round, with a round indusium, which is attached from the middle of the underside and placed over the sporangium rather like an umbrella.

1 Fronds once-pinnate, the leaflet margins smooth or toothed (or with 1 lobe at the base of the leaflet in *P. kruckebergii*) 2
1 Fronds once- or twice-pinnate, at least the lower primary leaflets pinnately lobed to pinnate 5

2(1) Small fern, the fronds less than 25 cm long; the lowest primary leaflet usually one-lobed .... *P. kruckebergii*
2 Medium-sized to large ferns; fronds usually longer than 25 cm; the primary leaflets not lobed 3

3(2) Lowest primary leaflets triangular, reduced in size; lf stalk to 10 cm .... *P. lonchitis*
3 Lowest primary leaflets lanceolate to lance-ovate, not much reduced in size; lf stalk longer than 10 cm 4

4(3) Indusium with hairs on the margin; scales on lf blade ovate, persistent ... . *P. munitum*
4 Indusium toothed or not, but lacking hairs on the margin; scales on the lf blade lanceolate, deciduous .... *P. imbricans*

5(2)   Small fern, the fronds less than 25 cm long; leaflets to 2 cm long, ovate, usually with one lobe .... *P. kruckebergii*
5   Large fern, the fronds to 100 cm long; leaflets to 10 cm long, lanceolate, deeply pinnately lobed .... *P. andersonii*

Polystichum andersonii Hopkins—ANDERSON'S SWORD FERN

Anderson's sword fern can be easily distinguished from western sword fern, *P. munitum*, in having toothed leaflets that are themselves pinnately divided. The larger fronds also typically bear a vegetative bud on the midvein: when the old frond falls and lies flat on the ground, this bud may give rise to a new plant. The indusium has an irregular, toothed margin.

A rare fern in the Park, found on moist, shaded slopes and said to favor alder (*Alnus sinuata*) thickets. Collections have been made near the snout of the Nisqually Glacier; around Longmire; and in Van Trump Park at 3000 ft. Reported by Ed Alverson on a brushy slope on the trail to Comet Falls.

Polystichum imbricans (D. C. Eaton) D. H. Wagner ssp. *imbricans* IMBRICATE SWORD FERN

Once considered a smaller variety of the western sword fern, this fern has narrower, shorter fronds, typically less than 50 cm long; the leaflets overlap each other and each is somewhat folded over its midvein. The indusium may be toothed or not but lacks the marginal hairs of the indusium of *P. munitum*.

Much less common than the closely related *P. munitum*. Typically found on exposed cliffs and roadcuts, as in Stevens Canyon, in distinctly drier habitats than those favored by *P. munitum*. Also known at Mowich Lk and Eunice Lk.

Polystichum kruckebergii W. H. Wagner KRUCKEBERG'S HOLLY FERN

Described as a separate species only in 1966, this fern somewhat resembles P. *lonchitis* but has much shorter leaflets, to about 2 cm long, that are more deeply divided. The upper lobe at base of the midvein of the leaflet is conspicuously larger than the others. The leaflets are spine-tipped.

Rare in the Park. One herbarium collection was found in the course of this study, made by J. B. Flett on a talus slope at Owyhigh Lk. (The name used by Flett in 1903 was "P. *aculeatum lobatum* Roth.")

Polystichum lonchitis (L.) Roth—MOUNTAIN HOLLY FERN

Characterized by tough and leathery fronds 15–60 cm long. The fronds are relatively slender, to about 7 cm wide near the middle, on short, stout, scaly stalks (with leaflets, in other words, nearly to the base). The lower leaflets are almost triangular in outline while those at the middle are sickle-shaped; the teeth of the leaflets are uneven and spiny. The sori usually touch each other and the indusium is not toothed.

Uncommon but widespread, occurring on rocky slopes in open forests from middle elevations up to timberline; sometimes growing with *P. munitum*. Found near the summit of Gobblers Knob and on Mount Wow; at Mowich Lk; near Longmire; below the snout of the Nisqually Glacier; at Cayuse Pass and up the road to Tipsoo Lk; on Laughingwater Cr at 3000 ft; at Berkeley Park; on cliffs at St. Elmo Pass at 7415 ft; and above Deadwood Lk.

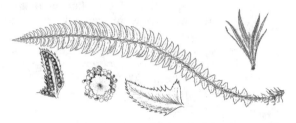

Polystichum munitum (Kaulf.) C. Presl
WESTERN SWORD FERN

With tough evergreen fronds to 150 cm or more growing in heavy clumps, this fern dominates many parts of the forest floor at lower elevations. The fronds are the widest of the genus in the Park: to 25 cm across near the middle. The leaflets are flat and do not overlap, characters by which this species may be distinguished from P. *imbricans*. The lf stalk and midvein are chaffy with persistent scales. On especially wide leaflets, the sori sometimes occur in more than one row. The indusium is fringed with short hairs.

Common and abundant, often dominating the understory in dense forests at lower elevations in all the Park's drainages, but also occurring up to the subalpine parks, as on the upper Paradise R at 6000 ft. Some specimens collected around Longmire feature leaflets with large, somewhat spreading teeth, similar in appearance to *P. lonchitis*.

Woodsia R. Br.—CLIFF FERN

*Woodsia*s are small, tufted, delicate ferns typically found on rocky slopes and cliffs, with once- or twice-divided lvs. The lvs are deciduous, but the stalks tend to persist on the short rootstock. The indusium is distinctive: cuplike and attached by its base, splitting into segments to expose the sorus. These ferns were not listed by Jones in his 1938 flora.

1   Lower side of blade both glandular and with glandless hairs ....
     .... *W. scopulina* ssp. *scopulina*
1   Lower side of blade glandular, hairs absent .... *W. oregana* ssp. *oregana*

Woodsia oregana D. C. Eaton ssp. *oregana* OREGON CLIFF FERN

The fronds are slender, twice-pinnate, and reach 25 cm long, with leaflets that are again lobed. The indusium splits into ragged or hairlike segments.

An uncommon fern of crevices, rocky ledges and talus slopes. Known from along Westside Rd at about 2500 ft; and on Tokaloo Spire at 7440 ft.

Woodsia scopulina D. C. Eaton ssp. *scopulina*— ROCKY MOUNTAIN CLIFF FERN

Somewhat larger than *W. oregana*, with once-pinnate fronds reaching 30–40 cm. The plant is finely hairy, with whitish hairs as well as stalked glands. The indusium segments are narrowly triangular.

Rare west of the Cascades in Washington and not previously reported for the Park. A collection in the Park herbarium was made on Hwy 706, growing in rock crevices of an east-facing wall above the road, 3.8 mi west of the Hwy 123 junction (that is, on the Cowlitz Divide overlooking the Ohanapecosh R). According to a note file at the University of Puget Sound, another collection was made by J. R. Slater on Westside Rd, 1 mi north of the main Park road, in 1963. The specimen has been lost, but the record almost certainly represents a misidentification of *W. oregana*, which is frequently seen in this area.

Polypodiaceae Bercht. & J. Presl
POLYPODY FAMILY

Polypodium L.—POLYPODY

These are ferns that are well-adapted to life off the ground, with slender, creeping, scaly rootstocks that enable the plants to grow on tree trunks and rock faces. The once-pinnate fronds are scattered along the rootstock. The sori are more or less round, relatively large, and lack indusia.

1   Upper surface of the midribs of the leaflets hairy; leaflets pointed; typically growing on logs and tree trunks .... *P. glycyrrhiza*

1   Upper surface of the midribs of the leaflets smooth; leaflets blunt; typically growing in crevices on rocky cliffs and outcrops 2

2(1) Sori circular, set close to the leaflet margin; fronds persisting two full years .... *P. amorphum*

2   Sori oval, midway between the midrib and the margin; fronds dropping during the second year .... *P. hesperium*

Polypodium amorphum Suksd.—IRREGULAR POLYPODY

Plants in the Park belonging to this species have long gone by the name *Polypodium hesperium*. The key differences are microscopic, and include features of the scales covering the creeping stem and the presence of structures called sporangiasters in the sori. The blade of the frond is seldom more than 10 cm long, only occasionally reaching 20 cm; the blade is about 4 cm at the widest point. The sori are circular and placed close to the margin of the leaflet.

The critical features distinguishing this species from *P. hesperium* are not useful in the field, and the easily seen features are not always sufficient to separate the two. Without the microscopic evidence, then, one can guess that one has found *P. amorphum* and be right much more often than not. See the *Flora of North America*, volume 2, for full details on these two species.

Common and widespread, on cliffs and in crevices in rocks, 2000-5000 ft. Three miles below Mowich Lk along road; at the summit of Gobblers Knob; along Westside Rd at 2700 ft; on the lower slopes of the Tatoosh Range; in the vicinity of Cliff Lk and Reflection Lks; on cliffs along the road at Cowlitz Divide; at the foot of Pinnacle Peak; and at Eunice Lk.

Polypodium glycyrrhiza D. C. Eaton—LICORICE FERN

Named for the licorice flavor of the rootstocks (although some plants are unpleasantly bitter). The scales of the rootstock measure 5–9 mm long. The lf stalk is short compared to the fronds, which are thin and widely scattered on rootstock; the blade is 15–50 cm long. The leaflets are lanceolate and tapered to a point. The sori are circular or oval and set close to the midrib.

Common, on mossy logs and tree trunks (preferring big-leaf maple); occasionally on mossy ledges; up to about 5000 ft but mostly below 3500 ft in humid valley bottoms. Along Westside Rd at 2700 ft; on the Rampart Ridge trail; near Longmire at 3000 ft; and 1 mi northwest of Chenuis Falls at 2000 ft. Less often seen in the White R valley.

Polypodium hesperium Maxon
WESTERN POLYPODY

A smaller fern than the licorice fern, with the fronds reaching 20 cm long and more crowded on the rootstocks; the blade of the frond is 5–7 cm wide. The scales of the rootstock are shorter, too: 3–5 mm long. The leaflets are oblong and rounded at the tips, and the stalk is nearly as long as the leafy part of the frond. The sori occupy most of the underside of the leaflet; they are oval and placed midway between the midrib and the margin. Sporangiasters are absent.

More common east of the Cascades, and known with certainty in the Park only in a few places: on rock outcrops along the lower reach of the trail from Longmire to Van Trump Park; in rock crevices near Ohanapecosh Hot Springs; and on cliffs at Tipsoo Lk.

Gymnosperms—Conifers

Conifers completely dominate the lower mountains and valleys around Mount Rainier and some of the finest conifer forests in the northwest, outside of the Olympic Peninsula, are found in the Park. Descriptions of the Park's forests tend to be extravagant, like this one by C. Frank Brockman (1949):

> Like a brilliant diamond whose sparking beauty is enriched by contrast with surrounding emeralds, the majesty of Mount Rainier is enhanced by the verdure of the forests that clothe the lesser ranges at its base.

The conifers of the Park are all evergreens (the deciduous western larch, *Larix occidentalis*, approaches the Park on the eastern slopes of the Cascade Divide but does not cross Chinook Pass). The leaves resemble scales or needles and true flowers are not present. Instead, the wind-pollinated plants produce male pollen and female ovules in separate cones; in this book they are called the "pollen cones" and the "seed cones," respectively. Juniper and yew differ in having male and female cones on separate plants. For most, the seeds develop on scales in woody cones (in juniper and yew, the seed is enclosed in a fleshy, berrylike structure).

In many species, the bark of the trunk is distinctively patterned and can provide an aid to identification—important when the lowest branches of an old-growth specimen may be thirty meters above the ground.

1	Lvs opposite or whorled	Cupressaceae
1	Lvs alternate or in bundles	2

2(1)	Seeds borne singly in fleshy, reddish, berrylike structures	Taxaceae
2	Seeds several to many, borne in woody cones	Pinaceae

Cupressaceae Rich. ex Bartl.—CYPRESS FAMILY

A diverse family, best differentiated from the other groups of conifers by the fact that the lvs are scalelike and opposite on the branches. Male "flowers" are borne in very small cones in the axils of the leaves; female cones are woody, generally small, and rounded to elongated (in *Juniperus* the cone is berrylike).

1	Prostrate, spreading shrub	*Juniperus communis* var. *montana*
1	Upright trees	2

2(1)	Cones more or less spherical; lvs green to bluish green; the branchlets drooping	*Chamaecyparis nootkatensis*
2	Cones egg-shaped or elongated; lvs yellowish green; the branchlets spreading	*Thuja plicata*

Chamaecyparis Spach—CEDAR

Chamaecyparis nootkatensis (D. Don) Spach
ALASKA YELLOW CEDAR, NOOTKA CEDAR

Trees to 50 m tall, with thin, gray-brown, scaly bark that peels in short strips (in contrast to the bark of western red cedar). The tree has a "weeping" appearance: the leader, or topmost shoot of the tree, droops to one side and the branchlets hang from the major branches. The branchlets are flattened and densely clothed with green or bluish green, 4-ranked, scalelike lvs. The tips of the lvs are sharp and spread from the branchlet. The pollen cones are roundish and very small; the seed cones are round and about 1 cm in diameter, composed of 4–6 thick, woody scales, green and glaucous when young and reddish brown at maturity. Mature trees develop heavy, flaring buttresses at the base, but old trees on exposed ground at timberline may be no more than spreading shrubs.

A runner-up in the "big tree" competition is an ancient cedar growing in the Ipsut Cr valley 0.5 mi northeast of Ipsut Pass; trees in this grove are at least 1200 years old. Large, old trees are also found at Narada Falls.

A common forest tree on rocky slopes, 3000-7000 ft, best-developed below about 5000 ft. On Gobbler's Knob near the summit; at Mowich Lk; on the North Puyallup R at 3500 ft; along the main Park road between Longmire and Paradise; at Bench Lk; and at Tipsoo Lk.

Juniperus L.—JUNIPER

Juniperus communis L. var. *montana* Aiton
COMMON JUNIPER

A species of variable stature, growing as a shrub in the Park, with the trunk as well as the branches trailing on the ground and over rocks, reaching no more than about 50 cm high but up to 2.5 m across. The bark is thin, reddish brown, and scaly. The lvs are 4–8 mm long, crowded on the branches, and narrowly lanceolate with sharp tips. Pollen and seed cones are borne in the axils of the lvs on separate plants. The male cone is round and 4–5 mm in diameter, while the scales of the seed cone become fleshy and form a roundish berry, 5–8 mm in diameter, blue-black in color and coated with a white bloom.

Common, on exposed ridges and rocky slopes, 2500-7500 ft. Found on ridges above Spray Park; at Longmire; at Panorama Point; at Bench Lk; on Eagle Peak at 5700 ft; on Plummer Peak; on the trail between Tipsoo and Dewey Lks; and along ridges above Sunrise.

Thuja L.—ARBORVITAE

Thuja plicata Donn. ex D. Don
WESTERN RED CEDAR

A tree that reaches about 60 m tall in the Park; the base of the trunk usually develops flaring buttresses and the branches spread widely. The thin, fibrous bark is reddish to grayish brown and easily peels in long strips. Young branchlets are conspicuously flattened; the lvs are paired on the branches and do not appear 4-ranked. The individual lvs are more or less oval, pointed at the tip, and flattened against the branchlet. The pollen cones are very small and borne singly near the ends of the branches, while the seed cones are rather narrowly egg-shaped, with about 10 woody scales, 1–1.5 cm long, and are borne in small clusters.

The tallest western red cedar listed by Van Pelt in *Champion Trees of Washington State* (1996) grows along the Ohanapecosh R at Hwy 123: it's 234 ft tall and relatively young.

Common and abundant on moist or wet ground, up to 3500 ft. Near Cataract Falls on Carbon R road at 3000 ft; at 3000 ft on the Puyallup R; around the Nisqually entrance; at Longmire; and at Ohanapecosh.

Pinaceae Lindl.—PINE FAMILY

Most conifers in the Park belong to this family. They are medium-height to tall trees, although at timberline some grow as shrublike krummholz. The lvs are needlelike, much longer than wide, and not at all scalelike. They are single and alternate on the branches, or in bundles of 2–5. The seed cones are mostly long and slender, with numerous woody scales. Two winged seeds develop on each scale and each scale is paired with a covering bract. The pollen cones are small and roundish and produce abundant pollen from numerous stamens.

1   Lvs in bundles of 2–5 .... *Pinus*
1   Lvs borne singly on the branches 2

2(1) Lvs with woody bases that persist as peglike structures or raised rims on the twigs when the lvs have fallen 3
2   Lvs lacking such woody bases; twigs smooth or nearly so after the lvs have fallen 4

3(2)   Lvs sharp-tipped, squarish in cross section .... *Picea*
3   Lvs rounded at the tip, flattened to angled in cross section .... *Tsuga*

4(2)   Seed cones pendulous, falling intact .... *Pseudotsuga*
4   Seed cones erect on the branches, the scales deciduous and falling
       separately (cones rarely found intact) .... *Abies*

(Stands of western larch, *Larix occidentalis* Nutt., are found just east of Chinook
Pass in Yakima County, but no example of the species has been observed in the
Park.)

Abies Mill.—FIR

Other conifers may be called "firs," but only *Abies* species are correctly
so named. The most conspicuous characteristic of the genus is that the
seed cones are erect on the branches at the top of the tree. As the cone
matures, the scales fall away, beginning at the tip, dispersing the winged
seeds; whole cones, therefore, are rarely found on the ground. The bark
is generally grayish, thin to thick, and smooth to somewhat scaly or
shallowly furrowed; the young twigs are hairy. The leaves are flattened,
with 2 rows of white bands (the tiny stomata) lengthwise on the
underside; 2 bands of stomata are also found on the upper surface of
some species. The leaves twist so that they run along the sides of the
branches and in some species they also curve upward. When the leaves
fall, they leave a flush, circular scar. Each scale of the seed cone is
covered by a papery brown bract, and the length of the bract relative to
the length of the scale is an important feature.

1   Upper surface of the lvs whitish (a hand lens reveals two lines of white
       stomata) 2
1   Upper surface of the lvs green, lacking white stomata 3

2(1)   Lvs flattened; bracts of the cones concealed; subalpine trees or shrubs ...
       .... *A. lasiocarpa* var. *lasiocarpa*
2   Lvs angled in cross section; bracts exserted; forest tree of lower elevations
       .... *A. procera*

3(2)   Lvs in 2 horizontal ranks, alternating longer and shorter on each side of the
       twig (the top of the twig thus not covered with lvs) ....
       .... *A. grandis* var. *grandis*
3   Lvs clustered on the top half of the twig, not 2-ranked, of more or less
       equal length (the top of the twig covered with lvs) .... *A. amabilis*

Abies amabilis Douglas ex Forbes—SILVER FIR, LOVELY FIR
Reaching about 60 m with a pointed crown and with thin, gray bark that
is fairly smooth until old age, when shallow furrows and scales develop.
The lvs are 2–3 cm long. The lvs are blunt at the tip, dark green on the
upper surface and with 2 white bands of stomata on the lower surface.
They are clustered on the upper half of the branch and curve outward

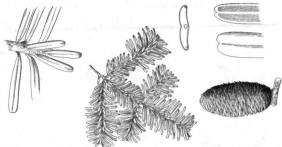

and upward. The seed cones are purple-brown, 8–15 cm long, and covered with drops of resin when young; the bracts are hidden by the scales.

A dominant tree in forests throughout the Park, 3000 and 5000 ft, occasionally higher, coming into greatest prominence above the lower-elevation zone dominated by western hemlock and Douglas fir. Said by C. Frank Brockman to be one of the most widely distributed trees in the Park. It can be seen along the main Park road from Longmire up to 3500 ft; at Paradise; along the Sunrise road; in Lodi Canyon at 5500 ft; and in Cold Basin at 5100 ft; also prominent at Cayuse Pass and Mowich Lk.

Abies grandis (Douglas ex D. Don) Lindl. var. *grandis*—GRAND FIR

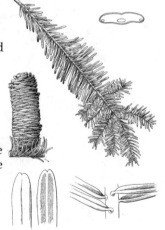

Although Jones listed a maximum height of 100 m for this species, no tree of such height is known in the Park. Typically, grand fir grows to 25–60 m and has a rounded crown. The bark of mature trees is grayish brown to reddish brown, thickish and furrowed, the furrows long and fairly close together. The lvs are 2-ranked on tides of the branches, giving the branches a flat appearance. They are green and alternate long and short along the branch, ranging in length from 1.5 to 5 cm. The lf tips are notched or blunt and have rows of white stomata only on the lower surface; the upper surface is flat except for a groove along the middle. The seed cones are 8–12 cm long, greenish brown, and resinous; the bracts are shorter than the scales.

Common in river valleys generally below 2500 ft, occasionally to 3000 ft. Less frequently seen than the other firs. Collections have been made on the North Puyallup R, along on Westside Rd at 3000 ft, and near Longmire; also known from the lower reach of the White R.

Abies lasiocarpa (Hook.) Nutt. var. *lasiocarpa*
SUBALPINE FIR

The smallest of the true firs in the Park, reaching about 20 m at the lower edge of its range, and just 5–10 m near timberline. The crown is narrow and spirelike, and the trunk usually carries the straight branches to near the ground. The bark is gray, thin, and rather smooth. The lvs are 1–3 cm long, curve upward on the branch, and have blunt tips; the upper surface has two faint rows of white stomata. The seed cones are purplish and 5–10 cm long; the bracts are shorter than the scales.

The classic spirelike conifer at higher elevations, 4500-6000 ft, encircling the mountain: near the Nisqually Glacier, at Paradise, at Chinook Pass and at Sunrise. The lower branches have a tendency to root where they touch the ground, giving rise to new trunks and, over time, forming a small grove.

Abies procera Rehder—NOBLE FIR

Capable of reaching about 60 m tall in the Park; this species differs from *A. amabilis* in having a rounded crown and from *A. grandis* by its bluish green lvs. The bark of mature trees is thick, deeply furrowed, and reddish brown. The lvs are notably 4-sided in cross section, 2.5–3.5 cm long, and curved upward, with stomata on the upper side. The seed cones are greenish brown and 10–20 cm long; the bracts are much longer than the scales and protrude from the cone, bending over the tips of the scales.

A common forest tree at middle elevations, 3000-5000 ft, often occurring as the dominant tree in small, dense stands in cool, moist locations. Collections are known from Longmire; from along the Glacier Vista trail; above Bench Lk; and at Sunrise. Especially large specimens can be found at Nahunta Falls.

Picea A. Dietr.—SPRUCE

The spruces are easily distinguished from the other conifers of the Park: the needles occur singly on the branches and each has a woody, peglike base that persists after the needle has fallen. The leaves are also very sharp at the tips. Both species in the Park have thin, scaly, gray bark. The seed cones hang from the branches and are egg-shaped to oblong.

1   Lvs flattened, with lengthwise lines of whitish stomata on the upper surface .... *P. sitchensis*

1   Lvs angled in cross section, with stomatal lines on both surfaces .... *P. engelmannii*

Picea engelmannii Parry ex. Engelm.
ENGELMANN SPRUCE

Trees in the Park are smaller than typical for the species, reaching about 20 m. The young twigs are hairy and densely clothed with green needles. The needles are about 2.5 mm long, are 4-sided in cross section, and look as if they have been brushed forward toward the branch tip. The seed cones are 5–10 cm long and yellowish brown; the scales are thin, papery, and have an irregular margin at the tip.

A characteristic tree of the mountains east of the Cascade Divide, it is infrequently seen on the north side of the Park, on moist montane slopes and ridges, 4500-6500 ft. Collections noted in the White R V above the campground and at Sunrise; on Chenuis Mtn at 5900 ft; and, surprisingly, at 2200 ft along Ipsut Cr. Floyd Schmoe, in *Our Greatest Mountain* (1925), notes that at "Summerland it is plentiful. In the Nisqually V a few isolated trees grow, and on Rampart Ridge . . . at 4900 feet a solitary individual stands." A hybrid with *Picea sitchensis* is said to occur in the Park.

Picea sitchensis (Bong.) Carrière
SITKA SPRUCE

To about 50 m tall, with a massive trunk that is usually free of branches most of its length. The twigs are hairless and less densely leafy than *P. engelmannii*. The lvs are yellowish green and 1–1.5 cm long. The seed cones are 6–10 cm long and reddish brown; the scales are toothed at the tip.

A tree of coastal forests that is very rare on the west side of the Park: in the Nisqually, Carbon, and Puyallup R valleys below 2500 ft.

Pinus L.—PINE

In the pines the leaves are in bundles of 2–5, each bundle in the axil of a short, thin bract. In all the other conifers in the Park, the leaves are single or paired on the branches. The bundled leaves tend to remain together even after they have fallen. The branches are generally whorled on young trees. The species vary greatly in size and stature, from tall forest trees to stunted krummholz at timberline. Pollen cones are clustered at the tips of the branches. The seed cones tend to be heavy and woody; sometimes they're carried in whorls on the branches. They take two years to mature and, in *P. contorta*, persist on the branches for many years.

1	Lvs 2 or 3 per bundle	2
1	Lvs 5 per bundle	3

2(1)	Lvs in bundles of 2	*P. contorta* var. *murrayana*
2	Lvs in bundles of 3	*P. ponderosa* var. *ponderosa*

3(2)	Cones ovate, to 8 cm long; timberline tree	*P. albicaulis*
3	Cones cylindric, more than 10 cm long; forest tree of lower and middle elevations	*P. monticola*

Pinus albicaulis Engelm.—WHITEBARK PINE

Whitebark pine assumes a variety of forms. Sometimes a forest tree to 15 m tall at the lower edge of its range, it is more typically seen on open slopes and ridge tops in its krummholz form, twisted by strong winds and bent low by snow, in age no more than a low, rounded shrub. The bark is thin and whitish, giving the tree its common name. The lvs are 4–6 cm long, stiff, dark green, and in bundles of 5. The seed cone is purple-brown, egg-shaped, 4–8 cm long, with very thick scales that are pointed at the tip; they fall before opening. The large seeds are sought by Clark's nutcracker as well as small animals, and the cones are often found only after they have been torn apart.

Common tree on open ridges above 5000 ft on the north and east sides of the Park, although it also grows in upper Paradise Park and on the ridge north of the Cowlitz Glacier. Collections noted from around Sunrise, Burroughs Mtn, and Plummer Peak.

Pinus contorta Douglas ex Loudon var. *murrayana* (Grev. & Balf.) Engelm.
LODGEPOLE PINE

A small tree, often with a gaunt appearance, reaching 5–15 m tall, with dark-gray, thin, scaly bark. The species epithet *"contorta"* refers to the twisted lvs, in bundles of 2, that are 3–5 cm long and dark yellow-green. The seed cones are dark brownish green, rather slender, curved, and persist on the branches for many years; they typically open to release the seeds only when subjected to fire. The scale tip is knoblike, with a stout prickle.

Found mostly on the north side of the Park in the White R drainage, but prominent also at Longmire along the Trail of the Shadows; and up to 5700 ft at Mystic Lk. C. Frank Brockman (1949) speculated that the Longmire trees were accidentally introduced there when hay from the Longmire ranch at Yelm, where lodgepole pine grows, was carried up to the springs at Longmire.

Pinus monticola Douglas ex D. Don
WESTERN WHITE PINE

The tallest of the pines in the Park, with the classic form: spirelike with spreading branches. It can grow to 50 m tall and has dark gray bark that splits into thin, squarish scales or small plates. The lvs are 4–10 cm long, light blue-green, and flexible, in bundles of 5. The seed cone is cylindrical, slender, and somewhat curved, 10–25 cm long. The scales are thin and lack a prickle. The cones are produced in large numbers.

Found between 1700 and 5000 ft, but most abundant between 3500 and 4500 ft. Noted in the White R valley, at Longmire, and just inside the Ohanapecosh entrance. It grows more strongly on poorer soils than do other conifers and is able to successfully compete with Douglas fir and western hemlock in such places. The species is subject to the fungal disease white pine blister rust, which is spread by bark beetles and kills infected trees.

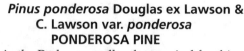

Pinus ponderosa Douglas ex Lawson & C. Lawson var. *ponderosa*
PONDEROSA PINE

Trees in the Park are smaller than typical for this species but still can reach about 40 m tall, with sturdy branches. The bark is distinctive: it breaks into largish yellow-brown plates, separated by darker fissures. The lvs are in bundles of 3 (although paired lvs occasionally are seen) and are yellow-green, flexible, and 1.5–2.5 cm long. The seed cones are yellow-brown at maturity, egg-shaped, and 7–15 cm long; the scales are heavy and each bears a sharp prickle.

Extremely rare in the northeast portion of the Park. Collections have been made along the Dewey Lk trail and along the road to Sunrise at 4000 ft. Although the species can be found on gravelly prairies between Tacoma and Olympia, the occasional trees in the Park represent outliers from east of the Cascade Divide, where ponderosa pine is very common.

Pseudotsuga Carrière—DOUGLAS FIR

Pseudotsuga menziesii (Mirb.) Franco var. *menziesii*
DOUGLAS FIR

The tallest trees in the Park are Douglas firs, and heights of 75 m are known. Mature trees have very thick, deeply fissured, grayish brown to reddish brown bark and spreading branches. The lvs are borne singly on the branches and are slender, straight, blunt at the tips, and 2–3 cm long; they are dark green on the upper side, lighter below from 2 rows of grayish stomata. They leave a slightly raised base on the branch when they fall. The seed cone is 5–10 cm long, cylindrical and slender, with 3-toothed bracts that are much longer than the thin scales.

Abundant, and frequently the dominant tree in forests below 3500 ft. Also occurring as a component of mixed forests to 4500 ft. The highest recorded collection was made at the head of Stevens Canyon at 4700 ft. Very large trees are found along the lower Ohanapecosh R, as at the Grove of the Patriarchs, and in the Mowich R valley, while thousand-year-old trees grow around the Cougar Rock campground, their relatively small size belying their great age.

Tsuga Carrière—HEMLOCK

Somewhat similar to the spruces, but far more common in the Park and easily distinguished by the leaves, which are blunt or rounded at the tip and which leave slightly raised but not peglike leaf scars when they fall. Another characteristic of the hemlocks is that the leader, the shoot at the top of the tree, droops to one side. The bracts of the seed cones are hidden by the scales. Each species is an important component of forests within its elevational range, so much so that the two of the three forest zones in the Park are named for hemlocks: the *Tsuga heterophylla* and the *Tsuga mertensiana* Zones.

1   Lvs flattened, 2-ranked on the sides of the twigs, of unequal length; forest tree of low and middle elevation .... *T. heterophylla*

1   Lvs angled in cross section, growing around the twig, of more or less equal length; subalpine trees .... *T. mertensiana*

Tsuga heterophylla (Raf.) Sarg.—WESTERN HEMLOCK

Growing to 50–60 m tall, with relatively short, spreading to pendent branches in a pyramidal crown. The bark of mature trees is rather thin and reddish brown, with broad ridges separated by narrow fissures. The lvs are straight and flattened, and vary in length, from 5 to 20 mm. They appear to grow in 2 rows, along the sides of the branches, and are dark green above, whitish below. The seed cones are egg-shaped and 1.5–2.5 cm long, with thin scales.

Very common in lower-elevation forests, up to about 4500 ft; said by C. Frank Brockman (1947) to be the most abundant conifer in the Park. Trees in the lower valley of the Carbon R and at the Grove of the Patriarchs on an island in the Ohanapecosh R are of huge stature. A few specimens can be found around Reflection and Bench Lks. The tree is host to a mistletoe, *Arceuthobium tsugense*. Infected branches become distorted, forming a "witch's broom."

Tsuga mertensiana (Bong.) Carrière
MOUNTAIN HEMLOCK

A tree 10–30 m tall, with a spirelike crown and short branches that sweep upward. The bark is purplish brown to dark reddish brown, thin, in narrow ridges. The thickened, rather than flat, lvs are 1–2 cm long and arranged around all sides of the branches; they are bluish green when young, with bands of whitish stomata on all sides. The seed cones are slender and more or less elliptical, 3–7 cm long, with thickish scales.

Found on open slopes from about 4000 ft up to timberline at about 7500 ft, chiefly on the west and south sides of the Park (on the north and east sides, mountain hemlock is largely replaced in this habitat by whitebark pine). It often follows ridge lines.

Taxaceae Gray—YEW FAMILY

Taxus L.—YEW

Taxus brevifolia Nutt.—WESTERN YEW

A small tree or large shrub, reaching 10 m but seldom growing with a straight and erect trunk; the branches often droop. The bark is reddish brown and thin, and shreds easily. The lvs are flat, 1–2 cm long, sharp-pointed, and arranged in 2 ranks on the branches. They are dark green on the upper surface and lighter on the lower, but lacking lines of whitish stomata. Trees are either male or female. The pollen cones consist of a few small bracts surrounding the stamens while the tiny female ovule develops into a bright red, fleshy, berrylike structure (an aril) that surrounds the small, woody seed.

Not uncommon, but often hidden from view in forests in deep valleys, to about 3000 ft. Perhaps most likely to be seen by visitors at the Cougar Rock campground.

Angiosperms—Flowering Plants
Dicotyledonous Plants (Dicots)

An angiosperm is a plant that bears true flowers and whose seeds are formed within an ovary. In contrast, the gymnosperms (the conifers) lack flowers and bear their seeds in cones. The angiosperms are made up of two groups: dicotyledons and monocotyledons. The dicots are easily differentiated from the monocots, at least as far as plants in the Park are concerned:

	Dicots	Monocots
Cotyledons (seed leaves)	2	1
Leaf venation	Mostly net-veined	Mostly parallel-veined
Petals, sepals, stamens	Typically borne in sets of 4 or 5, rarely 3	Typically borne in sets of 3

Key to the Families of Dicots

1 Trees or shrubs, with woody stems 2
1 Annual or perennial herbs, without woody stems (although some perennials may be slightly woody in the lower stem or have woody rootstocks) 3

2(1) Petals absent from the fls; sepals present or absent, the sepals sometimes brightly colored and petal-like .... Group 1 (below)
2 Petals present; sepals present but occasionally very much reduced in size .... Group 2 (p. 68)

3(1) Petals absent from the fls (but the sepals sometimes brightly colored and petal-like) .... Group 3 (p. 69)
3 Petals present 4

4(3) Petals distinct (or essentially so and united no more than a very short distance at the base) .... Group 4 (p. 70)
4 Petals united part or all of their length .... Group 5 (p. 72)

Group 1: Trees or shrubs, petals lacking

1 Fls in catkins 2
1 Fls not in catkins 4

2(1) Lvs pinnately lobed; only the male fls in catkins; fruit an acorn (a nut held in a scaly cup) .... Fagaceae (*Quercus*) (see "Doubtful and Excluded Species," p. 469)
2 Lvs not lobed; male and female fls in separate catkins; fruit a capsule or nutlet (but the nutlet not held in a scaly cup) 3

3(2) Lvs ovate, toothed; fruit a large or small nutlet (that may appear to be merely a hard seed); male and female fls on the same plant .... Betulaceae

3 Lvs linear to lanceolate or oblanceolate, toothed or not; male and female fls on separate plants; fruit a many-seeded capsule .... Salicaceae

4(1) Fls greenish, in small umbels; ovary inferior .... Rhamnaceae (*Rhamnus*)

4 Fls colored, solitary or in racemes; ovary superior 5

5(4) Fls yellow; lvs with spine-tipped teeth; fruit in clusters .... Berberidaceae

5 Fls purplish; lvs short, thick, turned under along the margins, not toothed; fruit solitary .... Empetraceae

Group 2: Trees or shrubs, petals present

1 Lvs opposite 2
1 Lvs alternate 8

2(1) Ovary inferior (remains of the style can be seen at the top of the fruit) ....
.... Caprifoliaceae
2 Ovary superior (the top of the fruit does not bear remains of the style) . 3

3(2) Petals united most of their length; fls irregular; plants low or tufted 4
3 Petals distinct; fls regular; stature various, but not low or tufted 5

4(3) Plant with a mintlike odor; stems square in cross section; fruit breaking into 4 nutlets .... Lamiaceae (*Monardella*)
4 Plant with an indistinct odor; stems round in cross section; fruit a many-seeded capsule .... Scrophulariaceae (*Penstemon*)

5(3) Lvs palmately veined and lobed, the margins toothed .... Aceraceae
5 Lvs pinnately veined but not lobed; the margins toothed or not 6

6(5) Stamens 8–50 .... Hydrangeaceae
6 Stamens 4–7 7

7(6) Fls clustered at the ends of the stems into heads (the heads subtended by large, petal-like bracts) or in flat-topped clusters (these lacking such bracts)
.... Cornaceae
7 Fls in small clusters along the stems in the axils of the lvs ..... Celastraceae

8(1) Petals united most of their length 9
8 Petals distinct, at most united only at their bases 10

9(8) Individual fls greenish, tubular, small, grouped in small, roundish heads; the heads in racemes or spikes .... Asteraceae (*Artemisia*)
9 Individual fls larger, colorful and showy, often bell-shaped; never in heads
.... Ericaceae

10(8) Stamens 2 or 4; plants low and heatherlike; lvs small, with inrolled margins
.... Empetraceae
10 Stamens 5 to many; plants various; if low, then not with the above characteristics 11

11(10)	Stamens 10–20 or more .... Rosaceae
11	Stamens 5 (or occasionally 4) 12
12(11)	Lvs pinnately veined; fls in small umbel-like clusters in the axils of the lvs .... Rhamnaceae
12	Lvs palmately veined; inflorescence various, but not of axillary umbels 13
13(12)	Stem densely spiny but leafy only at the top; lvs prickly on the underside and 10 cm or more broad .... Araliaceae
13	Stem prickly or not, leafy along most of its length; the lvs not prickly and rarely more than 10 cm broad .... Grossulariaceae

Group 3: Annual or perennial herbs, petals absent

1	Plant aquatic, submerged all or much of the growing season 2
1	Plant terrestrial (or parasitic on the branches of conifers); if of wet places, then not submerged 3
2(1)	Lvs opposite .... Callitrichaceae
2	Lvs whorled .... Hippuridaceae
3(1)	Plant without green color in the lvs or stems; the stems colored with red and white stripes .... Ericaceae (*Allotropa*)
3	Plant with normal green lvs 4
4(3)	Plant parasitic on the branches of conifers, rooted in the branch of the host .... Viscaceae
4	Plant not parasitic, rooted in the soil 5
5(4)	Plants tall, erect, with stinging hairs on the stems; fls in drooping racemes .... Urticaceae
5	Plants of various stature, lacking stinging hairs; inflorescence various 6
6(5)	Sepals 3, the tips very long and tapered; fls hidden beneath the lvs at the base of the plant... Aristolochiaceae
6	Sepals more than 3 (but absent in Achlys); fls conspicuous 7
7(6)	Sepals absent; the fls notable for the long and showy white stamens .... Berberidaceae (*Achlys*)
7	Sepals present.. 8
8(7)	Sepals white, petal-like; fls in a dense spike; lvs pinnately divided .... Rosaceae (*Sanguisorba*)
8	Plants without the above combination of characteristics....................... 9
9(8)	Plants with creeping stems and opposite lvs; sepals 4; stamens 8 .... Saxifragaceae (*Chrysosplenium*)
9	Plants without the above combination of characteristics..................... 10
10(9)	Stamens many; pistils 2 to many .... Ranunculaceae
10	Stamens 1–10; pistil 1 11

11(10) Lvs opposite, not toothed; fruit a capsule with 2 to many seeds ....
.... Caryophyllaceae

11 Lvs alternate, sometimes toothed; fruit 1-seeded, usually an achene ..
.... 12

12(11) Plant tall, weedy, with broadly toothed lvs that have mealy particles on the underside; stamens 5; fruit a single seed enfolded by the calyx (called a utricle) .... Chenopodiaceae

12 Stature various; the lvs seldom toothed; stamens 3, 6, or 9; fruit an achene .... Polygonaceae

Group 4: Annual or perennial herbs, petals present and distinct

1 Ovary inferior or at least partly inferior 2
1 Ovary superior 6

2(1) Ovary only partly inferior, the calyx attached to the ovary less than 2/3 its length 3
2 Ovary fully inferior 4

3(2) Plant with slender, trailing partly (but often not obviously) woody stems .
.... Hydrangeaceae (Whipplea)
3 Stems not trailing and not at all woody .... Saxifragaceae

4(2) Lvs lobed or divided; fls in umbels .... Apiaceae
4 Lvs not divided or lobed; fls in spikes, racemes, or heads 5

5(4) Lvs whorled at the top of a short stem; fls in a dense head above the lvs, subtended by white, petal-like bracts; stamens 5 .... Cornaceae
5 Lvs along the length of the stem; fls in racemes or spikes; stamens 2 or 8
.... Onagraceae

6(1) Plant aquatic, the lvs floating on the water surface; fls yellow ....
.... Nymphaeaceae
6 Plants terrestrial; if in wet places, then never submerged 7

7(6) Plants lacking green lvs (the lvs reduced to white, yellow, pink, or brown scales) .... Ericaceae
7 Plants with normal green lvs 8

8(7) Fls strongly irregular and bilaterally symmetrical 9
8 Fls regular and radially symmetric (but modestly asymmetric in Pyrola [Ericaceae] and some Saxifragaceae) 11

9(8) Fl pealike, with 5 petals: the upper petal more or less flared and bannerlike and the lower 2 folded over the stamens and pistil forming a "keel" ......
.... Fabaceae
9 Fl not pealike; instead either 1 or more of the sepals or petals sacklike or spurred 10

10(9) Uppermost sepal hoodlike or spurred; fls blue or purple ....
.... Ranunculaceae
10 1 or more of the petals sacklike or spurred; fls light to dark pink ....
.... Papaveraceae

Key to the Dicots

11(8)	Sepals 2 .... Portulacaceae	
11	Sepals 3 or more 12	
12(11)	Sepals 4, petals 4, stamens 6.. Brassicaceae	
12	Fl parts not in this combination 13	
13(12)	Insectivorous, the lvs basal and covered by sticky, gland-tipped hairs .. .. Droseraceae	
13	Lvs various, but not adapted for catching insects 14	
14(13)	Lvs thick, succulent; fls numerous, in flat-topped clusters .... Crassulaceae	
14	Lvs thin, not succulent; if thickish, then the fls otherwise 15	
15(14)	Lvs opposite or whorled on the stems 16	
15	Lvs alternate on the stems or basal... 18	
16(15)	Fls bright yellow; stamens more than 15 .... Hypericaceae	
16	Fls colored otherwise; stamens 2–10, often 5 17	
17(16)	Pistils 2, each of just 1 chamber; tufted alpine plant with purple fls .... Saxifragaceae (*Saxifraga*)	
17	Pistil 1, of 2–5 chambers; stature and fls various .... Caryophyllaceae	
18(15)	Stamens more than 10 19	
18	Stamens 10 or fewer 21	
19(18)	Filaments of the stamens united to form a tube surrounding the pistil .... Malvaceae	
19	Stamens not united to form a tube 20	
20(19)	Sepals more or less united, forming a cuplike receptacle; stamens attached to the upper part of the inside of this receptacle (above the pistils) .... Rosaceae	
20	Sepals distinct; stamens attached beneath the pistils ... Ranunculaceae	
21(18)	Sepals and petals 6 .... Berberidaceae (*Vancouveria*)	
21	Sepals and petals 5 (or 4 in some of the Saxifragaceae) 22	
22(21)	Lvs with 3 heart-shaped leaflets .... Oxalidaceae	
22	Lvs otherwise 23	
23(22)	Sepals distinct; styles 5, united from near the base to the tips; style persistent on the nutlet as a coiled beak.......................... Geraniaceae	
23	Sepals united at the base; style 1 or 2 (rarely 3 or 4), not united and not persisting on the fruits as beaks 24	
24(23)	Lvs alternate, or sometimes tufted but then the stem with at least 1 lf; styles 2 (rarely 3 or 4) .... Saxifragaceae	
24	Lvs basal; stems leafless; style 1 25	
25(24)	Lvs thin, deciduous; petals united at the base into a short tube, with the free lobes turned backwards .... Primulaceae (*Dodecatheon*)	
25	Lvs thickish, evergreen; petals distinct and not turned backwards .... Ericaceae	

Group 5: Annual or perennial herbs, petals present and united

1   Plant aquatic; lvs with 3 leaflets that are held above the water .... Menyanthaceae

1   Plants terrestrial; if in wet places, then not submerged 2

2(1)   Plant lacking green lvs, parasitic on the roots of other plants ....Orobanchaceae

2   Plants with normal green lvs 3

3(2)   Lvs basal, slimy and able to capture small insects; fl 1 per stem ....Lentibulariaceae

3   Lvs various, but not insectivorous; fls almost always more than 1 per stem 4

4(3)   Ovary inferior 5

4   Ovary superior 9

5(4)   Fls in heads, each head with an involucre of bracts; individual corollas tubular and narrow, usually small; stamens 5 .... Asteraceae

5   Inflorescence various, but not of heads subtended by involucres; corollas not small and narrowly tubular; stamens various 6

6(5)   Trailing plant with small, evergreen, opposite lvs, the stems slightly woody; the pink fls in pairs on forked, leafless stalks .... Caprifoliaceae (*Linnaea*)

6   Annual or perennial herbs, without the above combination of characteristics 7

7(6)   Lvs whorled; ovary developing into a pair of nutlets .... Rubiaceae

7   Lvs alternate or opposite; fruit a capsule or an achene 8

8(7)   Lvs alternate; fls blue; fruit a capsule .... Campanulaceae

8   Lvs opposite; fls white; fruit an achene............................... Valerianaceae

9(4)   Fls irregular, bilaterally symmetric (weakly so in some species of Lamiaceae and Scrophulariaceae) 10

9   Fls regular, radially symmetric 13

10(9)   Fl pealike, with 5 petals: the upper petal more or less flared and bannerlike, the lower 2 folded over the stamens and the pistil, forming a "keel" .... Fabaceae

10   Fl not pealike, often 2-lipped.. 11

11(10)   1 or 2 of the petals sacklike or spurred at the base .... Papaveraceae

11   Corolla not sacklike or spurred at the base; instead, weakly or strongly 2-lipped 12

12(11)   Stem square in cross section; lvs all opposite; plant with a mintlike or distinctive odor .... Lamiaceae

12   Stem round in cross section; lvs alternate or opposite; distinctive odor absent, or at least never mintlike .... Scrophulariaceae

Key to the Dicots

13(9)	Lvs thick, succulent; fls in flat-topped clusters; petals only sometimes united a short distance at the base .... Crassulaceae
13	Lvs more or less thin; inflorescence various, but not a flat-topped cluster 14
14(13)	Stamens equal in number to and placed opposite the lobes of the corolla .... Primulaceae
14	Stamens various in number, alternating with the lobes of the corolla 15
15(14)	Stamens twice the number of the corolla lobes, generally 10 .... Ericaceae
15	Stamens 4 or 5 16
16(15)	Stamens 4; corolla of a papery texture .... Plantaginaceae
16	Stamens 5; corolla normal, not of a papery texture 17
17(16)	Pistils 2, developing into a pair of follicles; plants with a milky juice .... Apocynaceae
17	Pistil 1; fruit a capsule, nutlet, or berry; plants lacking a milky juice . 18
18(17)	Ovary 4-lobed, becoming 4 nutlets .... Boraginaceae
18	Ovary not 4-lobed; fruit a capsule or berry 19
19(18)	Style 3-branched .... Polemoniaceae
19	Style unbranched or with 2 branches 20
20(19)	Lvs opposite; plant without hairs; fl usually 1 per stem .. Gentianaceae
20	Lvs alternate; plants more or less hairy; fls several, in clusters 21
21(20)	Fruit a juicy berry; fls in small umbels; uncommon weeds with trailing stems .... Solanaceae
21	Fruit a capsule; fls in coiled ("scorpioid") cymes; mostly tufted or erect native plants .... Hydrophyllaceae

Aceraceae Juss.—MAPLE FAMILY

Acer L.—MAPLE

Medium-sized shrubs to tall, spreading trees, with opposite, palmately lobed, deciduous leaves. The numerous small flowers, borne in dense racemes or rounded clusters, are followed by paired, winged seeds called samaras.

1   Larger lvs more to 20 cm or more broad; fls in racemes; large trees .... *A. macrophyllum*
1   Larger lvs less than 15 cm broad; fls in rounded clusters; small trees or shrubs 2

2(1) Lvs 3–5-lobed; fls greenish .... *A. glabrum* var. *douglasii*
2   Lvs 7–9-lobed; fls reddish .... *A. circinatum*

Acer circinatum Pursh—VINE MAPLE

Generally growing as a tall, multitrunked shrub, 5–10 m tall. The lvs are roundish in outline, 6–12 cm across, and parted into 7–9 shallow, toothed lobes. The small fls, with purplish sepals and white petals, are quite attractive, but this maple is especially noted for its brilliant scarlet and gold fall colors. The wings of the samara are red, spread widely, and form a nearly straight line.

Very common along streams and in the forest understory, most often seen below 3000 ft, but reaching 5000 ft in river valleys; most abundant on the west and south sides of the Park. In the White R drainage, it can be found growing with *Acer glabrum* var. *douglasii*.

Acer glabrum Torr. var. *douglasii* (Hook.) Dippel
ROCKY MOUNTAIN MAPLE

Usually growing as a tall shrub in the Park, but sometimes taking the form of a small tree. The lvs are pentagonal in outline, parted into 5 lobes (occasionally just 3), and 3–8 cm broad. The sepals and petals are greenish yellow and the samaras diverge only slightly, forming a V shape while still united. The autumn lvs are predominantly bright yellow.

Uncommon, except in the northeast quarter of the Park. Typically found along streams and on rocky slopes, in more open and often drier places than *Acer circinatum*. Found on Westside Rd at the beaver dams; around Longmire; on Cougar Cr trail at 3000 ft; on Hwy 123 above Dewey Cr; between Cayuse Pass and Tipsoo Lk; and along the White R below the campground.

Acer macrophyllum Pursh—BIG-LEAF MAPLE

A massive tree, to 30 m tall with heavy, widely spreading branches. The lvs are deeply divided into 3–5 lobes and can reach 30 cm across. The greenish yellow fls are about 5 mm across and sweetly fragrant. Big-leaf maple has the largest samaras of the three species of the Park, with the wings about 4 cm long. Mature trees typically bear a heavy cloak of ferns and mosses on the trunks.

A common tree at the lowest reaches of the Nisqually and Ohanapecosh Rivers, where it grows in deep soils in the river bottoms. It grows as high as 3000 ft along Westside Rd at Fish Cr and on Tahoma Cr.

Apiaceae Lindl.
(formerly Umbelliferae)
PARSLEY FAMILY, CARROT FAMILY

Short to tall perennials (or biennials), generally with hollow stems, pinnately divided lvs, and small fls in flat-topped umbels. The bases of the stem lvs are usually expanded and sheathe the stalk. The umbels may be themselves composed of smaller umbels, called umbelets, each of which is on a special stem called a "ray." Each fl has 5 sepals and 5 petals, most commonly white or yellow. The inferior ovary develops into a dry fruit, a schizocarp, which at maturity splits into a pair of seeds held by a short, stemlike structure. The fruits are variously ribbed or winged. Many members of this family have distinctive odors.

1	Fls in a simple, compact umbel; fruit with hooked bristles	*Sanicula*
1	Fls in compound umbels; fruit smooth to short-hairy, but lacking hooked bristles	2
2(1)	Fruit elongated, at least 4 times longer than broad	*Osmorhiza*
2	Fruit various, but not elongated	3
3(2)	Fruit markedly flattened; some or all of the ribs of the fruit winged	4
3	Fruit oblong, blocky, or rounded, not flattened; ribs of the fruit not winged (but sometimes thickened)	7
4(3)	Lvs 2 or 3 times compound, the leaflets broad and coarsely toothed	5
4	Lvs 3 times compound, the leaflets narrowly divided or dissected	6
5(4)	Leaflets more than 10 cm broad; stems very stout, 1–3 m tall, long-hairy	*Heracleum*
5	Leaflets less than 5 cm broad; stems more slender, to 1.5 m tall, smooth to short-hairy	*Angelica*

6(4) Rays of the umbel fewer than 10; plants low, mostly less than 10 cm tall with basal lvs .... *Lomatium*

6 Rays of the umbel usually more than 10; plants taller, to 60 cm tall, the lvs basal or 1 or 2 on the stem .... *Ligusticum*

7(3) Plants of subalpine meadows and moist slopes; fls yellow .... *Tauschia*

7 Plants of lowland marshes, swamps, and streamsides; fls whitish 8

8(7) Umbels crowded, the rays of the umbelets 2 cm long or less .... *Oenanthe*

8 Umbels looser, the rays of the umbelets 2–8 cm long .... *Cicuta*

Angelica L.—ANGELICA

Both angelicas in the Park are stout perennials with leafy stems and reach 120 cm or more tall. The leaves are divided into 3 large divisions, each of which is further divided into broad, coarsely toothed, lanceolate to ovate leaflets. The flowers are white and numerous, in large, compound umbels. The fruits are both winged on the lateral margins and ribbed.

1 Midrib of the lvs prominently bent away from the stem .... *A. genuflexa*

1 Midrib of the lvs straight .... *A. arguta*

Angelica arguta Nutt.
SHARP-TOOTH ANGELICA

The most common angelica in the Park has lvs that are essentially flat, in contrast to the next species. The leaflets are 6–9 mm long. The mature fruit is 4–6 mm long with broad wings.

Wet, sunny places and boggy ground, mostly above 4000 ft, reaching about 5500 ft. Found at Spray Falls; in Paradise Park; below the Cowlitz Glacier; along Hwy 410 in wet ditches below Tipsoo Lk; in the meadow around Tipsoo Lk; and at 4200 ft on the trail from White R to Summerland.

Angelica genuflexa Nutt.
KNEELING ANGELICA

Easily distinguished by its unique, "kneeling" lvs. That is, the stalk of each lf is bent downward at the point of first division of the leaflets, each of which is 4–10 cm long; the primary leaflets themselves are also reflexed. The fruit is 3–4 mm long, with wings so broad that the fruit appears almost round in outline.

Scarce in the Park; found in swampy areas below 2500 ft in the Nisqually R drainage. Two collections were located: one from the Longmire swamp and one, made by O. D. Allen in 1891, from an unspecified place in the upper Nisqually V. Also reported at Berkeley Park.

Cicuta L.—WATER-HEMLOCK

Cicuta douglasii (DC.) J.M. Coult. & Rose—WESTERN WATER-HEMLOCK

A highly poisonous plant that grows 50–150 cm tall. The clustered stems are thickened at the base and a stack of small air chambers is found within the base of each stem. The lvs are 2 or 3 times pinnately divided into narrowly lanceolate leaflets. (A subtle but distinctive feature of the leaflet is that the veins branching off the midvein reach to the bases, rather than the tips, of the teeth of the leaflet.) The numerous whitish fls are in compound umbels, the whole about 6 cm across. The fruits are nearly round and wingless, with thickened ribs.

Widespread below 3000 ft, in swamps and wet meadows, often in standing water, as at Longmire. Also seen on boggy ground at Ohanapecosh and in the marsh at the Stevens Canyon entrance station. *Cicuta vagans* Greene, not a valid species but once listed for the Park, was said to differ from *C. douglasii*, based on minor characteristics of the leaflets. A number of early collections of *Oenanthe sarmentosa* (see below) were misidentified as *Cicuta vagans*.

Heracleum L.—COW-PARSNIP

One of the Park's most notable wildflowers, the tall and imposing cow-parsnip, with its wide, flat-topped clusters of small white flowers, is common on damp roadsides at low and middle elevations. The other species, giant hogweed, is one of the worst weeds to have reached the Park. The genus is characterized by large, coarsely divided leaves and large, flattened, winged seeds.

1   Plant 1–2 m tall; largest umbel less than 25 cm across .... *H. maximum*
1   Plant 2–3 m tall; largest umbel more than 50 cm across ....
    .... *H. mantegazzianum*

Heracleum mantegazzianum Sommier & Levier
GIANT HOGWEED

An oversized version of the next species, with heavier and larger stems and lvs. The stems are conspicuously flecked with red and may be completely reddish or reddish purple. The lvs are up to 100 cm long and divided into 3 major lobes, which have large and irregular teeth. The compound umbels are 50–80 cm broad, with very numerous white fls.

A Federal Noxious Weed (the only plant in the Park so designated), giant hogweed has been seen on roadsides in the vicinity of Kautz Cr in recent years, according to Park Botanist Regina Rochefort. Young plants might well be mistaken for the much more common cow-parsnip. The plant should not be handled: substances in the sap react strongly with sunlight, causing severe skin irritation.

Heracleum maximum W. Bartram
COW-PARSNIP

An eye-catching plant of impressive size, with coarse, hairy stems 1–2 m tall. The lvs are divided into 3 large, palmately lobed, coarsely toothed segments. The small, white fls are in very broad, compound umbels, 15 cm or more across. The fruits are flattened, more or less round in outline, and 8–12 mm long, with conspicuous ribs and broad wings on the lateral margins.

Very common on wet soils in meadows, along streams, and in roadside ditches, below about 5500 ft. At Mowich Lk; at Indian Henrys Hunting Ground; in the Longmire meadow; in wet places along the road through Stevens Canyon; on Westside Rd; on Laughingwater Cr trail at 4500 ft; and along the road south of the White R entrance.

Ligusticum L.—LOVAGE

Mid-sized perennials that grow from taproots. The leaves are mainly basal and divided into numerous small leaflets. Plants with pinkish flowers are occasionally seen, but white is the more common condition. The fruits are somewhat flattened to nearly round, with narrowly winged ribs.

1   Terminal umbel with 7–10 (or 15) rays; stems lacking lvs or with 1 small leaflet .... *L. grayi*
1   Terminal umbel with 15–30 rays; stems with 1 or more well-developed lvs .... *L. canbyi*

Ligusticum canbyi Coult. & Rose
CANBY'S LOVAGE

The less common of the two species in the Park and the taller, with stems 50–120 cm tall. The stems bear 1 or more lvs, usually near the midpoint. The compound umbel has, on average, 15–30 rays (each terminating in an umbelet), giving the plant a densely flowered appearance.

Found on rocky ground that is at least seasonally moist, above 4000 ft; most frequently seen on the north and east sides of the Park. At Chinook Pass; at Three Lks at 5000 ft; on Dewey Cr at 4000 ft; and at Goat Pass at 5000 ft. A number of collections made in the Park have been misidentified as *Ligusticum apiifolium*, a lowland plant whose fruits lack wings on the ribs.

Ligusticum grayi J.M. Coult. & Rose
GRAY'S LOVAGE

The 20–60 cm tall stems of this common species may lack lvs or, at most, show a single small leaflet. The compound umbels are smaller and less densely flowered, with 7–10 rays.

Common and widespread, in subalpine meadows and along streams, 5000-7000 ft. Known from Mowich Lk and Spray Park; on Mount Wow; at Indian Henrys Hunting Ground; on upper Butter Cr; in Paradise Park; in the meadow above Bench Lk; in the meadow on Shriner Peak; on Cowlitz Ridge at 7000 ft; along the road between Cayuse Pass and Chinook Pass and at Tipsoo Lk; at Sunrise; along the Burroughs Mtn trail; and at Eunice Lk.

As with the species above, collections of *L. grayi* have been misidentified as *L. apiifolium*. *Ligusticum tenuifolium* also has been listed for the Park. See "Doubtful and Excluded Species" p. 469.

Lomatium Raf.—BISCUIT-ROOT

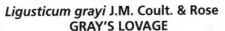

A genus of many species in the Pacific Northwest and most common in dry areas east of the Cascades. "Biscuit-root" refers to the thick, often tuberlike, taproot of the plant; many species were important sources of food for native peoples. The lvs are dissected into small leaflets. The fruits are flattened and have conspicuous lateral wings.

1   Lvs divided into long, threadlike segments; fls yellow ....
.... *L. triternatum* var. *triternatum*
1   Lvs divided into small, toothed leaflets; fls whitish .... *L. martindalei*

Lomatium martindalei (J.M. Coult. & Rose) J.M. Coult. & Rose—MOUNTAIN-PARSLEY

A plant of tufted habit, with stems 15–40 cm tall. The lvs are mostly basal, twice-divided, and bluish green; the ultimate lf segments are 1–1.5 cm long and toothed. The 3–6 small umbelets of whitish fls are on relatively long rays. The oval fruits are 8–15 mm long.

Widespread in the Park, on open rocky slopes and on cliffs, mostly above 5000 ft. Collected at Klapatche Park; on the saddle of Eagle Peak at 5900 ft; at 5000 ft on Mount Wow; by Sluiskin Falls; at Frozen Lk; on upper Butter Cr; on the trail to Burroughs Mtn from Sunrise; along the road from Cayuse Pass to Tipsoo Lk; and on the west slope of Naches Peak. Also collected once on dry ground near the bridge at Longmire.

Lomatium triternatum (Pursh) J.M. Coult. & Rose var. triternatum
NINE-LEAF MOUNTAIN-PARSLEY

Similar in size and stature to the preceding species, but with the leaflets 3 times divided into linear leaflets that are 2–8 cm long. The rays of the umbel number 5–15, with small umbelets of yellow fls. The fruits are 8–10 mm long.

Much less common than the above species; found chiefly on the west side of the Park, on open, rocky ground, 3000-5500 ft, often blooming soon after the snow melts. Found at Klapatche Park and on Mount Wow.

Oenanthe L.
WATER-PARSLEY

Oenanthe sarmentosa C. Presl
WATER-PARSLEY

With lax stems reaching 30–60 cm tall, this perennial has pinnately compound lvs and coarsely toothed leaflets, the largest segments 1–2 cm long. The small, white fls are crowded in 10 or so umbelets, the whole umbel 3–5 cm across. The fruits are nearly cylindrical, blunt at both ends, and 3–4 mm long, with prominent ribs.

Common, growing in shallow standing water in swamps and ponds, as well as along slow-moving streams; also along roads in wet ditches; below 3000 ft. Found on Westside Rd, in the swamps between Tahoma Cr and Longmire; at Longmire; and at Moraine Park. Early collections of this plant were often misidentified as *Cicuta vagans*.

Osmorhiza Raf.—SWEET-CICELY

Medium-sized herbs of wooded places, named "sweet" for the aromatic roots of some species. All are 30–60 cm tall and have leaves that are 3 times divided with toothed leaflets and widely spreading umbelets of relatively few flowers. An examination of the fruits is important in differentiating the species, as the shape of the stylopodium (base of the style), a small structure at the tip of the fruit, is critical.

1   Fruit hairless, blunt at the base .... *O. occidentalis*
1   Fruit bristly, the base slender and tapered to the ray 2

2(1)   Fls pink to purplish; the base of the style blunt, broader than high ....
.... *O. purpurea*
2   Fls whitish or greenish white; the base of the style conical, as high as broad
.... *O. berteroi*

Osmorhiza berteroi DC.
COMMON SWEET-CICELY

The most frequently seen of the Park's species and the most annoying, as the fruits are bristly on the ribs and easily become entangled in clothing. The fls are whitish to greenish white and followed by fruits that are 16–20 mm long, slightly wider above and tapered at the base, and topped with a conical stylopodium. The fruits are 12–25 mm long, linear or somewhat widened above. Very common in open forests and at roadsides below about 3000 ft. Around the Nisqually entrance at Sunshine Point and along Westside Rd; at Box Canyon; at the Ohanapecosh entrance; and in forests around the White R entrance.

Osmorhiza occidentalis (Nutt. ex Torr. & A. Gray) Torr.—WESTERN SWEET-CICELY

The least frequently seen species, and the one with the largest umbels: the umbelets are on rays 3–6 cm long, which spread widely. The stems are more clustered and upright than the other species. The fls are typically greenish white, but occasionally yellowish. The fruits are 12–18 mm long, hairless, rather cylindrical in shape, and blunt at the ray, with a conical stylopodium.

Uncommon in the Park, at elevations above 4500 ft in brushy places. Collected by O. D. Allen in the 1890s on Mount Wow; found also along the trail to Burroughs Mtn from Sunrise.

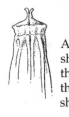

Osmorhiza purpurea (J.M. Coult. & Rose) Suksd.
PURPLE SWEET-CICELY

Although characterized by fls that are typically a purplish shade (or rarely greenish), the key distinction of the species is the stylopodium, which resembles a crown (much broader than high) at the top of the fruit. The fruits are 8–13 mm long, short-hairy, wider above, and tapered to the ray.

Frequently seen in open, moist forests above 4000 ft. Found 2 mi above Ipsut campground on the Mowich Lk trail; on Mount Wow at 4000 ft; at Paradise Park at 5000 ft; at Tipsoo Lk; and along the Interfork of the White R at 5000 ft.

Sanicula L.—SANICLE

Sanicula graveolens Poepp. ex DC.
SIERRA SANICLE

A common plant on dry, open slopes and bluffs in the Pacific Northwest, but rare in the Park; growing from a heavy taproot with mostly basal lvs that are pinnately divided into 3 major, toothed lobes, each lobe again lobed and toothed. The flowering stems reach 40 cm tall, with 3–5 small, globe-shaped umbelets in each umbel. The yellow fls are followed by oval fruits about 4 mm long and ornamented with hooked bristles.

Known in the Park only from the upper slopes of Mount Wow, where it grows on open, rocky ground.

Tauschia Schltdl.—TAUSCHIA

Tauschia stricklandii (J.M. Coult. & Rose) Mathias & Constance—STRICKLAND'S TAUSCHIA

A small species of unique appearance among the members of this family. The basal lvs arise from a branched taproot on long stalks and are divided once or twice into lanceolate to oval segments that are 8–30 mm long, thickish, and pale green. The umbel, on a stem reaching about 15 cm, contains 3–6 umbelets of tightly clustered, yellow fls. The fruits are nearly round and 2–3 mm long.

A species once thought to be endemic to Mount Rainier; discovered by O. D. Allen, and found in subalpine meadows and on moist slopes, 5000-6500 ft. At Spray Park; on Klapatche Ridge at 5500 ft and in St. Andrews Park; in Seattle Park; at Mystic Lk; in Grand Park (the only place where it is really abundant); and at Sunrise. The species is now known to occur in a limited part of the Columbia R Gorge in Oregon. Russ Jolley, in *Wildflowers of the Columbia Gorge*, has a photograph of a specimen found in 1980 in meadows at 3500 ft on upper Moffett Cr in the mountains south of Bonneville Dam.

Apocynaceae Juss.—DOGBANE FAMILY

This family is represented in the Park by a species that only barely enters the Park and by a weedy garden plant that is of doubtful persistence. Members of this family have opposite lvs and a milky juice in the stems. The flowers have 5 petals that are fused into a tube, at the base of which the 5 stamens are borne. The 2 pistils are united only at the single, enlarged stigma and develop into two elongated, podlike follicles.

1   Plant erect, branched; fls pinkish .... *Apocynum*
1   Plant with trailing stems; fls blue .... *Vinca*

Apocynum L.—DOGBANE

Apocynum androsaemifolium L. ssp. *androsaemifolium*
SPREADING DOGBANE

A very attractive low perennial with branched stems, mostly about 50 cm tall, and numerous opposite, oval lvs that are 3–8 cm long and tend to droop. The fls are pink and borne in cymes. Each corolla is bell-shaped, 5–8 mm long, with the short, free lobes curved backwards. The slender follicles are 6–8 cm long.

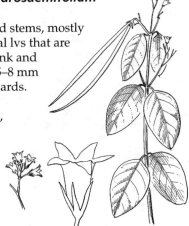

Found in a few places below about 3000 ft, on dry, sandy or gravelly soil in open places and at the margins of woods. Found at the Glacier View Bridge; at Ricksecker Point; at Nickel Cr at the roadside; and on the hillside above the Ohanapecosh entrance.

Vinca L.—PERIWINKLE

Vinca minor L.—PERIWINKLE

The common garden plant, with glossy, dark green lvs on long, arching stems that tend to grow into mounds. The fls are blue-violet, with a tube about 2 cm long and free lobes 3–4 mm long that spread sharply to form a flat face, 10–12 cm broad.

A trailing perennial introduced plant, reported to be in the Park but not verified in this study.

Araliaceae Juss.—ARALIA FAMILY

Oplopanax (Torr. & A. Gray) Miq.—DEVIL'S CLUB

Oplopanax horridus Miq.
DEVIL'S CLUB

A formidable shrub, with stems 1–4 m tall and heavily armed with yellowish spines. The lvs, clustered toward the tops of the stems, are 10–30 cm broad, palmately lobed with 5–7 major divisions, and toothed at the margins; the lvs and stalks are also spiny. The sole aralia in the Park has small fls in rounded umbels, each fl with 5 petals and 5 sepals that are fused at their bases; the pistil is 5-parted. The greenish yellow fls are borne in a tall, narrowly pyramidal raceme of small umbels that matures to an elongated cluster of red berries.

A common and conspicuous shrub of swampy ground and in wet forests; most frequently seen below 3000 ft, but occasionally to almost 5000 ft, as at Mowich Lk. Also from the Nisqually entrance up to Longmire; at 3500 ft in Cataract V; on Butter Cr; at Ohanapecosh; on Dewey Cr at the crossing of Hwy 123; along the White R in the campground area; at Green Lk; and throughout the Carbon R V. [Spelled *Oplopanax horridum* in many sources.]

Aristolochiaceae Juss.—BIRTHWORT FAMILY
Asarum L.—WILD GINGER

Asarum caudatum Lindl. var. *caudatum*
WILD GINGER

Spreading perennial, with stout rootstocks that have
the scent of ginger. The lvs are heart-shaped, deep
green, and 5–10 cm broad on slender stalks 10–15
cm tall. The solitary fls are borne on short stalks
at the base of the plant and will not be seen
unless the lvs are pushed aside. Petals are
absent; instead, the 3 purplish brown sepals
are united below to form an urn-shaped cup.
The free lobes of the sepals are long and very
slender ("caudate," or tailed).

Widespread but uncommon, below about
4500 ft in forests, especially in the south and
west sides of the Park. Found in the
Nisqually R V between the Park entrance
and the Glacier View Bridge; at Box Canyon; at Longmire; at
Ohanapecosh and the Grove of the Patriarchs; along Butter Cr; on wet
banks along the road south of the White R entrance; at the White R
campground; and in the Carbon R V below the snout of the glacier.
Along the main road just west of the Glacier View Bridge, a dense
growth of the plants lines a sunny, wet ditch.

Asteraceae Dumort. (formerly Compositae)
SUNFLOWER FAMILY, ASTER FAMILY

By far the largest family of plants in the Park, with numerous and
diverse members that are found in all habitats. Some of the worst weeds
in the Park are members of this family, as are many of the classic
meadow wildflowers.

The 47 genera are united by unique features of the inflorescence.
What might appear to be a single blossom is actually a head composed
of few to many small, individual flowers borne on a common receptacle.
One to many heads may be present on a plant. Each head is subtended,
or cradled, by one or more series of bracts, or phyllaries, which together
form the involucre; the phyllaries also may be called "involucral bracts."
A phyllary may be rather leaflike, papery, or even spiny. Two types of
individual flowers are found. Ray (or ligulate) fls are so named because
each has a corolla with a strap-shaped ligule (the whole resembling a
"petal" in the sense usually invoked by that word). Disk (or tubular) fls
have tubular corollas, without any petal-like extensions. A head is said
to be radiate when ray flowers are present; the ray flowers are usually
placed at the outer edge of the receptacle with disk flowers at the center,

but in some genera all the flowers are radiate. A head is discoid if all the flowers are disk flowers. In some genera, the individual flowers are subtended on the receptacle by chaffy bracts or scales. The individual flower has 5 petals united into a narrow tube, 5 stamens, and an inferior 1-celled ovary that develops into an achene. The style is 2-branched, and the shape of the branch tips is important at one point in the key to the genera. There is wide variation in the shape, size, and surface of the achene; a slender extension at the tip of the achene, called a "beak," may be present. The sepals have been modified in many genera into a ring of hairs, bristles, awns, or scales at the top of the ovary (and later the achene) called a "pappus"; the nature of the pappus is important in the keys that follow, but not all composites have a pappus.

Identification of the composites is made much easier by the use of 10-power magnification in order to discern the fine details of the phyllaries, pappus, and achenes. Careful dissection of the heads will also be required.

1	Plants thistlelike, the lvs and phyllaries variously spine-tipped	2
1	Plants not thistlelike; the lvs and phyllaries may be toothed, but not spine-tipped	4
2(1)	Plants with a milky sap	*Sonchus*
2	Plants with a clear, watery sap	3
3(2)	Marginal fls of the head enlarged	*Centaurea*
3	Fls of the head of similar size	*Cirsium*
4(1)	Fls all ligulate, the ligules 5-toothed; sap milky	5
4	At least some tubular fls present; ligules if present 2- or 3-toothed; sap watery	18
5(4)	Pappus absent	*Lapsana*
5	Pappus present	6
6(5)	Pappus of minute scales; fls blue	*Cichorium*
6	Pappus bristlelike to featherlike; fls variously colored, not blue	7
7(6)	Pappus featherlike	8
7	Pappus bristlelike	9
8(7)	Plant with slender, grasslike lvs on tall stems	*Tragopogon*
8	Plants with basal, toothed or lobed lvs, and leafless flowering stems	*Hypochaeris*
9(7)	Achenes flattened	10
9	Achenes round or angled	12
10(9)	Achenes beakless	*Sonchus*
10	Achenes beaked	11
11(10)	Involucre of 3–5 elongated phyllaries and many short basal ones; fls 5 per head	*Mycelis*
11	Involucre of several to many phyllaries of unequal lengths; fls 5 to many per head	*Lactuca*

12(9)	Lvs all basal; heads on leafless stalks	13
12	Basal and stem lvs present, the stem lvs typically reduced in size; heads generally on branched, leafy stems	16
13(12)	Phyllaries in 2 series, long and very short; achene minutely spiny at the upper end	*Taraxacum*
13	Phyllaries in several series; achenes not spiny	14
14(13)	Achenes beaked	*Agoseris*
14	Achenes beakless	15
15(14)	Pappus bristles brown, barbed	*Microseris*
15	Pappus bristles white, smooth	*Nothocalais*
16(12)	Fls white; heads nodding; lvs hastate (shaped like arrowheads, with the lower lobes turned back)	*Prenanthes*
16	Fls yellow or orange (white in *Hieracium albiflorum*); heads erect; basal lvs not hastate	17
17(16)	Pappus whitish; achene tapered at both ends	*Crepis*
17	Pappus brownish; achene tapered only below, or blunt at both ends .. *Hieracium*	
18(4)	Ray fls absent or inconspicuous (less than 1 mm long)	19
18	Ray fls present, generally conspicuous	36
19(18)	Pappus absent	20
19	Pappus present	24
20(19)	Each phyllary tipped with a slender spine	*Centaurea*
20	Phyllaries not spine-tipped	21
21(20)	Lvs pinnately dissected; weedy, often annual, generally odorous plants	22
21	Lvs variously lobed; native perennials, odor generally not distinct	23
22(21)	Heads solitary or few; receptacle hemispheric	*Matricaria*
22	Heads many, in a flat-topped cluster; receptacle flattish	*Tanacetum*
23(21)	Lvs deltoid-ovate, even to shallowly lobed	*Adenocaulon*
23	Lvs various, not deltoid-ovate, mostly deeply lobed to dissected .... *Artemisia*	
24(19)	Pappus of 2 barbed awns	*Bidens*
24	Pappus of slender bristles	25
25(24)	Involucre burrlike, the phyllary tips hooked	*Arctium*
25	Involucre various, but not burrlike	26
26(25)	Phyllaries papery or transparent; lvs woolly with white hairs	27
26	Phyllaries herbaceous, green; lvs variously short-hairy or hairless, never white-woolly	30
27(26)	Plants with a taproot	28
27	Plants fibrous-rooted, often with rootstocks or runners	29

28(27)	Annual or biennial, often unbranched; pappus bristles united at their bases .... *Gamochaeta*
28	Short-lived perennial, much-branched; pappus bristles not united .... *Gnaphalium*
29(27)	Basal lf rosettes present at flowering time; stem lvs much reduced .... *Antennaria*
29	Basal lf rosettes absent at flowering time; stem lvs about equal to lower lvs .... *Anaphalis*
30(26)	Plants annual .... *Conyza*
30	Plants perennial 31
31(30)	Receptacle bristly 32
31	Receptacle naked 33
32(31)	Phyllaries spine-tipped or the tips of the phyllaries fringed with short spines .... *Centaurea*
32	Phyllaries not spine-tipped .... *Saussurea*
33(31)	Lvs opposite .... *Arnica*
33	Lvs basal or alternate 34
34(33)	Fls mostly more than 20 per head; lvs deeply toothed .... *Senecio*
34	Fls mostly 5–15 per head; lvs untoothed to shallowly toothed 35
35(34)	Lvs white-woolly on the lower side .... *Luina*
35	Lvs hairless .... *Rainiera*
36(18)	Pappus absent 37
36	Pappus present 42
37(36)	Ray fls yellow .... *Madia*
37	Ray fls white 38
38(37)	Ligules less than 4 mm long; lvs pinnately divided into linear segments .... *Achillea*
38	Ligules more than 6 mm long; lvs, if pinnately dissected, then in broader segments 39
39(38)	Lvs obovate, shallowly toothed; lawn weed .... *Bellis*
39	Lvs longer than wide, variously dissected; chiefly roadside weeds ... 40
40(39)	Receptacle conic and chaffy at the center .... *Anthemis*
40	Receptacle flat, naked 41
41(40)	Ligules more than 10 mm long; heads solitary or few on leafless stems .... *Leucanthemum*
41	Ligules less than 10 mm long; heads numerous on leafy branches .... *Tanacetum*
42(36)	Pappus of slender bristles 43
42	Pappus of scales or awns 50
43(42)	Ray fls yellow 44
43	Ray fls blue, purple, or white 48
44(43)	Stem lvs opposite .... *Arnica*
44	Stem lvs alternate or lvs basal 45

Achillea L.—Yarrow

In his 1938 flora, G. N. Jones recognized two varieties of *A. millefolium*: var. *alpicola*, for high elevation plants of short stature with involucral bracts that have dark brown to nearly black margins; and *millefolium*, taller lowland plants with involucral bracts that are light brown. Some authorities, including *The Jepson Manual* (Hickman, 1993), describe *A. millefolium* as a highly variable polyploid complex and reduce the subspecies and varieties to synonymy with *A. millefolium*. Plants in the Park show intergradation in these features and there is sometimes no good correlation of these features with elevation. Still, Kartesz recognizes a number of varieties of this cosmopolitan species, and that practice is followed here.

Both varieties are short-hairy overall and have basal as well as stem lvs that are so finely divided as to appear fernlike or featherlike. The heads are arranged in flat-topped clusters and are numerous and small, about 5 mm tall, with 4 or 5 white or pinkish ray flowers. There are several yellow disk flowers. The achenes are flattened and lack a pappus.

The nonnative variety *millefolium*, frequently seen in western Washington, is apparently absent from the Park.

1   Plants of subalpine and alpine areas, 10–30 cm tall; phyllaries with margins dark brown .... var. *alpicola*
1   Plants of low to middle elevations, more than 30 cm tall; phyllaries with margins greenish to light brown .... var. *occidentalis*

Achillea millefolium L. var. *alpicola* (Rydb.) Garrett
YARROW

Tufted plants, with stems less than 30 cm tall. The phyllaries have dark brown margins. Common, often on dryish soil, 5000 -7000 ft in subalpine and alpine meadows. Found at Panorama Point; on upper Butter Cr; at Sunrise and on the Burroughs Mtn trail; at Frozen Lk, and at Goat Island Mtn.

Achillea millefolium L. var. *occidentalis* DC.
WESTERN YARROW

Mostly tall, lanky plants on low-elevation roadsides, but sometimes seen along trails. The stems can reach nearly 100 cm tall. The margins of the phyllaries are greenish to light brown.

Found at Longmire, the Glacier View Bridge, Ricksecker Point, and in the lower White R V. It can occur in disturbed places up to 5000 ft.

Adenocaulon Hook.—TRAIL PLANT

Adenocaulon bicolor Hook.
TRAIL PLANT, PATHFINDER

The lvs of this ubiquitous plant are green and smooth above but densely covered on the underside by white, woolly hairs. Supposedly, the lvs of the plant, which often grows at the sides of trails, could, by accident or design, be turned over to mark a trail. The lvs are 5–10 cm long, roughly triangular in outline and heart-shaped at the base with a lobed or wavy margin. The stems are slender and 30–100 cm tall, bearing open panicles of small heads. The involucres are about 3 mm high and 5 mm in diameter, and the heads are discoid with 5–10 whitish fls. The phyllaries are green, smooth, and reflexed at maturity. The achenes are glandular near the top and a pappus is absent.

Common and often seen along trails in forest openings below about 6500 ft. Collected at Longmire; on the Emmons Glacier trail at 4500 ft; at the Ohanapecosh and White R campgrounds; and in the woods around the Nisqually and White R entrances.

Agoseris Raf.—MOUNTAIN-DANDELION

Attractive dandelionlike plants. The lvs are entirely basal; the solitary heads are borne on leafless stems. The heads are composed entirely of ray fls, brightly colored and conspicuous. The pappus of the mature achene is of abundant soft, white bristles, and is attached to a slender beak. The stems and lvs secrete a bitter, milky juice when cut or broken.

1   Delicate annual; flowering stem less than 10 cm tall .... *A. heterophylla*
1   Perennials; stouter plants with stems at least 15 cm tall 2

2(1) Fls orange .... *A. aurantiaca* var. *aurantiaca*
2   Fls yellow 3

3(2) Beak about twice as long as the achene; basal lvs deeply dissected ....
      .... *A. grandiflora*
3   Beak very short or up to about the length of the achene; basal lvs not toothed .... *A. glauca* var. *dasycephala*

Agoseris aurantiaca (Hook.) Greene var. *aurantiaca*
ORANGE MOUNTAIN DANDELION

A taprooted perennial with lvs 10–20 cm long, narrowly oblanceolate with a few teeth along the margins. The head is 2–3 cm broad on a stem 10–30 cm tall, with orange to reddish orange ligules. The involucre is 15–20 mm high and woolly at its base. The beak of the achene is shorter than the body.

Common at roadsides and in open, dryish meadows throughout the Park, to at least 5000 ft. Known from Mowich Lk and Mtn Meadows; on Mount Wow; between Klapatche and St. Andrews Parks; along the of Nisqually R at the Glacier View Bridge; at the Canyon Rim overlook; at Reflection Lks; on Shriner Peak at 4000 ft; at Cayuse Pass and near the summit of Chinook Pass; on the road to the White R campground, as at Fryingpan Cr; at Sunrise and at Berkeley Park; and at Eunice Lk. The name "*Agoseris gracilens* (Gray) Kuntze" has been misapplied to plants from the Park with purplish spots on the involucral bracts.

Agoseris glauca (Pursh) Raf. var. *dasycephala* (Torr. & A. Gray) Jeps.
WOOLLY AGOSERIS

A stouter plant than *A. aurantiaca* and growing in harsher places. The stems rise no more than 15 cm above a tuft of thickish, untoothed, linear to oblanceolate, whitish-hairy lvs. The heads are 3–6 cm broad, with bright yellow ligules. The conspicuously woolly involucre is 12–16 mm high. The achenes are nearly beakless or have a beak no longer than the achene.

Very common on open, dry meadows and pumice fields, above 5000 ft. Collected at Sunrise; at Panhandle Gap at 6400 ft; in the Sourdough Mountains at 6800 ft; at the base of the Willis Wall; at Frozen Lk; above Mystic Lk at 6000 ft; at the base of Little Tahoma Peak; and on Burroughs Mtn. Plants from 8000 ft on Table Mtn are just 5 cm tall. Also reported from St. Andrews Park. This plant is one of the early colonizers of disturbed soils.

Agoseris glauca (Pursh) Raf. var. *monticola* (Greene) Q. Jones ex Cronquist has appeared on lists of Park plants, but its occurrence was not verified in this study. It is known, however, to occur as far north as Mount Adams. It can be distinguished from var. *dasycephala* by lvs that are lanceolate and pointed at the tips, as well as by less pubescence overall.

Agoseris grandiflora (Nutt.) Greene
LARGE-FLOWERED AGOSERIS

A taprooted perennial with basal lvs that are deeply dissected into linear lobes, the lobes pointing mostly outward and towards the tip of the lf. The lvs are oblanceolate in outline and 10–15 cm long, green and very finely hairy at 20x magnification. The flowering stems are 15–20 cm tall, smooth, but with a few loose hairs near the base and with short, thick hairs beneath the head. The head is about 20 mm high in fl (growing to around 30 mm long as the seeds mature), with phyllaries overlapping in 2 series: the outer are shorter, ovate-lanceolate, and tinged reddish, while the inner are lanceolate and become greatly lengthened in age. The ligules are yellow and about equal in length to the involucre. The pappus is of slender white bristles and the achene is 4–5 mm long, ribbed, and tapered to the tip, with a slender beak 10–15 mm long.

Found in 1998 in a few places along the road between Deer Cr and Deadwood Cr. First noted in the Park in this study; it would appear to be a new arrival in the Park.

Agoseris heterophylla (Nutt.) Greene var. *heterophylla*

A delicate little annual, from a slender taproot. The basal lvs are thinnish, oblanceolate, and short-hairy, up to about 8 cm long; the edges are usually not toothed, but more robust plants may be faintly toothed. Typically one stem per plant, but occasionally two, the stems 2-10 cm tall and very slender. The involucre is up to 1 cm high and is exceeded by the yellow ligules. The achene is topped by a slender beak 2-3 times the length of the achene, bearing a pappus of white bristles.

First seen in 1999 along the road to the White R campground, growing on the road shoulder and a nearby dry, rocky bank. First noted in the Park in this study.

Anaphalis DC.—EVERLASTING

Anaphalis margaritacea (L.) Benth. ex C.B. Clarke
PEARLY EVERLASTING

A rangy, weedlike plant, spreading by rootstocks to form loose clumps of stems, 60–100 cm tall. Most of the narrowly lanceolate lvs are on the stem; their upper surface is dark green, nicely contrasting with the white-woolly underside. The numerous heads are in a dense, flat-topped cluster. Each roundish head is 5–6 mm tall and the phyllaries are ovate, dry and papery, of a unique pearly white color. The fls are entirely discoid and dark yellow. The pappus is of short, fine, white bristles.

Widespread in the Park, from the entrance areas to 6000 ft. Typically along roads at lower elevations and on open, rocky slopes in subalpine regions, readily colonizing disturbed ground. Found on the road near Longmire; in the Paradise V to 6000 ft; in the Butter Cr Research Natural Area; at Ohanapecosh; near Tipsoo Lk; and on the Burroughs Mtn trail. High-elevation plants often have ligules tinged pinkish at base.

Antennaria Gaertn.—PUSSYTOES

A common group of plants, perennial and often spreading by runners to form low mats. The leaves and stems are variously short-hairy to woolly. The small heads are discoid, with male and female flowers found on separate plants. Male plants may occasionally be absent from a stand, the female plants then producing seeds without actual fertilization of the ovules. The shape, color, and texture (that is, whether papery or not) of the phyllaries are important in separating the species. The pappus is of fine bristles, and varies between male and female plants: The bristles of the pappus of female (pistillate) flowers are united at their bases into a ring, whereas the separate bristles of male (staminate) flowers are thickened or barbed.

1   Plants lacking runners and not forming mats ....*A. lanata*
1   Plants with runners, forming mats 2

2(1) Flowering stem glandular in the upper portion; heads in an open raceme
     .... *A. racemosa*
2   Flowering stem not glandular; heads in more or less dense, flat-topped clusters 3

3(2) Upper surface of the lvs green or sparsely long-hairy; lvs with 1–3 prominent veins 4
3   Upper surface of the lvs densely long-hairy; lvs with only 1 prominent vein 5

4(3)   Upper surface of the lvs sparsely long-hairy, at least when young ....
.... *A. howellii* ssp. *neodioica*

4   Upper surface of the lvs green and hairless, even when young ....
.... *A. howellii* ssp. *howellii*

5(4)   Phyllaries pinkish .... *A. rosea* ssp. *rosea*

5   Phyllaries at least in part greenish black to brownish black 6

6(5)   Tips of the involucral bracts pointed, the bracts greenish black throughout
.... *A. media*

6   Tips of the involucral bracts rounded, the bracts whitish at the tips, dark
brown or blackish brown on the lower portion .... *A. umbrinella*

Antennaria howellii Greene ssp. *howellii*
HOWELL'S PUSSYTOES

Differing from the variety *neodioica* in the absence of hairs from the upper surface of the lf; check young lvs, for those of var. *neodioica* can become nearly smooth in age. The lvs of var. *howellii* are also a little longer on the average, 2 cm compared to 1.5 cm.

Growing in similar habitats but evidently much less common in the Park than the following variety; observed once at the side of the road at the Ohanapecosh entrance.

Antennaria howellii Greene ssp. *neodioica* (Greene)
R.J. Bayer—FIELD PUSSYTOES

A mat-forming plant, with wide-ranging runners and obovate lvs about 1.5 cm long on tapering stalks. The slender stems reach 15–30 cm tall and are woolly but seldom glandular. The stem lvs are linear and not stalked, and the numerous small (8–10 mm high) heads are crowded into a cluster at the top of the stem. The phyllaries are narrowly lanceolate, brownish at the base and whitish at the papery, pointed tip.

Common on sunny, rocky slopes in forest openings and along road banks; mostly below 3000 ft on the west side of the Park, but also collected at 4500 ft on Sunrise Rd and along the road down the east side of the Cowlitz Divide. Once collected at an unrecorded elevation on Mount Wow. *Antennaria concolor* Piper was listed by St. John and Warren (1937); that name was evidently misapplied by O. D. Allen to plants of *A. howellii* ssp. *neodioica* collected on Mount Wow.

Antennaria lanata (Hook.) Greene—WOOLLY PUSSYTOES

Unique among the species in the Park, *A. lanata* forms tufted plants from heavy rootstocks, lacking runners or rootstocks and not forming mats. The stems are simple and reach 10–15 cm in height. The basal lvs are 4–10 cm long, oblanceolate, loosely woolly, and prominently 3-veined. The stem lvs are much reduced in size. The heads are crowded into a rounded terminal cluster. Female heads are 6–8 mm high, with dark brown to blackish phyllaries.

Common throughout the Park in alpine and subalpine areas, on rocky slopes and flats. In Spray Park; in the Paradise V to 7000 ft; on the upper reach of Butter Cr; on ridges above Seattle Park at 6500 ft; in Glacier Basin; at Sunrise and between Frozen Lk and Berkeley Park; at Goat Pass; at 5900 ft at Palisades Lk; at Chinook Pass; and in Glacier Basin. Of the five plants on a sheet made by Irene Creso, collected in Spray Park, two had 1 or 2 runners each; the dark involucral bracts of unusual specimens such as these may lead to confusion with *A. umbrinella*.

Antennaria media Greene—ALPINE PUSSYTOES

A mat-former, with long-hairy stems 3–8 cm tall. The runners are densely leafy, the lvs oblanceolate to spoon-shaped, 5–12 mm long, and profusely white-woolly. Leaves of the flowering stems are linear. The heads are crowded in a terminal cluster. Individual heads are 5–6 mm high. Both the inner and the outer phyllaries are pointed, long-hairy below and greenish black at the papery tip.

Not uncommon, on crevices in cliffs and among rocks on slopes above 6000 ft. Collected at Panorama Point; on Mount Fremont at 7200 ft; around Sunrise; and along the Burroughs Mtn trail. Also reported at Spray Park.

Antennaria racemosa Hook.—HOOKER'S PUSSYTOES

Easily distinguished by the open, elongated inflorescence, with the heads on slender stalks in a raceme or narrow panicle. Also taller than the other species, reaching 50 cm when growing in protected places. The upper stem is short-hairy but not woolly, and notably glandular. The basal lvs are densely hairy on the lower side and nearly smooth above, spoon-shaped to nearly round, on slender stalks, and 3–8 cm long; the stem lvs are linear. The heads are 5–8 mm high, with green or brown phyllaries.

Uncommon and found mostly on slopes in open woods, as on Mount Wow, but also noted along the Burroughs Mtn trail.

Antennaria rosea Greene ssp. *rosea* ROSY PUSSYTOES

The prettiest species, with pink heads on compact, mat-forming plants. The lvs are woolly on both surfaces, spoon-shaped to oblanceolate, and 2–3 cm long. The flowering stems reach 10–30 cm tall and bear 7–12 heads in a compact, terminal cluster. The heads are 5–6 mm high and the phyllaries are dull to bright pink (only rarely pale pink or whitish).

Common across the Park on hillsides or in dryish meadows, mostly above 5000 ft, but also at lower elevations on suitable sites, as along the Nisqually R above the Glacier View Bridge and at Silver Falls.

Collected at Indian Henrys Hunting Ground; just east of Reflection Lks; along the road from Cayuse Pass to Tipsoo Lk and above the lake; and along the trail below Summerland at 5400 ft.

Antennaria umbrinella Rydb.
DARK PUSSYTOES

Quite similar to *A. media* and distinguished chiefly by the phyllaries: in *A. umbrinella* they are rounded at the tip, while the papery portion of the inner phyllaries is whitish, or sometimes light brown.

Evidently rare in the Park and not previously reported. Known from collections originally labeled as "*A. alpina*" made "along road 2 mi from Paradise" (presumably west of Paradise), and in Paradise Park above of the Panorama Point.

Anthemis L.—DOG-FENNEL

Two weedy annual species, neither likely persisting for long nor spreading away from disturbed ground. The plants are short-hairy and have pinnately dissected lvs. The radiate heads have characteristic conical receptacles. The ray fls are white and a pappus is absent.

1   Lvs ill-scented; receptacle chaffy on the upper half ....*A. cotula*
1   Lvs not scented; receptacle chaffy from bottom to top ....
.... *A. arvensis* var. *arvensis*

Anthemis arvensis L. var. *arvensis*
CORN CHAMOMILE

Plants 20–40 cm tall, but sometimes sprawling on the ground, with branched stems. The heads are 2.5–3.5 cm across, with ligules 7–12 mm long.

Infrequent roadside weed, as at Longmire.

Anthemis cotula L.—STINKING CHAMOMILE

Usually less than 30 cm tall, erect with branched stems. The heads are smaller, 1–2.5 cm across with ligules 5–10 mm long.

Collected once in 1938 at the old Ohanapecosh ranger station.

Asteraceae

Arctium L.—BURDOCK

Arctium minus (Hill) Bernh.
COMMON BURDOCK

Coarse biennial, reaching about 1 m tall in the Park. The lvs are alternate on the stem, ovate or cordate, and up to 40 cm long. The numerous heads, 1.5–3 cm broad, are in a long, racemelike inflorescence. The heads are discoid, with purplish fls. The phyllaries are dull reddish, cobwebby, and armed with long, slender, hooked prickles. A pappus of numerous short bristles is present.

Uncommon weed of roadsides and waste ground, as at Ohanapecosh and Sunshine Point.

Arnica L.—ARNICA

The yellow-flowered plants that decorate the subalpine meadows and line the streams of the Park are often arnicas. With one exception, they have conspicuous yellow ligules. The heads often numerous in flat-topped inflorescences, set on medium-sized stems with opposite leaves. The ligules are usually blunt and toothed at the tip. The phyllaries are in 1 or 2 rows and a pappus of numerous fine bristles is present.

1	Heads with disk fls only	*A. parryi* ssp. *parryi*
1	Heads with disk fls and ray fls	2
2(1)	Stem lvs in 2–4 pairs; heads 5 or fewer per stem	3
2	Stem lvs in 5 or more pairs; heads 5 or more per stem	7
3(2)	Pappus of featherlike, light brown bristles	4
3	Pappus of barbed, whitish bristles	5
4(3)	Heads hemispheric; the lowest stem lvs usually the largest	*A. mollis*
4	Heads narrower, conical; the lvs at the middle of the stem the largest	*Arnica* x *diversifolia*
5(3)	Lvs lanceolate to oblanceolate, not toothed; lvs usually tufted at the base of the stem	*A. rydbergii*
5	Lvs broader, ovate to heart-shaped, toothed; lvs not tufted at the base of the stem	6
6(5)	Involucre with copious long white hairs; achene with short hairs from base to tip	*A. cordifolia*
6	Involucre hairless or with a few short hairs; achene hairless at the base and with hairs only on the upper part	*A. latifolia*
7(2)	Stems several from a somewhat woody base; lvs not toothed	*A. longifolia*
7	Stem usually rising singly from a rootstock; lvs toothed	*A. amplexicaulis*

Arnica amplexicaulis Nutt.—STREAMBANK ARNICA

With stems to about 60 cm tall rising from a creeping rootstock, this species is typically short-hairy to finely glandular-hairy, especially toward the top; occasional plants are nearly hairless. The lanceolate to ovate, toothed lvs are in 5–7 pairs, 4–10 cm long, and clasp the stem at the base; the lower lvs are smaller and often fall by flowering time. Typically there are 3–8 heads, each with an involucre about 12 mm high, arranged in a flat-topped cluster. The 8–12 ligules are 10–15 mm long and relatively broad. The pappus is light brown and featherlike.

Common, most frequently seen along streams below 6000 ft. Collected at Mtn Meadows; along Van Trump Cr; at Silver Falls; near Eunice Lk at 5400 ft; and near the saddle at Ipsut Pass.

Arnica cordifolia Hook.
HEARTLEAF ARNICA

A species with sparsely long-hairy stems reaching 40 cm tall, best distinguished by its toothed, heart-shaped lvs that are 4–10 cm long on stalks of about equal length. The 1–3 heads are on long, slender stalks. The involucre is about 15 mm tall and the phyllaries are densely hairy on the lower portion. The ligules are about 2 cm long and the pappus is white, of slender, barbed bristles.

Uncommon west of the Cascade Crest and known only from the east side of the Park. Collected at Sunrise and noted at Berkeley Park and on the Burroughs Mtn trail. An immature plant, not in flower, found on a bank at the confluence of Panther Cr and the Ohanapecosh R, had the distinctive lvs of this species.

Arnica x *diversifolia* Greene (pro. sp.)
HYBRID ARNICA

Perhaps the characteristics of *A. cordifolia* are the easiest to see in this plant: the lvs are ovate to heart-shaped, with the largest pair, about 8 cm long, at the middle of the stem. The upper lvs are generally not stalked, while the middle and lower lvs are stalked. The 1–3 heads are on long stalks. The involucre is rather narrow and widest at the top. The ligules are 1.5–2.5 cm long. The pappus is light brown and featherlike.

Rare in the Park. Thought to be a hybrid between *A. amplexifolius* or *A. mollis* and *A. cordata* or *A. latifolia*. Known from near Mowich Lk; at the Owyhigh Lks; on the south slope of Burroughs Mtn at 6600 ft; and at Berkeley Park.

Arnica latifolia Bong.
BROADLEAF ARNICA, MOUNTAIN ARNICA

With sparsely, finely hairy stems solitary or clustered on spreading rootstocks, this common species grows to about 50 cm tall (although notably shorter at higher elevations). The 2–4 pairs of ovate, toothed lvs reach about 4 cm long; those of the middle stem are usually not stalked. The 1–3 heads are set on long stalks. The involucre is 8–10 mm high, narrow, sparsely hairy with scattered glands or nearly hairless. Sometimes called the "daffodil arnica" for the bright yellow ligules which are about 1.5 cm long. The pappus is white, of barbed bristles.

Common and abundant across the Park, typically along streams and in meadows. Collected at the Glacier View Bridge; at Mowich Lk; at Indian Henrys Hunting Ground; along the river at Longmire; in Paradise Park; at Reflection Lks; on Olallie Cr at 4000 ft; on Dewey Cr at the road; on wet spots along Hwy 410 to Tipsoo Lk; on Sunrise Rd at 5500 ft; on Faraway Rock; at Berkeley Park at 6400 ft; and at Upper Palisades Lk. Plants from subalpine meadows are typically smaller in stature, with more numerous but smaller heads, and have called var. *gracilis* (Hitchcock and Cronquist, 1973). Kartesz does not maintain the varieties of *A. latifolia*, although Nelsa Buckingham does, in the *Flora of the Olympic Peninsula* (Buckingham et al., 1995).

Arnica longifolia D. C. Eaton—SPEARLEAF ARNICA

Easily distinguished from the similar but more common *A. amplexicaulis* by its untoothed lf margins. The species grows to about 50 cm tall, with several stems rising from a heavy, somewhat woody base. The lvs are in 5 or more pairs, lanceolate, and, like the stem, minutely glandular-hairy. The 3–5 heads are hemispherical, with glandular phyllaries. The ligules are 1.5 cm long; the pappus is yellowish brown and featherlike.

Rare. Known from Owyhigh Lks and Berkeley Park, where it grows on talus slopes; reported from Eunice Lk. Also found at 6600 ft on the south slope of Burroughs Mtn.

Arnica mollis Hook.—HAIRY ARNICA

This plant, 20–60 cm tall, has solitary stems rising from a branched rootstock. The 2–4 pairs of lvs are not stalked (or short-stalked at the base of the stem), not toothed to toothed, oblanceolate to elliptical and up to 15 cm long, with the longest lvs at the base of the stem. The 1–3 heads are up to 6 cm broad, on involucres 12–18 mm high. The phyllaries and stalks are glandular and hairy. The pappus is light brown and featherlike.

Common along small streams in subalpine meadows, although much less frequent there than *A. latifolia*. Found at Paradise and the Reflection Lks; in wet places along the road through to Stevens Canyon and around to the east side of the Cowlitz Divide; along the road below Tipsoo Lk; and along the trail to Dewey Lk. Also collected at Narada Falls, perhaps occurring there as a waif. Many early collections, especially those with relatively narrow lvs, were misidentified as *A. chamissonis* Gray.

Arnica parryi A. Gray ssp. *parryi*—NODDING ARNICA

The only arnica in the Park with disk fls alone, the 3–10 heads usually nod on long-hairy, glandular stems that reach 30–50 cm tall. The 3–4 pairs of lanceolate to ovate lvs are toothed and hairy on both sides. The involucres are 10–15 mm high, with a few scattered hairs. The pappus is featherlike and yellowish brown.

Uncommon, in mountain meadows, 5000–6000 ft (but also found below the White R campground). Collected on Mount Wow; on slopes above Sluiskin Falls; on Cowlitz Ridge at 6000 ft; at 6200 ft on a slope above Bear Park; from Deer Cr up to Chinook Pass and on Naches Peak; and at Sunrise and Frozen Lk. Also reported from Spray Park.

Arnica rydbergii Greene
RYDBERG'S ARNICA

A low plant, to 30 cm tall with just 2 or 3 pairs of small lvs on the stems. The lower lvs are not toothed, lanceolate to oblanceolate, about 10 cm long and tufted at the base of the stem. The 7–10 ligules are 1–2 cm long. The pappus is white and barbed.

Listed as "watch" in the *Endangered, Threatened, and Sensitive Vascular Plants of Washington* (Washington Natural Heritage Program, 1997).

Rare in the Park, this species generally ranges east of the Cascades. Collected on Mount Wow at 6000 ft; at Spray Park, where it grows on open subalpine slopes and in meadows; and on a ridge above Bear Park. Also reported from St. Andrews Park and Berkeley Park.

Artemisia L.—SAGEBRUSH

Perennial herbs or small shrubs in which the leaves and upper stems are finely white-hairy or densely clothed with matted, woolly hairs. The discoid heads are small and numerous, in panicles or racemes. The phyllaries are papery and overlapping, and a pappus is absent. Most species have strongly scented, silvery (from the hairs) foliage and tend to be found on rocky slopes and ridges. Only two of the Park's five species are frequently seen.

100

Asteraceae

1   Disk fls sterile, the achenes failing to develop; lvs dissected into nearly linear segments with smooth margins; heads in an open panicle ....
.... *A. campestris* ssp. *borealis*

1   Disk fls fertile, the achenes developing normally; plants without the above combination of other characteristics 2

2(1) Lvs all or mostly basal, the stem lvs reduced 3

2   Lvs all on the stem 4

3(2) Inflorescence a narrow spike; lf blade to 3 cm long ....
.... *A. furcata* var. *furcata*

3   Inflorescence a narrow panicle; lf blade longer than 3 cm ....
.... *A. arctica* ssp. *arctica*

4(2) Lvs densely long-hairy above and beneath ... *A. ludoviciana* ssp. *candicans*

4   Lvs densely long-hairy beneath, hairless to sparsely long-hairy above 5

5(4) Lvs about 1 cm wide (excluding the lobes); involucre wider than high .....
.... *A. tilesii* ssp. *unalaschcensis*

5   Lvs greater than 1 cm wide (excluding the lobes); involucre somewhat higher than wide .... *A. douglasiana*
(see "Doubtful and Excluded Species," p. 000)

Artemisia arctica Less. ssp. *arctica*—ARCTIC ARTEMISIA

Perennial herb with a woody base and stems 20–60 cm tall. Most of the lvs are at the base of the stem and are 3–10 cm long, sparsely long-hairy but becoming smooth in age, and pinnately dissected into lanceolate divisions mostly less than 1 cm long. The involucre is 4–7 mm high and the yellowish heads nod in a sparingly branched raceme.

Rare in the Park, found on a few rocky ridges on the west side: at 5500 ft at Ipsut Pass and on the north side of Tolmie Peak at 5800 ft.

Artemisia campestris L. ssp. *borealis* (Pall.) H. M. Hall & Clem. var. *scouleriana* (Hook.) Cronquist
PACIFIC WORMWOOD

Perennial herb, with several silky stems rising 30–70 cm from a basal rosette of lvs. The lvs are sparingly hairy and pinnately dissected into linear lobes. The basal lvs are about 8 cm long; the stem lvs are few and reduced in size. The numerous heads are borne in a tall, narrow panicle. The involucre is 2–3 mm high. The plant has no notable scent.

Rare, on dry subalpine slopes. Collected in Glacier Basin and at Burnt Park.

Artemisia furcata Bieb. var. *furcata*
THREE-FORKED ARTEMISIA

"Three-forked" for the lvs, which are palmately divided into three principle lobes, each of which is further divided into linear segments, about 7 cm long overall including the stalk. The lvs and stems are silvery-hairy. The heads are few, in a slender spike.

Rare in the Park, found on a few open ridges above 6000 ft, on the north and east sides of the mountain. Collected on ridges above Spray Park and Echo; on Mount Fremont at 7300 ft; at Knapsack Pass at 6000 ft; near Frozen Lk and on Burroughs Mtn.

Artemisia ludoviciana Nutt. ssp. *candicans* (Rydb.) D.D. Keck—SILVER WORMWOOD

A perennial, with several stems from a heavy, woody rootstock. The stems rise 50–100 cm tall and both the stems and lvs are long-hairy with white hairs. The lvs are pinnately divided into a few lanceolate lobes. The nodding heads are numerous in a tall, narrow panicle; short, simple lvs are interspersed among the lower panicle branches. The involucre is 3.5–4.5 mm high and woolly.

Common, on open talus slopes, on pumice flats, and gravelly places along rivers. Found just above Cayuse Pass on Hwy 410. Also collected on the "eastside trail 1 mile above road junction," possibly, that is, along the Ohanapecosh R near Olallie Cr.

Artemisia tilesii Ledeb. ssp. *unalaschcensis* (Besser) Hultén
ALEUTIAN WORMWOOD

Similar in appearance overall to *A. ludoviciana*, but the upper side of the lvs are green above and the heads are fewer and borne on longer-branched panicle. The involucre is notable for being wider than it is high, the reverse of *A. ludoviciana*.

Less common than *A. ludoviciana*, but found in similar habitats, chiefly on the west side of the Park. Known from between Klapatche and St. Andrews Parks at 5000 ft; at Ipsut Pass; on Mount Wow; at Tipsoo Lk; and on the Kotsuck Cr trail at 5000 ft.

Aster L.—ASTER

See "Doubtful and Excluded Species," p. 469, for *Machaeranthera canescens*.

Several genera have been split from the genus Aster, on the basis of a number of characteristics of the leaves, phyllaries, and overall habit, in the online version of Kartesz (1998). These include Canadanthus, Eucephalus, and Oreostemma. As used here, Aster has its traditional definition, and includes plants with leafy or leafless, branched or unbranched stems, and leaves that are toothed or not. The phyllaries overlap in 3–5 rows, may be sparsely hairy, and may be glandular. The ray flowers are blue to purple. The pappus consists of numerous slender hairs.

1	Phyllaries glandular; lvs toothed .... *A. modestus*	
1	Phyllaries short-hairy or hairless, not glandular; lvs generally not toothed 2	

2(1) Heads solitary on leafless stems; lvs basal ....*A. alpigenus* var. *alpigenus*
2   Heads several, on leafy, branched stems 3

3(2) Lower surface of the lvs hairy or wooly; phyllaries keeled ....
.... *A. ledophyllus* var. *ledophyllus*
3   Lvs hairless or, at most, finely short-hairy on the underside; phyllaries not keeled 4

4(3) Stem lvs clasping with earlike lobes at the base; outer phyllaries with greenish margins .... *A. foliaceus* var. *parryi*
4   Stem lvs attached by a narrow stalk; outer phyllaries with papery margins
.... *A. subspicatus* var. *subspicatus*

Aster alpigenus (Torr. & A. Gray) A. Gray var. *alpigenus*
ALPINE ASTER

Easily distinguished for having a single head per stem. The stem, less than about 20 cm tall, has a few small, narrow, bractlike lvs. The lower lvs are hairless, oblanceolate, and 3–8 cm long. The heads are 3–4 cm broad, with pale purple ligules. The phyllaries are linear and overlapping, 5–12 mm long. The pappus is of white bristles.

Widespread, 5000-8000 ft on stony slopes and in dryish meadows. Found at Spray Park; at Indian Henrys Hunting Ground; in upper Van Trump Park; throughout Paradise V; near Panorama Point; at 4500 ft on the north side of Pinnacle Peak; at Goat Pass at 5500 ft; in the meadow at the summit of Shriner Peak; on the trail to Burroughs Mtn and at Frozen Lk; at Tipsoo Lk; on the ridge above Seattle Park at 6500 ft; at Grand Park; and at Berkeley Park. Collections from Grand Park are noted "white phase," for the color of the fls. First collected in the area now within the Park by Tolmie in 1833.

Aster foliaceus Lindl. ex DC. var. *parryi* (D. C. Eaton) A. Gray—LEAFY ASTER

Reaching 20–60 cm in height, usually with a few branches. The lower stem lvs are 8–10 cm long, oblanceolate, and have a winged stalk. The stem lvs are smaller and clasp the stem at their bases.

The numerous heads are relatively few-flowered and 3–4 cm broad, with narrow, rose-purple ligules and yellow disk fls. The outer phyllaries are enlarged and some appear distinctly leaflike; they lack a papery margin. The pappus is white to pale yellow.

Common plant in subalpine meadows throughout the Park, mostly above 4000 ft. Collected at Mtn Meadows at 3900 ft; at Paradise; at Sunrise; and along Hwy 410 at 4500 ft. A widespread and variable species in western North America. Some plants from the eastside of the Park with wider bracts approach var. *canbyi* A. Gray (a variety from east and south of the Pacific Northwest). Only var. *parryi* is clearly verified for the Park by herbarium material, but the presence of var. *apricus* A. Gray is possible, given G. N. Jones's (1938) observation that "dwarf, thick-leaved specimens from high altitudes have been named var. *apricus*."

Aster ledophyllus A. Gray var. *ledophyllus* CASCADE ASTER

Growing 30–90 cm tall, with tufted stems, this aster has lanceolate-elliptical lvs 2–5 cm long that are hairless on the upper side but hairy beneath or wooly beneath. The lower lvs on the stem are smaller and often wither by flowering time. The heads are in a loose, flat-topped cluster. The phyllaries are narrowly lanceolate, 7–11 mm long, and keeled. The light to dark purple ligules are quite long, up to 2 cm. The pappus is of whitish bristles.

Common in subalpine meadows and moist open slopes, above 4000 ft, but best displayed above 5000 ft. Found at Ipsut Pass; on Mount Wow at 5000 ft; in the Butter Cr Research Natural Area; in Paradise Park; at the head of Stevens Canyon at 4500 ft; at 5500 ft at Goat Pass; on the trail to Dewey Lk; at Tipsoo Lk; across the meadows at Sunrise; on Burroughs Mtn; in Grand Park; and at Eunice Lk.

Aster modestus Lindl.
TALL NORTHERN ASTER

Unique among the Park's asters in having both toothed lvs and glandular phyllaries. The stems reach 40–90 cm tall, with leafy branches. The stalkless lvs are 4–8 cm long, lanceolate, with small teeth on the margins. The heads are numerous, 1.5–2 cm broad in a leafy, flat-topped cluster. The phyllaries are green throughout, little overlapping, and glandular. The ray fls are 10–12 mm long and deep purple. The pappus is brownish.

Common in moist meadows and along streams, 3000-5500 ft. Collected at Paradise Park, Tipsoo Lk, and at Longmire.

Aster subspicatus Nees var. *subspicatus*
DOUGLAS'S ASTER

Plants 30–100 cm or more tall, with branched stems and lanceolate lvs. Those of the stem are usually short-stalked (but if not stalked, then also not clasping at the base), lanceolate to oblong, and toothed. The heads are numerous and 1.5–2.5 cm broad, with light purple ligules and yellow disk fls. The phyllaries are firm and have papery margins. The pappus is a light reddish brown.

Said by earlier writers to be common about 3000 ft, on open, moist sites; no herbarium material was found during this study.

Bellis L.—DAISY

Bellis perennis L.—ENGLISH DAISY

The classic daisy, with a rosette of glossy, elliptical to obovate lvs and white-rayed heads, one per leafless stem. The yellow disk fls give this daisy its "eye." The phyllaries are hairy and a pappus is absent.

Common weed of lawns and along roadsides at low elevations, as at the Nisqually and Ohanapecosh entrances to the Park; at Sunshine Point; and at Longmire.

Bidens L.—STICKTIGHT

Bidens frondosa L.
LEAFY STICKTIGHT

Somewhat rank plants, with stems 50–80 cm tall, widespread branches, and opposite lvs. The lvs are 2–7 cm long and pinnately divided into 3–5 lobes, the leaflets lanceolate and sharply toothed. The ligules are golden-

105

yellow (but may be small and inconspicuous). The phyllaries are in 2 series, the outer longer and sparsely fringed with short hairs. The achene is narrowly wedge-shaped, 4-angled, with a pappus of two heavy, barbed awns.

An uncommon weedlike native plant along damp roadsides; collected at Ohanapecosh.

Centaurea L.—KNAPWEED

A genus of troublesome and highly invasive weeds. None were recorded in the Park in Jones's 1938 flora. The plants are usually much-branched, with alternate leaves. The heads are discoid and in some species the outer row of disk flowers is enlarged. The phyllaries are distinctly ornamented with a fringelike or spiny tip. The pappus is of short bristles or narrow scales.

1	Phyllaries spine-tipped	*C. diffusa*
1	Phyllaries not spine-tipped	2
2(1)	Lvs deeply and pinnately lobed	*C. biebersteinii*
2	Lf margins even or merely toothed	*C. nigra*

Centaurea biebersteinii DC.—SPOTTED KNAPWEED
Biennial, with pinnately divided lvs and stems to 1 m tall. The heads are numerous, in an open panicle, and 2–3 cm wide. The tips of the phyllaries are fringed with black hairs (hence, the involucre appears "spotted"). The disk fls are a light reddish purple, with a pappus of short bristles.

Collected by Park Botanist Regina Rochefort in Stevens Canyon, 1.8 mi east of Stevens Cr, on a talus slope on the downhill side of road.

Centaurea diffusa Lam.
TUMBLE KNAPWEED, DIFFUSE KNAPWEED
Biennial, reaching 120 cm in height and diffusely branched. The lvs are pinnately divided, the lower lvs 10–15 cm long. The heads are very numerous and relatively narrow, about 1.5 cm wide. The phyllaries are pale green and spine-tipped. The disk fls are few in each head and pale purple, occasionally nearly white. The pappus is of short, narrow scales, but sometimes is absent.

Seen only rarely, as in 1999 about 1.5 miles north of Cayuse Pass along Hwy 410.

Centaurea nigra L.—BLACK KNAPWEED

A perennial, 30–100 cm tall, softly hairy, with untoothed or
toothed lvs. The heads are few in number, in a more or less
flat-topped cluster. The phyllaries are blackish, fringed at
the tip with bristly teeth. The disk fls are numerous and
purple, in heads about 2 cm wide. The pappus is of
numerous, short bristles.

Collected once, in 1925, in the White R
campground, but evidently not persisting.

Cichorium L.—CHICORY

Cichorium intybus L.
CHICORY

A bright blue, cheerful
perennial weed, with rigid, spreading stems to
nearly 1 m tall in favorable situations. The basal
lvs, in a rosette, are deeply divided into angular
lobes. The heads, 3–4 cm broad, are stalkless on
the upper branches and composed of ray fls only.
The pappus is a low crown of scales, in 2 or 3
series.

An occasional weed in the Park, known from a
collection made at the roadside near the Shriner
Peak trailhead. Likely also to turn up in the
Nisqually and Ohanapecosh entrance areas.

Cirsium Mill.—THISTLE

Fortunately, only two weedy thistles are found in the Park, and neither is
common away from disturbed ground at low elevations, although
Cirsium edule, the native thistle, is an aggressive plant in its own right.
All have spine-tipped lvs and phyllaries, and sometimes spiny wings on
the stems as well. The alternate lvs are coarsely toothed. The heads are
globular, small to quite large, with numerous pinkish to purplish disk
fls. The pappus is of white, feathery bristles.

1   Plants with male and female fls on separate plants; phyllary spines to 1 mm
       long; fls reddish purple .... *C. arvense*
1   Plants with stamens and pistils in the same fl; phyllary spineslonger than 1
       mm; fls purple 2

2(1)   Margins of the lf stalks running down the stem and thus the stem "winged";
       weedy plant of roadsides and waste ground .... *C. vulgare*
2   Stems not thus winged; native, in open woods and on banks ..... *C. edule*

Cirsium arvense (L.) Scop.
CANADA THISTLE

A perennial with rather slender stems 1–2 m tall. The involucres are 1–2 cm high and bear the shortest thorns of the Park's thistles. The disk fls are usually a light reddish purple; male and female fls are borne on separate plants. Infrequent weed, typically at roadsides, below 3000 ft. Collected at Longmire and on the "Eastside R trail, above fork of the road" (possibly meaning along the Ohanapecosh R north of the Stevens Canyon Entrance Station).

Cirsium edule Nutt.
CAYUSE THISTLE, EDIBLE THISTLE

Usually growing as a biennial, but sometimes persisting additional years, 50–200 cm tall, with heavy, leafy stems. The lvs are woolly when young, but soon become mostly smooth. The involucres are 3–4 cm high, and the phyllaries are spine-tipped and densely webbed with hairs. The fls are purple.

Common and widespread in open forests, along roads, on open hillsides, to about 6000 ft. Along Westside Rd; between Klapatche and St. Andrews Parks; in Paradise Park; along the road through Stevens Canyon; on ridges around Moraine Park; about 1 mi west of Tipsoo Lk; in Glacier Basin; and at Eunice Lk.

The native *Cirsium brevistylum* Cronquist grows across western Washington and has been mistaken for *C. edule* elsewhere. It has not been found in the Park, but it would not be surprising to see it at the lower elevations, given the ease with which the wind spreads a thistle's feathery seeds. The two are easily distinguished: the styles of *C. edule* extend 3 mm beyond the opening of the fl, while in *C. brevistylum* (= "short-styled") they are as long as or no more than 1 mm longer than the fl tube.

Cirsium vulgare (Savi) Ten.
BULL THISTLE

Much more robust than Canada thistle, but of about the same height. The lvs are tipped with stronger spines and the stalk of the lf runs down the stem as a spiny wing. The involucres are 2.5–4 cm high, short-hairy, and heavily armed. The fls are purple.

Common weed along roadsides and on open, disturbed ground: along Westside Rd, at Longmire, and at Ohanapecosh.

Conyza Less.—CONYZA

Conyza canadensis (L.) Cronquist var. glabrata (A. Gray) Cronquist
HORSEWEED

Similar in appearance to the fleabanes of the genus *Erigeron*, this tall, slender weed is an annual with a taproot. Stems reach 50–100 cm tall, with numerous narrow, untoothed lvs. The heads are very numerous, in an open panicle, with involucres 3–5 mm high. The ligules are whitish and very short, best seen with low-power magnification. The pappus is of capillary bristles.

A roadside weed, common in western Washington but rarely seen in the Park. Collected once along the road to the White R campground.

Crepis L.—HAWKSBEARD

Crepis capillaris (L.) Wallr.
SMOOTH HAWKSBEARD

Annual or biennial, with a basal rosette of lanceolate, toothed lvs and stems 30–90 cm tall. The lvs of the stem are progressively smaller towards the top and clasp the stem with earlike lobes. The inflorescence is a panicle, with spreading, curving branches and a modest number of heads, each of which is bright yellow, of ray fls only, and 1–1.5 cm broad. The phyllaries are lanceolate, hairy as well as glandular, and about 7–8 mm long. The pappus is of numerous white capillary bristles.

Common roadside weed, ranging up to about 5500 ft; only rarely seen in meadows. Collected at Longmire and Paradise, and common around the Nisqually entrance. Plants on poor soils on the Kautz mudflow are much reduced in size.

Erigeron L.—DAISY, FLEABANE

Similar to the asters (and their related genera), these are annual, biennial, or perennial herbs with alternate leaves and heads borne in panicles or flat-topped clusters, or occasionally solitary. The phyllaries are in 1 or 2 series (3–5 series is typical for the asters). The ligules are numerous and narrow (in asters, they tend to be fewer and broader), with numerous yellow disk flowers. The pappus may be of simple bristles or may be double, consisting of bristles and an outer row of narrow scales.

1	Ray fls yellow	*E. aureus* var. *aureus*
1	Ray fls blue, purple, or white	2
2(1)	Ligules very short and inconspicuous, less than 4 mm long, erect	3
2	Ligules conspicuous, 6 mm or more long, spreading	4
3(2)	Plants 30–60 cm tall; heads several to many per stem .	*E. acris* ssp. *politus*
3	Plants less than 30 cm tall; heads 1 to a few per stem	
	.... *E. acris* ssp. *acris* var. *debilis*	
4(2)	Pappus of the disk fls in 2 series (long, slender bristles and short, stout hairs); pappus of the ray fls of short, stout hairs only	*E. strigosus*
4	Pappus of the disk fls and ray fls alike (long, slender bristles, or with both long bristles and short hairs)	5
5(4)	Lvs dissected into 3–5 lobes	*E. compositus*
5	Lvs toothed or not, but not divided into lobes	6
6(5)	Ligules 2–3 mm wide; phyllaries glandular	
	.... *E. peregrinus* ssp. *callianthemus* var. *callianthemus*	
6	Ligules to 1 mm wide; phyllaries long-hairy	7
7(6)	Ligules to 10 mm long; lvs toothed ... *E. philadelphicus* var. *philadelphicus*	
7	Ligules 10–20 mm long; lvs not toothed	8
8(7)	Stems and lvs mostly smooth, the latter with 1 prominent vein	
	.... *E. speciosus* var. *speciosus*	
8	Stems and lvs short-hairy, with 3 prominent veins	
	.... *E. subtrinervis* var. *conspicuous*	

Erigeron acris L. ssp. *acris* var. *debilis* A. Gray
BITTER DAISY

The least showy of the genus (along with the following subspecies), with short and inconspicuous pinkish ligules. The stems are 5–20 cm tall, with oblanceolate basal lvs and smaller, lanceolate lvs on the upper stem. The stalks and phyllaries are sparsely glandular-hairy and the pappus is brown. The heads may be solitary, or 2–3 per stem.

Infrequent, on open rocky ridges at timberline to about 7000 ft. Collected at McNeely Peak; at 5800 ft at Owyhigh Lks; at Sunrise. Also found near the mouth of Fish Cr.

Erigeron acris L. ssp. *politus* (Fr.) Schinz & Keller—BITTER FLEABANE

A lowland version of the above subspecies, growing to be 30–60 cm tall. The heads are numerous, and borne in an open panicle; the heads don't open as widely as those of var. *debilis*.

Occasional in dryish places in open woods below 4000 ft, as along the lower reach of Westside Rd. A collection was made at Longmire in the 1930s.

Erigeron aureus Greene var. *aureus* GOLDEN DAISY

A delightful alpine plant, growing from a compact base of short stems and reaching about 15 cm tall. The lvs are basal, obovate, and 2.5–5 cm long; stem lvs are few and narrow. The heads are solitary, 2–2.5 cm across, and have woolly, dark-brownish phyllaries and yellow ray and disk fls. The pappus is double, of bristles and scales.

Widespread and abundant on pumice slopes and flats, 7000-8500 ft, especially on the north and east sides of the mountain. Found at Spray Park; in Paradise Park on the Pebble Cr trail at 6850 ft; on the ridge above Seattle Park at 7000 ft; at Sunrise; at Frozen Lk; on the Burroughs Mtn trail; and at Berkeley Park.

Erigeron compositus Pursh DWARF MOUNTAIN FLEABANE

A dwarfed alpine, from a stout, woody stem base, with a few stems 5–10 cm tall. The lvs are unique among the erigerons in the Park: the lvs are hairy, mostly basal, and have long and slender stalks, with blades divided into 3 segments, each of which is again 3 times divided. The ligules are whitish to light pink, but are sometimes absent.

Uncommon, on stony flats and slopes, 4500-9000 ft. Collected on a moraine at Emmons Glacier and at Summerland 5500 ft and reported from ridges around Spray Park.

Erigeron peregrinus (Banks ex Pursh) Greene ssp. callianthemus (Greene) Cronquist var. callianthemus—MOUNTAIN DAISY

The common daisy of subalpine meadows, with unbranched short-hairy stems 30–60 cm tall. The basal lvs are alternate, stalked, spoon-shaped, and 5–20 cm long. The stem lvs are smaller and lanceolate. The heads are 2.5–4 cm broad, usually 1 per stem, with lavender to rose-purple ligules. The phyllaries are narrow, spreading, and minutely glandular. The pappus is simple and of whitish to light brown, fine bristles.

Abundant in subalpine meadows and along streams throughout the Park. Found above 3000 ft, but most common and impressive above 5000 ft. At Mowich Lk; at Indian Henrys Hunting Ground; along the Nisqually R at the Glacier View Bridge; across Paradise Park; in the Butter Cr Research Natural Area; at Silver Falls; between Cayuse Pass and Tipsoo Lk; at Chinook Pass; in lower Seattle Park; in the meadows at Sunrise up to Burroughs Mtn; and at Berkeley Park. One of the early fls to bloom at Paradise.

Erigeron philadelphicus L. var. philadelphicus PHILADELPHIA DAISY

A weedy plant, native to eastern North America, with stems 30–90 cm tall. The basal lvs are oblanceolate, toothed, short-hairy, and 3–10 cm long; the stem lvs become smaller and more lanceolate, with clasping bases. The inflorescence is a flat-topped cluster, with numerous small heads, 1.5–2.5 cm broad. The ligules are very numerous, pink. The pappus is white and simple.

Uncommon. Jones describes this as a plant of gravelly banks and shores." Collections are known from Ohanapecosh, where the species thrives at the edges of grassy areas surrounding the hot springs. The plant can also be found at the roadside at Kautz Cr.

Erigeron speciosus (Lindl.) DC. var. speciosus—SHOWY DAISY

A hairless, tufted perennial 30–50 cm tall with lanceolate, untoothed lvs. The 3–5 heads per stem are arranged in an open, flat-topped cluster. The ligules are very numerous and violet. The pappus is double, of bristles and scales.

Uncommon, found in a few widely scattered places in the Park: on a slope above Lk Allen on rocky soils at 5600 ft and in the Sunrise area at 6800 ft.

Erigeron strigosus Muhl. ex Willd. var. *strigosus*
DAISY FLEABANE, ROUGH-STEMMED DAISY

Annual or biennial, with stems 30–60 cm tall. The stems and lvs are short-hairy, the hairs on the stems usually lying flat. The lower lvs are narrowly spoon-shaped and 4–8 cm long; the stem lvs are smaller and linear. The 6–10 heads, each 1–2 cm broad, are in an open, flat-topped cluster. The phyllaries are narrow, sparsely hairy, and 3–4 mm long. The pappus of the disk fls is double, while that of the ray fls is single.

Evidently uncommon in the Park. One collection is known, made on dry ground along the White R road, presumably near the campground.

Erigeron subtrinervis Rydb. ex Britton var. *conspicuous* (Rydb.) Cronquist
THREE-NERVE DAISY

Quite similar in appearance to *E. speciosus*, but stems, lvs, and involucres are long-hairy. The stem lvs have 3 major veins.

Known in the Park from a few collections made on the lower slopes of Mount Wow along Westside Rd, where it prefers sunny roadcuts and open rocky slopes. Plants growing in shallow pockets of soil on rocks here reach just 10 cm in height.

Eriophyllum Lag.
ERIOPHYLLUM

Eriophyllum lanatum (Pursh)
J. Forbes var. *lanatum*
WOOLLY YELLOW DAISY, OREGON SUNSHINE

A very showy plant, with gray-green foliage and stems and large, bright yellow heads. The stems are tufted, to about 50 cm tall; it and the lvs are loosely woolly. The lvs are pinnately lobed or divided. The heads, one per stem, are 3.5–3 cm broad, with 6–15 ligules. The involucre is 10–12 mm high and the pappus is of 4–12 small, translucent scales.

Very common and conspicuous, growing on sandy or gravelly soils and on rocky banks, 2000-6000 ft. Along the upper reaches of Westside Rd; on the ridge between Mount Wow and Gobblers Knob; on the main Park road at Christine Falls; on the Ipsut Pass trail; in Paradise Park; in the Butter Cr Research Natural Area; along the highway through Stevens Canyon; near Ohanapecosh; and on the trail between Tipsoo and Dewey Lks.

Collections made on the ridge between Mount Wow and Gobblers Knob have very showy fls, with each ligule about 10 mm wide. Called "woolly-sunflower" in some books.

Gamochaeta Weddell—CUDWEED

Gamochaeta ustulata (Nutt.) Holub
PURPLE CUDWEED

Perennial herbs with lvs 2–8 cm long that are not divided, toothed, or stalked; the entire plant is densely woolly. The stems reach 30 cm tall. The heads, of yellowish disk fls only, are about 4–5 mm high and grouped on the branches of narrow panicle. The phyllaries are ovate, overlapping, brownish, and glossy near tip but woolly below. The pappus is of slender white bristles that are united at their bases and fall from the achene as a unit.

Uncommon weed found along roads and in open wooded areas, on dry gravelly soil.

Gnaphalium L.—CUDWEED

Gnaphalium microcephalum Nutt.
SLENDER CUDWEED

Somewhat bowed stems, 1–4 from a taproot, reach 20–70 cm tall, with a short, open panicle of numerous heads. The stems and lvs are whitish with woolly hairs. The linear-oblong lvs are 3–7 cm long below, shorter above, and run down the stem a short distance at their bases. The heads are of disk fls only and a dull yellow color. The phyllaries are overlapping in 3 distinct series, papery, and whitish. The pappus is of white bristles that are not united at their bases.

Inconspicuous but not uncommon, on dryish roadsides in open places, mostly below 4000 ft on the west and south sides of the Park.

Hieracium L.—HAWKWEED

Perennial herbs with a milky sap, the hawkweeds have, with one common exception, heads of bright yellow to orange ray flowers; disk flowers are absent. The phyllaries are in 1–3 series and are worth a look with a hand lens, being ornamented variously with hairs and glands. The pappus is of stiff, brown bristles. Two are weedy species, one of which successfully dominates the roadside at several places in the Park.

Hieracium longiberbe Howell, a "monitor" status plant listed by the Washington Natural Heritage Program, is said by the Program to occur in the Park, but the sole report in the Program's database points to a location called "East Slope Crystal Mtn," outside the Park boundary. Even this is certainly a case of mistaken identity, for *H. longiberbe* is a plant endemic to the Columbia R Gorge.

1	Fls white	*H. albiflorum*
1	Fls colored	2
2(1)	Plants with runners	3
2	Plants without runners	4
3(2)	Fls orange-red	*H. aurantiacum*
3	Fls yellow	*H. caespitosum*
4(2)	Lvs hairless; stems leafless	*H. gracile* var. *gracile*
4	Lvs short-hairy or hairy; stems with one or more lvs	4
5(4)	Lvs coarsely toothed; phyllaries densely stalked-glandular and bristly	*H. atratum*
5	Lvs not toothed; phyllaries glandular but sparsely if at all bristly	5
6(5)	Plants with long, bristlelike hairs throughout; lvs not glaucous	*H. cynoglossoides*
6	Plants variously short-hairy, but lacking bristlelike hairs above; lvs usually glaucous	*H. scouleri* var. *scouleri*

Hieracium albiflorum Hook.—WHITE HAWKWEED

A slender perennial, 40–80 cm tall, with unbranched stems, which are long-hairy near the base and hairless or almost so above. The lower stem lvs are short-hairy and oblanceolate, on slender stalks, 10–15 cm long. The upper lvs are smaller, lanceolate to elliptical, and not stalked. The heads are numerous and grouped at the top of the stems; each is about 1 cm broad and has white ligules. The phyllaries are linear-lanceolate, with a few simple hairs or hairless.

Very common throughout the Park, mostly below 5500 ft, but occasionally higher, along roadsides, in open, dryish woods, and on old burns. Seen along the main road between the Nisqually entrance and Narada Falls; at Spray Park; in the Butter Cr Research Natural

Area; through Stevens Canyon on to Ohanapecosh and White R; at Chinook Pass; and at Sunrise and Berkeley Park.

Hieracium atratum Fr.—POLAR HAWKWEED

A tall plant, biennial or perennial, reaching 150 cm. The lower lf blades are up to 12 cm long and taper to a short, notably long-hairy stalk; the upper lvs number 1 or 2. Both the upper and lower surfaces of the lvs are short-hairy with longish white hairs (the upper less so), and the margins are shallowly to deeply toothed. The lower stem is sparsely long-hairy, while starlike hairs are prominent near the top and in the inflorescence. Some plants show a few black, stalked glands near the top. The very numerous heads are in an open panicle; each is about 2.5 cm broad, with light to dark yellow ligules. The phyllaries show various combinations of white starlike hairs and black stalked glands, sometimes appearing from a distance to be nearly black.

Plants at Christine Falls are characterized by lvs that are all basal with clearly defined stalks; the upper lvs are represented by much-reduced bracts at the forks of the inflorescence.

A roadside weed that became established in the Park around 1975 and that has spread since then. Found on lower Westside Rd; at Christine Falls; at the Canyon Rim overlook; and at Ricksecker Point, where large numbers grow on the road banks. In 1998, it showed up in large numbers along the road between Longmire and Cougar Rock and in 1999 it had reached Narada Falls. There is at least one report of its occurrence in Stevens Canyon, which has not yet been substantiated, although the plant can be found at 3200 ft along the road on the east side of the Cowlitz Divide. It is also known from along the highway at Copper Cr, just west of the Nisqually entrance. Identified through a key to *Hieracium* in Washington prepared by Geraldine Allen for the Washington State Noxious Weed Control Board.

Hieracium aurantiacum L.—ORANGE HAWKWEED

An aggressive weed that spreads by runners and rootstocks. Most easily recognized by its reddish orange heads, which are in an umbel-like inflorescence and number 5–30. The stems, which can reach almost 100 cm tall, and the lvs, which are basal, are coarsely hairy.

A weed known from along Hwy 123 north of Ohanapecosh, at about 3000 ft. Evidently a recent arrival in the Park, it was noted in 1997 by Park Botanist Regina Rochefort.

Hieracium caespitosum Dumort.—MEADOW HAWKWEED

A weedy perennial which spreads rapidly by slender runners. The stems are 30-60 cm tall. Leaves lanceolate to oblanceolate, and roughly hairy; two smaller lvs are typically found on the stem. 10-20 heads are carried in a rather dense, flattish panicle. The ligules are yellow and about 8 mm long, slightly exceeding the involucre; heads up to 1 cm broad. The phyllaries are covered with coarse, black, stalked glands and whitish hairs.

First noted in the Park in 1999, when a large number of plants were found established on the road shoulder of Hwy 410, 1 mi north of the turnoff to the White R campground.

Hieracium cynoglossoides Arv.-Touv.
HOUND'S-TONGUE HAWKWEED

A slender perennial, with stems 30–60 cm tall that are long-hairy below and short-hairy near the top. The untoothed lvs are long-hairy and bristly, and about 12 cm long. The lower lvs are oblanceolate and tapered at the base to the stalk, while the upper lvs are smaller, lanceolate, and not stalked. The heads are numerous, 10–40, in a panicle. The heads are yellow, about 1 cm high and to 1.5 cm broad. The phyllaries are beset with black glandular hairs and a few whitish longer hairs.

Infrequent. Collected a few times in the northeast quarter of the Park: at Chinook Pass; at 5000 ft on crags above Goat Pass; near Tipsoo Lk; near Owyhigh Lks; at Summerland; and on the trail to Burroughs Mtn. Also collected once on the southwest side of the mountain, on a ridge between Iron and Copper Mtns. One collection at the University of Washington herbarium, originally designated *H. flettii*, approaches the condition of *Hieracium albertinum* (a plant of eastern Washington) in which the involucres bear copious bristles but almost lack glands. A collection made by J. B. Flett above the Owyhigh Lks is quite similar.

Hieracium gracile Hook. var. *gracile*
SLENDER HAWKWEED

A small plant, to no more than 30 cm tall, with basal lvs and slender, leafless stems. The lvs are 3–8 cm long, spoon-shaped to oval, and even or minutely toothed, mostly hairless. The small heads number 4–7, in a racemelike inflorescence; each is about 1 cm high and broad, with yellow ligules. The phyllaries are narrow and pointed, with black hairs and black hairlike glands.

Very common on dryish ground in subalpine meadows, 5000-7000 ft. Found at Spray Park and Mowich Lk; on St. Andrews Ridge at 6760 ft; along roadsides from Ricksecker Point, through Paradise V and down to the Silver Forest area; in the Butter Cr Research Natural Area; at Tipsoo Lk and Chinook Pass; along the White R road up to Sunrise; and at Berkeley Park.

Hieracium scouleri Hook. var. *scouleri*
SCOULER'S HAWKWEED

Small, like *H. gracile*, but with larger heads on leafy, long-hairy stems. The lower lvs are oblanceolate, stalked, 4–10 cm long, not toothed, and somewhat glaucous on the upper surface. The upper lvs are not stalked and lanceolate. The heads number 7–20 and are 1.5–2 cm broad. The phyllaries are glandular, long-hairy with white hairs, and with minute starlike hairs.

Known in the Park from one collection at the University of Washington, made at an unspecified place in Stevens Canyon by F. A. Warren in the early 1930s. St. John and Warren (1937) report "*H. flettii*" from rocky slopes at Indian Henrys Hunting Ground, but the vouchers have not been found; it seems likely that this actually represents *H. cynoglossoides*, which has been found in this area (see above).

Hulsea Torr. & A. Gray—HULSEA

Hulsea nana A. Gray—DWARF ALPINEGOLD

A dwarfed, taprooted perennial with glandular-hairy stems and lvs. The lobed lvs are thickened, up to 5 cm long, and arranged in a basal cluster. Each stem bears a single yellow head, about 3 cm broad. The phyllaries are lanceolate and the pappus is of translucent scales that are joined at their bases.

Listed as "watch" in *The Endangered, Threatened, and Sensitive Vascular Plants of Washington* (Washington Natural Heritage Program, 1997).

Widely scattered in the Park on pumice fields, 6000-9000 ft, but nowhere especially abundant: in Paradise Park at 6500 ft; at Panhandle Gap; on the moraine of Fryingpan Glacier at 7600 ft, and at the base of Little Tahoma Peak.

Hypochaeris L.—CAT'S-EAR

Hypochaeris radicata L.—ROUGH CAT'S-EAR

A rank perennial weed, with a basal rosette of pinnately lobed lvs up to 15 cm long, and a milky sap. "Rough" indicates the harsh hairs of the lvs. The branched, leafless stems can reach nearly 50 cm tall. The heads are numerous, of yellow ray fls only, and about 2–3 cm broad. The phyllaries are nearly linear and overlapping, and the achene has a long beak at the top of which is a pappus of feathery white hairs.

A common and abundant weed along roads, in lawns, and in campgrounds; only occasionally in open woods at low elevations.

Lactuca L.—LETTUCE

Two weedy species of lettuce are occasionally seen in the Park. *Lactuca serriola* seems not to have become very well established (in contrast to *Mycelis muralis*, the wall lettuce listed below). The plants have a milky sap, alternate and pinnately divided or lobed leaves, heads of ray flowers only, and beaked achenes that sport a feathery pappus.

1   Lf margins prickly; achenes with a long, threadlike beak .... *L. serriola*
1   Lf margins not prickly; achenes with a very short, stout beak .... *L. biennis*

Lactuca biennis (Moench) Fernald
TALL BLUE LETTUCE
An biennial native with stems reaching about 2 m tall. The lvs are toothed but not prickly. The heads are whitish to blue, very numerous in a narrow panicle. The pappus is brown.

Uncommon in open woods, mostly on the west side of the Park. Collected at the trail to Lk Allen on the lower slope of Mount Wow.

Lactuca serriola L.
PRICKLY LETTUCE
An annual or biennial weed that can grow 1–1.5 m tall. The stems and lvs are prickly and the heads are of yellow ray fls. The pappus is white.

A weed, known in the Park from two collections, one made in 1895 by Allen in the upper Nisqually V and one made in 1938 at the Ohanapecosh Ranger Station. It is likely to appear only sporadically, perhaps not persisting more than a year or two at a time.

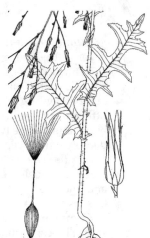

Lapsana L.—NIPPLEWORT

Lapsana communis L.
NIPPLEWORT
A taprooted annual weed, with a tall, slender, branched stem and a milky sap. The lower lvs are mostly ovate and stalked, toothed, up to about 10 cm long, and generally hairless. The upper stem lvs tend to be narrower and not stalked. The heads, of ray fls only, are yellow and very numerous. There is no pappus.

A weed of disturbed ground, usually in somewhat shaded places, as at Longmire. Very abundant in the White R V, chiefly at roadsides, but also on the banks of the river in relatively undisturbed but open woods.

Leucanthemum Mill.—DAISY

Leucanthemum vulgare Lam.
OXEYE DAISY

A tall, nonnative perennial, with alternate lvs that are lobed, toothed, and spoon-shaped in outline. The upper stem lvs are smaller and the stem terminates in a single head, of classic daisy form: a broad center of yellow disk fls and numerous white rays, the heads 3–6 cm broad. The overlapping phyllaries are ovate, greenish at the center and papery at the margins. There is no pappus.

Very common and frequently abundant weedy wildflower. Usually found along roadsides, but also occasionally in meadows and other open places, mostly below 5000 ft.

Luina Benth.—LUINA

Luina hypoleuca Benth.—SILVERBACK LUINA

A very attractive, mound-forming perennial that grows from a woody rootstock. The numerous stems reach 30 cm tall and the plant may be up to 1 m across. The alternate lvs are elliptical to ovate, not stalked, and little reduced in size toward the tops of the stems. The best feature is the lvs: bright green and shining above, densely white-woolly on the underside. The heads are of disk fls only, cream-colored, about 2 cm broad, and borne in umbel-like clusters. The phyllaries are linear and thickly short-hairy. The pappus is of soft, white bristles.

Common on dry, rocky banks, roadcuts, and on stony slopes, to about 6000 ft. At Sunshine Point; along Westside Rd to about 5000 ft on Mount Wow; at the Glacier View Bridge; at Ipsut Pass; at Longmire along the Nisqually R; on cliffs in the Cowlitz Canyon; in Stevens Canyon, where it is especially well-displayed; at Silver Falls; on cuts along the road between Cayuse Pass and Tipsoo Lk; on Eagle Peak; and at Eunice Lk.

Madia Molina—TARWEED

Small, slender annual plants with opposite, untoothed, softly to roughly hairy lvs. The stems, leaves, and involucre are glandular and frequently aromatic. The phyllaries are curved to enclose the achenes of the ray flowers. Both the ray and the disk flowers are yellow. No pappus is seen in these two species.

1   Involucre less than 2–3 mm high; fertile disk fl 1 .... *M. exigua*
1   Involucre 6–10 mm high; fertile disk fls 2–12 .... *M. gracilis*

Madia exigua (Sm.) A. Gray—LITTLE TARWEED

The stems are just 10–30 cm high, with lvs 1–4 cm long and very small heads on long, slender stalks. The ray fls number 3–8, and have minute ligules; there is only one disk fl.

Infrequent; said to prefer dry ground in open woods or along trails and roads at low elevations. One collection is known, made near the Glacier View Bridge.

Madia gracilis (Sm.) D.D. Keck & J. Clausen ex Applegate—SLENDER TARWEED

Growing to about 50 cm tall in the Park, with lvs 3–10 cm long and heads borne in a racemelike inflorescence. The 3–9 ray fls have ligules about 5 mm long and the disk fls number 2–12.

Found among low weeds on a dry roadside about 1 mi south of the White R entrance and between Tahoma and Kautz Creeks. Not previously reported for the Park. The plant has a pleasant, citrus-spicy odor.

Matricaria L.—WILD CHAMOMILE

Matricaria discoidea DC. PINEAPPLE-WEED

A small annual plant, to about 20 cm tall, with pinnately dissected lvs 3–4 cm long, scented of pineapple. The heads are cone-shaped, 5–10 mm high, yellow, and of disk fls only. The phyllaries are oval and papery. There is no pappus.

A common plant along roadsides below about 3000 ft; native to northwestern North America, but behaving as a weed in the Park.

Microseris D. Don—MICROSERIS

Microseris borealis (Bong.) Sch.-Bip.
NORTHERN MICROSERIS

A taprooted perennial with a milky sap and basal lvs. The lvs are narrowly lanceolate, usually minutely toothed, hairless, and 6–15 cm long. The heads are solitary on leafless stems, of ray fls only, 2–3 cm broad, and bright yellow. The phyllaries are long-tapered, somewhat overlapped, in 2–3 series. The pappus is of brownish, slender, barbed bristles.

Listed as "sensitive" in *The Endangered, Threatened, and Sensitive Vascular Plants of Washington* (Washington Natural Heritage Program, 1997).

Common in subalpine meadows across the Park, generally on wetter ground. Collected at Mountain Meadows; near Mowich Lk at 4900 ft; at Eunice Lk; and on open slopes around Sunrise, down to about 5000 ft.

Mycelis Cass.—WALL-LETTUCE

Mycelis muralis (L.) Dumort.—WALL-LETTUCE

A biennial herb, 30–70 cm tall, growing from a taproot, with slender, invasive rootstocks and a milky sap. The basal and stem lvs are deeply cut into wide pinnate lobes; the broad terminal lobe may be described as "ivylike." Each starlike head has 5 yellow ray fls and is about 1.5 cm broad; disk fls are absent. The phyllaries are slender and long, 1.5–2 mm long, forming a narrow tube. The achene has a short beak and a pappus of long white bristles.

Frequent weed of shady places at roadsides and in open woods; sometimes spreading into undisturbed places in forests. Found around the Nisqually entrance and at Sunshine Point; along Westside Rd; on the Kautz Cr service road; at the meadow edge at Longmire; in the forest uphill of the Ohanapecosh entrance; in the hot springs area at Ohanapecosh; and in open woods south of the White R entrance. In 1999, it was seen at Narada Falls. Prior to fieldwork for this study, the plant was known only from a single collection made in 1965 by Robert Wakefield at the roadside in the "Big Trees" area between Tahoma Cr and Longmire.

Asteraceae

Nothocalais (A. Gray) Greene
PRAIRIE DANDELION

Nothocalais alpestris (A. Gray) K.L. Chambers
SMOOTH MOUNTAIN DANDELION

A taprooted perennial with a milky sap and basal lvs. The lvs are hairless, spoon-shaped to lanceolate, 5–15 cm long with a few slender teeth (that often point toward the base of the lf). The heads are solitary on the leafless stems, with numerous yellow ray fls, and about 2–3.5 cm broad; disk fls are absent. The phyllaries are in 2 series, the outer series ovate-lanceolate and overlapping. The phyllaries sometimes sport small purple spots. The achene is slender but lacks a beak; the pappus is of long white bristles.

Very common in moist subalpine meadows and on open, moist slopes, 5000-6000 ft, most frequently seen on the north and east sides of the Park. Collected on Mount Wow; at Mowich Lk; in lower Paradise Park; at Goat Pass at 5000 ft; near Clover Lk; at Tipsoo Lk and Chinook Pass; in the Sunrise meadows; on the trail to Burroughs Mtn; and at Upper Palisades Lk. Plants from lower Paradise Park, near the road junction, are notable for having dark-spotted phyllaries.

Petasites Mill.—COLTSFOOT

The two varieties of the common species *P. frigidus* differ markedly in appearance. First to appear, very early in the spring soon after the snow melts, is a flowering stem bearing numerous heads in a flat-topped cluster. The stem is clothed in short, woolly white hairs and several heavy bracts. Heads bear ray and disk flowers, with lanceolate phyllaries that are hairy at the base. The pappus is of white bristles. The leaves, on long stalks borne directly on the rootstock, follow the flowering stem; they are greenish and mostly hairless on the upper side, whitish-woolly on the underside.

1   Plants of subalpine meadows; lvs ovate to triangular in outline ....
.... *P. frigidus* var. *nivalis*
1   Plants of low to middle elevations; lvs round in outline ....
.... *P. frigidus* var. *palmatus*

Petasites frigidus (L.) Fr. var. *nivalis* (Greene) Cronquist
SWEET COLTSFOOT

The stems of this variety reach about 30 cm tall and bear heads with whitish to pinkish ray fls and pink disk fls. The lvs are more or less triangle-shaped in outline, coarsely toothed and lobed, and 5–12 mm broad.

123

Frequently seen along small alpine streams in open meadows. Collected at Mowich Lk and Spray Park; in Paradise Park, where it is one of the first fls to bloom; at Tipsoo Lk; in Seattle Park at 5500 ft; in Glacier Basin at 7000 ft; and in Berkeley Park. The species was named from plants first collected on the north side of the mountain in 1889 by E. L. Greene.

Petasites frigidus (L.) Fr. var. *palmatus* (Aiton) Cronquist—ARCTIC COLTSFOOT

The lowland variety is taller, the flowering stem reaching about 40 cm. The heads are duller in color, whitish to pinkish. The lvs are roundish in outline and heart-shaped at the base, 10–40 cm broad.

Very common on moist banks, along streams, and in wet places on roadsides, up to about 5000 ft, although somewhat higher on the south side of the Mtn. Found along the main road from the Nisqually entrance to Narada Falls; at Longmire; at Ohanapecosh; at Indian Bar at 6000 ft; in Seattle Park at 5500 ft; and in the lower White R V. Blooms as early as late March at the Nisqually entrance.

Prenanthes L.—RATTLESNAKE-ROOT

Prenanthes alata (Hook.) D. Dietr. WESTERN RATTLESNAKE-ROOT

A perennial herb with a milky sap, growing upright to about 60 cm tall. The stems are leafy and sometimes branched. The lvs have a toothed, triangular blade and a stalk that is winged along its margins; the upper lvs are not stalked and smaller. The heads are of white ray fls only, in an open, flat-topped cluster. Each head is about 2 cm broad and nods on a short stalk. The phyllaries are linear, greenish, and finely hairy. The pappus is of stiff, brown bristles.

Infrequently found on rocky slopes in open forests. One herbarium collection was located, made along the Carbon R ".25 mi above Olsen's cabin," a location just above 3000 ft near Cataract Cr.

Rainiera Greene—RAINIERA

Rainiera stricta (Greene) Greene
TONGUE-LEAF RAINIERA

A large, coarse perennial, with leafy stems to nearly 100 cm tall. The lvs are oblanceolate, 15–35 cm long, stalked, not toothed, and hairless. The upper lvs on the stem are reduced in size and not stalked. The inflorescence is a dense raceme 10–30 cm long. The heads each narrow and about 1 cm broad and of 5 disk fls; ray fls are absent. The 5–7 phyllaries are linear and short-hairy, and the pappus is of brownish bristles.

Listed as "watch" in the *Endangered, Threatened, and Sensitive Vascular Plants of Washington,* under the name *Luina stricta* (Washington Natural Heritage Program, 1997).

Found infrequently on the north and east sides of the Park, at subalpine elevations in open woods, on dryish slopes, and in meadows. Collected at Spray Park; at Goat Pass; at the foot of the Cowlitz Glacier; near the summit of Chinook Pass; on Sunrise Ridge along the road; on Burroughs Mtn; and at Windy Gap at 5000 ft. Also seen on the west slope of Naches Peak.

Rainiera stricta was the name used by early botanists, later subsumed in *Luina.* It's most pleasing to see that Kartesz restores the name *Rainiera,* even if this is perhaps not the loveliest plant to share a name with the mountain. The species was first collected on Mount Rainier in 1889 by E. L. Greene.

Saussurea DC.—SAWWORT

Saussurea americana D. C. Eaton
AMERICAN SAWWORT

The regularly, closely toothed lf margins give this plant its common name. Growing 50–120 cm tall, the lower lvs are ovate and about 15 cm long. The lvs become smaller and more lanceolate further up the stem. The purple heads are numerous in a compact, flat-topped cluster, of disk fls only. The phyllaries are ovate, short-hairy, and in several series. The double pappus consists of an inner ring of bristles that are united at the base an an outer ring of shorter scalelike bristles.

Uncommon, found scattered in subalpine meadows, mostly above 5000 ft. Known from Mount Wow; at the foot of the Cowlitz Glacier; around the Owyhigh Lks; at Chinook Pass; and at Berkeley Park.

Senecio L.
GROUNDSEL, BUTTERWEED, RAGWORT

Perennial herbs from rootstocks or taproots, with alternate, toothed or pinnately divided lvs. (The weedy species are annuals or perennials.) Two of our species are discoid, but most have yellow ligules. The phyllaries are typically slender and pointed. The pappus is of soft, white, slender bristles. Some are weedy, but others are quite handsome plants. The common names listed above have been applied almost at random to senecios in the United States. I follow Mary Fries's advice in using "senecio" for the native species in the Park.

1	Lower stem lvs more or less equal to upper stem lvs	2
1	Stem lvs reduced upwards	6
2(1)	Ligules absent	*S. vulgaris*
2	Ligules present	3
3(2)	Lvs toothed but not lobed	*S. triangularis*
3	Lvs lobed to dissected	4
4(3)	Lvs pinnately dissected	*S. jacobaea*
4	Lvs toothed or lobed	5
5(4)	Lvs obovate, shallowly toothed; alpine plants	*S. fremontii* var. *fremontii*
5	Lvs lanceolate, deeply toothed to lobed; lowland plants	*S. sylvaticus*
6(1)	Lvs narrowly triangular, the blade squared off at the base	*S. triangularis*
6	Lvs variously shaped, but not triangular	7
7(6)	Plants hairless, at least at flowering time	8
7	Plants short-hairy at flowering time	11
8(7)	Heads discoid	*S. indecorus*
	(see "Doubtful and Excluded Species," p. 469)	
8	Heads radiate	9
9(8)	Heads solitary on the stems	*S. cymbalarioides*
9	Heads several to many on the stems	10
10(9)	Lower lvs cut into pinnate lobes, the terminal lobe largest	*S. flettii*
10	Lower lvs obovate to nearly round	*S. streptanthifolius* var. *streptanthifolius*
11(7)	Stems 10–30 cm tall, from an elongated, woody base; heads often nodding on the stalks	*S. elmeri*
	(see "Doubtful and Excluded Species," p. 469)	
11	Stems 20–60 cm tall, from a short crown, woody base absent; heads on erect stalks	12
12(11)	Ray fls yellow	*S. integerrimus* var. *exaltatus*
12	Ray fls whitish or cream-colored	*S. integerrimus* var. *ochroleucus*

Senecio cymbalarioides H. Buek
CLEFT-LEAF SENECIO

A delicate, hairless little plant, to 30 cm tall but frequently less. The blades of the lower lvs are 3–4 cm long, roundish to obovate, coarsely toothed, with stalks longer than the blades. The upper stem lvs are 2–4 cm long, oblong, lobed, and not stalked. The solitary head is about 3 cm broad.

A rare plant of wet meadows, 4000-5000 ft. Collected at Chinook Pass; on the Interfork of the White R; and in Glacier Basin.

Senecio flettii Wiegand—FLETT'S SENECIO

Plants 15–20 cm tall and hairless, from a slender rootstock. The basal lvs are pinnately lobed, the lobes themselves rounded and toothed, and about 4–12 cm long. The 1 or 2 stem lvs are much smaller. The 2–4 heads are in a compact, flat-topped cluster. Each head is about 2 cm broad, with 2–4 ligules.

Rare in the Park; known from an open rocky slope on Naches Peak at 5800 ft and, according to C. V. Piper, on the Cowlitz Chimneys at an unspecified elevation.

Senecio fremontii Torr. & Gray var.
fremontii—FREMONT'S SENECIO

A tufted and dwarf plant, less than 20 cm tall. The plant is hairless, with lvs that are mostly stalkless on the stems, thickish, obovate to spoon-shaped, coarsely toothed and occasionally lobed as well, less than 5 cm long. The heads are usually 2–3 per stem and the phyllaries are dark-tipped.

Not uncommon on talus slopes and in the rocky soil of moraines, 5500-9000 ft. Collected at Spray Park; on the upper reaches of the Paradise R at 6000 ft; on the moraine between Stevens and Paradise Glaciers; at Panorama Point; below the Cowlitz Glacier; on the moraine at Emmons Glacier; at Sunrise; and on the Fryingpan Glacier moraine.

Senecio integerrimus Nutt. var. *exaltatus* (Nutt.) Cronquist
WESTERN SENECIO

A perennial, 10–60 cm tall, greatly varying in size depending on its habitat, with a single flowering stem. The lower lvs are tapered to a stalk, thickish, with matted woolly hairs but becoming smooth with age, oblanceolate and 3–15 cm long. The stem lvs are much reduced. The heads are numerous in a compact, flat-topped cluster, each about 1 cm broad, with yellow ligules. The phyllaries are black-tipped.

Common on rocky slopes and open ridges, rocky slopes and ridges above 6000 ft. Known from Mount Wow and Sunrise.

Senecio integerrimus Nutt. var. *ochroleucus* (A. Gray) Cronquist—WESTERN SENECIO

Cream-colored or whitish ligules distinguish this variety.

Listed for the Park on the basis of two collections: one made by O. D. Allen at 6000 ft on Mount Wow and an unnumbered collection made by J. B. Flett on "grassy slopes" at 6500 ft.

Senecio jacobaea L.—TANSY RAGWORT, STINKING WILLIE

A biennial, short-hairy weed, to 100 cm tall with an unbranched stem, which is about equally leafy most of its length. The lvs are pinnately dissected and 5–15 cm long. The ligules are 4–10 mm long and the phyllaries are lanceolate and black-tipped.

A noxious weed, rarely seen in the Park and known from one collection made at a roadside at Sunshine Point.

Senecio streptanthifolius Greene var. *streptanthifolius* ROCKY MOUNTAIN SENECIO

A small and slender, hairless plant, with a single stem and mostly basal lvs, from a slender rootstock. The basal lvs are obovate to roundish, toothed, and taper to the long, slender stalk; the upper lvs are smaller and may be lobed. The 3–5 heads are in a small, flat-topped cluster, each about 1.5–2 cm broad.

St. John and Warren (1937) give a location of "rocky slopes, Indian Henrys Hunting Ground." In general, it is a plant of middle to high elevations on open rocky ridges and one collection from the Park, bearing no specific geographical information, is known.

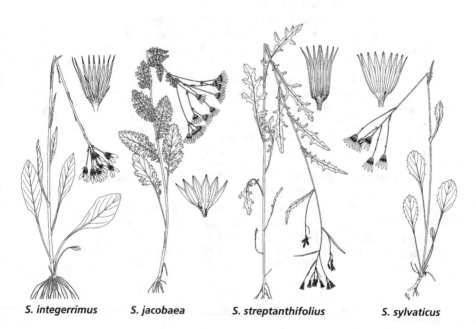

S. integerrimus   S. jacobaea   S. streptanthifolius   S. sylvaticus

Senecio sylvaticus L.—WOODS GROUNDSEL

A short-hairy weedy annual, similar in appearance to S. *vulgaris*, but the main stem is branched and reaches 30–90 cm tall, with grayish lvs that are sharply pinnately lobed. The heads are small with inconspicuous ligules up to 2 mm long and phyllaries that are green at the tip.

A common weed on disturbed ground, on old burns, and at roadsides.

Senecio triangularis Hook. ARROWLEAF SENECIO

A stately plant, 50–100 cm tall, with a leafy stem. The lvs are narrowly triangular, regularly toothed on the margins, heart-shaped or squared-off at the base, with a short stalk. The heads are numerous, in a fairly compact, flat-topped cluster, and each is about 2 cm broad. The phyllaries are blackish and hairy at the tips.

Common and often forming small colonies on streambanks and in meadows. Known from Longmire, but generally found above 4000 ft. Collected at Mowich Lk and Spray Park; at Indian Henrys Hunting Ground; in Van Trump Park; in Paradise Park; in the Butter Cr Research Natural Area; on Dewey Cr at the highway; at Tipsoo Lk and on Naches Peak; at the White R campground; and at Berkeley Park and Seattle Park.

Senecio vulgaris L. OLD MAN, COMMON GROUNDSEL

A low-growing annual weed, with branched stems less than 30 cm tall. The lvs are 1–5 cm long, pinnately lobed with sharp teeth. The heads are narrow and cylindrical, less than 1 cm broad, of disk fls only. The phyllaries black-tipped. Called "old man" for the puffy white pappus of the mature heads.

Very common weed at low elevations along roads, in planted areas, and in other disturbed places. Only rarely found in open forests.

Solidago L.—GOLDENROD

Perennial herbs, with alternate, undivided, toothed leaves. The heads have yellow disk and ray flowers. The phyllaries are lanceolate, in several series, papery at the base and greenish at the tip. The pappus is of slender, white bristles.

1   Lowland plants; stems about equally leafy above and below; lvs lanceolate .... *S. canadensis* var. *salebrosa*

1   Alpine or subalpine plants, with obovate to oblanceolate basal lvs and smaller stem lvs 2

2(1)   Taller plants, to 30 cm; If stalks fringed with short hairs ....
.... *S. multiradiata* var. *scopulorum*

2   Dwarf plants, to 10 cm tall; If stalks lacking hairs ....
.... *S. simplex* ssp. *simplex* var. *nana*

Solidago canadensis L. var. *salebrosa* (Piper) M. E. Jones—CANADA GOLDENROD

The only lowland goldenrod in the Park, with stems 50–100 cm tall, more or less equally leafy. The lvs are lanceolate, 6–12 cm long, and toothed. The very numerous heads are arranged in an elongated, pyramid-shaped panicle. The heads are about 5 mm broad, with short ligules.

Frequent in moist meadows below about 3000 ft. Found in the Nisqually R valley; along the road north of Ohanapecosh; and in the area of the White R entrance.

Solidago multiradiata Aiton var. *scopulorum* A. Gray NORTHERN GOLDENROD

A slender plant, to 30 cm tall, with oblanceolate lvs which are gradually reduced in size and pointed at the tips toward the top of the stem. The lvs are 2–8 cm long, even or minutely toothed; the lf stalks are fringed with short hairs. The heads are in an open, flat-topped cluster, each with about 13 ray fls. The phyllaries are in 2 overlapping series.

Common in subalpine meadows and on moist open slopes, 5000-7500 ft, although also seen at Silver Falls and the Canyon Rim overlook. Collected on the upper slopes of Mount Wow; in Paradise Park up to Panorama Point; on Eagle Peak at 5800 ft; on Pinnacle Peak; at the White R campground; at Sunrise and Frozen Lk; and at Chinook Pass.

Solidago simplex Kunth ssp. *simplex* var. *nana* (A. Gray) G.S. Ringius—DWARF GOLDENROD

To 10 cm tall, with a leafy stem, the lvs reduced upward. The lvs are up to 5 cm long, spoon-shaped to oblanceolate, thickish, and coarsely toothed; the margins of the lf stalk not fringed with hairs. The few heads are in a short and compact, racemelike inflorescence. The heads number fewer than 10 and the phyllaries overlap little.

Frequent on talus slopes and open ridge tops, 6000-7100 ft. Collected in the Paradise–Muir corridor; on slopes above Seattle Park; on Burroughs Mtn; and on Mount Fremont.

Sonchus L.—SOW-THISTLE

Weedy annuals, with a milky sap and ray flowers only (and therefore not true thistles), growing 50–100 cm tall. The hairless plants have toothed or prickly auriculate leaves, yellow ligules in heads that are 1.5–2.5 cm broad, and a pappus of soft, white bristles.

1   Lvs with spiny teeth; stem lvs earlike, with rounded lobes clasping the stem .... *S. asper* ssp. *asper*

1   Lvs toothed but lacking spines; earlike, clasping lobes of the stem lvs pointed .... *S. oleraceus*

Sonchus asper (L.) Hill ssp. *asper*
PRICKLY SOW-THISTLE

Notably prickly lvs 5–20 cm long distinguish this species. The achene is oval and smooth except for the lengthwise ribs.

Occasional weed on roadsides below about 4000 ft. It also grows on mineralized ground in the meadow at Longmire. Seen on Westside Rd; on the Cougar Cr trail burn; and at Longmire.

Sonchus oleraceus L.—COMMON SOW-THISTLE

The lvs of this species are toothed but lack spines, and are 5–25 cm long. The achene is narrowly obovate, ribbed lengthwise, and minutely roughened.

An uncommon weed of waste ground and along roads. In the Park herbarium is a plant collected along the road to Mowich Lk at 3500 ft, with lvs that are somewhat prickly-margined as in *S. asper*, but with the characteristically roughened achenes of *S. oleraceus*.

Tanacetum L.—TANSY

A pair of weedy perennials, strongly aromatic, with pinnately lobed to dissected leaves. The heads are numerous, with overlapping, papery phyllaries. The pappus is a short, scalelike crown.

1   Heads radiate; ligules white, disk fls yellow .... *T. parthenium*

1   Heads discoid, yellow .... *T. vulgare*

Tanacetum parthenium (L.) Sch.-Bip.—FEVERFEW

Coarsely short-hairy with branched stems reaching about 50 cm. The lvs are up to about 8 cm long, pinnately lobed, the lobes toothed or again divided. The heads are in an open, leafy panicle, each about 2 cm broad with 10–12 white ligules and yellow disk fls.

Occasionally reported as a weed of roadsides and waste ground, but perhaps not persisting for long. Neither herbarium specimens nor plants in the field were found in the course of this study.

Tanacetum vulgare L.
COMMON TANSY

Plants 50–100 cm tall, nearly hairless but glandular on the lvs. The stems are often branched, with stalkless or short-stalked lvs 10–20 cm long, the latter 3 times divided into small, linear, sharp leaflets. The very numerous yellow heads, each about 1 cm broad, are in large, flat-topped clusters.

Widespread but only infrequently seen; a noxious weed of roadsides and waste ground. Recorded at Sunshine Point, Longmire, and Paradise; also found along the service road at Kautz Cr and along the road south of the White R entrance to the Park.

Taraxacum Hall—DANDELION

Taraxacum officinale Weber ssp. *officinale*
DANDELION

A perennial with a milky sap and a basal rosette of hairless, pinnately lobed lvs, the lobes again toothed. The bright yellow head, of ray fls only, is solitary, 2–3.5 cm broad, and on a naked stem. The phyllaries are in 2 series, the outer series bent backward at the tips. The pappus is a parachutelike tuft of slender white bristles; the achene is usually olive-brown.

Common weed on waste ground, on roadsides, along trails, and in openings in woods, up to at least 5000 ft. Also occasionally in subalpine meadows.

Plants found away from waste places, especially in subalpine meadows, should be examined carefully, for at least two native dandelions may yet be found in the Park in such places. They are superficially similar to the weedy species described above. *Taraxacum officinale* ssp. *ceratophorum* (Ledeb.) Schinz & Thellung, the horned dandelion, has less deeply lobed lvs and inner bracts with hornlike processes at the tips, and the achenes are straw-colored to brown. The inner bracts of *T. eriophorum* Rydb. lack "horns" and the achenes are reddish to reddish brown.

Tonestus A. Nelson
GOLDENWEED

Tonestus lyallii (A. Gray) A. Nelson
LYALL'S GOLDENWEED

A low-growing perennial herb, with leafy stems reaching 20 cm tall in favorable places. The lvs are spoon-shaped to oblanceolate, finely glandular-hairy, and 1–5 cm long. The head is solitary, 2–3 cm broad, with numerous yellow ray and disk fls. The phyllaries are lanceolate and glandular, while the pappus is of hairlike white bristles.

Fairly common on alpine ridges and slopes, to almost 8000 ft. Collected on the upper slopes of Paradise valley and on Burroughs Mtn.

Tragopogon L.—SALSIFY

Two weedy, short-lived perennials, with a milky sap, differing in the color of the flowers. Both are 30–100 cm tall, with long, slender, almost grasslike leaves. The stems and lvs are notably glaucous. The heads are large, 5–10 cm broad, of ray fls only. The phyllaries are in 1 series, linear-lanceolate, and about 4–6 cm long. The pappus is of feathery bristles, about 2 cm long and parachutelike (that is, united for a distance above the achene).

1　Ligules purple .... *T. porrifolius*
1　Ligules yellow .... *T. dubius*

Tragopogon dubius Scop.
YELLOW SALSIFY, GOAT'S-BEARD
Uncommon weed along roadsides up to about 4000 ft. Found at the Canyon Rim overlook; at the roadside above Dewey Cr; and about 1 mi south of the White R entrance gate.

Tragopogon porrifolius L.—SALSIFY
Occasional roadside weed, with a collection known from the Longmire meadow.

Berberidaceae Juss.
BARBERRY FAMILY

A family of three genera in the Park. Although greatly differing in appearance, the key features uniting them include a superior ovary of 1 pistil, 6 petals, 6 or 9 sepals, and 6 stamens (although *Achlys* lacks petals and sepals and has 9–13 stamens). The leaves are alternate.

1   Shrubs with woody stems; lvs pinnate and bearing marginal spines.........
.... *Berberis*

1   Herbs; lvs compound but not pinnate, the margins not bearing spines .. 2

2(1) Lvs divided once; fls in a spike, petals absent................................*Achlys*

2   Lvs divided two or three times; fls in a panicle, petals present ..:..............
.... *Vancouveria*

Achlys DC.—VANILLA-LEAF

Perennial herbs with spreading rootstocks, forming patches in favorable places. The light green lvs are roundish in outline and divided into 3 lobes; each lobe is triangular to fan-shaped, with rounded teeth. The leaves are borne more or less horizontally on tall, rigid stalks, which rise singly from the rootstock. The flowering stems reach above the lvs and bear numerous white flowers. Sepals and petals are absent, and what is seen are the 9–13 conspicuous, long, white stamens. The fruit is a short, curved achene. The dry leaves have the sweet scent of vanilla.

For many years, two species went under the single name of *A. triphylla*. *Achlys californica* was described in 1970. The distinctions between the species are better seen in regions to the south, in Oregon and California, and not all of the plants in the Park can be satisfactorily assigned to one or the other.

The key follows that in *The Jepson Manual* (Hickman, 1993). Young plants are not always easily separable, and the measurements refer to lvs of mature plants. Hybrids, with mixed characters, are known to occur between the two species but have not been documented for the Park.

1   Central leaflet 6–8-toothed, 7–16 cm long and 8–17 cm wide; stamens 4–5 mm long; berry brownish .... *A. californica*

1   Central leaflet 3-toothed, 4–11 cm long and 4–8 cm wide; stamens 3–4 mm long; berry reddish purple .... *A. triphylla*

Achlys californica Fukuda & Baker—VANILLA-LEAF
Besides the features noted in the key, the lvs are separated on the rootstock by a distance averaging 8–10 cm and the plants reach 30–50 cm tall. The ovary is 1.5–2 mm long.

Apparently less common in the Park than the following species, and more likely to occur at the lower elevations. Found in the area of the Nisqually entrance; along the main road up to Tahoma Cr; at

Ohanapecosh; and along the road under tall shrubs between Stevens Cr and Bench Lk. Not noted in the White R drainage during the present study. Near Bench Lk, the plants are reduced in size but the lvs still show the "*californica*" pattern.

Achlys triphylla (Sm.) DC.—VANILLA-LEAF
In this species, the internodes on the rootstock average 2.5–5 cm and the plants reach 20–40 cm tall. The ovary is 1–1.5 mm long.

Common and often abundant in forests and along roadsides and trails, up to about 4500 ft. Collected in the area of the Nisqually entrance, up the road to Narada Falls; at Spray Park; in Van Trump Park; on the Rampart Ridge trail; and very common along the Ohanapecosh, White, and Carbon Rivers to about 4000 ft.

Berberis L.—OREGON GRAPE

Shrubs with woody stems and pinnately compound evergreen leaves that have sharp, spiny margins. The leaflets are set at angles on the midvein of the leaf. The yellow flowers are borne in racemes at or near the ends of the branches, and are followed by fleshy, glaucous, dark blue-purple berries.

Kartesz names this genus *Mahonia* rather than *Berberis*. However, A. T. Whittemore argues for *Berberis* in his treatment of the Berberidaceae for the Flora of North America, and that use is followed here.

1   Tall shrub, generally more than 1 m tall; leaflets 5–9, pinnately veined .... *B. aquifolium*
1   Low shrub, to 0.5 m tall; leaflets 11–19, palmately veined ....*B. nervosa* var. *nervosa*

Berberis aquifolium Pursh
OREGON GRAPE
Plants 1–3 m tall with erect stems. The lvs are divided into 5–9 spiny, dark green, oval to ovate leaflets that are rather widely spaced on the stems; the leaflets are notably glossy on the upper surface. The inflorescence is a cluster of racemes, each 5–8 cm long.

Uncommon, on rocky slopes on the east side of the Park. Collected at 5500 ft near Cayuse Pass, but mostly at lower elevations.

Berberis nervosa Pursh var. *nervosa*
DWARF OREGON GRAPE

Less than 50 cm tall, with stout, erect stems. The lvs are closely spaced and rise upward from the stems. They consist of 11–19 leaflets that are spiny, glossy on both surfaces, and lanceolate to ovate. The fls are in dense clusters of short racemes.

Quite common in forests mostly below 3000 ft; tolerant of a good deal of shade but flowering best in openings. Found throughout the Park; especially frequent around Longmire and in the lower White R Valley.

Vancouveria C. Morren & Decne.
INSIDE-OUT FLOWER

Vancouveria hexandra (Hook.) C. Morren & Decne.—INSIDE-OUT FLOWER

"Inside-out" for the appearance of the fl: the white sepals and petals are sharply reflexed (turned backwards) from the pistil and stamens. This is a perennial herb growing from a slender rootstock, with twice-compound lvs, 10–40 cm long overall, on slender stalks. The leaflets are more or less round and coarsely 3-lobed. The fruit is a follicle.

Infrequent in the Park. Known from more or less open, mixed woods in the Nisqually entrance area and at Sunshine Point, although the plant probably ranges higher in the Nisqually V.

Betulaceae Gray—BIRCH FAMILY

Trees or shrubs, with alternate, undivided, toothed leaves, which have straight veins. The flowers are unisexual: the male flowers are in pendulous catkins, with 2–10 stamens per bract and the female flowers are in clusters or catkins. They appear on the same plant. The fruit is a large to small nut (that may appear to be merely a hard seed).

1   Fls few, in small clusters; the fruit nutlike and enclosed in a papery husk . ... *Corylus*

1   Fls many, in catkins; the fruit a small winged nutlet, lacking a husk .... *Alnus*

Betulaceae

Alnus Mill—ALDER

Very common trees and shrubs in the Park, preferring streamsides and other moist habitats. The male catkins are long, slender, and bear numerous small flowers. The female catkins bear fewer flowers and at maturity take the shape of short, woody cones.

1   Trees; catkins appearing before the lvs; lower side of the lvs paler green the upper .... *A. rubra*

1   Shrubs; catkins appearing with the lvs; lvs about equally green on both sides .... *A. viridis* ssp. *sinuata*

Alnus rubra Bong.—RED ALDER

One of the earliest plants to flower in the spring, often before the snow melts. A tall tree, reaching 20 m in favorable places, with a narrow crown and smooth bark. The oval lvs are 5–15 cm long, singly toothed, somewhat inrolled at the margins, and appear well after the fls. The male catkins are 10–15 cm long and the female cones are 15–25 mm long.

Dominant tree along low-elevation stream and river bottoms, below about 3000 ft, on wet, deep soils; occasionally extending up onto moist hillsides.

Alnus viridis (Vill.) Lam. & DC. ssp. *sinuata* (Regel) A. Löve & D. Löve
SITKA ALDER, SLIDE ALDER

A much-branched shrub, to 5 m tall, although the stems are often bent down by the weight of the winter snow, an effect well-displayed along the road in Stevens Canyon below Bench Lk. The fls appear in the spring at the same time as the lvs. The lvs are doubly toothed, have flat margins, and are rounded at the base. The male catkins are 12–15 cm long and the female cones are 12–15 mm long.

Common shrub along streams, at lake shores, and on moist open slopes, 2500-5000 ft. Collected along the trail to Indian Henrys Hunting Ground; at Longmire; in Paradise Park; along the trail up the White R near the campground; and in the Butter Cr Research Natural Area.

Corylus L.—HAZELNUT

Corylus cornuta Marshall var. *californica* (A. DC.) Sharp
CALIFORNIA HAZELNUT

Shrubs with several stems, reaching 1–5 m tall. The lvs are doubly-toothed, flat at the margins, and somewhat cordate at the base. The fls appear before the lvs early in the spring. The male catkins are about 10

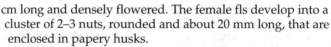

cm long and densely flowered. The female fls develop into a cluster of 2–3 nuts, rounded and about 20 mm long, that are enclosed in papery husks.

Frequent in open woods and on rocky slopes, below 4500 ft. Especially common in Stevens Canyon; also collected at Narada Falls and about Ohanapecosh.

Boraginaceae Juss.—BORAGE FAMILY

A small family with just one species that is a prominent part of the wildflower show at Mount Rainier: the striking bluebells, *Mertensia paniculata*. Most of the borages in the Park are small and several are weedy. Most of the species are roughly hairy, with alternate, undivided lvs and fls mostly in scorpioid cymes—that is, 1-sided racemes that uncoil like a scorpion's tail as the flowers mature. There are 5 sepals, which may be fused or free; the 5 petals of the corolla are fused into a tube below with 5 free lobes above. A magnifying glass helps to show the intricate, crestlike appendages at the bases of the free lobes. There are 5 stamens and 1 style, which is attached at the base of the deeply 4-parted ovary and develops into 1–4 nutlets.

1	Fls blue	2
1	Fls white, yellow, or orange	4
2(1)	Fl with a long, funnel-shaped tube, from the top of which the lobes spread gradually	*Mertensia*
2	Fl with a short tube, from which the lobes spread at nearly right angles	3
3(2)	Fls 1 or 2 per lf axil; calyx greatly expanded in fruit	*Asperugo*
3	Fls in a racemelike cyme; calyx not expanded in fruit	*Myosotis*
4(1)	Fls white	*Cryptantha*
4	Fls yellow or orange	*Amsinckia*

Amsinckia Lehm.—FIDDLENECK

With characteristic bright yellow to orange flowers, the fiddlenecks have no look-alike species in the Park. Bristly, annual plants to about 40 cm tall, with branched, leafy stems. The leaves are oblanceolate to nearly linear, 5–8 cm long. The sepals are bristly and the flowers are small, just 1–3 mm broad; the free lobes spread at right angles to the tube. The two species are of only casual occurrence in the Park.

1	Fl yellow, the corolla tube hidden by the calyx	*A. intermedia*
1	Fl orange (or yellow-orange), the tube exserted from the calyx	*A. menziesii*

Amsinckia intermedia Fisch. & C. A. Mey.
RANCHER'S FIDDLENECK

This variety has fls in which the free lobes of the petals are at the top of the calyx, the tube entirely hidden.

Rarely seen in the Park on roadsides. Reported by Jones in 1938, but not seen during in this study.

Amsinckia menziesii (Lehm.) A. Nelson & J. F. Macbr.—MENZIES'S FIDDLENECK

Here, the tube is exserted beyond the top of the calyx.

A collection from near Longmire is known, made by F. A. Warren in the 1930s. Not mentioned by Jones and probably not persisting in the Park.

Asperugo L.—MADWORT

Asperugo procumbens L.—MADWORT

Not mentioned by Jones in his 1938 flora, this is a weak, clambering annual weed with roughly hairy stems to about 100 cm long and lvs that are sometimes opposite. The fls are few in number, borne in the axils of lvs or in the forks of the stem. The lvs are oblanceolate, to about 2.5 cm long, and strongly rough-hairy. The sepals are fused for about half their length. The fls are bell-shaped, blue, and 2–3 mm long.

Collected twice at Longmire, once in the 1930s and once in 1965, in the area of the former horse barn. It is doubtful that it persists but is instead reintroduced from time to time.

Cryptantha Lehm. ex G. Don—CRYPTANTHA

A very large genus, difficult to differentiate, but with only two representatives in the Park. Small-scale characteristics of the nutlets separate *Cryptantha* from the other borages, but in the Park they are easily distinguished by their small, white flowers. Both are small, annual herbs, less than 30 cm tall and bristly-hairy. The lower stem leaves are opposite, the upper alternate, 1–3 cm long.

1   Fl conspicuous, the free lobes spreading to about 5 mm wide .... *C. intermedia*
1   Fl nearly hidden by the calyx, only about 1 mm broad .... *C. affinis*

Cryptantha affinis (A. Gray) Greene
SLENDER CRYPTANTHA

The fl here is so small that the long, straight, ascending hairs of the calyx partially obscure it. Known from a herbarium collection made 1 mi below the White R campground. Originally labeled *C. muriculata* (A. DC.) Greene by F. A. Warren, but the fls are very small, just 1–2 mm across, and the hairs of the calyx are straight; no nutlets are present. Perhaps only a waif in the Park; the plant's typical range is east of the Cascades. C. V. Piper (1916) says that J. B. Flett also collected *C. muriculata* on the "Goat Mountains"—that is, on Mount Wow—but that collection has not been located.

Cryptantha intermedia (A. Gray) Greene
COMMON CRYPTANTHA

Although the calyx is beset with long, straight, spreading hairs, the free lobes of the corolla are held above them. The appendages at the bases of the lobes are usually yellow, giving the fl a delicate little "eye."

Occasional on dry, open ground along roads and trails at low elevations. A collection in the herbarium at the University of Washington, made by F. A. Warren "near Ohanapecosh," is labeled *C. ambigua*, although this is clearly incorrect.

Mertensia Roth—BLUEBELLS

Mertensia paniculata (Aiton) G. Don var. borealis (J. F. Macbr.) L. O. Williams—TALL BLUEBELLS

A tall perennial, frequently to almost 1 m, with several stems in a tuft. The plant is mostly hairless and glaucous. The lvs are ovate-lanceolate to nearly elliptical, stalked, and sometimes with hairs on the lower side; the upper lvs are generally not stalked. The fls nod in scorpioid cymes; each is bell-shaped and about 10–14 mm long, pinkish in the bud and becoming blue to blue-violet. The sepals are about 4 mm long, glaucous, with a fringe of short hairs on the margins.

Common throughout the Park in moist, open places and in thickets; on rocky stream banks, at the borders of meadows, and on talus slopes, 2500-6500 ft. Between Mowich Lk and Lee Cr;

abundant on Westside Rd at about 2800 ft; around Longmire; on the road bank uphill from the Glacier View Bridge on up to Ricksecker Point; on the trail between Klapatche and St. Andrews Parks; near Ohanapecosh; on Hwy 410 at Dewey Cr; at 6200 ft on Brown Peak; at Tipsoo Lk and Chinook Pass; in Glacier Basin; near Sunrise; on the Burroughs Mtn trail; in Berkeley Park; at 4300 ft between White R and Summerland; and at Eunice Lk.

Myosotis L.—FORGET-ME-NOT

All three species are weeds in the Park and the first two are quite inconspicuous. Annuals or short-lived perennials, with softly hairy leaves and stems. The cymes are usually several per stem. The flowers are blue, quite small in the first two species, and have prominent white or yellow appendages at the bases of the free lobes, which spread at right angles from the tube. The calyx is roughly hairy, the sepals united below.

1   Fl 5–8 mm broad; introduced garden plant, doubtfully persisting .... *M. sylvatica*
1   Fls 1–2 mm broad; weeds of grassy waste places 2

2(1)   Fl stalk in fruit equal to or longer than the calyx .... *M. arvensis*
2   Fl stalk in fruit much shorter than the calyx .... *M. stricta*

Myosotis arvensis (L.) Hill—FIELD FORGET-ME-NOT

Rough, short-hairy annual, with stems to about 20 cm tall, usually branched near the base and with lower lvs about 2 cm long. The branches bear fls to near the base and a few leafy bracts are found in the lower part of the inflorescence. The fls are tiny, 1–2 mm in diameter, and best examined with a hand lens.

One collection was found in the Creso Herbarium at Pacific Lutheran University, made "10 miles down from Paradise," presumably around Longmire. Perhaps not persisting in the Park.

Myosotis stricta G. K. Link ex Roem. & Schult. SMALL FORGET-ME-NOT

Very similar to the species above, differing chiefly in the length of the fruit stalks, noticeable when the fruits are mature. It also tends to be lower-growing, seldom more than 15 cm tall. First noted in the Park in the course of this study and found in several places in the Park on dry, rocky roadcuts (which are, nevertheless, well-watered early in the growing season). Along the road between Tahoma Cr and Kautz Cr, and in Stevens Canyon.

Myosotis sylvatica Ehrh. ex Hoffm.
WOODLAND FORGET-ME-NOT

Taller and more robust than the above species, biennial or perhaps longer-lived. The branched stems may be up to 50 cm long and are long-hairy. The lvs are oblanceolate, to 10 cm. The fls are 5–8 cm long on stalks that are longer than the calyx at maturity. The calyx bears hooked hairs.

Reported by G. N. Jones to grow on wet ground around buildings at Longmire. Evidently a garden escape and probably not persisting.

Brassicaceae Burnett (formerly Cruciferae)
MUSTARD FAMILY

The former name of the family, Cruciferae, means "bearing a cross," a reference to the 4 petals that are arranged crosslike at right angles to each other. The inflorescence is a simple or branched raceme, or sometimes compact and headlike. The flowers have 4 free petals and 4 free sepals, 1 pistil with 2 chambers and podlike at maturity, and 6 stamens. The family includes annuals, biennials, and perennials. A number of species are weeds, but many more are distinctive parts of the native flora of the Park. The following key begins with an examination of the mature fruit. Sometimes the eventual shape of the fruit may be guessed well before maturity through an examination of the ovary. When keying out material lacking usable fruits, one will usually have success by following paths beginning at couplet 2 and at couplet 10 and comparing the results.

1	Fruit a silique: long, slender, and 4 or more times longer than broad 2
1	Fruit a silicle: short, oval to heart-shaped, generally no more than 2 times longer than broad 10
2(1)	Petals yellow 3
2	Petals white, pink, or purplish 8
3(2)	Silique with a prominent, beaklike style .... *Brassica*
3	Style, if prominent on the silique, not beaklike 4
4(3)	Plants hairless 5
4	Plants short-hairy 6
5(4)	Stems strongly angled; basal lvs not forming a rosette .... *Barbarea*
5	Stems round in cross section; basal lvs forming a rosette .... *Rorippa*
6(4)	Petals more than 1 cm long; hairs on lvs flattened and forked ... *Erysimum*
6	Petals less than 1 cm long; hairs various, but not flattened 7

7(6) Hairs on the lvs branched; stigma not divided .... *Descurainia*
7 Hairs on the lvs simple, unbranched; stigma 2-lobed .... *Sisymbrium*

8(2) Lvs mostly hairless .... *Cardamine*
8 Lvs short-hairy.. 9

9(8) Hairs of lvs mostly forked or branched; lvs partly to fully pinnate ....
.... *Smelowskia*
9 Hairs of lvs simple; lvs untoothed to toothed .... *Arabis*

10(1) Fls yellow 11
10 Fls white to pink... 13

11(10) Seeds 1 per chamber; flowering stems leafy .... *Lepidium*
11 Seeds 2 or more per chamber; flowering stems not leafy 12

12(11) Silicles elliptical to nearly linear, flattened .... *Draba*
12 Silicles roundish in outline, conspicuously inflated .... *Physaria*

13(10) Seed 1 per chamber; upper stem lvs clasping .... *Lepidium*
13 Seeds 2 or more per chamber; stem lvs absent or present but not clasping
.... 14

14(13) Plants with basal lvs only 15
14 Plants with basal lvs and stem lvs 16

15(14) Lvs mostly pinnately lobed; petals undivided.......................... *Teesdalia*
15 Lvs not toothed; petals divided into two lobes .... *Draba*

16(14) Basal lvs more or less not toothed; corners of the silicle rounded ....
.... *Thlaspi*
16 Basal lvs deeply toothed; corners of the silicle pointed .... *Capsella*

Arabis L.—ROCKCRESS

Biennials or perennials, from a taproot and often branched above the short stem. The basal leaves typically form a rosette and have stalks; the stem leaves are not stalked and in some species clasp the stem. The fruit is a long and very slender silique.

1 Mature siliques pendulous on the stem .... *A. holboellii* var. *retrofracta*
1 Mature siliques erect or ascending on the stem 2

2(1) Stem lvs clasping; plants mostly taller than 40 cm from a basal tuft of lvs
.... 3
2 Stem lvs attached directly to the stem but not clasping; plants low, branched, mostly less than 30 cm tall 5

3(2) Basal lvs narrowly oblanceolate .... *A. drummondii*
3 Basal lvs widely oblanceolate to obovate 4

4(3) Fls cream-colored; stem lvs glaucous ....*A. glabra* var. *glabra*
4 Fls white; stem lvs not glaucous ....*A. hirsuta* var. *glabrata*

5(2) Fls white; basal lvs blunt... *A. furcata* var. *furcata*
5 Fls rose-purple; basal lvs sharp-pointed .... *A. lyallii* var. *lyallii*

Arabis drummondii A. Gray
CANADIAN ROCKCRESS

Usually a biennial (but sometimes finishing its life cycle in one year), usually with a single stem 30–90 cm tall. The basal lvs are 2–4 cm long, narrowly oblanceolate and tapered to the stalk, untoothed or with a few small teeth, and with a few flattened, short hairs. The stem lvs are smaller and lanceolate and clasp the stem. The fls are 8–10 mm long, numerous, and white or pink. The siliques are 4–10 cm long and held erect and close to the stem.

Frequent in the Park, on rocky banks and slopes, 3000-6000 ft. Collected along the road between Longmire and Paradise at 3500 ft; at Sunrise; in Glacier Basin; and 1.5 mi above the White R campground.

Arabis furcata S. Watson var. *furcata*
CASCADE ROCKCRESS

A tufted perennial, with 1 or more stems 10–40 cm tall, and generally hairless. The basal lvs are spoon-shaped or obovate, not toothed, and 2–5 cm long; the stem lvs are not stalked. The fls are white, 4–8 mm long, and held more or less erect.

Rare in the Park and limited to dry, rocky ridges on the east side, as on talus slopes at Owyhigh Lks.

Arabis glabra (L.) Bernh. var. *glabra*
TOWER MUSTARD

A weedlike biennial, with spreading, forked or branched hairs. The stem is usually single, reaching about 100 cm tall, glaucous and hairless near the top, hairy below. The basal lvs are 2–5 cm long, oblanceolate, toothed or pinnately lobed, and short-hairy; the stem lvs are smaller and lanceolate and clasp the stem. The fls are cream-colored and 5–7 mm long. The fruits are erect and 4–10 cm long.

Infrequent, usually growing in open, disturbed places on gravelly soil along roadsides; also found in open woods to about 5000 ft. Known from Stevens Canyon; on Kotsuck Cr trail at 5000 ft; along Hwy 410 just above Cayuse Pass; and on the road 1.2 mi south of the White R entrance.

Arabis hirsuta (L.) Scop. var. *glabrata* Torr. & A. Gray
HAIRY ROCKCRESS

Usually biennial, with coarse, simple or forked hairs. The stem is generally single and 15–50 cm tall. The basal lvs are oblanceolate, 2–5 cm long, and untoothed or shallowly toothed. The stem lvs are not stalked, lanceolate to obovate, and clasp the stem. The fls are 4–6 mm long and white, and the fruits are 2–5 cm long and erect.

Infrequent; typically found on rocky cliffs or gravelly bars in open places, 2500-5000 ft. A collection was made near the trail along Tahoma Cr at 3000 ft.

Arabis holboellii Hornem. var. *retrofracta* (Graham) Rydb.
HOLBOELL'S ROCKCRESS

A biennial with both simple, short hairs and flat, starlike hairs. Several stems usually rise from a branched base, 30–60 cm tall. The basal lvs are narrowly oblanceolate, with shallow teeth, and 1–5 cm long; the stem lvs are lanceolate, with a clasping base. The fls are 7–8 mm long and pale purple. The fruits are 4–6 cm long, hanging downward on straight stalks. Commonly found on gravel bars, on moraines, and on talus slopes, 2000-5000 ft. Known from the west side of the Glacier View Bridge and 0.5 mi below White R campground. Plants with nearly white fls are occasionally seen.

Arabis lyallii S. Watson var. *lyallii*
LYALL'S ROCKCRESS

A small, tufted perennial, with several stems 5–20 cm tall. The basal lvs are 1–3 cm long, oblanceolate, not toothed, and short-hairy with branched hairs; the stem lvs are narrowly lanceolate and clasp the stem. The rose to purplish fls are 7–8 mm long. The pods are 2–5 cm long and held erect.

Quite common on talus slopes and flats above 5000 ft, mostly on the east side of the Park, to as high as 9000 ft on the south side of the mountain. Found near Panorama Point; on the Kotsuck trail at Goat Pass at 5500 ft; at about 5600 ft on Naches Peak; at Sunrise and along the trail to Burroughs Mtn; and on Tokaloo Spire at 7500 ft.

145

Barbarea Aiton f.—WINTERCRESS

Barbarea orthoceras Ledeb.
AMERICAN WINTERCRESS

A hairless biennial or perennial, 30–90 cm tall. The basal lvs are few in number and pinnately lobed, with smaller, paired leaflets beneath the larger terminal leaflet, to about 8 cm long. The stem lvs are reduced upward, similarly lobed but stalked and somewhat clasping the stem. The inflorescence is branched, with many pale yellow fls 4–5 mm long. The fruits are 2–4 cm long, spreading or ascending.

Uncommon, but found widely through the Park, mostly below about 4000 ft, on open, wet ground. A collection was made at Longmire.

Brassica L.—MUSTARD

1   Stem lvs glaucous and clasping the stem ....*B. rapa* var. *rapa*
1   Stems lvs green, not glaucous, not clasping the stem at their bases ....
.... *B. nigra*

Brassica nigra (L.) W.D.J. Koch—BLACK MUSTARD

A more delicate weed than field mustard, in the Park reaching about 60 cm tall. The basal as well as the stem lvs are toothed and lobed, roughly hairy, and green - never glaucous. The stem lvs are attached to the stem by short stalks. The fls are yellow, with petals about 5 mm long. The fruits are up to 2 cm long and held erect, close to the stem.

First found in the Park in 1999, on the side of the road about 1 mile south of the White R entrance gate.

Brassica rapa L. var. *rapa*—FIELD MUSTARD

A rank-growing annual weed, with branched stems to 100 cm tall. The basal lvs are toothed or lobed, 10–20 cm long, and short-hairy with simple hairs; the stem lvs are not toothed, glaucous and without hairs, and clasp the stem. The fls are yellow and 6–10 mm long. The fruits are 3–7 cm long, ascending to spreading, and have a beaklike style at the tip.

A rare roadside weed; known in the Park from a 1938 collection made at Ohanapecosh. It is likely that this plant is introduced from time to time and fails to persist more than a few seasons.

Brassicaceae

Capsella Medik.—SHEPHERD'S PURSE

Capsella bursa-pastoris (L.) Medik.
SHEPHERD'S PURSE

A small annual weed with stems that are branched and can reach 50 cm tall, but are often less when growing on poor soil. The plant is short-hairy with both simple and branched hairs. The basal lvs are in a rosette and are 3–6 cm long, oblanceolate, stalked, and pinnately toothed. The stem lvs are smaller, lanceolate, and somewhat clasping. The white fls are about 2 mm long. The fruits are heart-shaped and 6–8 mm long.

Common weed along roads and on open waste ground, with other low herbs, below 3000 ft, as at Sunshine Point and Longmire. It was observed once in 1932 at Tipsoo Lk and may well turn up in other unusual places from time to time.

Cardamine L.—BITTERCRESS

Including species that were formerly called *Dentaria*, these are delicate plants, with basal leaves that are often quite different from the stem leaves. The annual species are often weedy, while the perennials have lovely white or pink flowers. The plants are mostly hairless and the fruit is a silique.

1	Lvs all simple	*C. bellidifolia* var. *bellidifolia*
1	Lvs all, or at least those of the upper stem, lobed or compound	2
2(1)	Plants from a perennial, tuberlike rootstock; lvs arising from the rootstock mostly simple; fls pink	*C. nuttallii* var. *nuttallii*
2	Plants annual or perennial, with slender, creeping rootstocks or taprooted; lower lvs compound; fls mostly white	3
3(2)	Petals 8–14 mm long; leaflets mostly 3	*C. angulata*
3	Petals 2–6 mm long; leaflets (at least of the stem lvs) more than 3	4
4(3)	Plants taprooted, annual or biennial; raceme long and slender	5
4	Plants perennial, with a horizontal rootstock; raceme various	6
5(4)	Basal lvs none or only a few, not arranged in a rosette	*C. pensylvanica*
5	Basal lvs many, in a rosette	*C. oligosperma* var. *oligosperma*
6(4)	Raceme short, compact, and umbel-like	*C. oligosperma* var. *kamtschatica*
6	Raceme slender, elongated, not at all umbel-like	7
7(6)	Basal lvs not toothed or palmately lobed	*C. brewerii* var. *orbicularis*
7	Basal lvs pinnately lobed	*C. occidentalis*

Cardamine angulata Hook.—ANGLED BITTERCRESS

Tall plants, from slender rootstocks, often growing in patches, the stems reaching 60 cm tall. The lower lvs are compound, with 3 leaflets. The leaflets are up to 5 cm long, more or less ovate with a 1 or 2 pairs of lateral teeth; the leaflets of the upper lvs are more slender and may be stalkless. The petals are white and 8–12 mm long. The fruits are 12–25 mm long, on ascending stalks.

Uncommon, reaching the Park near the Nisqually entrance and extending up to about 5,000 ft. Collected at Ethania Falls on St. Andrews Cr.

Cardamine bellidifolia L. var. *bellidifolia*
ALPINE BITTERCRESS

A tufted perennial, from a very short, branched basal stem, reaching 3–8 cm overall. The lvs are ovate to elliptical, toothed but undivided, 3–12 mm long on short stalks. The inflorescence is a few-flowered head resembling an umbel, and the white petals are 3–5 mm long. The fruits are 15–35 mm long and erect, as if in a bundle.

Rare, on moraines and talus slopes, above about 7000 ft. Known from the moraine below Russell Glacier and on Burroughs Mtn. A collection made by J. B. Flett was made at 7500 ft on the ridge between North and South Mowich Glaciers.

Cardamine brewerii S. Watson var. *orbicularis* (Greene) Detling
BREWER'S BITTERCRESS

A perennial, from a very short rootstock, with unbranched, erect stems 15–45 cm tall. The basal lvs are few, with a larger terminal lobe above a pair of smaller lobes, or occasionally undivided and roundish in outline. The stems are leafy, the lvs stalked and 3-lobed. The leaflets are shallowly toothed. The fls are numerous, with white petals 3–4 mm long, followed by erect fruits about 2 cm long.

Only occasionally found in the Park, on wet ground in forests below about 4,000 ft. O. D. Allen collected the plant "partly submerged in swamps, upper Nisqually V."

Cardamine nuttallii Greene var. *nuttallii*
NUTTALL'S BITTERCRESS

Distinctive for the small (2–5 mm thick and elongated) tubers on the rootstock. The basal lvs rise directly from the rootstock and are on long stalks; the lf blades are round in outline and shallowly lobed. The slender stem is less than 20 cm tall and the stem lvs are divided into 3–5 more or less pointed lobes. The fls are few in number and pink, the petals 10–13 mm long. The fruits are 2–3 cm long on ascending stalks.

Infrequent in moist woods below about 2000 ft, blooming early in the spring and often not seen by visitors. O. D. Allen collected the plant in the "upper valley of the Nisqually."

Cardamine occidentalis (S. Watson ex B. L. Rob.) Howell
WESTERN BITTERCRESS

The rootstock of this perennial species is short and at the base of the stem is enlarged or tuberlike . The stem is 20–40 cm tall and often leans over and sometimes roots at the nodes of the lower lvs. The basal lvs are few and pinnately divided into 7–11 toothed lobes, with the terminal lobe the largest and about 1 cm broad. The stem lvs are smaller, with fewer lobes and shorter stalks. The white petals are 4–6 mm long. The fruits are 2–3 cm long, on ascending stalks.

Apparently rare in the Park, growing on rather wet, open ground on subalpine flats. Known from Berkeley Park.

Cardamine oligosperma Nutt. var. *kamtschatica* (Regel) Detling
SIBERIAN BITTERCRESS

A perennial, with a basal rosette of lvs up to about 8 cm long and pinnately divided into 7 or 9 rounded leaflets. The several stems are 10–40 cm tall and bear a few 3–5 narrow-lobed lvs. The white petals are 2–3 mm long and crowded into a short raceme, less than about 2 cm long and resembling an umbel. The fruits are erect, crowded, and 2–3 cm long.

Fairly common in subalpine areas, growing along small streams. Collected at 6000 ft in Glacier Basin, and reported from Mowich Lk.

Cardamine oligosperma Nutt. var. *oligosperma*
LITTLE WESTERN BITTERCRESS

Very similar to the above variety, but growing as an annual or biennial and somewhat shorter, the stems usually well under 30 cm tall. The raceme is more than 3 cm tall and is not at all umbel-like. The fruits are 1–2 cm long.

Common and often weedlike, in campgrounds, along roads, in open woods, preferring wet ground, and on mossy rocks, mostly below 4000 ft; especially abundant around the Nisqually and Ohanapecosh entrances.

Cardamine pensylvanica Muhl. ex Willd.
PENNSYLVANIA BITTERCRESS

Annual or biennial, with stems often branched, 30–60 cm tall. The basal lvs are few and disappear early. The stem is leafy, the stalked lvs up to about 6 cm long and divided into 7–11 narrow leaflets. The white petals are 2–4 mm long. The fruits are about 2.5 cm long, on spreading stalks.

Common on wet ground in open forests and on damp cliffs, below 3000 ft, as between Tahoma and Kautz Creeks.

Descurainia Webb & Berth.
TANSY-MUSTARD

Annuals, or occasionally biennials, with branched or starlike hairs on the stems and leaves, yellowish flowers, and linear siliques. The 2-lobed stigma separates this genus from the similar *Sisymbrium*. "Tansy" refers to the 2- or 3-times pinnate lvs, resembling those of *Tanacetum*, a genus of the aster family.

1   Plant with stalked glands; fruits 10–20 mm long ..... *D. incana* ssp. *viscosa*
1   Plants lacking glands; fruits 15–30 mm long .... *D. sophia*

Descurainia incana (Fisch. & C. A. Mey) Dorn ssp. *viscosa* (Rydb.) Kartesz & Gandhi—MOUNTAIN TANSY-MUSTARD

The slender stems are 30–90 cm tall, typically branched, short-hairy, and glandular, especially in the inflorescence. The lvs are up to 10 cm long, twice pinnately divided, with the lf segments linear; the upper lvs are smaller, and once divided. The petals are about 2 mm long. The fruits are linear, straight, and 1–2 cm long.

Found chiefly on the east side of the mountain, especially in the area of the White R campground, on dry, rocky slopes and banks.

Descurainia sophia (L.) Webb in Engl. & Prantl
HERB SOPHIA

To about 60 cm tall, with a branched stem that is short-hairy but not glandular. The lvs are 3 times pinnately divided, into narrowly lanceolate segments, to about 8 cm long. The petals are 2.5–3 mm long and the fruits 1.5–3.5 cm long, usually curved.

Uncommon weed, mostly around buildings and at roadsides in dry places up to 5000 ft. Also collected once on Theosophy Ridge at 5500 ft.

Draba L.—DRABA

With one exception in the Park, these are tufted, cushion- or mat-forming perennials. None is common and several species are known from just one or a few occurrences on the east side of the mountain. Most have yellow flowers and siliques that are flattened and linear to elliptical. Close inspection of the hairs of the leaves is important in distinguishing the species, and worth the attention for the intricate forms the hairs take.

1	Plants annual; petals deeply 2-lobed ....*D. verna*
1	Plants perennial; petals round to slightly notched 2
2	Petals white .... *D. lonchocarpa* var. *lonchocarpa*
2	Petals yellow 3
3(2)	Flowering stem leafy ....*D. aureola*
3	Flowering stem leafless, or with 1 or 2 lvs at its base 4
4(3)	Lvs with "double pectinate" hairs (i.e., complex structures that resemble two combs placed side to side along their backs), especially on the underside 5
4	Lvs with simple to star-shaped hairs 6
5(4)	Lvs to 1.5 mm broad, the short hairs pressed to the surface .... *D. oligosperma*
5	Lvs 2 mm broad, the hairs looser, not flattened ....*D. incerta*
6(4)	Midrib prominent on the lower surface of the lf; lvs to 1.5 mm broad .... *D. paysonii* var. *treleasii*
6	Midrib not prominent on the lower surface of the lf; lvs 2–4 mm broad .. .. *D. ruaxes*

Draba aureola S. Watson—GOLDEN DRABA

A beautiful little plant, with dense, tufted basal lvs and stems that are 5–10 cm tall. The leafy stems are unique among the drabas in the Park. The lvs are oblanceolate, blunt at the tip, 10–20 mm long, short-hairy on the underside with starlike hairs and on the upper with simple or branched hairs. The fls are numerous, with yellow petals 4–5 mm long. The fruits are elliptical and 10–15 mm long, on spreading stalks.

Listed as "watch" in *The Endangered, Threatened, and Sensitive Vascular Plants of Washington* (Washington Natural Heritage Program, 1997).

Scattered on pumice flats and stony slopes around the mountain, above about 7000 ft, ranging up to 10000 ft; it can be found around Camp Muir. Also reported from Spray Park.

Mount Rainier is the only station for the plant in Washington; it also occurs on a few volcanic peaks in Oregon and on Lassen Peak in northern California.

Draba incerta Payson
YELLOWSTONE DRABA

A tufted perennial, with old lvs clothing the base of the stems. The lvs are narrowly oblanceolate, 5–10 mm long, and short-hairy with starlike hairs. The stems are 10–20 cm tall, with a few yellow fls in a short raceme. The fruits are lanceolate, 4–8 mm long.

On rocky ridges above about 6500 ft, chiefly on the east side of the Park. Collected on Burroughs Mtn at 7200 ft and on Goat Island Mtn at 7000 ft.

Draba lonchocarpa Rydb. var. *lonchocarpa*
SPEAR-FRUITED DRABA

The only native white-flowered *Draba* of high elevations in the Park, with tufted stems less than 10 cm tall. The lvs are 5–15 mm long, oblanceolate, and short-hairy with starlike hairs. The fls are crowded into a short inflorescence. The fruits are linear, 10–15 mm long, and sometimes twisted.

Infrequent, in places similar to those in which the previous species is found, but ranging between about 5000 and 8000 ft. A collection in the Park herbarium was made at Frozen Lk at 6800 ft. Also known from Owyhigh Lks.

152

Draba oligosperma Hook.—FEW-SEEDED DRABA

A very small, cushion-forming plant, with stems less than 10 cm tall. The lvs are densely crowded, linear-oblanceolate, and short-hairy chiefly with double-pectinate hairs. The yellow fls are few, in a short, rather loose raceme. The fruits are less than 1 cm long, ovate or oblong-ovate.

Seen infrequently on the east side of the Park, 8000-10,000 ft.

Draba paysonii J. F. Macbr. var. *treleasii* (O. E. Schulz) C. L. Hitchc.—PAYSON'S DRABA

A perennial forming loose cushions, with the fl stems rising only a few centimeters above the top lvs, altogether less than 6 cm tall. The lvs are nearly linear, 4–8 mm long, with a complex arrangement of hairs: the margins are fringed with stiff simple and forked hairs, the underside with long and branched hairs, and the upper side with shorter, simple or forked hairs. The fls are few, in a short raceme, and the fruits are ovate, about 5 mm long.

Apparently rare in the Park. The plant favors rocky crevices and talus slopes, 6000-8000 ft and has been collected on Burroughs Mtn at 7000 ft.

Draba ruaxes Payson & H. St. John WIND RIVER DRABA

A loosely tufted plant with rather long, leafy branches and flowering stems that reach only 2–4 cm above the lvs. The lvs are about 1 cm long, oblanceolate, with a variety of hairs—simple, forked, and branched. The yellow fls are few, in a compact inflorescence. The fruit is nearly oval and 5–8 cm long.

Listed as "watch" in *The Endangered, Threatened, and Sensitive Vascular Plants of Washington,* as *D. ventosa* var. *ruaxes* (Washington Natural Heritage Program, 1997).

Known in the Park from a single collection made among rocks "one-half mile northwest of Frozen Lk" (originally labeled *D. glacialis*). Hitchcock and Cronquist (1973) said that this species occurs in Washington only on Glacier Peak.

Draba verna L.—SPRING DRABA

An annual, with basal lvs and a slender, leafless stem. The lvs are 1–3 cm long, oblanceolate, with simple, forked, and starlike hairs. The fls are fairly numerous, in a dense headlike raceme that becomes longer as the fruits mature. The petals are 2–2.5 mm long, and deeply divided into 2 elongated lobes. The fruit is 5–10 mm long and elliptical to ovate.

An early-blooming native plant, but possibly spread into the Park by human activity. First noted in the Park during this study, from a rocky cliff along the road between Tahoma Cr and Kautz Cr. It may be expected to spread fairly widely in the Park below about 3000 ft.

Erysimum L.—WALLFLOWER

Erysimum arenicola S. Watson var. *torulosum* (Piper) C.L. Hitchc. CASCADE WALLFLOWER

A biennial herb, with a short raceme of big, bright yellow fls. The first year's basal lvs are 3–8 cm long, coarsely toothed, and sparsely hairy. The stems reach about 40 cm tall and are leafy with smaller, untoothed lvs. The yellow petals are 15–18 mm long and spread widely, the fl therefore up to 1.5 cm broad. The siliques are squarish in cross section, 6–8 cm long, and twisted.

Common, on rocky slopes, 5000-7000 ft. Collected in Klapatche Park; on Gobblers Knob; near the saddle of Pinnacle Peak; on Mount Fremont; below Interglacier; at Sunrise; and at Eunice Lk.

Lepidium L.—PEPPERGRASS

Two insignificant weedy annuals, seldom seen among the grasses and other weeds with which they usually grow. The leaves in the inflorescence clasp the stem.

1	Petals white	*L. campestre*
1	Petals yellow	*L. perfoliatum*

Lepidium campestre (L.) R. Br.—FIELD PEPPERGRASS

Plants 20–50 cm tall, with a leafy stem and a branched inflorescence. The basal lvs are 5–7 cm long, oblanceolate, not toothed, and hairy. The stem lvs are reduced in size upwards and clasp the stem with projecting lobes. The white fls are 2 mm long. The fruits are about 5 mm long, ovate in outline, with a narrow wing on the margins.

Uncommon weed, seen along roads and in the Longmire area. A collection was once made in the "grassy bed of Ipsut Cr."

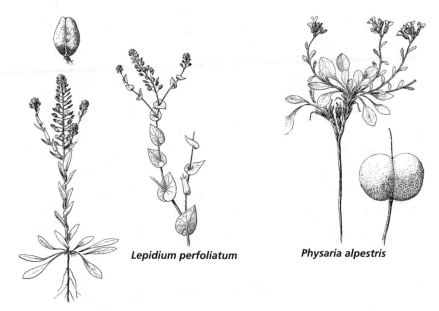

Lepidium perfoliatum

Physaria alpestris

Lepidium campestre

Lepidium perfoliatum L.—CLASPING PEPPERGRASS

Plants 20–60 cm tall, with a leafy stem and a branched inflorescence. The plant may be finely hairy below, but is hairless on the upper part. The basal lvs are 2 or 3 times pinnately divided into well-spaced, linear segments while the upper lvs are broadly lanceolate to ovate, not toothed, and "perfoliate" (completely surrounding the stem at their bases). The yellow petals are about 1.5 mm long. The unwinged fruits are 4 mm long, shaped rather like a diamond with rounded corners.

Another weed, collected once at Longmire, near the old corral, and perhaps not persisting.

Physaria (Nutt. ex Torr. & A. Gray) A. Gray—TWINPOD

Physaria alpestris Suksd.—ALPINE TWINPOD

A perennial, with several short stems from a somewhat woody base, 5–15 cm tall. The spoon-shaped lvs are 4–8 cm long and not toothed, whitish from the dense, starlike hairs; the lvs of the flowering stems are small and lanceolate. The fls are few, with yellow petals 7–9 mm long. The two chambers of the fruit are round in outline but notched at the top and bottom where they join, about 15 mm high, and both flattened and inflated.

Listed as "watch" in *The Endangered, Threatened, and Sensitive Vascular Plants of Washington* (Washington Natural Heritage Program, 1997).

Uncommon, growing on talus slopes and rocky ridges, above 7000 ft.

Rorippa Scop.—YELLOWCRESS

Rorippa curvisiliqua (Hook.) Bessey ex Britton var. *curvisiliqua*
WESTERN YELLOWCRESS

A native annual or biennial, with branched stems that typically creep a short distance along the ground before rising, to about 40 cm long. The lvs are up to about 5 cm long, narrowly oblanceolate in outline, deeply toothed, and clasp the stem. The petals are pale yellow and 2–2.5 mm long. The fruits are 5–10 mm, linear, and curved.

Common on moist, open ground and along streams up to 4500 ft, as around Longmire and at Chenuis Lks. Plants at Ghost Lk are less than 10 cm tall, with very small lvs.

Sisymbrium L.—TUMBLE MUSTARD

Annual weeds, with hairy lvs and stems, the hairs simple. The basal leaves are stalked and pinnately divided, with a larger terminal lobe and paired lobes below; the stem leaves are smaller, not stalked, with a few narrow lobes. The flowers are yellow, in many-flowered racemes. The fruits are linear and slender, on short, stout stalks.

1   Mature siliques beaked, flattened against the stem; petals 3 mm long .... *S. officinale*

1   Mature siliques spreading, not beaked; petals 6–8 mm long .... *S. altissimum*

Sisymbrium altissimum L.
TUMBLE MUSTARD

A coarse plant, with branched stems reaching 60–150 cm tall. The pale yellow petals are 6–8 mm long. The fruits are 6–10 cm long and spread widely.

A weed on dry, waste ground, uncommon in the Park. Known from Longmire.

Sisymbrium officinale (L.) Scop.—HEDGE MUSTARD

More slender than the above species, but also branched and reaching nearly 100 cm tall. The yellow petals are about 3 mm long. The fruits are 1–1.5 mm long and held closely against the stem.

Another weed, collected in 1965 by Robert Wakefield for his baseline weeds study, on a roadside near Ohanapecosh.

Smelowskia C. A. Mey.—SMELOWSKIA

Low, cushion-forming perennials with tufted stems. The leaves are pinnately divided, and both the leaves and stems are clothed with whitish hairs. The flowers are in short racemes, on stems less than 15 cm tall with 2 or 3 leaves. The petals are white and the fruit is a silique. Mount Rainier lies near the southern limit of the distribution of *S. calycina* and the northern limit of *S. ovalis*.

1   Basal lvs with long, stiff hairs on the stalk; siliques 5–10 mm long .... *S. calycina* var. *americana*
1   Basal lvs lacking long, stiff hairs on the stalk; siliques to 5–6 mm long .... *S. ovalis* var. *ovalis*

Smelowskia calycina (Stephan) C. A. Mey. var. *americana* (Regel & Herder) W.H. Drury & Rollins—ALPINE SMELOWSKIA

Besides the characteristics noted in the key, the petals in this species are 4–8 mm long.

Less common than the following species, found on talus slopes, 6000-10000 ft. Collections are known from about 10000 ft near Camp Muir and at 8500 ft above McClure Rock.

Smelowskia ovalis M. E. Jones var. *ovalis* SMALL-FRUIT SMELOWSKIA

In this species, the petals are smaller, 4–5 mm long.

Common, on rocky slopes and ridges, 7000-10000 ft. An exceptional group of plants grows at the "Beehive" on Cowlitz Cleaver, at 11000 ft, probably the altitude record for a vascular plant on the mountain. Collected at Panorama Point; on McClure Rock; on Wapowety Cleaver at 9000 ft; near Frozen Lk; and on Burroughs Mtn. Also reported from above Spray Park.

Teesdalia Aiton f.—TEESDALIA

Teesdalia nudicaulis (L.) R. Br.—TEESDALIA

Leaves in a basal rosette only, 5–15 mm long with slender stalks, the blades oval, obovate, or pinnately lobed. The stems are leafless, 5–20 cm tall, with a relatively tall raceme. The petals are white, about 1 mm long. The fruits more or less oval, with a notch at the top and a narrowly winged margin, about 2.5 mm long.

An inconspicuous weed, of grassy, open waste ground; first found in the Park during fieldwork for this study. Found at the Sunshine Point picnic area and at Longmire.

Thlaspi L.—PENNYCRESS

Two plants of very different appearance, both are hairless with white flowers and have clasping stem leaves. The fruit is a flattened silicle.

1   Annual weed; fruits round in outline, broadly winged .... *T. arvense*
1   Alpine perennial; fruits heart-shaped, not winged ....
.... *T. montanum* var. *montanum*

Thlaspi arvense L.—FIELD PENNYCRESS

An annual with branched stems 15–40 cm tall. The basal lvs are few in number, oblanceolate, tapered to the stalk and about 6 cm long; the stem lvs are smaller and oblong. The petals are 4 mm long. The fruits are strongly flattened, round in outline but deeply notched at the top, and 8–12 mm long, with a broad, thin wing.

Fairly common weed along roadsides and around buildings, as at Longmire.

Thlaspi montanum L. var. *montanum*
MOUNTAIN PENNYCRESS

A tufted perennial, 10–20 cm tall, with several stems from a low, branched base. The numerous basal lvs are elliptical to obovate, untoothed or shallowly toothed, to about 2 cm long. The petals are 4–6 mm long. The fruits are obovate, with a shallow notch at the top, and 4–8 mm long.

Common on open subalpine flats and on alpine slopes, 5000-7000 ft, mostly on the west side of the Park. Collected between Klapatche and St. Andrews Parks; at Mowich Lk; on Mount Wow at 5000; on Gobblers Knob; and near the saddle of Eagle Peak.

Callitrichaceae Link
WATER-STARWORT FAMILY

Callitriche L.—WATER-STARWORT

Perennial aquatic plants, with slender stems and opposite leaves. The flowers are inconspicuous and unisexual, not stalked and usually subtended by 2 bracts in the axils of the leaves. Sepals and petals are absent. The male flowers have 1 stamen and the ovary of the female flower is deeply lobed, splitting at maturity into 4 fruits, which are less than 1 mm long.

An interesting filamentous alga, *Nitella gracilis* (Sm.) J. G. Agardh., or brittlewort, can be found forming mats in shallow ponds at low elevations. Its strands could be confused with the slender submerged leaves of *Callitriche*. The plant consists of a threadlike stem with regularly placed whorls of 6–8 slender branchlets, each of which is about 20 mm long. Male and female reproductive structures are borne in tiny reddish bodies near the bases of the branchlets; these become dark and more visible as they mature.

1   Plants with linear lvs, remaining submerged .... *C. hermaphroditica*
1   Plants with both linear submerged lvs and broader floating lvs 2

2(1) Fruit wider above the middle and notched at the tip ....
.... *C. heterophylla* ssp. *bolanderi*
2   Fruit "blocky" and more or less straight-sided, shallowly notched at the tip
.... *C. palustris*

Callitriche hermaphroditica L.—NORTHERN WATER-STARWORT

Remaining submerged throughout its period of growth, with stems reaching about 25 cm long. The lvs are linear, with 1 main vein, 5–25 mm long, and with a broad notch at the tip. The bases of the opposite lvs do not meet. The fls are not bracted. The fruits are round in outline, with a shallow notch at the top.

Rare, in running water of small streams at low elevations; not found in the major rivers of glacial origin. St. John and Warren (1937) report this species at Reflection Lks, a highly unusual location.

Callitriche heterophylla Pursh ssp. *bolanderi* (Hegelm.) Calder & Roy L. Taylor
GREATER WATER-STARWORT

Stems up to about 40 cm long bear two different types of lvs: the submerged lvs are linear, shallowly notched at the tip, and 5–20 mm long, with 1 main vein, while the floating lvs are somewhat shorter and up to 5 mm wide, more or less obovate, with 3 main veins. The bases of the opposite lvs meet at the stem to form thin ridges. The fruits are wider above the middle and notched at the top—that is, somewhat heart-shaped.

Found in the shallow ponds near the Nisqually entrance and reported at Green Lk. This species was not mentioned by earlier writers.

Callitriche palustris L.
SWAMP WATER-STARWORT

With longer stems, to 45 cm, that have both floating lvs and submerged lvs, this species is sometimes found stranded on mud. The lvs are of two types and scarcely distinguishable from those of *C. heterophylla*, except that the floating lvs are up to 3 mm wide. The fruits are about rectangular in outline, with rounded corners and a shallow notch at the top, thus overall blocky in appearance.

Rare, or at least easily overlooked. In shallow ponds and swamps. A collection was made at Sunrise Lk in 1936, but the species is generally found at lower elevations.

Campanulaceae Juss.—BELLFLOWER FAMILY

Campanula L.—BLUEBELLS

Perennial herbs with alternate leaves and a milky sap. The flowers are bell-shaped, with the petals fused into a tube below, with 5 free lobes; the 5 sepals are likewise united below with free, spreading tips; there are 3 stigmas. The top flowers in the inflorescence bloom first. The ovary is inferior and develops into a short, capsule, shaped like a top, which opens by lateral pores.

1   Style exserted from the corolla; fls to 1.5 cm long .... *C. scouleri*
1   Style as long as the corolla; fls mostly more than 2 cm long 2

2(1)   Common native plant; fls 2–3 cm long, nodding .... *C. rotundifolia*
2   Rare introduced plant; fls 3–4 cm long, facing out and up .. *C. persicifolia*

Campanula persicifolia L.—PEACH-LEAF BLUEBELLS

Perennial with leafy stems 40–60 cm tall. The basal and stem lvs are both narrowly oblanceolate, finely toothed, and hairless; the longest are about 12 cm long and lack stalks. The 4–8 fls are borne in a false raceme; each is broadly bell-shaped, 3–4 cm long and usually somewhat wider at the mouth, and light purple.

Established around the old superintendent's house at the Nisqually entrance and found in 1998 at about 3200 ft on the side of the road coming down the east side of the Cowlitz Divide.

Campanula rotundifolia L.—SCOTS BLUEBELLS

One of the most delightful wildflowers of the Park and a signature species of cliffs and rocky slopes in the subalpine zone. The lvs are of two distinct types: the basal lvs are ovate to roundish, on slender stalks and 2–5 cm long, while the stem lvs are linear, not stalked, and 2–7 cm long. The flowering stems are 10–40 cm tall and unbranched, with the fls in an open raceme. The fls are a light blue-purple, 2–3 cm long, and nod on long, slender stalks; the free lobes are relatively short, hiding the style.

Common in open places on talus slopes, crevices on cliffs, and on streambanks to 6500 ft. Especially notable on the roadside below Tipsoo Lk in the fall. Collected along the trail between Mowich and Eunice Lks at 5000 ft; at Spray Park; on roadcuts along Westside Rd; at Paradise Park; at the foot of Cowlitz Glacier; on Shriner Peak; on cliffs along the highway between Dewey Cr and Tipsoo Lk; at Sunrise and Berkeley Park; at Grand Park; and at Eagle Peak. Also seen on cliffs along the Ohanapecosh R near the campground.

Campanula scouleri A. DC.—SCOULER'S HAREBELL

A somewhat sprawling plant, with stems to 30 cm long. The lvs are ovate below, becoming lanceolate above, sharply toothed, stalked, and 3–8 cm long. The fls are light blue to almost white, 1–1.5 cm long, bell-shaped but with the free lobes about half the length of the fl and curved back sharply, exposing the style.

Common in somewhat open forests, 2000-5000 ft. On the trail to Gobblers Knob at 4900 ft; on the trail along Tahoma Cr; on the South Puyallup R near the road; along the road down the east side of the Cowlitz Divide; at Ohanapecosh; and in the White R canyon at 4000 ft.

Caprifoliaceae Juss.—HONEYSUCKLE FAMILY

Shrubs or vines with opposite leaves. The flowers are often irregular, with united petals and sepals. The ovary is inferior and there are 4 or 5 stamens.

1   Fls densely clustered in terminal cymes or umbels; fls more or less flat ... 2
1   Fls in pairs or short racemes or in axillary clusters; fls tubular or funnel-shaped 3

2(1)   Lvs pinnate .... *Sambucus*
2   Lvs simple or lobed .... *Viburnum*

3(1)   Plants with slender, creeping stems and glossy, evergreen lvs .... *Linnaea*
3   Plants upright shrubs or vines; lvs deciduous, not glossy 4

4(3)   Fls nearly regular; berries white .... *Symphoricarpos*
4   Fls irregular; berries red or black .... *Lonicera*

Linnaea L.—TWINFLOWER

Linnaea borealis L. ssp. *longiflora* (Torr.) Hultén—TWINFLOWER

A subshrub with evergreen lvs and long, slender, trailing stems that can reach 1 m or more in length. The lvs are carried on the running stems and on well-spaced upright stems that reach 6–10 cm tall. The glossy, bright green lvs are oval to obovate, 1–2 cm long on short stalks, with toothed margins. The pink fls are borne in pairs at the top of a forked, glandular stalk and are narrowly bell-shaped, 12–16 mm long and slightly irregular; they have the delicate scent of almond. The ovary develops into a dry nutlet clothed with short, hooked hairs.

Well-conserved herbarium specimens hold the color of the lvs and fls very well: material gathered in 1891 by O. D. Allen, in the Park herbarium, is remarkably fresh.

Very common and often carpeting tracts of forest floor, up to 4000 ft. Well-displayed around the Nisqually entrance, at Longmire, at Box Canyon, and at Ohanapecosh. Notable for cascading down mossy banks and low cliffs on Westside Rd.

Lonicera L.—HONEYSUCKLE

Two honeysuckles are found in the Park, one a vine and one a shrub. Both have simple, deciduous leaves. The corollas are irregular: 2-lipped with a spurred base. The fruit is a berry.

1   Fls in pairs in the axils of lvs; shrub .... *L. involucrata* var. *involucrata*
1   Fls in small clusters at the ends of stems; vine .... *L. ciliosa*

Lonicera ciliosa (Pursh) DC.
ORANGE HONEYSUCKLE

A slender vine that clambers and twists over other shrubs, reaching 8 m in length. The lvs have short hairs on the margins, hairless and somewhat glaucous on the underside, and 2–6 cm long. The orange fls are 2–2.5 cm long and borne in small clusters, on the ends of the stems, sitting above a pair of joined lvs. The berry is red.

Fairly common in open woods in the Nisqually drainage, to about 3500 ft. Collected at Sunshine Point, Glacier View Bridge, and at the Ohanapecosh entrance gate.

Lonicera involucrata (Richardson in Franklin) Banks var. *involucrata*
TWIN HONEYSUCKLE

Upright shrubs, reaching 3 m tall. The lvs are oval to broadly lanceolate, 5–15 cm long and usually somewhat hairy on the underside. The fls are paired in the axils of the upper lvs; they are 1–1.5 cm long, tubular, yellow-orange, and sit above a pair of broad reddish purple bracts. The berries are black.

Found in a few places on wet ground below about 4500 ft. Collected at Sunshine Point and at the edge of the Longmire swamp.

Another shrubby species, *Lonicera utahensis* S. Watson, has been reported on the trail between Chinook Pass and Sheep Lk, just east of the Park boundary. Its light yellow fls are in pairs in the axils of the upper lvs. The fl bracts are narrow and inconspicuous.

Sambucus L.—ELDERBERRY

Tall, erect shrubs with weak stems. The leaves are pinnately divided into large, sharply toothed lobes. The numerous flowers are regular, small, and crowded into terminal cymes. The flowers are creamy-white, nearly flat, with 5 exserted stamens, and typically have an unpleasant odor. The fruit is a berry containing 3–5 nutlets.

1   Berries blue-black; fls in flat-topped clusters .... *S. cerulea* var. *cerulea*
1   Berries red; fls in pyramidal clusters ....
.... *S. racemosa* ssp. *pubens* var. *arborescens*

Sambucus cerulea Raf. var. *cerulea*
BLUE ELDERBERRY

Sometimes growing as a small tree and reaching 10 m tall, but usually shorter and shrublike, with multiple stems. The 5–9 leaflets are each 8–12 cm long and hairless. The fls are 4–5 mm broad, in a flat-topped, umbel-like cyme. The berries are blue-black, with a whitish bloom on the surface.

Uncommon in the Park, on dry ground or open woods on hillsides, to about 3000 ft. Known from the lower Nisqually V, at Longmire, and at Ohanapecosh.

Sambucus racemosa L. ssp. *pubens* (Michaux) House var. *arborescens* (Torr. & A. Gray) A. Gray—RED ELDERBERRY

A shrub reaching 5 m tall, with 5–7 leaflets per lf, each 5–15 cm long. The lvs are short-hairy on the underside. The fls are 5–6 mm broad, in an elongated, pyramidal cyme and the fruit is red.

Common on wet ground in forests, at roadsides, and along streams, below 5000 ft. Collected along Westside Rd at 2300 ft; at the trailhead to Lk George at 4000 ft; at Longmire; near Reflection Lks; on the Kotsuck Cr trail at 4500 ft; in the Grove of Patriarchs; at Sunrise; and at Eunice Lk.

Symphoricarpos Duhamel—SNOWBERRY

Small shrubs, much-branched with many small twigs. The deciduous leaves are short-hairy. The pink flowers are bell-shaped and short, in dense clusters at the ends of the stems. The fruit is a berrylike drupe, white and waxy, containing 2 seeds. The fruits typically hang on the branches for several months into the fall or winter.

1   Upright shrubs; branchlets mostly hairless .... *S. albus* var. *albus*
1   Trailing shrubs; branchlets short-hairy .... *S. hesperius*

Symphoricarpos albus (L.) S.F. Blake var. *albus*—SNOWBERRY

A shrub to about 1 m tall. The lvs are oval and 2–5 cm long, and on vigorous stems may often have wavy or lobed margins. The fls are 4–5 mm long and hairy inside. The berries are mostly 8–10 mm broad.

Common on rocky banks and in dry, open woods, below about 3000 ft. Collected on the Ohanapecosh R trail and seen along the Park road where it crosses Backbone Ridge.

Symphoricarpos hesperius G. N. Jones
CREEPING SNOWBERRY

A trailing or only weakly erect shrub, the stems to 1.5 m long. The lvs are oval, 1–3 cm long, even or sometimes lobed. The 4 mm long fls may have a few short hairs inside and the berry is about 6–8 mm long.

Uncommon, in dry woods on stony ground. One collection was located, from 3000 ft near the White R ranger station.

Viburnum L.—VIBURNUM

Viburnum edule (Michaux) Raf.—HIGH-BUSH CRANBERRY

A multistemmed shrub 1–2 m tall. The lvs are sharply toothed, palmately veined, and roundish; many are 3-lobed at the end. The white fls are about 3 mm broad and are borne in small cymes. The 5 stamens are shorter than the petals, which spread widely above the tube. The bright red fruits are berrylike drupes that persist on the bushes into the winter.

Rare in the Park, below about 4000 ft along streams. Known from Mountain Meadows and Ohanapecosh.

Caryophyllaceae Juss.—PINK FAMILY

A family with a large number of species in the Park, only a few of which are significant as wildflowers. Many are weeds, while many of the natives are low-growing, nondescript plants. These species are annual or perennial herbs, with opposite, untoothed leaves. Petals usually number 4 or 5 but may be absent. In some genera the petals are clawed, with a broader blade; that is, a relatively wide blade tops a slender, vertical stemlike claw. The 4 or 5 sepals are usually free, but may be united into a tube. Most species have 10 stamens. The ovary is superior, with 2–5 styles, developing into a capsule.

See "Doubtful and Excluded Species," p. 469, for *Saponaria officinalis*.

1	Sepals distinct or almost distinct	2
1	Sepals united	10
2(1)	Papery stipules present	3
2	Stipules absent	4
3(2)	Lvs opposite; styles and valves of the capsule 3	*Spergularia*
3	Lvs apparently whorled; styles and valves of the capsule 5	*Spergula*
4(2)	Capsule cylindrical, curved near the top	*Cerastium*
4	Capsule egg-shaped to elliptical, not curved at the tip	5
5(4)	Styles 5, alternating with the sepals	*Sagina*
5	Styles mostly 3, opposite the petals	6
6(5)	Petals notched or deeply divided at the tip	7
6	Petals not notched, rounded at the tip	8
7(6)	Petals notched; plants glandular-hairy	*Pseudostellaria*
7	Petals deeply divided; plants perhaps hairy but not glandular	*Stellaria*
8(6)	Mature capsule opening with 3 teeth; plants perennial	*Minuartia*
8	Mature capsule opening with 6 teeth; plants annual or perennial	9
9(8)	Styles 2–3 mm long; seeds with a fleshy appendage	*Moehringia*
9	Styles 0.5–2 mm long; seed appendage absent	*Arenaria*
10(1)	Styles 3 or 5 (rarely 4)	*Silene*
10	Styles 2	11
11(10)	Fls set closely above by 1 or more pairs of bracts	*Dianthus*
11	Fls not set above bracts	*Vaccaria*

Arenaria L.—SANDWORT

Minuartia and *Moehringia*, which are discussed below, are closely related, and species in both those genera were once included in *Arenaria*. *Arenaria* is distinguished by the number of teeth by which the mature capsule opens and the length of the style, both features best seen with magnification. In *Arenaria* there are 5 sepals and 5 white, spreading petals. The former are not united and the latter are not notched at the tips. There are 10 stamens and the egg-shaped capsule opens by 3 valves (although the valves are divided at the tips, and so the capsule may appear to have 6 valves).

1	Weedy annual; lvs ovate to lanceolate	*A. serpyllifolia* ssp. *serpyllifolia*
1	Native perennial; lvs needlelike	*A. capillaris* ssp. *americana*

Arenaria capillaris Poir. ssp. *americana* Maguire—SLENDER MOUNTAIN SANDWORT

A matlike perennial with short branches. The lvs are linear, sharply pointed, sometimes curved, and 1.5–6 cm long. The flowering stems are numerous and can reach 30 cm tall, bearing a few pairs of lvs. The fls are in few-flowered, flat-topped clusters.

Common and often abundant, on stony flats and talus slopes, 5000-7000 ft. Collected at Paradise Park; on the trail between Tipsoo and Dewey Lks; at the saddle on Eagle Peak at 5900 ft; at Sunrise and on the Burroughs Mtn trail; at 6400 ft above Berkeley Park; and reported from around Spray Park.

Arenaria serpyllifolia L. ssp. *serpyllifolia*—THYME-LEAF SANDWORT

An annual weed, with the branched stems mostly sprawling on the ground, 10–30 cm long. The lvs are ovate, 4–7 mm long, with short, stiff hairs, and sometimes glandular as well. The fls are in slender stalks in open cymes. The petals are 2 mm long, somewhat shorter than the sepals.

Occurring only rarely as a roadside weed in moist places, as at Longmire.

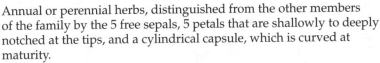

Cerastium L.—MOUSE-EAR CHICKWEED, CHICKWEED

Annual or perennial herbs, distinguished from the other members of the family by the 5 free sepals, 5 petals that are shallowly to deeply notched at the tips, and a cylindrical capsule, which is curved at maturity.

1   Native perennial; petals longer than the sepals .... *C. arvense* ssp. *strictum*
1   Weedy annuals or perennials; petals equal to or shorter than the sepals   2

2(1)   Fls clustered in heads; the fl stalks shorter than the sepals ....
       .... *C. glomeratum*
2   Fls in loose cymes; the fl stalks longer than the sepals ....
       .... *C. fontanum* ssp. *vulgare*

Cerastium arvense L. ssp. *strictum* (L.) Ugborogho FIELD CHICKWEED

An attractive plant, with prominent white fls. The stems are somewhat matted, hairy, and 15–40 cm tall. The lvs are linear to lanceolate, 2–3 cm long, with small bundles of lvs in the axils of the upper stem lvs. The bracts of the inflorescence have papery margins, while the stalks and sepals are minutely glandular. The petals are 12–15 mm long, deeply notched, and about 1 cm wide.

A collection made at Ipsut Pass by Dinni Fabiani approaches *C. berringianum*, with oblanceolate lvs and only a few bundled lvs in the axils, but it has the scarious-margined bracts characteristic of *C. arvense*. *Allen 237* from Mount Wow has a similar mix of features.

Quite common on open, stony slopes in dryish places, 3500-6000 ft.

Cerastium fontanum Baumg. ssp. *vulgare* (Hartman) Greuter & Burdet
MOUSE-EAR CHICKWEED

Generally a biennial, although sometimes persisting several years, short-hairy but not glandular, with matted stems 10–30 cm long. The dark green lvs are oval and 1–2 cm long. The cyme is few-flowered, the stalks much longer than the sepals. The white, shallowly notched petals are about as long as the sepals, 4–5 mm.

A common weed along roads and trails, to about 3000 ft, especially around Longmire and near the Ohanapecosh entrance.

Cerastium glomeratum Thuill.
STICKY CHICKWEED

Annuals, with stems erect and 10–20 cm tall, glandular-hairy throughout. The lvs are oval to narrowly ovate and 5–20 mm long. The fls are clustered into small heads at the tops of the stems, the stalks shorter than the sepals. The sepals are 5–6 mm long, the petals about the same.

An uncommon weed, seen chiefly in the Ohanapecosh area, at roadsides and in open woods.

Dianthus L.—PINK

Two non-native species are occasionally found in the Park. Unlike the much more common non-native silenes, the flowers bear only two styles and the calyx is not inflated. The petals are long, with a slender claw but are not ornamented with lateral lobes. The flower cluster is set above several pairs of leaf-like bracts and the sepals are fused into a ribbed tube. The capsule is tubular, opening by 4 valves.

Caryophyllaceae

1   Fls many in a dense head; plant hairless .... *D. barbat*
1   Fls several in a short, open cluster; plant short-hairy, at least in th
     inflorescence .... *D. armeri*

Dianthus armeria L.—GRASS PINK, DEPTFORD PINK
Usually an annual, but may live two years, with one erect stem 30-60 cm
tall. There is a tuft of linear, basal lvs up to 10 cm long; the stem lvs are
progressively smaller. The 3-6 fls are in a crowded cyme; the blade of the
petal is about 3 mm long, only slightly wider than the claw, and dark
pink.
     Collected in 1999 on the roadside along Hwy 410, about 1.5 mi south
of the White R entrance gate.

Dianthus barbatus L.—SWEET WILLIAM
A perennial, 30–60 cm tall, hairless throughout, from a
stout rootstock. The basal lvs are up to 10 cm long,
lanceolate to oblanceolate; those of the stem are somewhat
smaller and lanceolate. The fls are in congested heads,
variously white, pink, or violet; the blades of the petals are
6–10 mm long.
     Found a few times, many years ago, as a weed in the
Longmire area, having escaped from fl gardens in the
vicinity. Also collected once along the road 2.5 mi above
Longmire. It may not persist in the Park.

Minuartia L.—MINUARTIA, SANDWORT

Separated from *Arenaria*, in which genus these species were
once included. All are high-elevation perennials, with short,
linear lvs, and otherwise mostly similar except that in *Minuartia* the
capsule opens by 3 valves rather than 6.

1   Sepals blunt; fls mostly solitary on the stems .... *M. obtusiloba*
1   Sepals pointed; fls several per stem 2

2(1) Sepals 3-veined, sharp with a long point .... *M. nuttallii* ssp. *nuttallii*
2   Sepals 1-veined, sharp but not pointed .... *M. rubella*

Minuartia nuttallii (Pax) Briq. ssp. *nuttallii*—NUTTALL'S SANDWORT
Somewhat spreading, with the stems
reaching about 10 cm long, glandular-
hairy. The plant is notable for its
fragility, and the stems are apt to
shatter at the nodes of the lvs. The lvs
are linear, sharply pointed and 3–10 mm
long. The sepals and petals are about 4–5

e petals usually a bit shorter than the sepals. The mature
er than the sepals.
Park; found on dry pumice flats and on moraines, 7000-
y on the eastern side of the Park. Reported by Peter
t Cowlitz Chimneys and collected by St. John in the 1930s
k. Found in 1999 on the summit of Second Burroughs Mtn.

Minuartia obtusiloba (Rydb.) House
ALPINE SANDWORT

Densely tufted stems, trailing on the ground to about
10 cm long and forming mats. The lvs are linear,
minutely glandular-hairy, and 4–7 mm long. One
fl, with petals about 10 mm long, is borne on each
slender stem. The sepals are somewhat shorter
and tend to curve inwards over the petals. The
capsule is about as long as the sepals.

Fairly common on talus slopes and rocky flats,
6000-8000 ft. Collected at Sunrise at 6500 ft; at Frozen
Lk; on the Burroughs Mtn trail; and reported from around
Spray Park.

Minuartia rubella (Wahlenb.) Hiern
BOREAL SANDWORT

A tufted plant, growing into cushions about 10 cm
across. The lvs are finely glandular-hairy, sharply
pointed, and less than 1 cm long. The flowering
stems reach about 10 cm tall, with 2–5 fls in a loose
cyme. The petals are about 5 mm long, slightly
exceeding the sepals. The capsule is longer than
the sepals at maturity.

Common on talus slopes and rocky ridges,
from 5000 to about 7000 ft. Collections have been
made on the summit of Mount Wow and at 6500 ft at
Sunrise.

Moehringia L.—GROVE-SANDWORT

Moehringia macrophylla (Hook.) Torr.
LARGE-LEAF SANDWORT

Perennial, from a slender and spreading
rootstock, with several leafy stems 5–15 cm
tall, the stems and lvs generally finely
hairy. The lvs are 1–6 cm long and
lanceolate to oblanceolate. Each stem
bears 2–6 white fls on long stalks, in an
open cyme. The sepals are 4–5 mm long,
somewhat longer than the petals.

Common in moist, partly shaded places on roadcuts, banks, in thickets, and in open woods from the Park entrances to about 6000 ft. Found at Mowich Lk; around the Nisqually entrance and up Westside Rd; on Rampart Ridge; at Longmire; at the Butter Cr Research Natural Area; at lower Paradise Park; in Stevens Canyon; through the Ohanapecosh area; at the White R campground; and along the road from Cayuse Pass to Tipsoo Lk.

Pseudostellaria Pax—STICKY-STARWORT

Pseudostellaria jamesiana (Torr.) Weber & Hartman
STICKY-STARWORT

Separated from the stellarias of the Park for being glandular-hairy in the upper part and for having petals that are much longer than the sepals. The plant is 15–40 cm tall, the stems curved at the base and usually branched. The lvs are lanceolate, 2–10 cm long, and reduced in length above. The inflorescence is a leafy cyme. There are 5 petals and 5 sepals, the petals shallowly notched at the tips and 6–12 mm long. The capsule is roundish.

Apparently rare in the Park, with one collection known, made on a moist slope above Lk George (this collection in the Park herbarium was originally identified as *Cerastium arvense*). The species is typically found in Washington east of the Cascades.

Sagina L.—PEARLWORT

Annuals and perennials, growing as compact or spreading mats. The leaves are linear and sharply pointed. There may be secondary bundles of leaves in the axils of the main lvs. The flowers are borne on threadlike stalks. The sepals and white petals number 4 or 5 and are less than about 2 mm long. The mature capsule is conical to egg-shaped, about 3 mm long, and opening by 4 or 5 valves.

1	Plant annual, with very slender stems, without a basal rosette of lvs .... *S. decumbens* ssp. *occidentalis*	
1	Plant perennial, with stems more stout and with a basal rosette 2	
2(1)	Alpine plant; sepals 5, appressed to the capsule in fruit .... *S. saginoides*	
2	Weedy lowland plant; sepals mostly 4, spreading from the capsule in fruit .... *S. procumbens*	

Sagina decumbens (Elliott) Torr. & A. Gray ssp. *occidentalis* (S. Watson) G. E. Crow—WESTERN PEARLWORT

A native annual, with slender, weak stems up to 15 cm tall, without a basal rosette of lvs and seldom having bundles of lvs in the axils. The lvs are 5–20 mm long. The fl stalks are usually glandular-hairy and remain straight after flowering. There are 5 petals and sepals, and 5 or 10 stamens. The capsule is conical.

Fairly common on moist ground in open places, sometimes behaving as a weed on disturbed ground, up to about 4000 ft.

Sagina procumbens L.—PEARLWORT

A matted weed, with long and slender stems, to about 15 cm long. There is a basal rosette of lvs and the stem lvs have bundles of smaller lvs in the axils; the major lvs are 3–10 mm long. The fls are axillary, on hairless stalks 5–20 mm tall that curve at the top at maturity. There are usually 4 sepals, stamens, and petals, although occasionally plants are seen that lack petals. The capsule opens by 4 valves.

Common weed, along roadsides and on waste ground below about 3000 ft. Found along Westside Rd and in the Nisqually entrance area; at Longmire; at Box Canyon; around Ohanapecosh; and at the roadside south from the White R entrance, reaching Fryingpan Cr.

Sagina saginoides (L.) H. Karst.
ALPINE PEARLWORT

A hairless, matted perennial with slender, ascending stems and basal rosettes of lvs. The linear lvs are 5–15 mm long. The flowering stems reach about 10 cm tall, and the slender, hairless stalks are 1–3 cm long. There are 5 sepals and petals, and 10 stamens. The capsule is conical.

Uncommon, on moist flats among rocks, 5500-7000 ft. Known from 6000 ft along the Paradise R and at 6000 ft at the head of Stevens Canyon. Also reported also from Spray Park.

Silene L.—CAMPION, CATCHFLY

A large genus, well-represented in the Park by both weedy and native species, including among the latter some fine wildflowers. The natives are perennials, with simple or branched stems. A conspicuous feature is the ribbed and sometimes inflated calyx, formed by the 5 fused sepals. The petals are exserted from the calyx, with a slender claw and broader blade, and are ornamented at the junction of the blade and claw with paired scalelike or pointed auricles and smaller lateral lobes. There are 10 stamens and 3 styles (5 in one weedy species). The capsule is egg-shaped. Most of the species are glandular-hairy, at least in the upper parts.

1   Styles 5 .... *S. latifolia* ssp. *alba*
1   Styles 3 2

2(1) Plants annual 3
2   Plants biennial or perennial 4

3(2) Plant short-hairy throughout; fls opening in the evening .... *S. noctiflora*
3 Plant glandular below the inflorescence, otherwise hairless; fls opening in the day .... *S. antirrhina*

4(2) Calyx hairless .... *S. acaulis* var. *exscapa*
4 Calyx short-hairy 5

5(4) Weedy biennial; calyx with a few hairs on the edges; plants glaucous .....
.... *S. vulgaris*
5 Native perennials; calyx more densely hairy; plants not glaucous 6

6(5) Calyx 5–8 mm long .... *S. menziesii*
6 Calyx more than 10 mm long 7

7(6) Plants low and matted, with flowering stems to about 10 cm tall; basal lvs to 2 cm long .... *S. suksdorfii*
7 Plants tufted, with flowering stems at least 15 cm tall; basal lvs more than 3 cm long 8

8(7) Stem short-hairy, but not glandular .... *S. douglasii* var. *douglasii*
8 Stem short-hairy and glandular, at least on the upper portion 9

9(8) Each stem with 2–4 fls, typically borne singly; auricles 2, blunt; blade with 4 lobes.. *S. parryi*
9 Each stem with 8–10 fls, often paired; auricles 4, pointed; blade with 2 forked lobes.. *S. oregana*

Silene acaulis (L.) Jacq. var. *exscapa* (All.) DC. & Lam. CUSHION PINK, MOSS CAMPION

Tufted and growing to form cushions 10–60 cm across. The lvs are 5–10 mm long, densely crowded on the stems, linear and fringed with hairs on the margins. The pink or pink-purple fls are solitary on short stalks, shallowly notched at the tip. The calyx is hairless and 4–7 mm long. The petals are 8–12 mm long, with a blade about 4 mm wide, not much wider than the claw; there are 2 very small scalelike auricles at the base of the blade.

Common on moraines, talus slopes, and rocky ridges, 6000-8000 ft; most abundant on the east side of the Park. Collections known from the cliffs above Spray Park; and at 7000 ft on Burroughs Mtn.

Silene antirrhina L.—SLEEPY CATCHFLY

A weedy annual, with a few erect stems 15–40 cm tall. A characteristic sticky band is found on the upper stem, just below the branches of the inflorescence, but the plant is otherwise hairless. The lvs are 2–6 cm long and linear to lanceolate. The white or pink fls are in cymes, on slender stalks. The petals are about 10 mm long, the blade tapering to the claw and shallowly notched at the tip.

An uncommon weed in the Park, growing on dry, disturbed ground below 4000 ft.

Silene douglasii Hook. var. *douglasii*
DOUGLAS'S CATCHFLY

A tufted perennial, with several stems that are curved at the base and unbranched, 20–50 cm tall. The plant is densely fine-hairy but not glandular. The lvs are narrowly lanceolate and 5–8 cm long. Flowers are borne 2–4 on each stem. The calyx is 10–13 mm long. The petals are white, or rarely pinkish, and 12–16 mm long. The blade is 2-lobed and narrower at its base than the top of the claw; the 2 scalelike auricles are narrow and about 1 mm long.

Fairly common and widespread in the Park, growing on rocky slopes and ridges above 5000 ft. Collected on Mount Wow; at 5000 ft on the trail between Mowich and Eunice Lks; at Sunrise and at Frozen Lk; at Mystic Lk at 6000 ft; and on the trail to Fremont Peak at 7000 ft.

Silene latifolia Poir. ssp. *alba* (Mill.) Rendle & Britten
WHITE CAMPION

The only silene in the Park with 5 styles, it was formerly placed in a separate genus, *Lychnis*. This is usually a tall, robust perennial, glandular in the upper parts, with a number of stems, to 100 cm or more tall. The sweet-scented fls are numerous, in leafy cymes, and open in the evening; male and female fls are borne on separate plants. The petals are white, about 15 mm long, deeply 2-lobed with 2 small scales at the base of the blade. The 5 styles are exserted from the female fls. The calyx becomes balloonlike as the capsule matures.

First collected by Robert Wakefield in 1965 in his "exotic plants" survey, at Longmire in the horse barn area; most likely not persisting long. Found once in the course of this study, along Hwy 410 about 1 mi south of the White R entrance.

Silene menziesii Hook. var. *menziesii*
WHITE CATCHFLY

A perennial with somewhat lax stems, reaching 10–30 cm tall, short-hairy and somewhat glandular in the inflorescence, including the calyx. The lvs are lanceolate and 3–8 cm long. The fls are in a leafy cyme, with white petals 6–8 mm long; the blade is much shorter than the claw.

Fairly common in brushy places and open woods, to about 4000 ft. Collections have been made at Longmire (*Warren 1711*) and on the road 2.5 mi above Longmire (*Warren 1570*).

Silene menziesii var. *viscosa* may occur in the Park: it is distinguished from var. *menziesii* by stems that are glandular-hairy below the inflorescence. One old collection, *Warren 1570*, does have such glandular hairs. On the other hand, in 1993, J. K. Morton annotated these same two collections at the University of Washington as *S. menziesii*, with no varietal designations. The varietal status of plants in the Park will need to await Morton's treatment of *Silene* in the *Flora of North America*.

Silene noctiflora L.
NIGHT-FLOWERING CATCHFLY

A glandular-hairy annual, 50–100 cm tall, blooming at night. The lvs are lanceolate to oblanceolate, 6–12 cm long, smaller toward the top of the stem. The inflorescence is an open cyme, with the fls on short stalks. The calyx is 2–2.5 cm long, narrow but expanding as the capsule matures, long-hairy, with heavy ribs and long, slender free lobes. The petals are white, 2.5–3 cm long; the blade is deeply divided into 2 spreading lobes, with 2 small scales at the base.

A weed, recorded by St. John and Warren (1937), based on *Warren 1658*, collected at Longmire.

Silene oregana S. Watson—OREGON CATCHFLY

Tufted, with stems 15–25 cm tall. The stems are short-hairy and glandular on the upper half. The basal lvs are 3–5 cm long, narrowly obovate, hairless, and not toothed on the margin. There are 2–3 pairs of smaller, short-hairy stem lvs. The calyx is 15 mm long and glandular, and cylindrical (before the capsule expands). The petals are white, 14–15 mm long overall. The blade is 2-lobed, but the slender lobes are deeply divided; each lobe is forked at about its midpoint. There are also 2 linear side lobes and 4 pointed auricles (making this the most intricate fl of the Park's *Silene* species).

A plant typically found on the east side of the Cascade Range, and first reported for the Park in this study. It can be found on the Naches Peak trail, on a ridge top close to the southern end of the "loop" trail, growing in a seasonally dry meadow.

A collection, labeled *Silene parryi*, made by C. L. Landes in the 1930s at Chinook Pass at 5000 ft, looks something like *S. oregana*, but the fls are poorly preserved and no firm determination is possible.

Silene parryi (S. Watson) C.L. Hitchc. & Maguire
PARRY'S CATCHFLY

A tufted perennial with mostly basal lvs, the stems 20–50 cm tall and finely short-hairy below and glandular above. The basal lvs are narrowly oblanceolate and 3–7 cm long, smaller and narrower upwards on the stem. Flowers are borne 2–4 in a narrow, loose, unbranched inflorescence. The ribs of the calyx are glandular; the petals are white and 12–15 mm long. The blade of the petal is deeply divided into 2 narrow lobes, each of which has a narrower lateral lobe about as long, thus appearing more or less 4-lobed, with 2 blunt auricles 1–2 mm long.

Fairly common above 5000 ft on rocky slopes and dry meadows. Known from around Spray Park; along the trail between Klapatche and St. Andrews Parks; on Tolmie Peak; at Berkeley Park and at Eunice Lk. A St. John collection from the 1930s made on Bald Rock is labeled *S. scouleri* but actually seems to be *S. parryi*.

Silene suksdorfii B. L. Rob.
CASCADE CATCHFLY

A low, tufted perennial, forming mats, the flowering stems to only about 10 cm high. The basal lvs are less than 2 cm long, more or less linear and blunt at the tip; there are 1 or 2 pairs of stem lvs. Flowers are borne 2 or 3 on each stem, on glandular stalks. The ribs of the calyx are densely clothed with purplish hairs. The white petals are 15–20 mm long. The blade is deeply divided, with 2 nearly linear auricles at its base; there are no lateral lobes.

Occasional on talus slopes and along ridges above 6000 ft, on the east side of the Park, but also reported from Spray Park. Collections known from the base of Little Tahoma Peak; on Plummer Peak at 6400 ft; at Sunrise; and on Fremont Peak at 7200 ft.

Silene vulgaris (Moench) Garcke
BLADDER CAMPION

A branched perennial, with stems to 50 cm high, hairless except for a few hairs on the calyx. The lvs are ovate-lanceolate, 3–5 cm long; the stem lvs are narrower and shorter. The white fls are numerous in open cymes; unlike *S. latifolia* ssp. *alba*, each fl has both stamens and a pistil. The petals are 15 cm long, white; the blade is deeply divided into 2 divergent lobes, with 2 nearly obsolete auricles at the base.

Uncommon weed of roadsides and waste places, perhaps not persisting long when it does enter the Park. Herbarium collections have been made in the Ohanapecosh area.

Spergula L.—SPERGULA

Spergula arvense L. var. *sativa* (Boenn.) Rchb.
CORNSPURRY

A small annual weed, with stems reaching 10–30 cm tall. The stem lvs are opposite, with small bundles of lvs in the axils, so that it appears the lvs are in whorls. Papery stipules are present, linear and 2–4 cm long. The fls are white, on slender stalks in open cymes, with 5 petals about 3 mm long, slightly shorter than the 5 sepals.

A roadside weed, known from the Longmire area.

Spergularia (Pers.) J. Presl & C. Presl
SANDSPURRY

Small weedy annuals, similar in appearance to *Spergula*, above, but with 3 styles (and, therefore, 3 valves of the capsule) and leaves that do not appear to be whorled. The flowers are pink or sometimes whitish, and have 5 sepals and petals, 2–10 stamens, and prominent papery stipules. The leaves are linear.

1   Small bundles of lvs in the axils of the major lvs; seeds brown ..... *S. rubra*
1   Lvs without such bundles; seeds black .... *S. diandra*

Spergularia diandra (Guss.) Murb.
ALKALI SANDSPURRY

A delicate plant with a few slender stems, to about 15 cm tall. The lvs are fleshy, 1–2.5 cm long, linear, and pointed at the tip. The fls are solitary or in small cymes and have 4–8 stamens.

A weed of roadsides and disturbed ground. Collected once in the Park at 4500 ft "along Naches Rd" (that is, along Hwy 410?).

Spergularia rubra (L.) J. Presl & C. Presl—RED SANDSPURRY

A matted annual, sometimes persisting more than a year, with prostrate stems reaching about 20 cm long. The lvs are 5–10 mm long; 2 or more lvs are bundled in the axils of the stem lvs. The fls are solitary on short, threadlike, glandular stalks at the tops of the stems. The stamens number 6–10.

A common and widespread weed along roads, trails, and in meadows. Collected at Longmire and at Sunrise; also observed on the roadside below Chinook Pass.

177

Stellaria L.—STARWORT

A genus of annuals or perennials, perhaps best defined by features seen in other members of the pink family that it does not have: stipules are absent, the petals are not clawed, extra bundles of leaves are not present in the leaf axils, and the capsule is not curved. They are mainly lax-stemmed, with flowers solitary or in small cymes. There are 5 free sepals, 5 white petals that are deeply divided into 2 lobes (or sometimes none), 3–10 stamens, and a capsule that is egg-shaped or somewhat cylindrical, with six valves.

1   Plants annual 2
1   Plants perennial 3

2(1)   Plants hairless; lvs mostly basal .... *S. nitens*
2   Plants with short hairs in lengthwise lines on the stems; stems leafy ....
                          .... *S. media* ssp. *media*

3(1)   Petals longer than the sepals 4
3   Petals shorter than the sepals, or petals absent 5

4(3)   Common native species; fls solitary in the axils of the lvs ....
                         .... *S. longipes* var. *longipes*
4   Uncommon weed; fls in a loose cyme .... *S. graminea* var. *graminea*

5(3)   Fls solitary in the axils of the lvs; lf margins minutely wavy .... *S. crispa*
5   Fls few to many in clusters; lf margins even 6

6(5)   Upper stems generally hairless; sepals 3.5–4.5 mm long, about as long as the mature capsule .... *S. borealis* var. *sitchana*
6   Upper stems short-hairy; sepals 1.5–3 mm long, shorter than the mature capsule .... *S. calycantha*

Stellaria borealis Bigelow ssp. *sitchana* (Steudel) Piper
BOREAL STARWORT

A sprawling perennial, with stems 15–50 cm long and hairless. The lvs are narrowly lanceolate to ovate and 2–8 cm long. The fls are a few to many in small cymes at the tops of the stems. The stalks turn downward as the capsule ripens. There are usually 5 petals, but these may be absent; the sepals are 3.5–4.5 mm long. The petals, when present, are shorter than the sepals.

Evidently scarce in the Park. It has been collected at Longmire and is chiefly associated with alder, 2500-3500 ft. Other collections are also known from the Nisqually V and on sandy ground in the bed of Tahoma Cr.

Stellaria calycantha (Ledeb.) Bong.
NORTHERN STARWORT

Perennial, with prostrate or weakly ascending stems, 5–25 cm long and short-hairy above. The lvs are elliptic to ovate, or in some plants lanceolate, and 3–25 mm long. The inflorescence is a few-flowered cyme, or sometimes the fls are solitary in the axils. The fl stalks often become curved in fruit. The sepals are 1.5–3 mm long and the petals are frequently absent. The petals, when present, are shorter than the sepals.

Fairly common in subalpine meadows, 5000-7000 ft. Found at Van Trump Park; at 5500 ft at Paradise Park; and at Seattle Park.

Stellaria crispa Cham. & Schltdl.
CRISPED STARWORT

A weak perennial, with prostrate stems that can reach 40 cm long and become matted. The ovate lvs are 5–25 mm long, the lower lvs on short stalks and the upper not stalked. The lf margins are "crisped"—that is, wavy but not toothed. The fls are solitary in the axils on slender stalks that reach 1–2 cm long. The sepals are 2.5–4 mm long, with papery margins.

Common, according to Jones, on moist ground in woods below about 3000 ft, but easily overlooked. It has been collected in the Nisqually valley, and was found twice during this study, clambering across moss on a boggy patch of ground, uphill from the Ohanapecosh entrance and on a muddy bank of Fish Cr at 3000 ft

Stellaria graminea L.—LESSER STARWORT

Sprawling perennial, with slender stems 10–50 cm long; the stems are square in cross section, smooth or with minute hairs on the edges. The lvs are 1–3 cm long, narrowly lanceolate, often with short hairs on the margin near the base. The inflorescence is a loose cyme, with slender and widely spreading branches, with a pair of small, papery bracts at the base of each branch. The petals are about 5 mm long, somewhat longer than the sepals, and divided nearly to the base. The sepals are lanceolate, 3-nerved, with a papery margin; some of the sepals have long hairs on the margins.

A weed, found in 1998 at Longmire near the Administration building. First reported in the Park in this study.

Stellaria longipes Goldie ssp. *longipes*
LONG-STALKED STARWORT

A mostly erect perennial, with stems 10–30 cm tall. The lvs are linear to narrowly lanceolate, 1–4 cm long, and sharply pointed. The inflorescence is a cyme with erect branches and several fls. The petals are longer than the 3–5 mm long sepals.

Rare in the Park, growing in damp places on streambanks and in meadows. Cited by Peter Dunwiddie in his 1983 dissertation and included on the National Park Service flora, but not by Jones or St. John and Warren in their earlier floras.

Stellaria media (L.) Vill. ssp. *media*
COMMON CHICKWEED

A mostly prostrate annual herb, with branched stems to 40 cm long; a single line of short hairs runs down the stem between each adjacent pair of lvs. The lvs are 5–25 mm long and ovate; the margin of the lf is fringed with hairs near the stalk. The fls are mostly solitary in the axils. The sepals are 3–4.5 mm long; the petals are shorter.

Very common weed on waste ground and along roadsides, especially around the Nisqually entrance and in the Ohanapecosh area.

Stellaria nitens Nutt.—SHINING CHICKWEED

A small and slender annual, with erect stems 5–10 cm tall; the lvs and stems are glossy, hence "shining." The lvs are 5–10 mm long, ovate to ovate-lanceolate, stalked toward the base of the stem but not above. The petals may be absent; if present, they are shorter than the 3–4 mm long sepals.

Uncommon, on open, dry ground and in meadows; most often seen at roadsides. Found at Sunshine Point and elsewhere in the Nisqually R V.

Vaccaria Wolf—COWCOCKLE

Vaccaria hispanica (Mill.) Rauschert
COWCOCKLE

A coarse annual weed, with stems 25–70 cm tall, hairless and glaucous throughout. The lvs are lanceolate to oblanceolate, 2–8 cm long; the pairs of opposite lvs are joined by narrow ridges on the stem at their bases. The fls are numerous, in flat-topped cymes. The rose-pink petals are clawed, with a broad blade that is notched at the top, overall about 12 mm long. The sepals are fused and the calyx becomes inflated in fruit.

A weed, only occasionally seen on disturbed ground at lower elevations, as at Longmire.

Celastraceae R. Br.—STAFF-TREE FAMILY

Paxistima Raf.

Paxistima myrsinites (Pursh) Raf.
MOUNTAIN-BOXWOOD

A low-growing, mound-forming shrub, generally not more than 50 cm tall, with opposite, evergreen lvs. The lvs are dark green, on very short stalks, elliptical to ovate, and toothed on the margin above the midpoint. The flattish, maroon fls are in small clusters in the axils of the lvs. They have 4 sepals, 4 petals, and 4 stamens, and the ovary is half-surrounded by a disk upon which the stamens are borne. The fruit is a small, asymmetrical capsule.

Common and often abundant, in dry to moist places, on rocky slopes and at the edges of meadows, to about 6000 ft. Known from Mowich Lk; around the Nisqually entrance; at 4000 ft along Westside Rd; on Rampart Ridge; at Longmire; at Paradise Park; at Box Canyon; on Kotsuck Cr trail at almost 6000 ft; in Stevens Canyon; at Ohanapecosh; along the road from Cayuse Pass to Tipsoo Lk; at 5600 ft on Naches Peak; in the White R V around the campground; and on the trail to Burroughs Mtn. [Often spelled *Pachistima*.]

Chenopodiaceae Vent.
GOOSEFOOT FAMILY

Chenopodium L.—PIGWEED, GOOSEFOOT

Chenopodium album L. var. album
LAMB'S QUARTERS

A tall, branched annual herb, 50–100 cm tall, with alternate lvs. The stalked lvs are ovate to diamond-shaped, irregularly toothed, and 2–4 cm long; the underside is covered with grayish, mealy particles. The stalkless fls are in terminal or axillary spikes; each is about 2 mm broad, with 5 short sepals and 5 stamens; petals are absent.

Uncommon weed of waste places. Noted at Longmire and Ohanapecosh.

Clusiaceae Lindl. (formerly Hypericaceae)
ST. JOHNSWORT FAMILY

Hypericum L.—ST. JOHNSWORT

Perennial herbs with opposite leaves that are dotted with glands. The flowers have 5 broad, rounded, yellow petals and 5 lanceolate sepals; the stamens number 15–50 and may be united at their bases in several groups. The fruit is a capsule.

1   Upright, weedy plant; sepals black-dotted on the margins . *H. perforatum*
1   Matted native plant of bogs and meadows; sepals not black-dotted ....
.... *H. anagalloides*

Hypericum anagalloides Cham. & Schltdl.—TINKER'S PENNY

Spreading from runners and becoming matted, with stems 5–25 cm long. The lvs are 3–10 mm long, roundish and clasping the stem, marked with greenish glandular dots. There are several fls in each cyme; the yellow petals are often shaded a salmon color and 2–4 mm long. The fruit is 1-chambered.

Common on wet ground in meadows and boggy places, 2500-6000 ft. Collected at Mowich Lk; in the Longmire meadow; below Sluiskin Falls at Paradise Park; at 5000 ft at Three Lks; and at Berkeley Park.

182

Hypericum perforatum L.—KLAMATH WEED, COMMON ST. JOHNSWORT

Erect plant, with numerous branched stems from a heavy taproot, reaching 30–100 cm tall. The lvs are 1.5–2.5 cm long, linear to oblong, with dark glandular dots; the sepals and petals bear similar dots. The fls are very numerous on stiff branches in a leafy cyme. The bright yellow petals are 8–12 cm long. The fruit is 3-chambered.

Common and noxious roadside weed. Found throughout the Park below about 5000 ft, as at the Nisqually entrance; at Longmire; in Stevens Canyon; around Ohanapecosh; and along Hwy 410, from the White R entrance, reaching 4500 ft below Cayuse Pass.

Cornaceae (Bercht. & J. Presl) Dumort.—DOGWOOD FAMILY

Cornus L.—DOGWOOD

Three species are found in the Park: one a perennial herb, one a shrub, and one a tree. The plants have simple, untoothed, opposite leaves (whorled in bunchberry). The flowers are crowded into cymes or heads. In two species, the flowers are subtended by an involucre of white, petal-like bracts, while the actual petals and sepals are small and greenish. The flower parts are in fours and the fruit is a berrylike drupe.

1   Plants appearing to be herbaceous, with a whorl of lvs at the top of a short stem less than 20 cm tall .... *C. unalaschkensis*
1   Plants trees or tall shrubs 2

2(1)   Floral bracts large and conspicuous .... *C. nuttallii*
2   Floral bracts absent .... *C. sericea* ssp. *sericea*

Cornus nuttallii Audubon ex Torr. & Gray PACIFIC DOGWOOD

Small trees, 6–12 m tall, with spreading crowns. The lvs are 8–12 cm long, obovate, sharply pointed at the tip and tapered to the stalk, bright green above and grayish green below. The fls are crowded into a hemispherical head, with 4–6 white obovate bracts that are 4–7 cm long. The fruits are bright red, about 10 mm long.

Found as isolated trees in open woods around Ohanapecosh. Also found at 3000 ft on the Laughingwater Cr trail. O. D. Allen made a collection in the 1890s in the "upper Nisqually V," but the precise location was not confirmed during this study.

Cornus sericea L. ssp. *sericea*—CREEK DOGWOOD

A shrub 2–4 m tall, the young branches with bright red bark. The lvs are ovate to oval, 3–8 cm long, and short-hairy on the underside. The fls are in terminal, flat-topped clusters about 8–10 cm across. The fruits are white and 7–9 mm long.

Occasional at the fringes of meadows and swamps, along streams, and in moist places at roadsides, to about 4500 ft, as at the Nisqually and Ohanapecosh entrances to the Park; also at Longmire.

Cornus unalaschkensis Ledeb. BUNCHBERRY

A perennial herb, with stems rising from branched, widely spreading rootstocks, usually forming patches. Each stem is 10–20 cm tall, with 4–7 more or less oval lvs in a false whorl at the top of the stem, each lf 2–8 cm long. A pair of smaller lvs is generally found at the middle of the stem. The fls are crowded in a head above the whorled lvs and are greenish white to greenish purple. Close below, and far more conspicuous, are the 4 large, white petal-like bracts. The fruits are orange-red and about 4 mm long.

Common, widespread, and often dominating the forest understory in mountain woods, to about 3500 ft. Collections have been made along Tahoma Cr; at the Twin Firs trail; around Longmire; at Box Canyon; at Ohanapecosh; at 2100 ft on the Laughingwater Cr trail; at the White R entrance and at the White R campground.

Crassulaceae DC.—STONECROP FAMILY

Sedum L.—STONECROP

Perennial herbs with short, thick stems and succulent, stalkless leaves; there is usually a basal rosette of leaves in addition to the leafy stem. The flowers are in cymes. The 5 sepals are united at their bases and the petals may be partially fused as well. There are 10 stamens in these species; *S. integrifolium* is unisexual, with male and female parts in separate flowers. The pistils develop into erect or divergent follicles.

1	Fls purplish; lvs flattened	*S. integrifolium* ssp. *integrifolium*
1	Fls yellow; lvs thick and succulent..	2
2(1)	Lvs on the flowering stems opposite	*S. divergens*
2	Lvs on the flowering stems alternate	3

3(2)   Lvs spoon-shaped, broadest above the middle, bright green ....
.... *S. oreganum* ssp. *oreganum*
3   Lvs linear to lanceolate, broadest below the middle, glaucous . *S. rupicola*

Sedum divergens S. Watson
SPREADING STONECROP

The name refers to the follicles, which spread widely as the seeds mature, although the plant itself also grows in a spreading fashion. The stems are leafy and reach 10 cm long. The lvs are bright green, not glaucous, and sometimes tinged red, broadly obovate and very thick, 7–9 mm long and widest above the middle. The lvs of the flowering stem are opposite, and somewhat shorter and flatter. The yellow petals are 6–7 mm long and not at all fused, with a lengthwise groove. The follicles diverge widely at maturity.

Common and abundant on cliff faces and on stony slopes, 4000-8000 ft, although also found near Longmire. Collections noted from Mowich Lk; Indian Henrys Hunting Ground; at Narada Falls; at Box Canyon and through Stevens Canyon; at 5800 ft on the Kotsuck Cr trail; on the upper slopes of the Butter Cr Research Natural Area; at Chinook Pass; along the trail between Tipsoo and Dewey Lks; along the Sunrise Rd; between Frozen Lk and Berkeley Park; on the trail to Clover Lk; at 8000 ft on Mount Ruth; at 6000 in Glacier Basin; on Pinnacle Peak; and at Eunice Lk.

Sedum integrifolium (Raf.) A. Nelson ssp.
integrifolium—KING'S-CROWN

A perennial with very thick stems that are branched at the base, reaching 5–10 cm tall. A basal rosette is absent and the stems lvs are alternate, flat, and obovate, even or sometimes toothed above the middle. The petals of the unisexual fls are purple and 3 mm long; the sepals are also purplish, about 2 mm long. The follicles are erect but curved outward at the tips.

Rare in the Park. Two collections are known: at 7000 ft on Mount Fremont, on a rocky

185

slope; and in the Sunrise area. The name *"Rhodiola integrifolium* Raf. ssp. *integrifolium"* has recently been proposed by Kartesz for this plant.

Sedum oreganum Nutt. ssp. *oreganum*
OREGON STONECROP

A perennial with strong rootstocks and basal rosettes of lvs and flowering stems are 10–20 cm tall. The lvs are 6–18 mm long, spoon-shaped to obovate, flattish, and bright green, those on the flowering stems are alternate. The yellow petals are fused a short distance at their bases, and 8–10 mm long. The follicles are erect.

An uncommon plant in the Park, found on cliff faces, road cuts, and on rocky slopes below about 5500 ft. Found at the Nisqually entrance; along Westside Rd; at Longmire; along the road below Ricksecker Point; through Stevens Canyon and along the east side of Cowlitz Divide; along the road between Cayuse Pass and Tipsoo Lk; along the Summerland trail; and at Eunice Lk.

Sedum rupicola G. N. Jones
LANCE-LEAF STONECROP

Distinguished by its glaucous, narrowly cigar-shaped lvs, both in basal rosettes and alternate on the flowering stems. The lvs are 3–7 mm long and are easily dislodged from the stems. The stems are 5–10 cm tall and the yellow petals are 4–6 mm long. The follicles are erect.

Fairly abundant on the north and east sides of the Park, on talus slopes and rocky ridges, 6000-7500 ft, as on Burroughs Mtn.

Droseraceae Salisb.
SUNDEW FAMILY

Drosera L.—SUNDEW

Drosera rotundifolia L. var. *rotundifolia*
ROUND-LEAVED SUNDEW

A perennial herb with a basal rosette of lvs, modified with long, reddish, sticky hairs on the roundish blade and capable of capturing small insects. The stalks are long and slender and the lvs overall are 3–7 cm long. The single flowering stem is 10–30 cm tall, with several white fls in a 1-sided raceme. There are 5 sepals and petals, and numerous stamens; the petals are 3–4 mm long. The fruit is a small oblong capsule.

Known from the Park only in the meadow at Longmire, growing on boggy ground.

Empetraceae Gray
CROWBERRY FAMILY

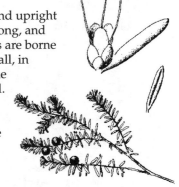

Empetrum nigrum L. ssp. *nigrum*
BLACK CROWBERRY

A low, heatherlike shrub, with creeping stems and upright branches. The lvs are thick and linear, 4–6 mm long, and rolled under at the margins. Male and female fls are borne on separate plants. The purplish fls are very small, in clusters of 2 or 3 in the axils of the upper lvs. The fruits are berrylike, black drupes, 4–6 mm broad.

Found in a few places in the Park, 6000-8000 ft on open, rocky slopes and ridges, forming attractive, low mounds. Collected near McClure Rock at 7100 ft; on Burroughs Mtn; at 6800 ft at Frozen Lk; above Seattle Park at 6500 ft; and at Moraine Park. Also reported from Spray Park.

Ericaceae Juss. (including Monotropaceae and Pyrolaceae)—HEATH FAMILY

A large and diverse family of trees, shrubs, and perennial herbs, which at one time or another has been broken into separate families, including the Monotropaceae (for small herbs lacking chlorophyll), the Pyrolaceae (small herbs with green leaves and separate petals), and the Vacciniaceae (shrubs with an inferior ovary). For this reason, the key to the dicots took several routes to arrive at this family.

Shrubby members of the family are prominent in subalpine meadows, including the heatherlike *Cassiope, Harrimanella,* and *Phyllodoce.* Salal is a dominant shrub in low-elevation forests, and white rhododendron is conspicuous in the shrub cover on higher slopes. A number of genera, some common and some quite rare, lack chlorophyll and are entirely red, yellow, or white in color. These obtain their sustenance through mycorrhizal fungi that provide a complex, bridging relationship between the plants and nearby trees, transferring sugars, nutrients, and water among the partners. The huckleberries, *Vaccinium,* provide a wealth of edible berries, and *Moneses uniflora,* or wood nymph, would be near the top of most people's lists of the loveliest wildflowers in the Park.

The numerous species are united in having simple leaves, 4–5 sepals that are usually free, 4–5 petals that may be free or fused, 8–10 stamens, and 1 style. In most species, the anthers open by terminal pores.

1   Ovary inferior; lobes of the calyx persistent on top of the berry (huckleberries) .... *Vaccinium*

1   Ovary superior; lobes of the calyx, if persisting, at the base of the fruit .. 2

2(1)   Fruit fleshy, berrylike 3
2   Fruit a dry capsule 5

3(2)   Bark of the branches rough, grayish; fruit a capsule surrounded by the fleshy calyx .... *Gaultheria*
3   Bark of the branches smooth and reddish; fruit a dry, berrylike drupe or many-seeded berry 4

4(3)   Trees; fruit surface "warty" .... *Arbutus*
       (see "Doubtful and Excluded Species," p. 469)
4   Shrubs; fruit surface smooth .... *Arctostaphylos*

5(2)   Herbs lacking green lvs 6
5   Herbs or shrubs with green lvs 11

6(5)   Stems slender, not fleshy or succulent; style elongated and conspicuous, often curved .... *Pyrola*
6   Stems fleshy; style straight, mostly included in the fl 7

7(6)   Stems red-and-white striped; petals absent .... *Allotropa*
7   Stems colored otherwise; petals present 8

8(7)   Corolla urn-shaped, the petals united; anthers bearing hornlike appendages .... *Pterospora*
8   Corolla mostly funnel-shaped, the petals distinct (except in *Hemitomes*); anthers lacking appendages 9

9(8)   Petals fused, the inner surfaces hairy; anther filaments hairy .. *Hemitomes*
9   Petals distinct, the inner surfaces and the filaments hairy or not 10

Allotropa Torr. & A. Gray—CANDYSTICK

Allotropa virgata Torr. & A. Gray ex A. Gray
CANDYSTICK

Growing as a single thick stem, lacking green lvs, colored red and white in vertical strips (rather like a peppermint stick), 30–50 cm tall. The stem bears scalelike, whitish lvs about 3 cm long. The fls are 5–6 mm long, in a dense raceme; there are 5 whitish to brownish free sepals, no petals, and 10 purple stamens. The fruit is a dry, 5-chambered capsule.

Uncommon in mixed fir woods. Collections have been made on Tahoma Cr at an unspecified elevation; at Longmire; and on Laughingwater Cr. Also observed on the Wonderland Trail up Rampart Ridge; on the trail along the west side of the river between Ohanapecosh and Silver Falls; and in the forest at the White R entrance.

Arctostaphylos Adans.—MANZANITA

Shrubs, with shiny reddish brown bark on the older branches, tough, evergreen leaves, and panicles or racemes of urn-shaped flowers (rather bell-like, that is, but nearly pinched closed at the mouth). The 5 sepals are partly fused and the 5 petals are fully united except for small free tips at the mouth of the flower. There are 10 stamens and the fruit is a dryish berrylike drupe holding several stony nutlets. *Arctostaphylos nevadensis* and *A. uva-ursi* can be found growing together on Mazama Ridge on the cliff overlooking Louise Lk.

1   Upright shrub; young branches with dense grayish, woolly hairs ....
.... *A. columbiana* ssp. *columbiana*
1   Prostrate shrubs; young branches smooth or merely short-hairy 2

2(1) Lvs rounded at the tip; berry bright red; plant prostrate .... *A. uva-ursi*
2   Lvs pointed at the tip; berry dull red; plant mounded ....
.... *A. nevadensis* ssp. *nevadensis*

Arctostaphylos columbiana Piper ssp. *columbiana*—BRISTLY MANZANITA

An erect shrub 1–3 m tall, with many branches. The bark of the young branches is densely clothed with short hairs, with a few longer bristles as well. The oval lvs are 2–5 cm long, gray-green, and short-hairy on the upper and lower surfaces. The white fls are numerous in compact, short-hairy panicles. The corollas are 6–7 mm long and the fruits are reddish brown, somewhat flattened, and 6–8 mm broad.

Found in a few places in the Park, most easily observed in lower Stevens Canyon. It also occurs on rocky slopes in the upper Nisqually V and in the Tatoosh Range, and is known from around Ohanapecosh.

In the Cascade Range and in the Olympic Mountains, where this species occurs in proximity with *A. uva-ursi*, occasional hybrids are found, which have gone by the name *Arctostaphylos* x *media* Greene. The hybrids, which are not yet known from the Park, feature lvs closer to *A. uva-ursi* on a shrub of the stature of *A. columbiana*, with branches that are short-hairy rather than grayish-woolly.

Arctostaphylos nevadensis Gray ssp. *nevadensis* PINEMAT MANZANITA

A mounded shrub, reaching about 50 cm tall; the spreading branches to 100 cm long and finely short-hairy. The lvs are 1–3 cm long, oblanceolate to obovate and pointed at the tip, bright green and shiny on both sides, and somewhat finely hairy when young. The fls are white, in very short

190

racemes (the inflorescence appearing therefore to be almost rounded). The corolla is 5–6 mm long and the red fruit is globose, about 7 mm broad.

Common on rocky slopes, roadcuts, and in dry woods, 2500-6000 ft. Collected on Gobblers Knob; on Rampart Ridge at the Longmire overlook; at Cougar Rock; at Nickel Cr on the main road; above Bench Lk; in Stevens Canyon and on the east side of the Cowlitz Divide; at Ohanapecosh; near Tipsoo Lk; at the White R campground; above Clover Lk; and along the Sunrise Ridge Rd to about 5000 ft.

Arctostaphylos uva-ursi (L.) Spreng.
BEARBERRY, KINNIKINNICK

A prostrate shrub, with the branches reaching to 100 cm long, rooting as they spread. The lvs are oblanceolate to obovate and rounded at the tip, 1–3 cm long, dark green on the upper side, pale green below, and mostly hairless. The fls are white or pinkish in short, dense racemes. The corolla is 4–5 mm long, and the fruit is globose and 8–10 mm broad.

Frequent through the Park to almost 8000 ft, growing on dryish ground among rocks or on sandy or gravelly slopes and banks; also used in Park landscaping, as around Longmire. Collected above Spray Park; on Rampart Ridge; on rocks at 8000 ft near Nisqually Glacier; on the south curve of the road across Cowlitz Divide; at Ohanapecosh; on the Cougar Cr trail at 3000 ft; on a dry bank where the road crosses Fryingpan Cr; and at Berkeley Park at 6600 ft.

Cassiope D. Don—CASSIOPE

Cassiope mertensiana (Bong.) G. Don var.
mertensiana—WHITE MOUNTAIN-HEATHER

A signature plant of subalpine meadows and, with pink heather, a dominant element of the shrub cover in these places. A low-growing, spreading shrub, with branches reaching 20–40 cm tall and densely clothed with scalelike, opposite lvs 2–4 mm long. The fls are solitary on slender, nodding stalks, rising from the axils of the upper lvs. The sepals are reddish; the bell-shaped corolla is pure white and 5–6 mm long. The fruit is a rounded capsule, 2–3 mm long.

Widespread and abundant in subalpine meadows, 5000-8000 ft, although best-developed around 5500 to 6500 ft. Collected at Mowich Lk and Spray Park; at Indian Henrys Hunting Ground; throughout Paradise V, especially in the Edith Basin; at Reflection Lks; in the meadow above Bench Lk; in the Butter Cr Research Natural Area; at Goat Pass; by Upper Palisades Lk; at Tipsoo Lk; on Burroughs Mtn; at Berkeley Park; and at Eunice Lk.

Chimaphila Pursh—PIPSISSEWA

Evergreen perennials, with short, woody stems. Each stem is topped by a loose whorls of toothed leaves. The inflorescence is a few-flowered raceme. The flowers have 5 sepals, 5 petals, 10 stamens, a short, blunt style, and a deeply grooved capsule. The flowers are pleasantly fragrant.

1   Fls 1–3 on stalks less than 5 cm long; lvs elliptic .... *C. menziesii*
1   Fls 5 to about 10 on stalks more than 5 cm long; lvs oblanceolate ....
.... *C. umbellata* ssp. *occidentalis*

Chimaphila menziesii (R. Br.) Spreng.
LITTLE PIPSISSEWA

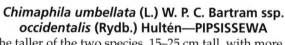

Smaller plants than the next species, the stems reaching 8–15 cm tall, with 1 to a few lvs per node on the stem. The lvs are lanceolate to ovate, 1–3 cm long, and a dull dark green. The 1–3 white fls nod on slender stalks, the petals spreading and curving backwards and becoming pinkish in age. The capsule is 5–7 mm broad.

Notably less common that *C. umbellata*, found in open coniferous woods below about 4000 ft. Known from the Nisqually entrance area; along Westside Rd; at Longmire; at 2500 ft on the Laughingwater Cr trail; at the Owyhigh Lks; and at the White R campground. An unusual occurrence was noted on the Kautz Cr mudflow, where it grows under alder. Both species of *Chimaphila* grow together in the White R Valley.

Chimaphila umbellata (L.) W. P. C. Bartram ssp.
occidentalis (Rydb.) Hultén—PIPSISSEWA

The taller of the two species, 15–25 cm tall, with more lvs on the stem. The lvs are elliptical to oblanceolate, bright green and shiny, and 2–4 cm long. The 4–8 light pink fls on nodding stalks are broadly bowl-shaped and 10–15 mm broad. The capsule is 6–8 mm broad.

Common on slopes in coniferous forests, to about 5000 ft. Known from the Nisqually entrance area; at Longmire and on Rampart Ridge; around Ohanapecosh; in the White R Valley; and at the Owyhigh Lks. It, too, grows under alder on the Kautz Cr mudflow.

Gaultheria L.—WINTERGREEN

Evergreen shrubs, two of which are prostrate ground covers, with simple, leathery, dark green, toothed leaves. The flowers are bell-shaped or urn-shaped, with 5 partly fused sepals, 5 united petals, and 10 stamens. The fruit is a capsule but becomes berrylike as the maturing calyx becomes fleshy and surrounds the capsule.

1   Upright, densely branched shrubs; fls in racemes; berry dark blue .... *G. shallon*
1   Creeping shrubs, with short, ascending branchlets; fls solitary; berry red 2

2(1) Lvs to 15 mm long; calyx smooth; subalpine, 5000–7000 ft .. *G. humifusa*
2   Lvs 20–40 mm long; calyx hairy; in forests below 4500 ft ...... *G. ovatifolia*

Gaultheria humifusa (Graham) Rydb.—ALPINE WINTERGREEN

The trailing stems rise from a rootstock and reach 10–15 cm long. The stems are mostly hairless. The oval lvs are about 15 mm long and rounded at the tip, toothless or sometimes toothed near the tip. The fls are white and about 5 mm long, on short, leafy vertical stems. The sepals are hairless and about as long as the corolla. The fruit resembles a round berry, red, and 5–6 mm broad.

   A less frequent, subalpine relative of the following species; typically found in meadows and on moist ground, 5000-7000 ft or higher. Collections are known from Mowich Lk, Paradise Park, and along the trail between Tipsoo and Dewey Lks. It has been reported from Clover Lk, Berkeley Park, and Eunice Lk.

Gaultheria ovatifolia Gray—SLENDER WINTERGREEN

A creeping shrub, with prostrate branches that run to about 30 cm long. The stems have spreading hairs. The lvs are ovate and toothed on the margins, 2–4 cm long and pointed at the tip. The fls are pinkish and about 4 mm long, on leafy stems 10–15 cm tall. The sepals are reddish-hairy and much shorter than the corolla. The red, berrylike fruit is 5–6 mm broad.

   Common below about 4500 ft in forests and at the edges of bogs and ponds. Found along the Carbon R below the glacier; along Kautz Cr, on the mudflow under alder; along the trail up

193

Rampart Ridge; at Box Canyon and Nickel Cr; at the Grove of Patriarchs; at 3000 ft on the Cougar Cr trail; at the White R entrance, in the White R campground and to about 4500 ft on Sunrise Rd; and on the Owyhigh Lks trail. At Nickel Cr, this wintergreen forms an attractive and interesting mixed ground cover with *Linnaea borealis*. The plant is able to colonize sunny, dry roadsides, as at the White R entrance and around Ohanapecosh. The berries, infrequently produced, are said to be "spicy."

Gaultheria shallon Pursh—SALAL

A shrub that can reach 2 m tall but is often less than 1 m. The branches are hairless, with grayish bark. The dark green, tough lvs are 5–10 cm long, roundish to ovate, sharply pointed, and toothed on the margin. The fls are in arching, 1-sided racemes. The corolla is white, bell-shaped, and 8–10 mm long. The fruit is about 10 mm broad, globose, short-hairy, and purple-black.

Abundant and widespread below about 4000 ft, where it prefers dryish woods, flourishing after disturbances. For all its abundance, it has only rarely been collected: vouchers are known from Longmire and Ohanapecosh.

Harrimanella Coville—BELL-HEATHER

Harrimanella stelleriana (Pall.) DC.
ALASKA CASSIOPE

A heatherlike shrub similar in appearance, when not in fl, to *Empetrum*. The plant grows as a spreading mat, with flowering stems to about 15 cm high. The rounded, linear lvs are arranged around the stem, spread widely, and are about 1 cm long. The fls are most often solitary at the ends of the stems, shaped like an open bell, and about 6–7 mm long, with spreading free lobes. The corolla is white, and the reddish sepals are united only at the base and are 2–3 mm long. The fruit is a dry, rounded capsule 4–5 mm long.

Rare in the Park, growing on open, moist ground, 5000-7500 ft. Found at Mowich Lk and Spray Park; near Pinnacle Peak; near Panorama Point; at 7300 ft on Mount Fremont; at Berkeley Park; and at Eunice Lk. The species reaches its southern limit of distribution at Mount Rainier.

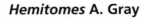

Hemitomes A. Gray

Hemitomes congestum A. Gray
GNOME-PLANT

Lacking green lvs, this odd, usually yellowish pink plant grows as a stout, short stem, about 15 cm tall, typically with most of its length buried in the duff. The stem is clothed with scalelike lvs and bears at the top a short, dense, spikelike cluster of fls. The fls are 8–15 mm long, narrowly bell-shaped, and held erect above 2 bracts. There are 2–4 sepals, 4 petals, and usually 8 stamens. The inner surface of the petals and the filaments of the stamens are hairy; the anthers are 2 mm long and open by lengthwise slits.

Rare, or at least very seldom seen, perhaps partly due to its habit of rising only a short distance above the duff; in dense coniferous forests below about 3500 ft. Three herbarium collections were located in this study: along the Mowich R at 3400 ft on the trail to Mowich Lk, in "woods at Longmire springs," and at 3000 ft on the trail up Laughingwater Cr. Reliably reported from the Ipsut Cr campground and along the trail to Summerland from White R Rd. The author found several healthy clumps of plants immediately inside the White R entrance. These had an attractive pink color, from the bracts. Similar clumps can be found at about 3200 ft near the road on the east side of the Cowlitz Divide. Gnome-plant may also grow along the trail on the west side of the river between Ohanapecosh and Silver Falls: in 1996, several just-emergent stems were found destroyed by a misplaced foot when they were yet too young to be identified with certainty.

Kalmia L.—LAUREL

Kalmia microphylla (Hook.) A. Heller
ALPINE LAUREL

An evergreen shrub of varying height and appearance. Lower-elevation plants are upright, branched shrubs reaching 40 cm tall; the relatively narrow, oblong lvs are 2–3 cm long, and have incurved margins. At higher elevations, plants are more or less prostrate or mounded, with slender, spreading branches and relatively broad lvs less than 2 cm long. Both versions have saucer-shaped pink fls in a loose rounded cluster at the top of the stem; 12–20 mm broad in

195

lowland plants, 10–12 mm broad in higher-elevation plants. There are 10 stamens, each of which is held within a small pocket on the corolla and "tripped" by the touch of an insect to release the pollen. The fruit is a 5-chambered, reddish capsule.

Infrequent, growing in sunny places in boggy meadows, on borders of streams and at lake shores, 4000-8000 ft. Known from Mountain Meadows and Mowich Lk; Spray Park; Edith Basin; close to the Nisqually Glacier at 7000 ft; in the meadows at Reflection Lks; the meadow above Bench Lk; Seattle Park at 6500 ft; Palisades Lk; Berkeley Park; and Mystic Lk.

K. microphylla and *K. polifolia* (also known as *K. occidentalis*) were recognized by Jones in his 1938 flora. The first name was given to plants of subalpine meadows while the second was applied to lower-elevation plants of sphagnum bogs. The taxa were said to be distinguished by lf size and overall stature. The nomenclature used in this book follows Ronald J. Taylor, who writes in *Mountain Plants of the Pacific Northwest* that "variation appears continuous [between *K. microphylla* and *K. occidentalis*], with intergradation at intermediate elevations." A collection made at 4000 ft on Mowich Lk road is labeled *K. occidentalis*; its largest lvs are just 2 cm long, while an Allen collection from the upper Nisqually V has lvs up to 3 cm long. These, and other herbarium collections at the Park, the University of Washington, and Pacific Lutheran University, support Taylor's view that just the one species is found in the Northwest.

Ledum L.—LABRADOR TEA

Ledum groenlandicum Oeder
BOG LABRADOR TEA

An erect, evergreen shrub, 0.5–1.5 m tall, with spreading branches. The lvs are narrow and oblong to lanceolate, 2–6 cm long, and fragrant. The underside of the lf is covered with rust-colored feltlike hairs, and the margin is strongly incurved. The white fls are numerous, in terminal, umbel-like clusters. Each is about 1.5 cm broad and has 5 petals that are not united and spread widely; there are also 5 partly fused sepals and 5 or 10 long stamens. The stigma is 5-lobed and the fruit is a 5-chambered capsule 5–7 mm long.

Uncommon, in swampy areas below about 3000 ft, as at the Longmire swamp.

Based upon a cladistic study of the rhododendrons, Kron and Judd (1990) proposed that *Ledum* should be considered a subsection of the genus *Rhododendron*. Kartesz has not adopted this usage, but that designation is used in a 1997 book edited by Sarah Spear Cooke, *A Field Guide to the Common Wetland Plants of Western Washington and Northwestern Oregon*, where this shrub is called *Rhododendron groenlandicum*.

Menziesia Sm.—MENZIESIA

Menziesia ferruginea Sm.
FOOL'S HUCKLEBERRY

A medium-sized to tall shrub with very much
the appearance of some of the Park's
huckleberries, although the fruit is a dry
capsule. The branches spread widely and the
plant may reach 2–3 m tall. The lvs are
deciduous, toothed, oblanceolate, and sparsely
hairy. The fls are in umbel-like clusters along the
upper stems. The corolla is pinkish to reddish
yellow and urn-shaped, about 8 mm long, on a
glandular-hairy stalk.

Frequent in shrubby areas in open woods, especially at the edges of
groves of trees, above about 4000 ft. Known from Mowich Lk; at the
Ipsut Cr campground; around Narada Falls; at Ricksecker Point; above
Bench Lk; in the Butter Cr Research Natural Area; at 4500 ft on the
Kotsuck Cr trail; at the White R campground; and at Green Lk.

Moneses Salisb. ex S. F. Gray—SINGLE DELIGHT

Moneses uniflora (L.) A. Gray ssp. reticulata
(Nutt.) Calder & Roy L. Taylor
WOOD NYMPH

A small perennial herb, from a creeping rootstock,
with mostly basal lvs. The lvs are roundish,
sharply toothed, and 5–25 mm long, on short
stalks. A single fragrant blossom nods at the top of
a stem that reaches 5–15 cm tall. The spreading
petals are 8–12 mm long and waxy white. The style
is straight and about 4 mm long, with an enlarged
stigma; the green ovary gives the fl an "eye." The fruit
is a rounded, dry capsule 6–10 mm broad. Out of
bloom, the plant closely resembles *Pyrola minor*.

Comparatively rare, in deep forests, to about 5500 ft.
Collections are known from Ipsut Cr (at an unspecified elevation); 0.5 mi
above the White R campground; at 3500 ft at the foot of the road to
Sunrise; and at 5500 ft on the trail from White R to the Owyhigh Lks.
Also reported from the Ipsut Cr campground, the White R campground,
and an unspecified location in the "upper valley of the Nisqually."

Monotropa L.—MONOTROPA

These are the most common of the leafless members of the Ericaceae,
especially *M. hypopithys*. The flowers have distinct petals and nod at the
top of an unbranched stem, which straightens as the seed capsule

matures. Both plants become black as they age. The petals lack the fringed margins of *Pleuricospora* and are not united like the petals of *Hemitomes*. The fruit is a small, roundish capsule.

1   Fls 1 per stem; plants waxy white .... *M. uniflora*
1   Fls several per stem; plants brownish to dull red, rarely yellowish ....
     .... *M. hypopithys*

Monotropa hypopithys L.—PINESAP

Usually found in clumps of several stems, 10–40 cm tall, with insignificant, scalelike lvs. The plants are typically a muddy yellowish brown, but individuals showing striking colors are occasionally found mixed with duller plants, including bright reds and clear yellow-browns; plants of a pale cream color are rare. The raceme has 3–20 fls, each 1–2 cm long.

Quite common, below about 4500 ft in old coniferous forests. Collected at Mowich Lk; at the Nisqually entrance area; on Rampart Ridge at about 3500 ft; at Longmire; at Ohanapecosh; in the White R V; and around the Carbon R entrance.

Monotropa uniflora L.
INDIAN-PIPE

Also typically found in clumps of stems, this species differs in being a cold, waxy white color. Damaged parts of the plant rapidly turn black. The stems are 10–30 cm tall, with a single, nodding fl 1–2 cm long.

Fairly common in coniferous woods, often in the vicinity of vine maple, below 2500 ft. Only a few herbarium collections were located: on the slope above the Sunshine Point campground, on Tahoma Cr near the main road, and up the hillside from the Ohanapecosh entrance to the Park. A clump estimated to contain 300 stems was found on the hillside above Sunshine Point.

Orthilia Raf.—SIDEBELLS

Orthilia secunda (L.) House
SIDEBELLS WINTERGREEN

Similar to the pyrolas, but with a distinctive inflorescence: the fls are on only one side of an arched stem. Growing from a spreading rootstock and able to form small patches, the lvs of each plant are mostly basal, roundish to oval, toothed, light green, and 1–3 cm long. The stem is slender and 8–20 cm tall, with 5–15 fls in a raceme at the top of the stem. The fls are bell-shaped, pale green in color, and 6–8 mm long. The straight style is longer than the corolla. The fruit is a round, dry capsule.

Widely scattered in the Park and locally abundant, to about 5500 ft. Found at Mowich Lk; at 3000 ft along Ipsut Cr on the trail to Lk James; on Rampart Ridge; along the Nisqually R above the Glacier View Bridge; at the Stevens Canyon entrance station; in the White R campground and along the road to Sunrise to about 4500 ft; on the trail to Mystic Lk 0.5 mi below Moraine Park; and along the Carbon R at the Big Rock Slide.

Phyllodoce Salisb.—MOUNTAIN HEATHER

These are the common "heathers" of the subalpine meadows of the Park, although not the same as the classic heather of Europe. These low shrubs grow as spreading mats, with many short, upright stems. The stems typically reach 10–40 cm tall, depending on elevation and exposure. The evergreen leaves are about 1 cm long, with the glandular margins recurved and the underside appearing grooved. The flowers are borne at the tops of the stems and usually nod on glandular stalks. The fruit is a roundish, dry capsule.

1   Corolla rose-pink, hairless on the outside .... *P. empetriformis*
1   Corolla yellowish, glandular-hairy on the outside .... *P. glanduliflora*

Phyllodoce empetriformis (Smith) D. Don
PINK HEATHER

The rose-pink, bell-shaped fls of this species are 6–8 mm long, with the free lobes of the petals turned back. They may be solitary or in small umbel-like clusters. The style is prominently exserted, and the capsule is 2–4 mm long.

Common and abundant in subalpine meadows and on moist slopes, 5000-8000 ft throughout the Park (rarely as low as 4000 ft, as on Rampart Ridge). Collections have been made at Mowich Lk and Spray Park; at Van Trump Park; at Indian Henrys Hunting Ground; across Paradise Park; at Reflection Lks; in the meadow at the summit of Shriner Peak; at Chinook Pass; at Sunrise and along the trail up Burroughs Mtn; at Berkeley Park; and at Eunice Lk. White-flowered plants are occasionally seen.

Phyllodoce glanduliflora (Hook.) Coville
YELLOW HEATHER

Here, the fls are urn-shaped, with a narrow mouth; the style is concealed. Each is 6–8 mm long, a pale greenish yellow and sticky with many small glands. The capsule is about 4 mm long.

Found in similar habitats to pink heather, but less frequently seen, with a preference for somewhat drier and stonier ground, to 8000 ft. Collected at Spray Park; near the saddle of Pinnacle Peak; at Paradise Park; along the Burroughs Mtn trail; and at Frozen Lk.

Pleuricospora A. Gray
FRINGED-PINESAP

Pleuricospora fimbriolata A. Gray
FRINGED-PINESAP

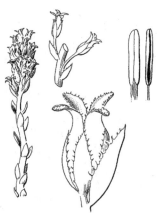

Lacking green lvs and growing as a single, rather stout and fleshy stem with a few leaflike scales, the plant reaches 6–12 cm tall and is a creamy yellow color when fresh. Usually all that appears above the duff of the forest floor are the fls, which are crowded in a raceme. Each is 8–10 mm long, with 4 free sepals and usually 4 free petals. The petals and sepals are hairless, with minutely toothed margins (and hence "fringed"). The anthers on the 8 stamens open by lengthwise slits and are about 3 mm long. The fruit is a fleshy, berrylike capsule.

Listed as "watch" in *The Endangered, Threatened, and Sensitive Vascular Plants of Washington* (Washington Natural Heritage Program, 1997). This represents a downgrade from the 1994 edition, where the plant had a "sensitive" status.

Not previously reported from the Park. Known from just one location, on the slope across from the entrance to Sunshine Point campground, along the path of the old cable trail. About 50 stems were seen here in 1996, forming a circle nearly 4 ft across. Fewer stems appeared in 1997.

Pterospora Nutt.—PINEDROPS

Pterospora andromedea Nutt.—PINEDROPS

Another species lacking green lvs, but quite different in appearance from the others. Tall, strong, reddish brown stems, sometimes in loose clumps, reach from 30 to nearly 100 cm tall and may persist through the winter. The stems are glandular-short-hairy and bear numerous, small, scalelike lvs on the lower half. The raceme occupies the top half of the stem, bearing urn-shaped fls on short, nodding stalks. The fls are whitish when young but soon turn reddish brown, and are 6–7 mm long. The 10 stamens are hidden within the fl and the fruit is a dry capsule.

Fairly common in open coniferous woods, sometimes on rather dryish ground, below 3500 ft. Collected at Spray Park; on Ipsut Cr; along the trail from Tahoma Cr to Indian Henrys Hunting Ground; at Longmire; around Ohanapecosh; on Olallie Cr at 2500 ft; in forests between the White R entrance and the White R campground; at 3400 ft on the trail to Windy Gap; and at 3000 ft along the Carbon R.

Pyrola L.—WINTERGREEN

Mostly small, evergreen perennial herbs, found in rich soil on the forest floor and spreading by slender rootstocks. The leaves are chiefly basal, but 1 or a few leaflike bracts may be found on the flowering stems. The flowers are in racemes and are roughly saucer-shaped, with 5 free sepals and 5 free petals; 2 of the petals usually form something of a hood over the stamens, which are turned up at the ends. The style may be straight or curved. The fruit is a dry capsule. The genus *Orthilia* was once included with *Pyrola*; it has fls arranged in a 1-sided raceme.

(Plants leafless and lacking chlorophyll have been called *Pyrola aphylla*, which is not a true species at all. In fact, several species of *Pyrola* are known to occur in this condition, and in the Park both *P. picta* and *P. chlorantha* are found in leafless forms. This key will work most of the time to identify such plants.)

1	Style straight	*P. minor*
1	Style bent or curved	2
2(1)	Fls greenish white to cream-colored; sepals pointed; lvs dark green and more or less mottled with white lines on the veins	*P. picta*
2	Fls variously colored; sepals pointed or rounded; lvs not mottled and not a deep green	3
3(2)	Petals greenish white to cream-colored	*P. chlorantha*
3	Petals pink	4

4(3) Lvs with small teeth, squared off at the base; sepal 3.5 mm long or longer .... *P. asarifolia* ssp. *bracteata*

4 Lvs not toothed, heart-shaped at the base; sepals less than about 3.5 mm long .... *P. asarifolia* ssp. *asarifolia*

Pyrola asarifolia Michx. ssp. *asarifolia*
PINK WINTERGREEN

Both subspecies of *P. asarifolia* are relatively large plants, with thickish, basal lvs 3–8 cm long and stems 10–40 cm tall. The fls are also large, up to 1.5 cm across, with pink to rose-red petals 7–9 mm long. The style is curved downward and the fruit is a dry, rounded capsule. The subspecies *asarifolia* has lvs that are heart-shaped at the base, not toothed on the margins, and with a dull luster. The fls tend to be a lighter pink shade.

Distinguishing the subspecies by lf characteristics is not always easy: some plants at Nickel Cr, for example, have shallowly heart-shaped lf bases. A collection from just outside the southwest corner of the Park at Bear Prairie had scalloped lf margins, with the lf round at apex, squared off at lower end.

Not very frequently seen, compared to the following subspecies. Found in moist forests below about 3000 ft on the south side of the Park, as at Ohanapecosh and at Box Canyon.

Pyrola asarifolia Michx. ssp. *bracteata* (Hook.) Haber
PINK WINTERGREEN

The subspecies *bracteata* has lvs that meet the stalk more or less squarely at the base, with shallowly toothed margins, and a glossy luster on the upper surface. The fls are more brightly colored, especially where the plants are growing in sunny places; light to deep rose-red is typical.

Common on ground in forests below 4000 ft. Known from along the Carbon R between the entrance and the Ipsut Cr campground; on the lower slopes of Tumtum Peak; on the Kautz mudflow; at Longmire; in the Cougar Rock campground; at Nickel Cr on the roadside; and between the White R entrance and the White R campground. The plant grows especially vigorously at the shoulder of the road just inside the White R entrance on Hwy 410, where the brilliantly colored fls are easy to spot while driving by.

Pyrola chlorantha Sw.—GREEN-FLOWERED WINTERGREEN

A smaller pyrola, with stems 10–20 cm tall. The lvs are on long stalks, thick and dark green, rounded in outline, and 1–3 cm long. The fls are few in number, generally no more than 5–8, and greenish white or cream-colored, about 8 mm long. The style is curved and longer than the petals.

A reddish-colored, leafless form is occasionally seen. In the forest at the White R entrance, leafless plants grow alongside normal ones.

Fairly common but perhaps easily overlooked, in deep forests below about 4000 ft, often growing on decayed logs. Collections are known from along Upper Tahoma Cr; on the Green Lk trail; on the Laughingwater Cr trail at 2500 ft; and at 4000 ft on the road to Sunrise. Also observed on the Kautz Cr mudflow (an ecologically interesting occurrence) and at Longmire.

Pyrola minor L.
LESSER WINTERGREEN

Another small species, with rounded to oval, toothed, basal lvs on slender stalks, overall 1–4 cm long. The stem is 10–20 cm tall, with 5–20 white to pink fls about 1 cm broad. The style is straight and shorter than the petals, but several collections made at Longmire had straight styles 4–5 mm long, well above the usual length for the species, which is about 2 mm. Out of bloom, the plant closely resembles *Moneses uniflora*.

Rare in the Park, growing in mossy places in forests below about 3000 ft. Most often seen in the Ohanapecosh area, especially at Silver Falls; also known from Longmire and in the forest from the White R entrance up to the White R campground.

Pyrola picta Sm.—WHITE-VEIN SHINLEAF

Easily distinguished from the other species by its lvs: these are 2–6 cm long, ovate to oval (some plants have narrower lvs), thick, and dark green, with whitish markings following the major veins. The degree of this marking varies and rarely it is not in evidence, but some plants display quite striking patterns, reminiscent of the rattlesnake orchid, *Goodyera oblongifolia*. The fls are numerous, on stems reaching 15–30 cm tall. Each is 1–1.5 cm broad, with greenish white petals. The style is somewhat longer than the petals and curved.

Common below about 5000 ft, generally in dryish woods. Collected at Mountain Meadows; along Westside Rd at 2600 ft; on Rampart Ridge; at the Cougar Rock campground; above 4500 ft along the Nisqually R; and throughout the Ohanapecosh area; evidently absent from the White R V. Most leafless pyrolas found in the Park are *P. picta*, and leafless and normal plants sometimes grow in close proximity.

Rhododendron L.—RHODODENDRON

Rhododendron albiflorum Hook. var. *albiflorum*—CASCADE AZALEA, WHITE RHODODENDRON

A deciduous shrub 1–2 m tall, with widely spreading branches. The lvs are 3–8 cm long, elliptical to oblanceolate, and sparsely short-hairy. The slightly irregular white fls are borne in small clusters at the top of the previous year's growth, below the current season's lvs; they are shallowly bell-shaped and about 2 cm broad. The 10 stamens are curved and fruit is an oblong capsule.

Common shrub of open slopes, at the edges of meadows, and on brushy ground, above about 3500 ft. Known from Mowich Lk and Spray Park; on Rampart Ridge at 3700 ft; at Ricksecker Point; at Paradise Park; in the Butter Cr Research Natural Area; on the slopes above Bench Lk; near Chinook Pass; at Sunrise; at the Owyhigh Lks; and at Eunice Lk.

There are reports that *Rhododendron macrophyllum* G. Don, the rose-bay rhododendron and the state fl of Washington, is to be found in the Park. In his 1906 *Flora of the State of Washington*, C. V. Piper lists under the synonym *Rhododendron californicum* a collection made by O. D. Allen in June of 1893, presumably from Allen's usual study area in the southwest corner of the Park. This collection was not located and the National Park Service flora does not include it. Piper did not mention the species in his 1902 list of the Park's flora.

There is a recent reliable report that a colony of *R. macrophyllum* plants grows on the western approach to Mount Beljica, about 2 mi west of the Park, in the Copper Cr drainage, north of the Nisqually entrance.

Vaccinium L.—HUCKLEBERRY, BLUEBERRY

A diverse group of shrubs, many of which are eagerly sought for their tasty fruit. The species range in size from low and matted to tall and strongly branched; some have evergreen leaves while the deciduous species add color to the Park in the fall. The white to pinkish flowers are solitary in the axils of the leaves (except for *V. ovatum*) and urn-shaped or bell-shaped, with the petals united most of their lengths while the sepals are at least partly united and persist on the fruit. The short, free tips of the petals curve backward at the mouth of the blossom. The fleshy fruits range in color from red to blue-black to black, often with a glaucous sheen.

See "Doubtful and Excluded Species," p. 469, for a discussion of *V. oxycoccos*, the swamp cranberry.

1   Lvs evergreen; fls in racemes .... *V. ovatum*
1   Lvs deciduous; fls solitary or in clusters of 2 or 3 2

2(1)   Berries red (occasionally purplish); younger branches strongly angled and bright green; lvs mostly much less than 30 mm long 3
2   Berries blue to black; younger branches dull green to gray, not strongly angled; lvs seldom less than 25 mm 5

3(2)   Tall shrub, more than 1 m tall; lvs not toothed, thin .... *V. parvifolium*
3   Low shrubs, to about 40 cm tall; lvs finely toothed, thickish 4

4(3)   Branches broomlike, erect, dense, and parallel; lvs less than 15 mm long; berry always red, less than 5 mm broad .... *V. scoparium*
4   Branches not broomlike, although usually erect and not widely spreading; lvs 10–30 mm long; berry dark red or more usually blue to purplish, and more than 5 mm broad .... *V. myrtillus*

5(2)   Calyx deeply lobed and persisting as the berry grows; fls in small clusters in the axils of the upper lvs .... *V. uliginosum*
     (see "Doubtful and Excluded Species," p. 469)
5   Calyx shallowly lobed or apparently not lobed; fls solitary in the axils 6

6(5)   Low shrub, seldom more than 40 cm tall; lf margins toothed above the midpoint .... *V. deliciosum*
6   Taller shrubs, seldom less than 50 cm tall; lf margins variously toothed but not limited to being toothed only above the midpoint, occasionally untoothed 7

7(6)   Lf margins markedly toothed from base to tip; tips of the lvs long-pointed .... *V. membranaceum*
7   Lf margins nearly toothless to weakly toothed, generally below the midpoint; tips of the lvs short-pointed to rounded 8

8(7)   Lvs usually sparsely short-hairy, at least along the veins, to 6 cm long; corolla "squatty-globular"; its stalk straight or nearly so, at least 10 mm long and wider below the fruit .... *V. alaskense*
8   Lvs to about 4 cm long, hairless; corolla "stretched-globular"; the stalk curved, less than 7 mm long, not notably wider below the fruit ....
     .... *V. ovalifolium*

Vaccinium alaskense Howell —ALASKA BLUEBERRY

A tall, upright shrub, with stems 0.5–2 m tall and somewhat angled twigs. The lvs are 2.5–6 cm long, elliptical or ovate-elliptic, the margin typically minutely toothed toward the base of the lf. The fls are described as "bronzy-pink" and are about 7 mm long, roundish but broader than long and therefore appearing somewhat flattened. The berries are bluish black, not glaucous, and 7–10 mm broad. Ed Tisch and T.R. Ogilvie, writing in 1991 in issue 9 of the *Botanical Electronic News*, discuss ways to distinguish *V. alaskense* from *V. ovalifolium* is by appearance and by taste: "You can tell them apart with your eyes closed. *V. ovalifolium* has tasty fruit and *V. alaskense* is tasteless, or even tarty."

Two collections examined in the course of this study key convincingly to *V. alaskense* in Hitchcock and Cronquist (1973). These are: *Allen 220a* at the University of Washington, from the "upper valley of the Nisqually" and originally determined by Allen to be *V. ovalifolium*; and *Heller 1498* at the University of Washington, from along the river above Longmire, which was initially identified by him as *V. deliciosum*. Also reported from the Butter Cr Research Natural Area by Martha Cushman. Kartesz reduces *V. alaskense* to synonymy with *V. ovalifolium*. However, controversy remains, especially among Canadian botanists, over the delineation of *Vaccinium* in the Pacific Northwest, and the status of *V. alaskense* as a recognizable species is maintained here. It is certainly a rare plant in the Park compared to the other members of the genus. [Spelled "*alaskaense*" in some references.]

Vaccinium deliciosum Piper—RAINIER BLUEBERRY

One of the shorter huckleberries, less than 30 cm tall, with round, widely spreading branches and matlike growth of the main stems. The lvs are oval to obovate, mostly toothed on the margins above the middle, somewhat glaucous on the underside, and 2–3 mm long. The pinkish fls are nearly round in profile, 5–6 mm long. The black, glaucous berries are 6–8 mm broad.

Very common and much sought after for its fine fruit. Found in subalpine meadows, often forming dense patches, 5000-8000 ft. Found at Mowich Lk and Spray Park; at Indian Henrys Hunting Ground; at Van Trump Park; at Paradise; in the Butter Cr Research Natural Area; on the Kotsuck Cr trail at 5000 ft; around Tipsoo Lk; at Seattle Park; and at Eunice Lk. The type specimen for the species was collected in 1896 in the Nisqually V by O. D. Allen.

Vaccinium membranaceum Douglas ex Hook.
THINLEAF HUCKLEBERRY

Medium-sized to tall, with spreading branches and somewhat angled twigs, 0.5–1.5 m tall. The lvs are ovate to oblong-ovate, with a long, tapered tip, light green but not glaucous, finely toothed along the whole margin, and 1–4 cm long. The fls are roundish in outline but wider than long, pink to greenish pink, and 4–5 mm long; they contain a sweet drop of nectar. The berries are black, 7–10 mm broad.

Abundant in coniferous forests, 3500-5500 ft; occasionally in open brushy places. Known from Mowich Lk; at 4100 ft on the trail to Lk George; at 5000 ft at Ipsut Pass; on the main road at the Canyon Rim overlook; at Paradise; in the Silver Forest; at 3500 ft on the Kotsuck Cr trail; on the Naches Peak loop; and at Fryingpan Cr on the trail to Summerland at 4500 ft; at the White R entrance; at Eunice Lk; and at Green Lk.

Vaccinium myrtillus L. var. *oreophilum* (Rydb.) Dorn
DWARF BILBERRY

A low-growing plant that tends to form mats, the upright stems at the most 20–30 cm tall. The twigs are greenish and sharply angled. The lvs are more or less ovate, thickish, dark green, 1–3cm long, and sharply toothed. The fls are nearly roundish in outline and about 5 mm long. The berry ranges in color from dark red to bluish and is 5–8 mm broad.

Only two herbarium collections were found during this study, made by Irene Creso, housed in the herbarium at Pacific Lutheran University, from along the trail between Lee Cr and Spray Park, on moist, open ground. The material includes no fls or fruit, but the lvs and branches are characteristic of *V. myrtillus*; the collections are clearly not *V. scoparium*. The plant is actually much more common, especially in the northeast part of the Park. It grows in the forest at the White R entrance, and in the White R campground it tends to occupy the ground right under the "skirts," or low branches, of conifers. It can also be found on the north slope below Cayuse Pass. In his flora, Jones evidently did not distinguish between *V. myrtillus* and *V. scoparium*. The berries have little flavor.

Vaccinium ovalifolium Sm.
EARLY BLUEBERRY

An upright shrub with sharply angled twigs, reaching 1–2 m tall. The lvs are 2–4 cm long, ovate-elliptic, and glaucous on the underside, not toothed or rarely with a few fine teeth on the margin. The pink fl is egg-shaped and 6–8 mm long. The berries are 6–9 mm broad, black and glaucous.

Common in fir forests, 4000-5500 ft, mostly on the south and west sides of the

Park, but occasionally as low as Longmire. Collected at Mowich Lk; at about 4000 ft on the lower slopes of Mount Wow; on the trail along Tahoma Cr on the way to Indian Henrys Hunting Ground; at Longmire; at Cougar Rock campground; near Reflection Lks; on the Kotsuck Cr trail at 4000 ft; near Cayuse Pass; on the trail between Tipsoo and Dewey Lks; and at Green Lk.

Vaccinium ovatum Pursh var. *ovatum*
EVERGREEN HUCKLEBERRY

The only evergreen *Vaccinium* in the Park, this is a tall shrub, reaching 2–3 m, with mostly erect branches. The lvs are ovate to elliptic-ovate, sharply toothed, dark green, shiny on the upper surface, and 1–4 cm long; the lvs are 2-ranked on the branches. The pink fls are crowded into short racemes, bell-shaped and 5–7 mm long. The berries are black and 7–8 mm broad.

   Common in coniferous forests at low elevations. Notable in the Nisqually and Ohanapecosh entrance areas; scarce at the White R entrance.

Vaccinium parvifolium Sm.
RED HUCKLEBERRY

An erect shrub, reaching 1–2 m tall, with sharply angled, light green branches and lvs. The lvs are more or less oval, not toothed, and 5–20 mm long. The round, pink fls are about 5 mm long, followed by tart, red berries 6–8 mm broad.

   Very common in low-elevation coniferous forests, usually where there is little competition from other shrubs, often sprouting on decaying stumps and logs. Collections noted from Ipsut Pass; at 4000 ft on Westside Rd; at Longmire; on the trail from Ohanapecosh to Silver Falls; in the White R V; at the trailhead below Windy Gap; and at Green Lk.

Vaccinium scoparium Leiberg ex Coville
GROUSEBERRY

A low shrub, 15–40 cm tall, with "broomlike" stems and branches; that is, the stems are strictly erect and the branches diverge but little from the stems. The lvs are 8–15 mm long, narrowly ovate and toothed on the margins. The fls are urn-shaped, about 4 mm long, and the bright red berries are 3–5 mm broad.

   Frequent on rocky slopes in open woods, or in thickets at the edges of meadows, 4000-7000 ft, mostly on the north and east sides of the Park.

Collections known from Mount Wow at 5000 ft; at Spray Park;
at 4000 ft on Rampart Ridge; at Paradise; at 5400 ft on Eagle
Peak; along the trail between Tipsoo and Dewey Lks; at
Chinook Pass; at 4750 ft on the trail between White R and
Summerland; in Glacier Basin at 7000 ft; and on Burroughs
Mtn at 7000 ft.

Fabaceae Lindl. (formerly Leguminosae) PEA FAMILY

A diverse family of plants, united by features of the flowers:
the 5 sepals are fused into a tube with free tips, the 5 free
petals that differ in size and shape (the upper petal is broad
and forms a banner, the 2 lateral petals are narrow and
winglike and enfold the 2 lower petals, which in turn surround
the single pistil and 10 stamens, forming a keel), and a fruit that
is a beanlike pod, usually straight but sometimes coiled. These
are annual or perennial herbs or shrubs, with compound leaves
bearing stipules, mostly pinnate, but palmate in some genera.

1 Upright shrubs; lvs much reduced in size or some stems nearly leafless ... *Cytisus*
1 Herbs or perennials (sometimes with somewhat woody bases), with normal lvs ... 2

2(1) Lvs trifoliate or palmately divided ... 3
2 Lvs pinnately divided (sometimes falsely trifoliate in *Lotus*, the lower pair of leaflets mimicking stipules) ... 6

3(2) Leaflets 5 or more ... *Lupinus*
3 Leaflets 3 ... 4

4(3) Fls in dense heads ... *Trifolium*
4 Fls in short or long racemes ... 5

5(4) Fls in short racemes; pods coiled ... *Medicago*
5 Fls in long racemes; pods egg-shaped ... *Melilotus*

6(2) Odd number of leaflets, including the terminal leaflet on the midrib 7
6 Even number of leaflets, the midrib ending in a tendril ... 9

7(6) Fls solitary or in umbels in the axils of the lvs; plants weedy annual or perennial herbs ... *Lotus*
7 Fls in short terminal or axillary racemes; plants perennial, sometimes with somewhat woody bases ... 8

8(7) Fls yellowish; calyx and pod short-hairy ... *Oxytropis*
8 Fls purplish; calyx and pod covered with long black hairs ... *Astragalus*

9(6) Style flattened, hairy along one side ... *Lathyrus*
9 Style round and slender, hairy only surrounding the tip ... *Vicia*

Astragalus L.—LOCOWEED

Astragalus alpinus L. var. *alpinus*
ALPINE MILK-VETCH, ALPINE LOCOWEED

A perennial herb, with 1 or 2 stems from a creeping rootstock. The lower lvs are pinnately compound with oval to obovate leaflets, overall to about 15 cm long. The leafy stems reach 20–30 cm tall and bear racemes of fls on long stalks in the axils of the lower lvs. The fls are purple with whitish wings and about 1.5 cm long. The calyx and pods are heavily covered with long, black hairs.

Rare in the Park, and known from one collection made above Bear Park on a rocky slope of Brown Peak. See the remarks on the Park's other locoweed, *Oxytropis monticola*.

Cytisus Desf.—BROOM

Cytisus scoparius (L.) Link var. *scoparius*
SCOT'S BROOM

A tall, deciduous shrub with erect, straight, slender, strongly angled branches 2–3 m tall. The lvs are small, 3-parted lower on the branches and simple above; proper lvs may be absent, especially on the upper branches. The bright yellow fls are solitary in the axils of the upper lvs (or nodes, where the lvs are absent), and about 2 cm long. The pod is hairless and stout, bearing 2–4 seeds.

A noxious, weedy shrub, sporadically occurring in the Park, generally along roads and open trails. Present at Sunshine Point and at Kautz Cr; also known from an old collection made by F. A. Warren on Tahoma Cr, 0.25 mi below the glacier.

Lathyrus L.—PEAVINE, WILD PEA

Climbing or straggling perennial herbs, with pinnately compound leaves that terminate in a tendril. The stipules are often prominent and leaflike. The flowers are borne in racemes in the axils of the upper lvs. In *Lathyrus*, the style is hairy only along the upper side, a feature that distinguishes the peavines from the similar vetches (*Vicia*).

1	Leaflets 2; plants weedy, chiefly on roadsides	2
1	Leaflets 4 or more; plants native, in a variety of habitats	3
2(1)	Leaflets and stipules linear to lanceolate; fls 10–15 mm long	*L. sylvestris*
2	Leaflets and stipules ovate; fls 15–20 mm long	*L. latifolius*

3(1)   Plants sparsely hairy; stipules narrow, much smaller than the leaflets ....
.... *L. nevadensis* var. *nevadensis*
3   Plants hairless; stipules ovate, nearly as large as the leaflets . *L. polyphyllus*

Lathyrus latifolius L.—EVERLASTING PEAVINE

A coarse and vigorous plant, with broadly winged
stems, rambling to 2 m long. The stipules of each lf
are 3–5 cm long and leaflike; the 2 leaflets are
lanceolate, 5–12 cm long, topped by a branched
tendril. The fls are relatively few, 20–25 mm long,
and pink to pinkish red. The pod is smooth and
about 6 cm long.

Appearing occasionally as a weed; likely not
persisting for long. One collection from Longmire
was found during this study.

Lathyrus nevadensis S. Watson
ssp. *nevadensis*
NUTTALL'S PEAVINE

The stems of this native pea
are angled but not winged and climb or clamber to
about 1 m. The lvs are 8–12 cm long, of 4–10 lanceolate
to oval leaflets, each 2–3 cm long with a branched tendril;
the stipules are very slender and about 1 cm long. The
racemes are 4–6-flowered; the fls light bluish violet in color
and 10–16 mm long. The straight pod is hairless and 3–4 cm
long.

Infrequent, in open woods below about 5000 ft in
the west and south sides of the Park, but also seen in
the lower White R V. Known from Mount Wow
and Lk Allen; along Westside Rd; and on Olallie
Cr at 4800 ft. [Kartesz uses "*L. nevadensis* ssp. *lanceolatus* (T. J. Howell)
Hitchc. var. *pilosellus* (M. E. Peck) Hitchc." However, S. L. Broich, who is
writing the treatment of the genus for the *Flora of North America*, uses the
name followed here.]

Lathyrus polyphyllus Nutt.
LEAFY PEAVINE

A robust native perennial, with stems to 1 m
long. The stems are angled but not winged. The
ovate-lanceolate stipules are often nearly as long
as the leaflets. The lvs are 14–18 cm long, with 12–
16 oval leaflets and a terminal branched tendril. The
fls are purplish and 16–20 mm long. The pod is
straight, smooth, and 7–8 cm long.

Said by G. N. Jones to be common at low elevations
in open woods, but only a single voucher is known, *Allen*

132, from "steep rocky hillsides, upper Nisqually V." The plant, in its typical form, can be seen along the Kautz Cr service road.

Lathyrus pauciflorus has been listed for the Park, but G. N. Jones notes that this was a result of a misinterpretation of *Allen 132*, "which belongs quite plainly to *L. polyphyllus.*" See "Doubtful and Excluded Species," p. 469.

Lathyrus sylvestris L.
SMALL EVERLASTING PEAVINE

Another vigorous weedy plant, with winged stems much like those of *L. latifolius*. The 2 leaflets are linear-lanceolate and 4–6 cm long; the stipules are very narrow and curved, 1–2 cm long. The reddish fls are about 15 mm long, borne on very long, slender stalks, 4–7 per raceme.

As with *L. latifolius*, this is a weedy pea, perhaps not long persisting when it does occur. A collection was made on river alluvium near Longmire and the plant is common around the shop areas at Kautz Cr.

Lotus L.—BIRDSFOOT-TREFOIL

None of these species was included by Jones in his 1938 flora, and each is listed here for the first time for the Park. Two are slender annual native plants, generally growing on disturbed ground in the Park, and one is an invasive perennial. The leaflets number 3–5, with a terminal leaflet rather than a tendril; the stipules are represented by small dark glands. The flowers are typically solitary on short to long bracted stalks in the axils of the upper leaves, although *L. corniculatus* has its flowers in umbels. The pod is straight and slender.

1   Plant perennial; fls in umbels .... *L. corniculatus*
1   Plants annual; fls solitary 2

2(1)   Calyx hairy, with teeth longer than the tube; leaflets mostly 3 ....
       ....*L. unifoliatus* var. *unifoliatus*
2   Calyx sparsely hairy, the teeth about equal to the tube; leaflets mostly 5 .
       .... *L. micranthus*

Lotus corniculatus L.—BIRDSFOOT-TREFOIL

Generally hairless perennial, with spreading stems 40–80 cm long, forming broad clumps. There are 5 obovate leaflets, each 10–15 mm long; the lower pair is borne next to the stem and might appear to be the stipules. The true stipules are reddish brown, glandlike membranes. The bright yellow fls number 3–8, in umbels borne on stalks 6–8 mm long in the axils of the lvs. The fl is 10–12 mm long.

First noted in the Park in 1998, growing in roadside ditches and other wet, disturbed places. Found at the Stevens Canyon entrance station, along Hwy 123, and here and there on the road to the White R campground.

Lotus micranthus Benth.
SMALL-FLOWERED LOTUS

Hairless (except for a few hairs on the calyx), with a weak stem 10–15 cm tall. The leaflets are 5–12 mm long and mostly obovate. The fl is pink to salmon, about 5–6 mm long; the banner is tipped forward.

Found in a few places on thin soils among low grasses: at the roadside by Tumtum Peak; around Ohanapecosh; between Dewey Cr and Cayuse Pass; and along the road south of the White R entrance. First noted in the Park in the course of this study.

Lotus unifoliatus (Hook.) Benth. var. *unifoliatus*
SPANISH-CLOVER

Hairy throughout and 10–20 cm tall. The leaflets are 1–2 cm long and lanceolate to elliptical. The fl is 5–9 mm long and yellow to pink; the banner is erect.

Known for the Park from one collection at the University of Washington, for which precise location information is not given. In 1998, the plant was seen as a weed in a planted bed at Longmire.

Lupinus L.—LUPINE

The lupine species of the Park have challenged generations of taxonomists, and the disposition of the species has varied tremendously. Four species are recognized here, one of which, a diminutive annual, has not been previously reported for the Park. The three perennial species are conspicuous members of open meadow communities, but also are found on sunny streambanks and on roadsides, and are among the most attractive wildflowers in the Park. Although varying in height, each has leaves that are palmately divided into slender leaflets and typical pea-shaped flowers in some shade of blue. The pods are oblong, flattened, and short-hairy.

1   Plants annual, less than 20 cm tall and lacking any kind of woody base; fls 5–7 mm long .... *L. polycarpus*

1   Plants perennial, taller than 20 cm; if less than 20 cm tall, then the plant matted, with stems that are woody at the base; fls more than 10 mm long 2

2(1)   Plants low and tufted; lvs silvery with dense silky hairs ....
.... *L. lepidus* var. *lobbii*

2   Plants with erect stems; lvs greenish, hairy but not silvery or silky 3

3(2)   Plants about 60 cm tall, robust; fls obviously whorled; keel lacking marginal
hairs .... *L. burkei* var. *burkei*
(see "Doubtful and Excluded Species," p. 469)

3   Plants less than 50 cm tall, more slender; fls not distinctly whorled or
irregularly whorled; keel fringed with short hairs on the upper margin .. 4

4(3)   Plants mostly taller than 40 cm, with straight, flattened hairs on the stems
and lvs; leaflets 3–6 cm long .... *L. latifolius* ssp. *latifolius*

4   Plants less than 40 cm tall, with dense, long, soft, spreading hairs (the hairs
often yellowish to reddish); leaflets 1.5–3.5 cm long ....
.... *L. arcticus* ssp. *subalpinus*

Lupinus arcticus S. Watson ssp. *subalpinus* (Piper & Robinson) D.B. Dunn—SUBALPINE LUPINE

A perennial herb, with hairy stems and lvs, from 15–40 cm tall, usually unbranched. The long-stalked lvs bear 5–7 leaflets, each oblanceolate and 1.5–3.5 cm long. The fls are numerous and blue to blue-violet, 10–16 mm long, borne in a tall raceme. The pod is about 4 cm long.

The common lupine between 5000 and 7000 ft, spreading across open slopes, in forest openings, and in meadows surrounding the mountain; replaced at lower elevations by *L. latifolius* ssp. *latifolius*. Collections have been made at Mowich Lk, Indian Henrys Hunting Ground, Van Trump Park, Paradise, Seattle Park, Tipsoo Lk, Sunrise, Burroughs Mtn, Frozen Lk, the lower slopes of Goat Island Mtn, Moraine Park, and Berkeley Park.

Lupinus volcanicus Greene was once used for plants with conspicuous yellowish to rusty-red hairs and smaller fls, although a good deal of variation can be seen in these characteristics. *Lupinus latifolius* var. *thompsonianus* (Smith) C.L. Hitchc. [or, according to Kartesz, *L. arcticus* ssp. *canadensis* (C. P. Sm.) D. Dunn] has been listed for the Park, on the basis of, it seems, a misidentification of *L. arcticus* ssp. *subalpinus*. A collection from the lower slopes of Mount Wow, made by O. D. Allen in the 1890s, was labeled *L. rivularis*; it is actually *L. arcticus* ssp. *subalpinus*, as is another Allen collection that was labeled "*Lupinus sericeus*, reduced mountain form."

Lupinus latifolius Lindley ex J. G. Agardh ssp. *latifolius*
BROAD-LEAF LUPINE

A robust perennial, with hairy stems that may be branched and 30–60 cm tall. The oblanceolate leaflets number 7 or 8 and are 3–6 cm long, occasionally longer. The blue or blue-violet fls are about 20 mm long, obscurely whorled in a tall raceme. The pod is 4 cm long and hairy.

Common and abundant below about 5000 ft, in open places in forests, on roadsides, in meadows, and on riverbanks. Collected around the Nisqually entrance; at Indian Henrys Hunting Ground; around Longmire and on Rampart Ridge; below Paradise; in Stevens Canyon; around Ohanapecosh; and in the White R Valley.

Plants along the lower Nisqually R sometimes key, half-convincingly in Hitchcock and Cronquist, to *L. rivularis* Douglas, but that lowland species cannot be considered verified for the Park. Plants from the Cayuse Pass area on the east side of the Park are sometimes not easy to distinguish from *L. arcticus* ssp. *subalpinus*. A collection made by Irene Creso, below Paradise, has leaflets up to 9 cm long, reminiscent of *L. polyphyllus* Lindl.

Lupinus lepidus Douglas ex Lindl. var. *lobbii* (A. Gray ex S. Watson) C.L. Hitchc.—DWARF LUPINE

"Silver lupine" would be an appropriate name for this striking plant. A low perennial, less than 15 cm tall, of matlike growth and arising from a stout, woody base. All parts of the plant are silvery with long, flattened hairs. The lvs are usually basal and 5–7-parted; the leaflets are oblanceolate and 5–10 mm long, on slender stalks. The inflorescence is a densely flowered, short, headlike raceme. The deeply fragrant fls are blue, with a white patch on the banner, and 10–12 mm long, although white-flowered plants are occasionally seen. The name used here is that tentatively proposed by Teresa Sholars for her treatment of the perennial lupines in the *Flora of North America*.

Common and often abundant on talus slopes, along rocky ridges, and on pumice flats, mostly above 7000 ft and reaching nearly 9000 ft. Found above Van Trump Park; in the upper reaches of Paradise Park; at Sunrise; around Frozen Lk; on Burroughs Mtn; and on the ridge above Moraine Park. [*Lupinus lyallii* Gray ssp. *lyallii* var. *macroflorus* Cox has been used for plants from the east side of the Park with larger fls (Cox, 1974).]

Lupinus polycarpus Greene—FIELD LUPINE

A small annual, with widely spreading branches, 10–20 cm high. The stems and lvs are short-hairy. The fls number 8–16 in an open raceme; they are not whorled. Each fl is 5–7 mm long on a stalk 1–2 mm long; the fl is blue, although the banner shows a patch of white. The banner tips forward toward the wings. There are 4–6 seeds in the short-hairy pods.

First found in 1996, at Sunshine Point. Now also found at Kautz Creek and in the lower White R Valley.

Medicago L.—MEDICK

Two rambling, weedy plants, short-hairy throughout or sometimes hairless, with 3-foliate, toothed leaves. The inflorescence is a densely flowered, axillary raceme. The fruit is a tightly coiled pod.

1 Fls blue-purple .... *M. sativa* ssp. *sativa*
1 Fls yellow .... *M. lupulina*

Medicago lupulina L.
BLACK MEDICK

An annual, with spreading stems to about 50 cm long. The leaflets are obovate and 1–1.5 cm long. The inflorescence is egg-shaped and 5–20 mm long. The fls are about 2 mm long, colored bright yellow. The pod is black.

Fairly common roadside weed, mostly in the Ohanapecosh area but also elsewhere at low elevations; once collected at 3600 ft at Round Pass.

Medicago sativa L. ssp. *sativa*—ALFALFA

The cultivated perennial forage plant, sometimes escaping in the Park. Generally growing into a mound, with stems 20–80 cm long. The leaflets are 1–1.5 cm long and lanceolate to obovate, with toothed stipules. The inflorescence is rather long, slender, and spikelike, with purple or blue-purple fls, which are 8–10 mm long. The pod is light brown.

Collected in the 1930s at Longmire, in the horse barn area. Unlikely to persist long following introduction.

Melilotus P. Mill.—SWEET-CLOVER

Weedy annuals, usually hairless, with tall, erect stems and 3-foliate leaves. The leaflets are toothed and oblanceolate to obovate. The flowers are in slender, long, spikelike racemes. The fruit is a short, egg-shaped pod, variously ridged or bumpy.

1   Fls 4–7 mm long, white .... *M. alba*
1   Fls 2–3 mm long, yellow .... *M. indica*

Melilotus alba Medik.—WHITE SWEET-CLOVER

Growing with numerous erect stems 1–2 m tall. The raceme can be up to 10 cm long. The pod is 3–5 mm long, with irregular crosswise ridges on the surface.

An occasional weed at roadsides.

Melilotus indica (L.) All.
YELLOW SWEET-CLOVER

The shorter of the two species, with stems reaching about 60 cm. The raceme reaches about 3 cm long. The pod is 2–3 mm long, ornamented with threadlike lines.

A rare weed of waste places.

Oxytropis DC.—CRAZYWEED

Oxytropis monticola Gray
NORTHERN YELLOW LOCOWEED

A tufted perennial, from a heavy, branched rootstock. The lvs are pinnately divided, with 8–11 pairs of lanceolate leaflets 4–10 mm long and a terminal leaflet rather than a tendril. The raceme is a short and headlike, on a stalk that reaches above the tops of the lvs. The cream-colored fls are about 15 mm long. The calyx has flattened black and white hairs and the ellipsoid, short-hairy pod is 1–2 cm long.

Known in the Park only from Mount Wow, where it grows on rocky ridges at about 5000 ft.

It's interesting that the Park's only other locoweed, *Astragalus alpinus* L. var. *alpinus*, also is highly restricted in its occurrence in the Park, found only on the slope of Brown Peak above Bear Park. *Astragalus* and *Oxytropis* are genera that rarely cross the Cascades. Mount Wow and its surrounding ridges, as well as the ridges in the northeast of the Park, are composed of volcanic rocks older than those found in the rest of the Park, and apparently they offer conditions required by these plants.

Trifolium L.—CLOVER

Leaves comprised of 3 leaflets help define this genus. The stipules are prominent. All clovers in the Park are perennials, with creeping or erect stems. The flowers are on naked stalks, arranged in more or less dense heads. The flowers turn sharply downward on their short stalks as the pod develops. In some species, the inflorescence sits above a bowl-like involucre of green bracts.

All *Trifolium* species known from the Park are introduced. The native *T. longipes* has been found just outside the Park on the east slope of Crystal Mtn, near Hen Skin Lk; the variety is not certain. In this key, it would come out as *T. hybridum*, which is a roadside weed; the features of *T. longipes* are noted below, for it may be expected in the Park itself, especially on Crystal Mtn and Naches Peak. There is also a report, unverified, that the plant was once found in a streamside meadow between Grand Park and Lk Eleanor.

1	Fls yellow	*T. dubium*
1	Fls white to pink or purplish	2
2(1)	Stems creeping; fls white	*T. repens* var. *repens*
2	Stems erect; fls pink to purplish	3
3(2)	Fl heads on stalks; calyx sparsely hairy	*T. hybridum*
3	Fl heads nearly stalkless; calyx densely hairy	*T. pratense*

Trifolium dubium Sibth.—LITTLE HOP-CLOVER

The only yellow-flowered clover at Mount Rainier, with slender stems reaching 5–30 cm long. The lvs are pinnately compound: the terminal leaflet is on a short stalk (in the other clovers in the Park, the lvs are palmately compound, the leaflets meeting at their bases). The stipules are lanceolate and not toothed. The obovate leaflets are 5–12 mm long and toothed. The heads are small, less than 8 mm broad, and the fls are 3 mm long. The pod is egg-shaped, about 3 mm long.

Frequent weed along roads and on open, disturbed ground at low elevations, as at Sunshine Point and at Ohanapecosh.

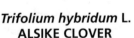

Trifolium hybridum L.
ALSIKE CLOVER

Usually a perennial, with spreading or erect stems 30–60 cm tall. The leaflets are obovate, sometimes blunt or notched at the tips, and 1–3 cm long; the stipules are ovate-lanceolate. The inflorescence is headlike, rounded, and 1.5–3 cm broad. The fls are pink, slender, and 6–10 mm long.

Common weed of roadsides at low elevations, especially in the White R Valley.

218

Trifolium longipes Nutt. will key here. It may be distinguished from *T. hybridum* by these features: it is a native perennial of subalpine slopes; the plant has a few slender stems and is less than 30 cm tall; the lvs are narrowly elliptical and blunt at the tips; the head is more oblong than rounded; and the whitish fls are 11–18 mm long.

Trifolium pratense L.—RED CLOVER

A tall, short-hairy perennial, with erect stems. The leaflets are 1–4 cm long, obovate, with small teeth. The head is bright red-pink, about 2 cm across, nearly stalkless above a false involucre of leaflike bracts; the fls are about 12 mm long.

Common at roadsides and on waste ground, to about 4000 ft. Found at Sunshine Point and up Westside Rd; at Longmire; at Ohanapecosh; in the White R campground; and in the Carbon R Valley.

Trifolium repens L. var. *repens*
WHITE CLOVER

A prostrate perennial, rooting along the creeping stems, which can reach about 60 cm long. The leaflets are obovate, toothed, and sometimes notched at the tips, and 1–2 cm long. The heads are rounded, about 1.5 cm broad. The white fls are 6–10 cm long.

Very common along roadsides, on disturbed ground, in lawns, and rarely in meadows, to about 5000 ft. Found at Sunshine Point and along Westside Rd; at Longmire; in Stevens Canyon; at Ohanapecosh; in the White R campground; and along the road below Tipsoo Lk.

Vicia L.—VETCH

A genus of pealike plants, differing from La*thyrus*, the true peas, in a characteristic of the style: in *Vicia*, a ring of short hairs encircles the tip of the style. Otherwise, the leaflets of the vetches of the Park tend to be narrower, the stipules smaller, and the flowers both smaller and narrower. The plants climb or clamber by mean of tendrils. The stems are angled (in some *Lathyrus* species, the stems are winged on the angles).

1   Fls 4–6 mm long; weedy annual .... *V. tetrasperma*

1   Fls 10–25 mm long; native or weedy perennials 2

2(1)   Fls 1–3 in the axils of the lvs; occasional weed .... *V. sativa* ssp. *nigra*

2   Fls several to many in racemes on slender stalks; common native ....
        .... *V. americana* ssp. *americana*

Vicia americana Muhl. ex Willd. ssp. americana—AMERICAN VETCH

A perennial with sparsely hairy stems that can reach 100 cm long. The leaflets are oval to linear-oblong, notched to blunt or pointed at the tip, smooth to short-hairy, and 1–3 cm long; the stipules are about 1 cm long and have several sharp lobes. The blue-purple fls number 4–9 in a 1-sided raceme, and are 1.5–2 cm long. The narrow pods are hairless and 2–4 cm long.

Common in open woods and brushy places, usually on drier ground; sometimes behaving in a weedy manner. Collected on the lower slopes of Mount Wow; at the roadside at Kautz Cr; in the White R campground area.

Plants from the Park have been identified as *V. americana* ssp. *minor* Hook. as well as var. *villosa*, neither of which can be considered verified for the Park. The variety *minor* has narrow, tough, densely short-hairy lvs while var. *villosa* has fls less than 15 mm long. Peter Dunwiddie included "var. *minor*," citing C. F. Brockman's 1947 book, but the latter does not actually mention varieties of *V. americana* in the Park.

Vicia sativa L. ssp. nigra (L.) Ehrh. COMMON VETCH

A plant similar in appearance to American vetch, but with 1–3 fls, stalkless in the axils of the upper lvs. The fls are pinkish purple to purple and about 2 cm long; the banner spreads more widely than the above species.

Not previously reported for the Park, this weed species is occasionally seen at low elevations, as at Sunshine Point and Longmire.

Vicia tetrasperma (L.) Moench SLENDER VETCH

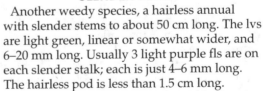

Another weedy species, a hairless annual with slender stems to about 50 cm long. The lvs are light green, linear or somewhat wider, and 6–20 mm long. Usually 3 light purple fls are on each slender stalk; each is just 4–6 mm long. The hairless pod is less than 1.5 cm long.

A weed, found in 1996 in planted beds at Longmire. First noted in the Park in this study. at Box Canyon; around Ohanapecosh; and at the roadside south from the White R entrance, reaching Fryingpan Cr.

Gentianaceae Juss.—GENTIAN FAMILY

In a 1901 letter written to C. V. Piper at the Washington State College in Pullman, J. B. Flett listed a number of plants he had collected at Mount Rainier, plants that Piper did not include in his 1901 "Mazama" article. One of these was *Gentiana sceptrum*, which Flett said grew "on the Carter Ranch near Longmire." Flett's collection, or collections, have not been found, and no writer on the Park's flora mentions the species. Piper himself seems not to have received specimens from Flett—unusual, given the amount of material Piper acquired from Flett over the years. In any case, Carter's Ranch, owned by pioneer Henry Carter, was located on Bear Prairie, just a mile or two south of the present Park boundary. *Gentiana sceptrum* is a plant of coastal areas and the Puget Lowlands and could conceivably turn up within the Park. It may be distinguished from *G. calycosa* by its lanceolate leaves that are 3–6 cm long and a cluster of several flowers at the top of each stem.

Gentiana L.—GENTIAN

Gentiana calycosa Griseb.—RAINIER PLEATED GENTIAN, MOUNTAIN BOG GENTIAN

Perennial hairless herb with opposite leaves and unbranched stems 5–30 cm tall. The ovate leaves are undivided, have untoothed margins, and are 1.5–3 cm long. A single flower tops each stem: 3–4 cm long and dark blue but dotted green on the interior. The 5 petals are united most of their length, with ovate free lobes. The fruit is a capsule.

A set of plants collected on Mazama Ridge and kept in the Park Herbarium shows "white, old rose, and blue color phases," according to the collector's label. The colors have somewhat faded, but the sheet is still striking. The plant is late to bloom, with the flowers sometimes not appearing until the huckleberry lvs have begun to color in the fall.

A common plant in moist subalpine meadows, 4500-8000 ft. Collected at Mowich Lk and Spray Park; at Indian Henrys Hunting Ground; at Klapatche Park; through Paradise Park and on Mazama Ridge; around the Reflection Lks; at Bench Lk; at the head of Stevens Canyon; at 4500 ft on the trail from White R to Summerland; at 6500 ft below Russell Glacier; around Sunrise; at Berkeley Park; at Windy Gap; and at Eunice Lk. Discovered in 1833 by Tolmie in the northwest corner of the Park.

Geraniaceae Juss.—GERANIUM FAMILY

A family of weedy plants, including some serious pests; none native to the Park. Annuals (*Geranium robertianum* may also be a biennial) with 5 free sepals, 5 free petals and 5 or 10 stamens. The 1 pistil is deeply divided into 5 lobes, each developing into a nutlet; the 5 styles are fused but break apart as the nutlets mature, remaining attached and sometimes becoming coiled.

1	Lvs pinnate; stamens 5	*Erodium*
1	Lvs palmate; stamens 5 or 10	*Geranium*

Erodium L'Her. ex Aiton—FILAREE

Erodium cicutarium (L.) L'Her. ssp. *cicutarium*
FILAREE

A weed of the planted beds at Longmire and first noted in the Park in this study. A short-hairy annual, with chiefly basal lvs that are oblanceolate in outline and pinnately divided into toothed leaflets. The stems are leafy and reach 10–30 cm long, typically sprawling on the ground. The fls are in umbels of 2–4 on long stalks in the axils of the upper stem lvs, and are about 1 cm broad with reddish purple petals. The style at maturity is 4–5 mm long and coiled.

Geranium L.—CRANESBILL

Hairy plants, with palmately divided leaves on slender stalks on erect or spreading stems. The rather showy flowers are in leafy or bracted cymes. The style is persistent on the nutlet as a stout or slender beak.

1	Lvs pentagonal in outline, deeply dissected; petals 7–10 mm long .... *G. robertianum*	
1	Lvs rounded in outline, shallowly divided but not dissected; petals to 5 mm long	2
2(1)	Nutlet smooth, with tiny hairs; stamens 5	*G. pusillum*
2	Nutlet wrinkled across its face, hairless; stamens 10	*G. molle*

Geranium molle L.—DOVE-FOOT GERANIUM

Softly hairy annual, with spreading stems about 30 cm long. The lf blades are round in outline, on slender stalks and 2–4 cm broad; each is divided into 7 or so major lobes, each of which has several teeth at the tip. The petals are about 4 mm long, reddish purple, and notched at the tip; there are 10 stamens.

Uncommon weed along roads and in waste places, as at Sunshine Point, and not reported for the Park before this study.

Geranium pusillum L.
SMALL-FLOWERED CRANESBILL

G. pusillum

Quite similar to *G. molle*, but the fls pink to violet with 5 stamens. Magnification is needed to discern the distinctive features of the nutlets.

A weed along roadsides and in waste places. Known in the Park from a Longmire collection.

Geranium robertianum L.
HERB ROBERT, STINKY BOB

An annual or biennial, with stems that can reach 50 cm long, typically sprawling over the ground. The lf blade is divided into five more or less diamond-shaped lobes, each of which is deeply toothed; the blade overall is 5-sided and 4–8 cm wide at its widest point. The fls are in an open cyme and the petals are pink to red-purple, with darker veins, 12–20 mm long.

A highly troublesome weed in the Puget lowlands that has recently become established in the Carbon R V at the Chenuis Falls trailhead, where it is the subject of eradication efforts. In 1999, the plant was found in the White R V, about 1 mi south of the entrance gate. Stinky Bob grows vigorously and crowds out other low-growing plants, especially in shaded places.

Grossulariaceae DC.—GOOSEBERRY FAMILY

The gooseberries and currants were formerly included in the Saxifragaceae. These are all shrubby plants and many (the gooseberries) bear prickles or spines on the branches; currants lack prickles or spines. The alternate leaves are palmately toothed and lobed and are often short-hairy and glandular. The flowers are in racemes in the axils of the lvs; each is saucer-shaped to tubular, with the calyx partly united with the ovary. The petals are shorter than the sepals (which may be colored and petal-like). The inferior ovary develops into a berry that may be covered with prickles and is usually juicy if unremarkable in flavor. In many species, the berry is glaucous, with a heavy white bloom that imparts a bluish color to the berry.

1	Spines absent from the lf nodes of the stems (currants)	2
1	Spines present at the lf nodes and sometimes along the stems (gooseberries)	7
2(1)	Ovary and berry hairless, not glandular; racemes hanging down; fls purple	*R. triste*
2	Ovary and berry with flat or stalked glands; racemes mostly erect; fls greenish to reddish	3

3(2) Glands of the ovary and berry flat; fls greenish; lvs with a foul smell .... *R. bracteosum*

3 Glands of the ovary and berry stalked; fls greenish or reddish; lvs with mild smell 4

4(3) Fls 10–15 mm long, the tube bell-shaped 5
4 Fls 4–6 mm long, the tube nearly flat 6

5(4) Fls dark pink to red, in 10–20-flowered racemes; lvs short-hairy but not glandular .... *R. sanguineum* var. *sanguineum*
5 Fls light green, in 4–8-flowered racemes; lvs glandular ... *R. viscosissimum*

6(4) Racemes hanging down; bracts of the raceme nearly as long as the fl stalks .... *R. acerifolium*
6 Racemes spreading to erect; bracts of the raceme much shorter than the fl stalks .... *R. laxiflorum*

7(1) Fls 5–15 in the racemes; stalk jointed just below the ovary .... *R. lacustre*
7 Fls fewer than 5 in the racemes; stalk not jointed 8

8(7) Style long-hairy; ovary and berry hairless ... *R. divaricatum* var. *divaricatum*
8 Style hairless; ovary and berry glandular or bristly 9

9(8) Ovary and berry bristly; fls greenish white .... *R. watsonianum*
9 Ovary and berry glandular; fls some shade of purple .... *R. lobbii*

Ribes acerifolium Koch
MAPLELEAF CURRANT

About 1 m tall, with spreading branches and lvs that are indeed "maplelike," with 5–7 wide, rounded lobes finely toothed on the margins and 3–7 cm broad. The inflorescence is a drooping raceme of about 10 greenish fls (look closely for the small, reddish petals). Each fl is less than 10 mm across and saucer-shaped. The filament of each stamen is broad at the base, distinctive among *Ribes* in the Park. The berry is 6–10 mm broad, black and glaucous, with a few stalked glands on the surface.

Common and widespread in the Park, as low as 4000 ft, although usually found above 5000 ft, near tree line. Known from Mowich Lk; at Spray Park; at 6000 ft in Van Trump Park; near Narada Falls on the Nisqually R; in the lower reaches of Paradise Park; at 5,000 ft on Kotsuck Cr trail; along Lodi Cr at 5300 ft; at 4000 ft on the trail along Laughingwater Cr; at Tipsoo Lk and Chinook Pass; and on the trail between Huckleberry Cr and Forest Lk at 5700 ft.

Ribes bracteosum Douglas ex Hook.
STINK CURRANT

Plants 2–3 m tall, with erect, unarmed stems. The lvs are mostly 7-lobed, 5–20 cm broad, the lobes toothed and sharp-pointed; the lvs are glandular on the underside and have a skunklike odor when crushed. The fls are nearly white, 20–40 in an erect raceme 5–25 cm long. The fls are small and saucer-shaped. The berry is blackish and glaucous, about 1 cm broad, and unpleasantly flavored.

Common on moist ground along streams and in swamps, to about 5000 ft. Found in Mountain Meadows; at 4900 ft at Mowich Lk; at the old beaver dams along Westside Rd; at the Deer Cr bridge; on Chinook Cr at 2500 ft; on the highway at Dewey Cr; and in the Butter Cr Research Natural Area.

Ribes divaricatum Douglas var. *divaricatum*
STRAGGLY GOOSEBERRY

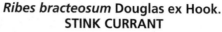

The arched stems of this 2–4 m shrub are heavily armed with stout thorns. The lvs have 3–5 toothed lobes, short-hairy at least on the underside and 2–6 cm broad. There are 3–5 fls in each lax raceme. The free portion of the reflexed sepals is purplish and 5–7 mm long while the petals are erect, white, and about 2 mm long. The berry is black, smooth, and about 1 cm broad.

Seen fairly often in woods in the Nisqually V below about 2500 ft, in low, wet areas.

Ribes lacustre (Pers.) Poir.
SWAMP GOOSEBERRY

Plants 1–2 m tall, with mostly ascending stems, armed with short spines; the young branches are bristly. The 3–7-lobed lvs are hairless, toothed, and 3–5 cm broad; the lf stalks and fl stalks are glandular. The 5–15 fls are in an erect raceme, greenish, and only about 7–8 mm broad. The black fruit is 5 mm broad, sparsely covered with stalked glands.

Frequent in moist woods, at the edges of swamps, and along streams (where it is less common than *R. bracteosum*). Found about the Nisqually and Ohanapecosh entrances to the Park; at Mowich Lk; at 3700 ft on the Rampart Ridge trail; at Longmire; along the trail up Tahoma Cr to Indian Henrys Hunting Ground; and on the road to Sunrise at 5000 ft.

Ribes laxiflorum Pursh
TRAILING BLACK CURRANT

Stems 1–3 m long, spreading or trailing on the ground; spines absent. The lvs are 5–10 cm broad, deeply lobed, toothed and hairless or sparsely glandular-hairy. The 5–10 fls are in a semi-erect raceme, with glandular stalks. The fls are saucer-shaped and about 1 cm broad, with purplish sepals and red petals. The berry is about 5 mm broad, black and glaucous, and glandular-bristly.

Uncommon in the Park, chiefly on the east side to 6500 ft, along streams in open woods, as at the White R campground. An O. D. Allen collection was said to have been made at "7000 feet, Goat Mountains," a topographic impossibility.

Ribes lobbii Gray
GUMMY GOOSEBERRY

A low shrub with thorny, spreading branches 1–2 m long. The lvs are 2–3 cm broad, 3–5-lobed with rounded teeth, and glandular-hairy on the underside. The lf stalks and fl stalks are heavily glandular. The lax inflorescence has 1–3 fls, each with a narrow tube and sharply turned-back sepals, the free lobes of which are about 10 mm long. The petals are erect, white, and 4–5 mm long. The reddish berry is 10–15 mm long, oblong, and heavily glandular-bristly.

Evidently rare in the Park; known from a collection (*Allen 28*) made in the 1890s along the upper Nisqually R.

Ribes sanguineum Pursh var. *sanguineum*
RED-FLOWERING CURRANT

A large, erect shrub with unarmed stems 2–3 m tall. The lvs are 2–7 cm broad, with 3–5 rounded lobes, whitish-hairy on the underside and finely toothed at the margin. Numerous reddish pink fls are borne in long, ascending to erect racemes. The fl is about 15 mm long, with a narrow tube and flaring sepals; the petals are 2–3 mm long, white, and erect. The berry is 5–8 mm broad, blue-black, glaucous, with numerous glandular hairs on the surface.

Widespread to about 3000 ft, but typically becoming crowded out as forests mature. Writing in 1925, Floyd Schmoe observed that "the lower few miles of the roads just within the Park are lined" with this shrub. At the end of the century, only a few shrubs can now be observed in the Nisqually and Ohanapecosh entrance areas. Well-displayed on rocky roadcuts along the road across the Cowlitz Divide.

226

Ribes triste Pall.
SWAMP RED CURRANT

Low-growing, with unarmed stems about 1 m long,
spreading or prostrate. The lvs have 3–5 toothed lobes and
are 3–6 cm broad. The saucer-shaped fls number 7–10, in
drooping racemes. The sepals are purplish and the short,
spreading petals are reddish purple. The immature ovary
and the berry are hairless; the red berry is 6–8 mm broad.

Rare on wet ground, 5000-6000 ft on the east side of the
Park. Collections are known from the area of Chinook
Pass and on "Crystal Mtn (Pierce County)."

Ribes viscosissimum Pursh—STICKY CURRANT

Stems spreading, unarmed, and 1–2 m long. The lvs are 5–
8 cm broad, with 5 rounded, toothed lobes, glandular-hairy
on each side. The 6–12 fls are borne in a spreading raceme.
Each fl is whitish or greenish white, with a narrow tube and
spreading sepals; the petals are small,
white, and erect. The berry is 10
mm broad, black, and
sticky with numerous
glandular bristles.

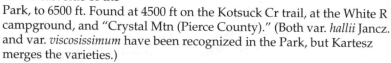

Infrequent, in
dryish, open woods
on the east side of the
Park, to 6500 ft. Found at 4500 ft on the Kotsuck Cr trail, at the White R
campground, and "Crystal Mtn (Pierce County)." (Both var. *hallii* Jancz.
and var. *viscosissimum* have been recognized in the Park, but Kartesz
merges the varieties.)

Ribes watsonianum Koehne—SPINY GOOSEBERRY

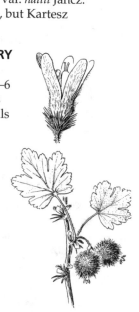

A shrub with spiny and erect to spreading stems 1–2 m
long. The lvs are 3–5 lobes, with shallow, coarse teeth, 3–6
cm broad and glandular-hairy on both sides. The 2–4 fls
are borne in short, lax racemes. The greenish white sepals
are 6–8 mm long, united below into a short tube and
flaring above; the petals are white and much shorter
than the sepals. The immature ovary and the purplish
berry are densely bristly, some of the bristles gland-
tipped.

Another *Ribes* of limited distribution in the Park,
known only from the White R V and along Sunrise
Ridge, where it grows on rocky slopes, 4500-5500 ft.

227

Hippuridaceae Link. (formerly Halorgidaceae)
MARE'S-TAIL FAMILY

Aquatic perennial herbs, found on streambanks and on very wet ground where standing water has receded. The plants grow from creeping rootstocks, with erect, unbranched stems that superficially resemble horsetails—the leaves are narrow and whorled, longest at the middle of the stem, and not toothed. The flowers are solitary in the axils of the leaves and lack petals and sepals; each consists of a single stamen and a single ovary. The flowers are wind-pollinated; the fruit is a 1-seeded achene, about 2 mm long.

1   Lvs 6–12 per whorl, 10–35 mm long .... *H. vulgaris*
1   Lvs 5–6 per whorl, 8–12 mm long .... *H. montana*

Hippurus L.—MARES-TAIL

Hippurus montana Ledeb. in Reichenb.
MOUNTAIN MARE'S-TAIL
Less than 10 cm tall, with lvs less than 12 mm long and 1 mm wide.
    Fairly common and locally abundant along small streams and in wet subalpine meadows. Collected in the lower Paradise meadows and at Reflection Lks.

Hippurus vulgaris L.
COMMON MARE'S-TAIL
Plants 10–40 cm tall, with lvs 10–35 mm long and 1–2 mm wide.
    Rather uncommon but also abundant in a few places, in swamps, ponds, and on lakeshores, below about 4000 ft; collected by O. D. Allen and C. V. Piper around 1895 at Longmire.

Hydrangeaceae Dumort.
MOCK-ORANGE FAMILY

A small family, formerly included in the Saxifragaceae, but distinguished by the woody stems (which may be obscure in *Whipplea*), opposite leaves, an inferior ovary, and 10–12 stamens.

1   Tall shrub with erect branches .... *Philadelphus*
1   Trailing plant; flowering stems to 10 cm tall .... *Whipplea*

Philadelphus L.—MOCK-ORANGE

Philadelphus lewisii Pursh var. *gordonianus* (Lindl.) Jeps.
LEWIS'S MOCK-ORANGE

Mock-orange is a tall shrub, reaching 2–4 m, with grayish bark that flakes easily. The deciduous lvs are opposite, ovate with a tapered point, and 3–5 cm long, usually toothed on the margins. The fls are in a raceme, white, sweetly fragrant (hence, "mock-orange"), and about 3 cm broad, with 4 sepals, 4 petals, numerous stamens, and a half-inferior ovary that develops into a woody capsule.

Scarce and found in a few places in the southwest part of the Park, in moist, open woods below 5000 ft. Found near Ipsut Pass; on Tumtum Peak; and at 4800 ft below Paradise Park on the Narada Falls trail.

Whipplea Torr.—YERBA DE SELVA

Whipplea modesta Torr.
YERBA DE SELVA

A prostrate and trailing subshrub, the slender stems seldom reaching more than 10 cm tall. The bark tends to peel on older stems. The lvs are evergreen, nearly stalkless, elliptical to ovate, and 2–4 cm long. The upper lf surface is hairy. The white fls are in a dense, short raceme, with 4–6 petals, each 3–6 mm long. The ovary is half inferior and the fruit is a segmented capsule.

Previously reported for Washington only on the northeast side of the Olympic Peninsula. It was noted by Richard Olmstead (Botany Department, University of Washington) along the trail between Longmire and Christine Falls, by way of Van Trump Park; the exact location is not recorded. Elsewhere in its range through Oregon and California, the plant prefers more or less open, dryish, rocky wooded areas, so it should be looked for, I suspect, along the lower part of Rampart Ridge.

Hydrophyllaceae R. Br.—WATERLEAF FAMILY

Perennial herbs, although some phacelias are woody at the base. The leaves are mostly alternate, toothed or not, and all parts of the plants are often roughly hairy. The flowers are typically in 1-sided, coiled cymes. The 5 sepals are partly fused and the 5 petals are as well, the corolla therefore bowl- to bell-shaped. There are 2 styles, partly to wholly united. The fruit is a dry capsule.

1   Delicate annual; fls solitary in the axils of the lvs .... *Nemophila*

1   Perennials, often coarse; fls in cymes 2

2(1)   Stamens included within the corolla; lvs kidney-shaped and toothed ....
.... *Romanzoffia*

2   Stamens longer than the corolla; lvs various, but not kidney-shaped and toothed 3

3(2)   Fls in elongated, coiled cymes .... *Phacelia*

3   Fls in congested, headlike cymes .... *Hydrophyllum*

Hydrophyllum L.—WATERLEAF

Low herbs, from a short rootstock. The stems are succulent and leafy, although the leaves are mainly basal. The inflorescence is a loose headlike cyme. The numerous flowers are bell-shaped, with long-exserted stamens. The style is 2-cleft.

1   Lobes of the lvs 3–7, the lvs about as long as wide; calyx lobes hairless on the back (but fringed with hairs on the margins) .... *H. tenuipes*

1   Lobes of the lvs 7–11, the lvs longer than wide; calyx lobes short-hairy on the back .... *H. fendleri* var. *albifrons*

Hydrophyllum fendleri (A. Gray) A. Heller var. *albifrons* (Heller) J. F. Macbr.
FENDLER'S WATERLEAF

The lvs of this middle-elevation waterleaf are roughly oval in outline and distinctly pinnately lobed, 10–20 cm long overall, and harshly short-hairy with coarsely toothed lobes. The stem is 20–30 cm tall, roughly hairy with downward-pointing hairs, its upper lvs reduced in size. The fls are white and 7–8 mm long.

Common in moist subalpine meadows, above about 4000 ft. Collections known from Mowich Lk; Van Trump Park; at 5000 ft on the Kotsuck Cr trail; in wet places along the road between Cayuse Pass and Tipsoo Lk; in the White R valley 1 mi above the campground; and at Eunice Lk. The species was first described from plants collected at Mount Rainier.

Hydrophyllaceae

Hydrophyllum tenuipes A. Heller
PACIFIC WATERLEAF

Differing from the above species in having lvs that
are roundish in outline, with a very short midrib
and less distinctly pinnate, 10–15 cm long, softly
hairy on both sides, and sharply toothed. The fls are
greenish white and about 7 mm long.

Common on wet ground in low-elevation
forests, most often under alder and maple, usually
in deciduous woods. Collections have been made in
the area of the Nisqually entrance and along the
lower reach of Tahoma Cr; also on Westside Rd.

Nemophila Nutt.
NEMOPHILA

Nemophila parviflora Douglas ex
Benth. var. *parviflora*
SMALL-FLOWERED NEMOPHILA

A small, delicate annual with sprawling
stems reaching about 20 cm long. The lvs
are mostly opposite on the lower part of
the stem, alternate above. The lvs are
pinnately lobed, with 5 coarsely toothed
major divisions. The fls are solitary on slender
stalks in the axils of the stem lvs, bowl-shaped,
white with a few black spots, and about 5 mm broad. The
style is 2-cleft.

Locally abundant in the Nisqually entrance area, at Sunshine Point
campground, and along Westside Rd.

Phacelia Juss.—PHACELIA

An intergrading, complex group of species, in which plants from low
and middle elevations are often difficult to identify with confidence. The
species in the Park are rough, hairy perennials (*P. sericea* is woody at the
base), with erect to spreading stems. The flowers are in coiled cymes in
an inflorescence that is compound and spikelike to racemelike. The style
is 2-cleft.

1	Alpine or subalpine plants, low and densely tufted	2
1	Plants of low to subalpine elevations, rarely above 6000 ft; plants with erect, leafy stems	3

2(1)	Lower lvs pinnately lobed; fls bluish violet	*P. sericea* ssp. *sericea*
2	Lower lvs undivided or with a pair of small lobes at the base; fls whitish	*P. hastata* var. *compacta*

3(1)   Basal lvs compound, with at least 2 pairs of lobes below the terminal blade; hairs of the stem coarse and stiff; plants robust, with erect stems to about 1 m tall .... *P. nemoralis* ssp. *oregonensis*

3   Basal lvs simple or with one pair of lobes; stem hairs not both coarse and stiff; plants generally less than 50 cm tall, the stems spreading or weakly erect 4

4(3)   Stems weakly erect; the inflorescence slender and elongated, with short lateral cymes; basal lvs usually with a lateral pair of lobes .... *P. mutabilis*

4   Stems curved at the base and spreading; the inflorescence broader and rather short, the cymes somewhat congested; basal lvs usually undivided .... *P. leptosepala*

Phacelia hastata Douglas ex Lehm. var. *compacta* (Brand) Cronquist—SILVER-LEAF PHACELIA

Low, tufted plants. The ascending stems are 10–30 cm long and clothed with stiff, straight hairs. The mostly basal lvs are 10–20 cm long, including the stalks, lanceolate, and silvery from the dense, flattened hairs. A small pair of lobes may be found at the base of the blade of the larger lvs. The fls are whitish to pale lavender, bell-shaped, and 4–7 mm long.

Jones, in his 1938 flora, evidently did not distinguish between this species and plants now called *P. leptosepala*.

Fairly common on dry, rocky flats and talus slopes, 4000-7000 ft. Collected on the upper slopes of Mount Wow; in drier parts of Paradise Park; at Chinook Pass; on the Interfork of the White R at 5000 ft; and at Sunrise. Reported from Mowich Lk and Eunice Lk.

Phacelia leptosepala Rydb. NARROW-SEPALED PHACELIA

A taller plant than *P. hastata* ssp. *compacta*, with stems 30–50 cm tall, and not tufted in habit; instead, the several stems curve near the base and spread as they rise. The larger basal lvs often have a pair of lobes at the base of the blade; the lf surface is roughly hairy but neither whitish or silvery. The fls are whitish, bell-shaped, and 5–7 mm long.

Uncommon in the Park, found between about 3000 and 5000 ft on open, dryish, stony ground. Found on Mount Wow; below the Nisqually Glacier; near the Cowlitz Glacier; along the road south of Cayuse Pass, at an unspecified elevation; at 4500 ft on the road to Sunrise; and at 3500 ft on the Carbon R, near the glacier.

Phacelia mutabilis Greene—CHANGEABLE PHACELIA

Sometimes a biennial, with several more or less erect, stiffly hairy stems 20–60 cm tall. The lower lvs are 5–20 cm long, oblanceolate, usually with a pair of lobes. The whitish to yellowish fls are 4–6 mm long.

Uncommon; typically found on seasonally moist ground in open woods, especially along roads and trails. Found on the lower part of Westside Rd; on the trail between Klapatche and St. Andrews Parks; at 3700 ft on Kautz Cr; on the Nisqually Glacier trail; at Goat Pass (possible—the specimen is incomplete); and 1 mi above the White R campground.

Phacelia nemoralis Greene ssp. *oregonensis* Heckard
SHADE PHACELIA

Quite distinctive for its size, the thick, erect stems 50–150 cm tall and clothed with dense, stiff hairs. The lower lvs are 5–25 cm long, ovate to lanceolate, and pinnately compound, with 3–7 lobes. The fls are in short, bracted cymes; each is bell-shaped, 4–6 mm long, and yellowish.

Rare in the Park: a plant of the Puget lowlands, it reaches the Park in the Nisqually V, where it grows in dryish, shaded places, typically along roads. Known from Sunshine Point and on the lower part of Westside Rd; also occurring as high as about 4500 ft near Ipsut Pass and reported by Peter Dunwiddie from the Ipsut campground.

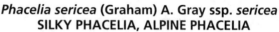

Phacelia sericea (Graham) A. Gray ssp. *sericea*
SILKY PHACELIA, ALPINE PHACELIA

The second easily distinguished phacelia in the Park. This is a tufted plant from a woody base, with silky-hairy stems and lvs, reaching 10–20 cm tall. The lvs are narrowly oblanceolate in outline and deeply cut into slender, pinnate lobes, 2–6 cm long. The fls are 5–6 mm long, bluish violet, in dense, oblong, headlike clusters; the stamens are much longer than the corolla.

Common on open rocky slopes, 7000-9000 ft. A collection in the Park herbarium was made at St. Elmo Pass.

Romanzoffia Cham.—MISTMAIDEN

Romanzoffia sitchensis Bong.
SITKA MISTMAIDEN

A delicate, mostly hairless perennial, with a "bulbous" base, formed where the expanded and thickened stalks of the basal lvs overlap. The stems reach about 25 cm tall and bear 1 or 2 small lvs. The basal lvs are kidney-shaped, lobed on the margin, and 1–3 cm broad. The

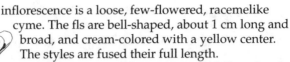

inflorescence is a loose, few-flowered, racemelike cyme. The fls are bell-shaped, about 1 cm long and broad, and cream-colored with a yellow center. The styles are fused their full length.

Uncommon, and limited to wet cliffs and rocks along streams. Collected in the Park on Mount Wow, on a slope above Lk George. It was also once found by Mary Fries in a slide area along the Carbon R trail east of Ipsut Cr.

Lamiaceae Lindl.
(formerly Labiatae)
MINT FAMILY

Perennial herbs (*Lamium* is an annual), with several key characteristics: the stems are square in cross section, the leaves are opposite, the flowers are mostly in whorls, and most species have a distinctive odor. The calyx is 5- or 10-lobed and the 5 petals are fused; the corolla is irregular and 2-lipped, of 4 or 5 lobes. There are 4 stamens and the style is 2- or 4-cleft. The fruit develops into 4 nutlets.

1	Stamens 2; fls nearly regular; plants odorless	*Lycopus*
1	Stamens 4; fls nearly regular to markedly 2-lipped; plant with a mintlike or other prominent odor	2
2(1)	Fls nearly regular, the two lips of equal length	3
2	Fls 2-lipped, the lips unequal in size	4
3(2)	Fls in whorls in the axils of the upper lvs or in terminal whorls	*Mentha*
3	Fls in a dense terminal head	*Monardella*
4(2)	Plants with slender, creeping, woody stems	5
4	Plants erect, annuals or perennials, lacking woody stems	6
5(4)	Fls on threadlike stalks in the axils; uncommon native	*Satureja*
5	Fls in terminal heads or short spikes; rare weed	*Thymus*
6(4)	Calyx 2-lipped, with unequal lobes	*Prunella*
6	Calyx regular or almost regular, the lobes or teeth nearly equal	7
7(6)	Inflorescence terminal, densely flowered; fls whitish	*Nepeta*
7	Inflorescence few-flowered, axillary (or if the inflorescence is terminal, then the fls in interrupted spikes); fls rose to purplish	8
8(7)	Plant tall, erect; native plant	*Stachys*
8	Plant sprawling or creeping; weeds	9
9(8)	Annual; fls red-purple	*Lamium*
9	Perennial; fls blue-violet	*Glechoma*

Glechoma L.—GROUND-IVY

Glechoma hederacea L. var. *micrantha* Moric.
GROUND-IVY

A creeping plant spreading by slender stems that are 30–40 cm long. The lvs are roundish to kidney-shaped, 1–3 cm across, with rounded teeth on the margin. The fls are in whorls in the axils of the lvs, few in number, with blue-violet corollas, marked with magenta spots at the throat. The corolla is 15–20 mm long and 2-lipped: the upper lip is 2-lobed, the lower 3-lobed. The odor is faintly unpleasant.

A weed of grassy places along roadsides and at the edges of thickets. Known from between the Nisqually entrance and Longmire, although likely to occur also around Ohanapecosh.

Lamium L.—DEAD-NETTLE

Lamium amplexicaule L.
DEAD-NETTLE

A sprawling annual, with slender stems 10–25 cm tall. The lvs are heart-shaped to ovate, 1–2 cm long, and coarsely toothed; the stem lvs are stalked while the lvs at the tops of the stems clasp the stalk and are reduced in size. The red-purple fls are in whorls in the axils of the upper lvs. The corolla is 12–18 mm long, with a slender tube and an abruptly expanded 2-lipped mouth: the upper lip is 2-lobed and hoodlike, while the lower lobes are turned down. The odor is unpleasant.

An uncommon weed in the area of the Nisqually entrance.

Lycopus L.—BUGLEWEED

Lycopus uniflorus Michx. var. *uniflorus*
NORTHERN BUGLEWEED

An odorless plant, from a tuberous root and with spreading runners. The unbranched stems are finely hairy and 20–60 cm tall. The lvs are short-stalked, lanceolate, and

235

sharply toothed, 2–6 cm long. The small (2–3 mm long), white fls are numerous, in the axils of the upper lvs. The corollas are 4-lobed and only slightly irregular; there are 2 stamens.

Known only from bogs and wet ground in the Ohanapecosh and Stevens Canyon entrance areas, below 3000 ft.

Mentha L.—MINT

Highly aromatic plants, with tall, erect stems, spreading rapidly by runners. The sharply toothed leaves are short-stalked or not and the flowers are in dense whorls at the tops of the stems. The corolla is slightly irregular, with the upper of the 4 lobes being larger, and in each of these species 4–5 mm long.

1   Fls whorled in the axils of ordinary lvs on the upper stems .. *M. canadensis*
1   Fls in spikelike inflorescences, the lvs close below much reduced in size 2

2(1)   Plants with a citrus odor .... *M. aquatica*
2   Plants with a peppermint odor .... *M.* x *piperita*

Mentha aquatica L.—WATER MINT

A tall, hairless plant, reaching 90 cm, with ovate to round-ovate, dark green lvs 1.5–3 cm long. The inflorescence is spikelike, with small bracts below the whorls of fls. The corolla is usually pink.

Said by earlier writers to be a weed occasionally found on wet ground. Herbarium collections were not found in the course of this study, and the plant perhaps has not persisted in the Park.

Mentha canadensis L.—FIELD MINT

Plants 20–60 cm tall, with short-hairy lvs and stems. The lvs are lanceolate, light green, and 2–8 cm long. The inflorescence is of widely separated whorls of dark pink fls; each whorl has 2 normal-sized lvs beneath it. The odor is rather sharp, suggestive of lemon and peppermint.

A native mint, known from the Longmire meadow.

Mentha x *piperita* L. (pro sp.)—PEPPERMINT

Plants 20–60 cm tall, hairless with purplish stems and lvs. The lvs are lanceolate or broadly lanceolate, 2–5 cm long. The fls are pink to purplish pink.

An occasional weed, probably escaped from cultivation, in the Longmire meadow. Said to be a hybrid between *M. aquatica* and *M. spicata*.

Monardella Benth.—MONARDELLA

Monardella odoratissima Benth. ssp. *discolor* (Greene) Epling—MOUNTAIN MONARDELLA

A perennial, the tufted stems woody at the base but not actually a shrub; stems 10–30 cm tall and short-hairy. The lvs are ovate to lanceolate, on short stalks, whitish-hairy, and 5–20 mm long. The pale violet fls are crowded into a headlike inflorescence about 2 cm wide, just above several purplish, oval bracts. The corolla is about 7 mm long, the upper lip erect and 2-lobed and the lower lip somewhat spreading with 3 linear lobes.

Frequent on the east side of the Park, 5000-7000 ft, on moraines and talus slopes. Collected at Cowlitz Glacier, Glacier Basin, and on the Emmons Glacier moraine at 7000 ft.

Nepeta L.—CATNIP

Nepeta cataria L.—CATNIP

A perennial with grayish green, short-hairy lvs and stems, the latter erect and to almost 1 m tall. The lvs are ovate, coarsely toothed on the margin, and 2–6 cm long. The fls are in dense spikes at the ends of the stems and as well as from the axils of the upper lvs. The corolla is 2-lipped, white with purple spots on the lower lip, and 7–9 mm long.

A weed, only rarely seen in the Ohanapecosh area.

Prunella L.—SELF-HEAL

A low plant, spreading by short rootstocks and able to carpet small areas. The flowering stems are short-hairy and terminate in a dense, oblong, headlike spike of blue-violet fls, with leaflike bracts below it. The lvs are chiefly basal, dark green, lanceolate to more or less ovate, and 2–6 cm long. The corolla is 10–14 mm long and 2-lipped: the upper lip is of 2 lobes that are arched and cover the stamens while the lower is of 3 spreading lobes.

1   Native subspecies; lvs on the middle stem about 3 times as long as wide, tapering at the base .... *P. vulgaris* ssp. *lanceolata*

2   Introduced subspecies; lvs on the middle stem 2 times as long as wide, rounded at the base .... *P. vulgaris* ssp. *vulgaris*

Prunella vulgaris L. ssp. *lanceolata* (W.P.C. Barton) Hultén
COMMON SELF-HEAL

In addition to the characteristics of the lvs, this subspecies has stems that are erect or ascending. Although native, this subspecies often grows in a "weedy" manner at roadsides. On moist, shaded to sunny ground in openings in forests, to about 4000 ft. Found at the Nisqually entrance; at Longmire; on the west side of the river at the Glacier View Bridge; in Stevens Canyon; at the Grove of the Patriarchs and Silver Falls; along the road south of the White R entrance; and to about 4000 ft on Sunrise Rd.

Prunella vulgaris L. ssp. *vulgaris*
COMMON SELF-HEAL

More apt to have stems that lie along the ground, rising only weakly.

Evidently rare in the Park; known from a collection made by Robert Wakefield in a 1965 "exotic plants" survey, along the main road 1.5 mi southwest of Longmire.

Satureja L.—SAVORY

Satureja douglasii (Bentham) Briq. in Engl. & Prantl
YERBA BUENA

An evergreen subshrub, the slender, somewhat woody stems running along the ground, and rooting at the nodes of the lvs, able to reach about 100 cm in length. The stem and lvs may be sparsely hairy. The lvs are ovate to roundish, broadly toothed and deeply veined, green above and purplish below, 5–20 mm long on short stalks. The fls tend to be on short stems that branch off the main runner. The white, 2-lipped fls are solitary in the axils of the lvs, on threadlike stalks, and 7–8 mm long.

The scent is reminiscent of spearmint, but less sweet, more complex, and somewhat musky. Plants from the Park are more compact than plants from the Puget lowlands, with shorter stems and less distance between the pairs of lvs.

Said by Jones, in 1938, to be common in low-elevation woods, and perhaps this was true earlier in the century when these forests were more open. Now the plant is seldom seen. Collected by O. D. Allen in the vicinity of the Nisqually entrance in the 1890s, most likely on a lower slope of Mount Wow.

Stachys L.—HEDGE-NETTLE

Stachys cooleyae Heller—COOLEY'S HEDGE-NETTLE

A rank-smelling perennial, 1–2 m tall, the stems sometimes purplish. The lvs are stalked, narrowly triangular to ovate, coarsely toothed, and 5–15 cm long. The fls are in loose, terminal clusters. The corolla is 2–3.5 mm long, 2-lipped, and rose-purple, spotted with white on the lower lip.

Common on wet ground around swamps, on stream banks, and other wet places, to about 3000 ft, as around Longmire, along the Ohanapecosh R, and along the lower White R.

Plants in the Cascade Mountains have been incorrectly called *Stachys ciliata* (a shorter plant from near the coast, with narrower lvs and paler pink fls).

Thymus L.—THYME

Thymus praecox Opiz ssp. *arcticus* (Dur.) Jalas
THYME

A creeping perennial, rooting along the stem. The lvs are mostly not stalked, oval, and 6–8 mm long. The stems rise to lift the terminal, crowded, spikelike inflorescence. The corolla is 2-lipped, purple, and about 4 mm long.

A collection in the Park herbarium was made by Robert Wakefield in his 1965 exotic plants survey, where the plant grew as a lawn weed at Longmire. It may not have maintained itself there.

Lentibulariaceae Rich.—BLADDERWORT FAMILY

Pinguicula L.—BUTTERWORT

Pinguicula vulgaris L.
COMMON BUTTERWORT

A most interesting insectivorous plant. The lvs are all basal, not toothed, oblanceolate to elliptical, with a short stalk, and a sickly yellow-green color. The lf margins curve inward and the upper surface is slimy-sticky, able to trap and hold small insects while they are digested. The corolla is 1.5–2.5 cm long, deep lavender with a whitish throat; it is 2-lipped with 2 lobes above, 3 below, and a flaring throat, which is prolonged backward into a spur. The fruit is a roundish capsule.

Listed as "review – group 1" in *The Endangered, Threatened, and Sensitive Vascular Plants of Washington* (Washington Natural Heritage Program, 1997).

Infrequent on wet cliffs and banks, distributed in the Park in a few places below 5000 ft: at Mountain Meadows; on cliffs along Van Trump Cr and at Comet Falls; at Indian Henrys Hunting Ground; at Cowlitz Park; and along the road 1 mi west of Tipsoo Lk.

Malvaceae Juss.—MALLOW FAMILY

Malva L.—MALLOW

Two insignificant, weedy plants, both annuals. The stems mostly sprawl on the ground and may reach 40 cm in length. The alternate, long-stalked leaves are palmately veined and typically somewhat lobed and toothed on the margin. The flowers are in clusters of 2 or 3 in the axils of the lvs. There are 5 sepals and 5 petals, and numerous stamens that are united below to form a tube surrounding the 10–15 styles. The ovary develops into a wheel-shaped fruit of 10–15 1-seeded segments (hence "cheeseweed," resembling a wheel of cheese).

1   Lvs on the upper stem divided into linear segments; fls 2–3 cm broad .... *M. moschata*
(See "Doubtful and Excluded Species," p. 469)

1   Lvs even or shallowly lobed; fls to 1 cm broad 2

2(1) Petals much longer than the sepals; individual sections of the wheel-like fruit smooth .... *M. neglecta*

2   Petals about equal to the sepals; individual sections of the fruit wrinkled .... *M. parviflora*

Malva neglecta Wallr.—DWARF MALLOW
The more attractive of the two species, with pink fls about 2 cm broad, the petals wedge-shaped, notched at the tip, and 10–13 mm long, much longer than the sepals. The lvs have 5–7 very shallow lobes and are 2–5 cm broad, with starlike hairs.

Occasional weed of waste ground, as at Longmire and Ohanapecosh.

Malva parviflora L.
CHEESEWEED
Similar to the above species, but the lvs more deeply 5–7-lobed and wider, up to 8 cm broad. The petals are white to pink and 4–5 mm long, about equal to the sepals.

A weed, collected once at Longmire.

Menyanthaceae (Dumort.) Dumort.
(formerly included in the Gentianaceae)
BUCKBEAN FAMILY

Menyanthes L.—BUCKBEAN

Menyanthes trifoliata L.—BUCKBEAN

"Buckbean" derives from "bog-bean," for the 3-parted lvs of this bog-dwelling perennial do look the lvs of a cultivated bean plant. The plants grow in clumps from thick rootstocks, directly upon which the lvs are borne. The 3 leaflets are elliptical to lanceolate and 5–10 cm long, lifted above the water on long stalks. The showy fls are numerous in a raceme on a leafless stem. Each is about 2 cm broad, the 5 petals united below and spreading above, white but tinged pink on the outside of the petals, and decorated on the inner surface with long white hairs. The fruit is an oval capsule.

Infrequent, growing in low-elevation swamps, shallow ponds, and at the edges of lakes, in water less than 50 cm deep. Collected at Longmire, Ohanapecosh, and near Silver Falls.

Nymphaeaceae Salisb.—WATER LILY FAMILY

Nuphar Sm.—POND LILY

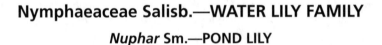

Nuphar lutea (L.) Sm. ssp. polysepala (Engelm.) E. O. Beal
YELLOW POND LILY

A perennial aquatic herb from a stout, heavy rootstock. The lvs float on the surface of the water on long, thick stalks, with heart-shaped blades 10–30 cm long. The fls are cup-shaped and about 10 cm broad. Most visible are the 6–12 bright yellow sepals; the petals are only about 1 cm long and are hidden by the numerous reddish stamens. The latter surround a broad, bright yellow, disklike stigma. The fruit is a large, oval, leathery capsule.

Found in the Park in a few shallow ponds below 4500 ft, such as the pond at the side of the road at the Stevens Cr entrance station; also found in a pond at 4000 ft near Lk James.

241

Onagraceae Juss.—EVENING-PRIMROSE FAMILY

A diverse family of annual and perennial herbs, with simple, mostly opposite leaves and slender stems. The flower parts are in 4s or 2s, with an inferior ovary.

1   Fl parts—sepals, petals, and stamens—in 2s .... *Circaea*
1   Fl parts in 4s 2

2(1) Seeds with a tuft of hairs at one end; sepals erect to spreading ....
.... *Epilobium*
2   Seeds lacking such a tuft of hairs; sepals turned back .... *Gayophytum*

Circaea L.—ENCHANTER'S NIGHTSHADE

Circaea alpina L. ssp. *pacifica* (Asch. & Magnus) P.H. Raven—ENCHANTER'S NIGHTSHADE

A low perennial with tuberous roots, forming colonies on the forest floor. The leafy stems are 10–50 cm tall, short-hairy near the top. The opposite lvs are 3–6 cm long, on short stalks, finely toothed on the margins, and somewhat short-hairy on the underside. The fls are in a terminal raceme, small and white, with 2 sepals that are turned back, and 2 notched petals about 2.5 mm long. The fruit is a pair of nutlets, clothed with hooked hairs.

Common in moist forests, below about 4000 ft, throughout the Park. Found at Mowich Lk; at the Nisqually entrance and along Westside Rd; in the Butter Cr Research Natural Area; around Ohanapecosh; in riverside woods along the lower White R; and at the Ipsut campground. Also collected at an unspecified location on the Reflection Lks trail.

Epilobium L.—WILLOW-HERB

The most diverse, in terms of a count of species, of any dicot genus in the Park, to which a confusing number of names have been given. Annuals or perennials, small and inconspicuous to tall, and with mostly opposite leaves. Many are attractive wildflowers, including fireweed. The flowers have 4 sepals, 4 petals, and 8 stamens; the stigma may be 4-lobed or nearly unlobed, and the long, slender capsule is 4-celled. The seeds bear a tuft of hairs at the end, aiding in wind dispersal. A number of species are hairy "in lines": from each side of the base of the stalk of a leaf is seen a thin line of small hairs running down the stem

Onagraceae

1   Fls yellow; petals 10–15 mm long .... *E. luteum*

1   Fls white, pink, or purple 2

2(1) Lvs alternate; fls more than 10 mm long, the fl tube not prolonged beyond the top of the ovary 3

2   Lvs opposite, at least on the lower stem; fls seldom more than 10 mm long, the fl tube prolonged beyond the top of the ovary 5

3(2) Plants 1–2 m tall; fls numerous in terminal racemes .... *E. angustifolium* ssp. *angustifolium*

3   Plants less than 0.5 m tall; fls few, in the axils of reduced upper lvs of the stem 4

4(3) Petals rounded at the tip, 15–30 mm long .... *E. latifolium*

4   Petals notched at the tip, 10–12 mm long .... *E. x pulchrum*

5(2) Plants annual; the skin of the lower stem dry and peeling away at maturity 6

5   Plants perennial; the skin not peeling 8

6(5) Plants 40–100 cm tall; lvs hairless .... *E. brachycarpum*

6   Plants to 30 cm tall; lvs more or less hairy 7

7(6) Inflorescence leafy, many-flowered and dense .... *E. densiflorum*

7   Inflorescence loose and sparsely flowered .... *E. minutum*

8(5) Turions (small, below-ground winter buds formed of fleshy, overlapping lf scales at the base of the stem) or basal rosettes of lvs present 9

8   Turions or basal rosettes absent 11

9(8) Stems mostly branched above the midpoint; petals 5–10 mm long .... *E. ciliatum* ssp. *glandulosum*

9   Stems mostly unbranched; petals 3–5 mm long 10

10(9) Lvs not stalked or short-stalked; capsule glandular .... *E. halleanum*

10   Lvs stalked; capsule hairy but not glandular ..... *E. ciliatum* ssp. *ciliatum*

11(8) Plants almost hairless; lvs glaucous 12

11   Plants more or less short-hairy throughout; lvs not glaucous 13

12(11) Stems 30–80 cm tall; upper lvs lanceolate, not overlapping .... *E. glaberrimum* ssp. *glaberrimum*

12   Stems less than 35 cm tall; upper lvs ovate and so closely placed on the stem as to overlap .... *E. glaberrimum* ssp. *fastigiatum*

13(11) Fls white .... *E. lactiflorum*

13   Fls pink to rose or purple 14

14(13) Stems erect, solitary or a few per plant 15

14   Plants tufted or matted, with numerous stems 16

15(14) Petals 4–8 mm long; lvs 2–4 cm long .... *E. hornemannii* ssp. *hornemannii*

15   Petals 3–5 mm long; lvs 5–7 cm long .... *E. ciliatum* ssp. *ciliatum*

16(15)   Plants hairless or with a few hairs in the inflorescence ... *E. oregonense*
16   Plants short-hairy, the hairs often in lengthwise lines on the stem ... 17

17(16)   Lvs pale green; fl buds erect; capsule club-shaped .... *E. clavatum*
17   Lvs bright green; fl buds nodding; capsule linear .... *E. anagallidifolium*

Epilobium anagallidifolium Lam.—ALPINE WILLOW-HERB

Normally reaching little over 10 cm tall, the tufted or matted stems are minutely hairy in lines and nodding at the top (a collection from Paradise was a full 20 cm tall). The hairless lvs are 6–15 mm long, oval to lanceolate, and light green, usually with a few small teeth. The inflorescence may be sparsely glandular. The petals are pink to rose-purple and 3–5 mm long. The capsule is 15–30 mm long, with a few small hairs.

Common in wet meadows and along small streams above 4000 ft, but most frequently seen above 5000 ft. Found at Mowich Lk; near Sluiskin Falls and elsewhere at Paradise; in the Tatoosh Mountains; at 4000 ft on Dewey Cr near the road; at Summerland at 5500 ft; at Tipsoo Lk; at Berkeley Park; at Three Lks.

Epilobium angustifolium L. ssp. *angustifolium*
FIREWEED

The most prominent member of the genus in the Park, with very leafy stems reaching 1–2 m tall and typically growing in clumps that rise from spreading rootstocks. The upper portion of the stem is grayish-hairy. The lvs are alternate, stalkless or short-stalked, lanceolate, and 8–15 cm long. The racemes are densely flowered and up to 40 cm long. The fls are 2–3 cm broad; the stigma is prominent and 4-lobed. The capsule is 4–9 cm long; it sometimes is colored silvery-gray on the upper side and purplish on the lower.

Common and often very abundant on favorable sites, below 5500 ft; often at roadsides, but best developed on burns and other large tracts of disturbed ground. Found at Mowich Lk; Longmire; at Paradise; in the Butter Cr Research Natural Area; through Stevens Canyon; around Ohanapecosh; and in the White R V.

The subspecies *circumvagum* Mosquin does not appear to be present in the Park, although it is known from Olympic National Park and is common to the south through Oregon and California. It reaches about 30 cm tall, with a small number of fls in the axils of reduced lvs or bracts at the top of the stem. It may be distinguished from the similar-looking *Epilobium latifolium* by its fl bracts, which are very much smaller than the stem lvs, and by its obscurely veined lvs.

Epilobium brachycarpum C. Presl
AUTUMN WILLOW-HERB

A tall and diffuse annual with branched stems, 30–80 cm tall. The skin at the base of the stem becomes dry and peels as the plant ages. The lvs are mostly alternate below but opposite above, narrowly lanceolate and "folded" along the midvein, nearly hairless, and 2–5 cm long. The white to rose-pink fls are in loose racemes on the paniclelike upper branches; the petals are 2–8 mm long. The capsule is slender and somewhat club-shaped, 10–15 mm long.

Infrequent on the east side of the Park, growing in sandy soil on slopes and along roads, above about 4000 ft.

Epilobium ciliatum Raf. ssp. *ciliatum*
COMMON WILLOW-HERB

A variable species. In ssp. *ciliatum*, overwintering basal rosettes of lvs are usually but not always present. The stems are slender, branched in the inflorescence, glandular-hairy in the upper part (the hairs both in lines and scattered), and may reach 100 cm tall. The lvs are lanceolate to elliptical and 1–10 cm long, with pronounced veins. The fls are pink to rose-purple, the petals 2–6 mm long. The fruit is very slender, hairy, and 15–100 mm long.

Fairly common on wet ground, in sunny places below 4000 ft. Observed along lower Westside Rd, at the Glacier View Bridge on the main road, and in open woods along the lower White R. Collected as "*E. leptocarpum*" by M. C. Huntley in the 1930s, below Chinook Pass.

Epilobium ciliatum Raf. ssp. *glandulosum* (Lehm.)
Hoch & P.H. Raven—COMMON WILLOW-HERB

Turions rather than basal rosettes of lvs are found in this subspecies (turions are short, fleshy offshoots at the base of the stem). Other differences: the lvs are ovate, the inflorescence is leafy, and the petals are 4–14 mm long.

Found in similar habitats as ssp. *ciliatum*, but to about 5000 ft. Known from Van Trump Park and near Longmire.

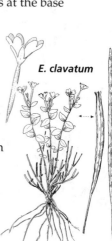

E. clavatum

Epilobium clavatum Trel.
CLUB-POD WILLOW-HERB

A tufted plant about 20 cm tall, the ascending stems with flattened hairs. The lvs are elliptical to ovate and 10–25 mm long. The fls buds are erect and the petals are 3.5–6 cm long. The capsule is 2–4 cm long, hairy, and club-shaped, widest at the tip.

Common, and not always easily distinguished from *E. anagallidifolium*, which shares its habitat. Found on talus slopes above 5000 ft, chiefly on the east side of the Park. Collected on the west slope of Naches Peak at 5800 ft and at Berkeley Park.

Epilobium densiflorum (Lindl.) Hoch & P.H. Raven
DENSE-FLOWERED WILLOW-HERB

An annual, with simple or branched stems to about 80 cm tall. The skin at the base of the stem peels as the plant ages and the stem is usually hairy in the upper portion. The lvs are mostly alternate, not stalked, narrowly lanceolate, and 1–8 cm long. The racemes are dense and the fls are white to pink, the petals 3–10 mm long. The cylindrical capsule is 4–10 mm long.

In a 1901 letter written to C. V. Piper, now in the archives of the library at Washington State University, J. B. Flett listed a number of plants he had collected at Mount Rainier but which Piper did not include in his 1901 "Mazama" article. One of these was *Boisduvalia densiflora* (the name under which this species is listed in Hitchcock and Cronquist), which Flett said grew on the northeast side of the mountain between 3500 and 4400 ft. Flett's collection or collections have not been found and no writer on the Park's flora mentions the species. Piper himself seems not to have been sent specimens from Flett. In any case, the Park lies within the range given by Hitchcock and Cronquist (1973) for *E. densiflorum*, and so the species is included here. It should be looked for in the lower valley of the White R.

Epilobium glaberrimum Barbey in Brewer & S. Watson ssp. *fastigiatum* (Nutt.) Hoch & P.H. Raven—GLAUCOUS WILLOW-HERB

A perennial with clumped, erect stems to about 35 cm tall, hairless and glaucous. The stems are unbranched and the clasping, lanceolate-ovate lvs are 1–3.5 cm long and closely placed on the stems so as to overlap. The pink to rose-purple petals are 3–7 mm long. The capsule is 2–7 cm long and may be sparsely hairy.

Infrequent and evidently less abundant than ssp. *glaberrimum*, growing on cliffs, talus slopes, and sometimes at roadsides and on gravelly stream banks, above about 4000 ft. Known from along Westside Rd; at Mowich Lk; along the Nisqually R from Longmire to the Glacier View Bridge; in the Butter Cr Research Natural Area; in Stevens Canyon; among rocks on roadcuts south from the White R entrance; and at Eunice Lk.

Epilobium glaberrimum Barbey in Brewer & S. Watson ssp. *glaberrimum*—GLAUCOUS WILLOW-HERB

This subspecies has quite a different look. The stems are taller, reaching 80 cm, and often branched. The lanceolate lvs are 2–7 cm long and well-spaced on the stem. The petals are 5–12 mm long.

More frequent than the preceding subspecies on the east side of the Park, growing mostly above 5000 ft. Collected near Bench Lk; below the summit at Chinook Pass; and in the White R Valley.

Epilobium halleanum Hausskn. HALL'S WILLOW-HERB

Known from one set of collections in the Park herbarium. A perennial about 40 cm tall, sparsely hairy on the upper stem as well as short-hairy in lines. The lvs are mostly ovate to oblong, stalkless or very short-stalked. The fls nod in the bud and the pink petals are about 4 mm long. The capsule is 3–5 cm long and sparsely glandular.

Uncommon, in moist subalpine meadows. Collected at the head of the Paradise R.

Epilobium hornemannii Rchb. ssp. *hornemannii* HORNEMANN'S WILLOW-HERB

Growing in loose clumps, with leafy, unbranched stems 10–40 cm tall. The short-stalked lvs are ovate, with spreading teeth, hairless, and 2–4 cm long. The inflorescence is few-flowered and usually nodding at the top. The pink petals are 4–8 mm long and the capsule is slender and sparsely hairy, 4–7 cm long.

Common on wet marshy ground in open places. Known at Paradise, in Stevens Canyon, and around Cayuse Pass.

Epilobium lactiflorum Hausskn. WHITE WILLOW-HERB

A perennial, with slender, tufted stems 10–30 cm tall that are usually hairless, sometimes minutely hairy in lines on the upper portion. The lvs are 2–5 cm long, narrowly ovate-lanceolate and remotely toothed. The inflorescence is glandular and may be nodding or erect. The petals are white and 3–7 mm long. The slender capsule is hairy and 5–10 cm long.

Fairly common in wet meadows and on moist ground in open woods, 5000-7000 ft. Found around Mowich Lk, in Berkeley Park, and at Eunice Lk.

247

Epilobium latifolium L.
RED WILLOW-HERB

A perennial that is somewhat woody at the base of the
stem, with spreading stems less than 40 cm tall and
usually glaucous overall. The racemes are short and
few-flowered, but the rose-purple fls are even larger
and more showy than those of fireweed and 3–5 cm
broad. The capsule is 3–10 cm long, short-hairy,
and narrowest at the base.

Uncommon, on stream banks and wet rocky
slopes. Known from Paradise Park and near the
terminus of Cowlitz Glacier, but probably more
widely distributed. A collection was made by J. B.
Flett at an unspecified elevation on Nickel Cr.

Epilobium luteum Pursh
YELLOW WILLOW-HERB

A loosely clumped perennial, 20–60 cm
tall, and hairy in lines on the upper stem. The lvs are
opposite, not stalked, hairless, 4–8 cm long and
toothed. The petals are some shade of yellow and 15–
18 mm long; the stigmas are 4-lobed. The capsule is 4–8
cm long and glandular-hairy.

Widespread and locally abundant in subalpine
meadows, mostly on the south side of the Park,
reaching 6500 ft. Collected at 4700 ft along Lee Cr 0.75
mi from Mowich Lk; at Indian Henrys Hunting
Ground; at Tahoma Cr at 3800 ft; at Paradise; on
Kotsuck Cr trail at 4500 ft; and as low as the Glacier
View Bridge.

Epilobium minutum Lindl. ex Lehm. in Hook.
SMALL-FLOWERED WILLOW-HERB

A slender annual, no more than 30 cm tall and short-
hairy throughout. The lvs are mostly opposite,
lanceolate, toothed, and 10–20 mm long. The
inflorescence is a few-flowered, bracted raceme. The
pink petals are 3–4 mm long and the stigma is 4-lobed.
The capsule is 2–2.5 cm long, slender and curved.

A lowland plant that is rare in the Park, growing on
dry soil in open woods and along roads. Found twice in
the course of this study: on an open slope uphill from
the Ohanapecosh entrance and at the roadside about 1 mi
south from the White R entrance. Listed in Piper's *Flora of
Washington* (1906), citing a collection made by O. D. Allen
in the upper Nisqually V.

Epilobium oregonense Hausskn.—OREGON WILLOW-HERB

A delicate perennial, with matted, hairless stems 5-20 cm tall. The lvs are opposite, not stalked, and ovate-lanceolate, widely spaced o the stem, and 10-20 mm long, marked with reddish veins. There are typically 1-3 fls per stem; the pink petals are 3-5 mm long. The capsule is 2-4 cm long, slender, and glandular-hairy.

An uncommon plant of wet meadows. Known from Berkeley Park.

Epilobium x *pulchrum* Suksd. (pro sp.)—HYBRID WILLOW-HERB

A hybrid species: *E. luteum* is one parent and the other is one of the smaller-flowered perennial species. The hybrid is a nearly hairless perennial, with stems 20-40 cm tall in small clumps. The lvs are opposite, not stalked or very short-stalked, toothed, and 2-3 cm long. The inflorescence is glandular, with a few fls held erect. The petals are rose-purple (or said to occasionally be white) and 10-12 mm long. The capsule is 2-3 cm long, very slender, and somewhat curved.

Said by early writers to be found infrequently along subalpine rivulets.

Gayophytum A. Juss.—GROUNDSMOKE

Gayophytum ramosissimum Torr. & A. Gray
GROUNDSMOKE

Similar in appearance to the smaller annual equilobiums, but more diffusely branched with alternate lvs. The key distinguishing feature is that the seed in *Gayophytum* does not have a tuft of hairs at the end. The stems are 20-60 cm tall and hairless. The lvs are linear and 2-4 cm long. The petals are white or pink and just 1-2 mm long, and the capsule is club-shaped and less than 1 cm long.

Rare in the Park and more common east of the Cascades. Known from a collection made on dry, sandy soil at the Owyhigh Lks at 5200 ft and reported by C.V. Piper at the foot of the Cowlitz Glacier. Possibly reaching a lower elevation in the White R Valley.

Orobanchaceae Vent.—BROOMRAPE FAMILY

Orobanche L.—BROOMRAPE

Orobanche uniflora L.—NAKED BROOMRAPE

A leafless, parasitic herbaceous plant, growing on the roots of the host. A single fl is borne on a purplish, stemlike stalk, which is 5-10 cm tall and glandular-hairy. The fl is tubular and curved, with fused petals, and 2-lipped, the upper lip 2-lobed and erect and the lower lip 3-lobed and spreading, 15-20 mm long, colored purplish but white at the throat. The fruit is an egg-shaped capsule, 6-8 mm long.

   Small and inconspicuous, but not uncommon. Parasitic on *Sedum* species and possibly on *Montia parviflora* as well. Most easily observed in crevices on the cliffs of Stevens Canyon; also in the area of the Glacier View Bridge.

Oxalidaceae R. Br.—WOOD-SORREL FAMILY

Oxalis L.—WOOD-SORREL

Perennial herbs, with scaly rootstocks and 3-lobed, heart-shaped leaves. There is usually a very short stem, but the leaves sometimes arise directly from the rootstock. The flowers are borne on hairy, leafless stems and have 5 free sepals and petals, with 10 stamens and a superior ovary. The fruit is a capsule.

1   Fls 1 per stem; petals 1-2 cm long .... *O. oregana*

1   Fls 2 or more per stem; petals less than 1-1.5 cm long .... *O. trilliifolia*

Oxalis oregana Nutt.—REDWOOD SORREL, OREGON OXALIS

The rootstocks of this species are horizontal and creeping. The leaflets are 2-4 cm long on stalks 15-25 cm tall. One fl is borne on each stem, usually below the level of the lvs; the petals are white with purplish veins and 1-2 cm long. The capsule is egg-shaped and 6-10 mm long.

   Common and often carpeting the forest floor, below 2500 ft in the west and south sides of the Park; apparently absent from the lower White R Valley. Especially abundant around the Nisqually entrance and along the lower reaches of Tahoma Creek. Seed pods are rarely seen.

Oplopanax horridus; Asarum caudatum;
  Agoseris glauca

Heracleum maximum; Arnica latifolia

Arnica longifolia; Cirsium edule

Aster alpigenus

Erigeron aureus; Erigeron peregrinus; Rainiera stricta

Eriophyllum lanatum; Solidago simplex

Campanula rotundifolia; Mertensia paniculata

Linnea borealis

Cornus unalaschkensis; Cassiope mertenisana;
  Phyllodoce empetriformis

Sedum oreganum; Lupinus arcticus

Lupinus latifolius; Lupinus lepidus

Rhododendron albiflorum

Phlox diffusa; Oxyria digyna; Polygonum bistortoides

Cistanthe umbellata; Dodecatheon jeffreyi

Anemone occidentalis; Aquilegia formosa; Caltha
  leptosepala

Delphinium glareosum; Aruncus dioicus; Holodiscus discolor

Luetkea pectinata; Potentilla flabellifolia

Spiraea splendens; Heuchera glabra

Pentaphylloides fruticosa

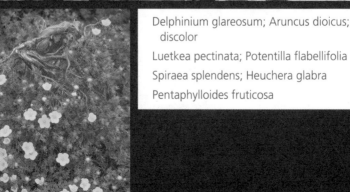

Saxifraga tolmiei; Castilleja hispida

Castilleja miniata; Castilleja
  parviflora; Mimulus lewisii

Pedicularis bracteosa; Pedicularis
  groenlandica

Castilleja rupicola

Pedicularis ornithorhyncha; Valeriana sitchensis; Viola orbiculata
Penstemon davidsonii var. davidsonii; Penstemon rupicola
Carex spectabilis; Clintonia uniflora; Erythronium grandiflorum

Lilium columbianum; Trillium ovatum; Veratrum viride

Xerophyllum tenax; Calypso bulbosa; Corallorhiza mertensiana

Erythronium montanum; Festuca viridula

Oxalis trilliifolia Hook.—THREE-LEAF WOOD-SORREL

The rootstocks of this species are vertically oriented and the plant does not form the colonies typical of *O. oregana*. The leaflets are 3-5 cm long and the stalks reach to 30 cm tall. Fls are borne 6-8 in an umbel and have white petals 8-14 mm long. The capsule is straight and slender, 2-3 cm long.

Much less common than the above species; also growing in deep woods, but said to prefer somewhat higher elevations, to about 3000 ft. Reported to sometimes grow with *O. oregana*, a phenomenon not seen in the course of this study. Collections have been made in the Longmire area and elsewhere along the lower Nisqually R. One Longmire collection carried 6 fls atop the stem.

Papaveraceae Juss. (including Fumariaceae) POPPY FAMILY

Perennial herbs, with bluish green, glaucous foliage and bitter juice in the stems and leaves (dangerous—best not to taste!). The flowers are irregular in the species in the Park, with one or more of the outer petals spurred at the base: there are 2 sepals, 4 petals, 6 stamens, and a 2-chambered pistil. The fruit is a capsule.

Opinion among botanists remains divided over whether to recognize the Fumariaceae and Papaveraceae as separate families; the treatment of J. L. Reveal is followed here. None of the "real" poppies occur in the Park, although it is certainly possible that the California poppy, *Eschscholzia californica*, might turn up as a weed on roadsides around the Nisqually, Ohanapecosh, or White R entrances.

1   Both outer petals spurred at the base; fl stems delicate, leafless .. *Dicentra*
1   Only 1 outer petal spurred at the base; fl stems robust, leafy ..... *Corydalis*

Corydalis DC.—CORYDALIS

Corydalis scouleri Hook. SCOULER'S CORYDALIS

A tall and vigorous plant, with branched stems often more than 1 m tall. The large lvs are alternate and 2 or 3 per stem, 3 times divided into lanceolate to oblong ultimate segments, the whole 40–50 cm wide. The fls are borne in a long raceme at the top of the stem; each is pink to rose and 2–3 cm long, including the single slender spur.

Common along streamsides below about 4000 ft, in more or less open forests. (Also prominent around Ashford on the approach to the Park from the west.) Collections known from the Nisqually entrance area; along Westside Rd and on the lower slopes of Mount Wow; on the main road above Longmire; and in the Carbon R V, where it is abundant.

Dicentra Bernh.—BLEEDING-HEART

Glaucous herbs with fernlike leaves and unusual flowers: the two outer petals are partly united, each with a spur. The common names suggest the shapes the blossoms take.

1   Fls 1 per stem; rare subalpine plant .... *D. uniflora*
1   Fls several per stem; common lowland plant ...... *D. formosa* ssp. *formosa*

Dicentra formosa (Haw.) Walp. ssp. *formosa*
WESTERN BLEEDING-HEART

Forms colonies by means of strongly spreading rootstocks. The lvs are basal, rising on long stalks and 2 or 3 times divided into angular segments, 20–50 cm long overall. Fls 4 or more, heart-shaped, pink to rose-purple, nodding in a loose raceme on arching stems; each is 14–18 mm long and the outer petals are both spurred at the base and spreading at the tips. The capsule is narrow and 15–20 mm long.

Common in low-elevation forests, especially near streams, in the Carbon and Ohanapecosh R Valleys. Known to reach 4500 ft on the trail from the Carbon R to Mystic Lk. Also known from the Butter Cr Research Natural Area.

Dicentra uniflora Kellogg
STEER'S-HEAD

A small plant, less than 10 cm high, from a cluster of tuberlike rootstocks. The lvs are 2 or 3 times divided into slender, oblong segments. A single pinkish fl about 15 mm long is borne on each short stem. Its outer petals curve back strongly (the horns of the steer) while the spurs are saclike. The capsule is ovate and about 10 mm long.

Known in the Park only in the area of Chinook Pass, blooming soon after the snow melts and often overlooked. Of much concern due to visitor impact in this area. This is said to be the only station for this plant west of the Cascade Divide.

Plantaginaceae Juss.—PLANTAIN FAMILY
Plantago L.—PLANTAIN

The family is represented in the Park only by two common and weedy species. Both are perennials, with basal leaves and leafless flowering stems. The inflorescence is a dense spike of small brownish flowers. The sepals and petals each number 4; the lobes of the petals have a papery texture. The 4 exserted stamens give the flower spike an interesting appearance: as the flowers mature, from the bottom of the spike upward, the stamens give the spike a "halo." The ovary is superior and develops into a small capsule.

1   Lvs lanceolate, tapered to an indefinite stalk .... *P. lanceolata*
1   Lvs elliptic to ovate, with a slender stalk .... *P. major* var. *major*

Plantago lanceolata L.—ENGLISH PLANTAIN
The 5–20 cm long lvs are lanceolate and held erect, surrounding the 25–45 cm tall flowering stem. The spike is up to 5 cm long and rather slender.

A common weed on waste ground and alongside roads; occasionally in dry meadows. Found in the Nisqually entrance area; up Westside Rd; at Longmire; at Ohanapecosh, including the hot springs area; and to 4200 ft in the White R Valley.

Plantago major L. var. *major* COMMON PLANTAIN
In this species, the ovate lvs, which are 5–20 cm long, tend to lie close to the ground, and form an ill-defined rosette. The flowering stem may reach 50 cm in height, with a spike up to 20 cm long.

As common as the above species, but preferring moister situations. Known from the Nisqually entrance area; at Longmire; at the top of Chinook Pass; and at Ohanapecosh, including the hot springs area. Also reported from Green Lk.

Polemoniaceae Juss.—PHLOX FAMILY

A family that includes a number of attractive wildflowers found in wet to dry subalpine meadows. Most are perennial herbs, with 5 sepals, 5 fused petals (showing a well-developed slender tube below the spreading free lobes), 5 stamens, and a 3-chambered ovary, with 3 distinct stigmas atop the 3 united styles. The fruit is a dry, 3-valved capsule.

253

1 Calyx tube eventually rupturing as the capsule matures; at least the lower lvs opposite 2

1 Calyx tube expanding with the maturing capsule and not rupturing; lvs alternate 4

2(1) Stamens attached at different levels in the tube; lvs not toothed ..... *Phlox*

2 Stamens attached at or about at the same level in the tube; lvs mostly divided 3

3(2) Lvs all opposite and palmately divided .... *Linanthus*

3 At least the upper lvs alternate, undivided or pinnately divided .... *Gilia*

4(1) Lvs pinnately compound, the individual leaflets distinct on the midvein of the lf .... *Polemonium*

4 Lvs pinnately to palmately divided, the lobes often toothed but not clearly distinct as leaflets .... *Collomia*

Collomia Nutt.—COLLOMIA

Our species include two annuals and a perennial. *Collomia* is separated from the other members of the family by a characteristic of the calyx: the fused sepals are not separated by intervening membranes, but rather are pleated and expand as the capsule matures. The capsule, therefore, does not rupture the calyx as it grows. The leaves are alternate and the plants are variously glandular and hairy.

1 Tufted alpine perennial... *C. debilis* var. *larsenii*

1 Lowland annuals 2

2(1) Plant erect and usually unbranched; lvs not toothed.................. *C. linearis*

2 Plant erect to somewhat sprawling, branched; lvs pinnately divided ....
.... *C. heterophylla*

Collomia debilis (S. Watson) Greene var. *larsenii* (A. Gray) Brand
ALPINE COLLOMIA

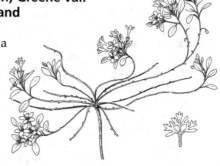

A tufted perennial, growing into a low cushion about 15 cm tall; the stems are branched and tend to sprawl. The lvs are palmately divided into 3 major lobes, each of these again divided into several slender lobes. Several pink to purplish fls are borne in a leafy, headlike cluster; each is about 10–15 mm long.

 Common on talus slopes and pumice flats, 6000-10000 ft, on the south side of the Mountain. Collected at Tolmie Peak; Spray Park; in upper Van Trump Park at 7150 ft; at Moraine Park at 6000 ft; at Frozen Lk; and at Berkeley Park.

Collomia heterophylla Hook.
VARIABLE-LEAF COLLOMIA

A low annual, with branched stems less than 20 cm tall. The lvs are pinnately lobed into irregular divisions; the upper stem lvs may be undivided. The fls are in small terminal clusters as well as in the axils of the upper lvs. The fl tube is 10–14 mm long and the free lobes another 3–4 mm long; light to dark pink, except for a white eye at the throat.

Widespread through the Park below about 4000 ft. Found at Sunshine Point; on the Kautz Cr mudflow; on Rampart Ridge and at Longmire; through Stevens Canyon; at the Ohanapecosh entrance; and in the White R Valley.

Collomia linearis Nutt.—NARROW-LEAF COLLOMIA

An annual with erect, usually unbranched stems up to 30 cm tall. The stalkless lvs are linear to lanceolate, untoothed (or the larger lower lvs with a few teeth), and 2–5 cm long. The 10–20 pink fls are crowded into a head at the top of the stem and each fl is 8–10 mm long.

Known from one collection, made at Longmire; said to grow in open woods and on gravelly banks.

Gilia Ruíz & Pav.—GILIA

Gilia formerly included the plants now in *Collomia* as well as a phlox, *P. gracilis*. What is left in *Gilia* are plants in which a membrane is present between the ribs of the calyx (which ruptures as the capsule matures) and in which the stamens are attached to the corolla tube at about the same level. Both of these species are small-flowered annuals. The 3-lobed style is exserted beyond the throat of the corolla.

1   Lvs not divided; fls scattered along open branches .... *G. capillaris*

1   Lvs pinnately divided; fls in dense terminal heads ....
.... *G. capitata* ssp. *capitata*

Gilia capillaris Kellogg—SLENDER GILIA

A diminutive plant, well worth a close examination. The glandular-hairy stems usually have a few branches and grow to 15 cm tall (but sometimes unbranched and less than 5 cm tall when growing on poor ground). The lvs are linear, not toothed, and 1–3 cm long; there is no basal rosette. The inflorescence is open, with 1–3 fls per branch. The corolla is 6–8 mm long, well-exceeding the length of the calyx, with a very narrow tube. The tube is yellow and glandular, while the throat and free lobes are blue.

Infrequent and most often found on poor soils, usually seen in low grasses along trails, as at Sunshine Point, on Westside Rd, and around Longmire. Not reported for the Park prior to the present study.

Gilia capitata Sims ssp. capitata
GLOBE GILIA

An annual, with erect, branched stems 20–80 cm tall, and pinnately dissected lvs. The blue fls are in dense, roundish heads at the tops of the stems. The corolla is 8–10 mm long.

Perhaps only casual in the Park. Included by G. N. Jones in his flora, without a specific location, and said to grow on "gravelly soil." St. John and Warren (1937) speak of observing the species at Longmire and refer to a collection made by J. B. Flett at Ohanapecosh.

Linanthus Benth.—LINANTHUS

Linanthus nuttallii (A. Gray) Greene ex Milliken ssp. nuttallii
NUTTALL'S LINANTHUS

A hairy perennial with tufted stems, forming broad clumps, the stems somewhat woody at the base and reaching 30 cm tall. The lvs are opposite, not stalked, 1–2 cm long, with 5 palmate-spreading lobes, the lvs therefore giving the appearance of being whorled. The fls are in dense terminal clusters. The yellow tube of the corolla is about 8 mm long, not exceeding the calyx, while the white, spreading lobes are about 4 mm long.

A common plant of dry meadows, 5000-6500 ft, known chiefly from the southwest side of the Park. Collections have been made on Mount Wow and Gobblers Knob; in St. Andrews Park at 5300 ft; at 6000 in Klapatche Park; and on Eagle Peak at 5500 ft.

Phlox L.—PHLOX

Phlox diffusa is perhaps the most eye-catching of the flowers one sees on dry, stony slopes heading down the road from Reflection Lks to Stevens Canyon. Phlox have opposite, untoothed leaves and solitary flowers. A key feature is that the 5 stamens are attached to the tube at different levels.

1   Lowland annual .... *P. gracilis* ssp. *humilis*
1   Tufted perennial, of alpine and subalpine areas 2

2(1) Lvs 5–13 mm long; stems mostly erect .... *P. caespitosa*
     (see "Doubtful and Excluded Species," p. 469)
2   Lvs 10–15 mm long; stems matted and spreading ....
     .... *P. diffusa* ssp. *longistylis*

Phlox diffusa Benth. ssp. *longistylis* Wherry—SPREADING PHLOX

"Diffuse" describes the habit of this plant: the leafy stems spread widely along the ground and old plants are matlike. The lvs are linear and 10–15 mm long. The fls, solitary and without stalks at the ends of the branches, open sky-blue and pass through pink to white as the blossom ages. The corolla tube is about 10 mm long and the lobes are 6–8 mm long. Common on talus slopes and rocky ridges, 5500-7500 ft. Found at Mowich Lk and Spray Park; at St. Andrews Park; at Paradise Park; at Reflection Lks; in the Butter Cr Research Natural Area; in the meadow at the top of Shriner Peak; on the Kotsuck Cr trail; around Tipsoo Lk; at Sunrise and up to Burroughs Mtn; and at Frozen Lk.

Phlox gracilis (Douglas ex Hooker) Greene ssp. *humilis* (Greene) H. Mason—ANNUAL PHLOX

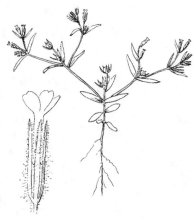

An erect, branched annual, with a slender stem 10–25 cm tall. The lvs are oblanceolate, or narrower toward the top of the stem, 1–3 cm long, and may also be alternate upwards on the stem. The fls are solitary in the axils of the upper lvs, pink, and 8–10 mm long.

Common on gravelly soil in open places, below about 5000 ft. Known from the Nisqually entrance area and up Westside Rd; in Stevens Canyon; in the White R V; and below the Carbon Glacier.

Polemonium L.—SKY-PILOT, JACOB'S LADDER

Three very appealing wildflowers, although the plants have a skunky odor. The flower tube is quite short compared to the other members of the family and the lobes spread to form a bell-shaped to shallow cup-shaped corolla. The flowers are clustered at the ends of the leafy stems, in loose to dense cymes. The leaves are alternate and pinnately divided.

1   Leaflets more than 20, overlapping on the midvein of the lf; corolla longer than wide, funnel shaped ....*P. elegans*

1   Leaflets fewer than 20, separated on the midvein of the lf; corolla about as wide as long, bell-shaped.. 2

2(1)   Tufted plants of high elevations; larger leaflets of the basal lvs to about 10 mm long ....*P. pulcherrimum* ssp. *pulcherrimum*

2   Looser, taller plants, mostly of low to middle elevations; larger leaflets of the basal lvs 20–30 mm long .... *P. californicum*

Polemonium californicum Eastw.—CALIFORNIA POLEMONIUM, SHOWY JACOB'S LADDER

Loosely tufted plants, with lax, spreading stems 15–30 cm tall and sparsely glandular-hairy. The lvs are 10–20 cm long, with oval leaflets each 1–3 cm long; the terminal three leaflets usually merge at their bases. The fls are in loose cymes and about 9–12 mm long and bell-shaped; the throat and tube are yellow, the lobes blue to blue-violet. The lobes of the calyx are much longer than the tube.

Typically found in woods and meadows up to the subalpine zone. Collections have been made in lower Paradise Park; in Glacier Basin at 5900 ft; and on a ridge above Bear Park. Common around Sunrise.

Polemonium elegans Greene SKY-PILOT, ELEGANT JACOB'S LADDER

A densely tufted and compact plant, with heavily glandular-hairy stems and lvs, 5–12 cm tall. The lvs are 4–7 cm long, with very many leaflets, each roundish and 2–5 mm long, overlapping on the midvein. The fls are in a dense cluster and the corolla is 12–15 mm long, funnel-shaped and blue-violet, yellow at the throat.

The species was discovered on Mount Rainier in 1895 by C. V. Piper. Plants growing on Mount Wow tend to be less sticky, with longer hairs and white to pale blue fls, were once called *P. viscosum* Nutt. var. *pilosum* Greenm. Common on high ridges, 7000-10000 ft, but occasionally lower, as near Tipsoo Lk and at the summit of Mount Wow.

Found above Spray Park; in upper Paradise Park at Panorama Point and above McClure Rock; on Wapowety Cleaver at 9000 ft; on the west side of Wilson Glacier at 8200 ft; at the summit of Chinook Pass; and on Burroughs Mtn. Also reported from the cliffs above Carbon Glacier, at an unspecified elevation.

Polemonium pulcherrimum Hook. ssp. *pulcherrimum*—LOW JACOB'S LADDER

A tufted and compact perennial, sparsely hairy and glandular, with erect to somewhat spreading stems less than 10 cm high. The are mostly basal, about 5 cm long with ovate to roundish leaflets 4–8 mm long; the terminal leaflet is distinct from the pair of leaflets just below it. The inflorescence is crowded, with white to blue fls; the corolla is bell-shaped and 8–10 mm long, with a yellow throat. The calyx lobes are about equal to the tube, or somewhat shorter.

   Common on cliffs and rocky ridges as well as on pumice flats, usually at lower elevations than *P. elegans*, 5000-6500 ft. Collected at Mowich Lk; at Indian Henrys Hunting Ground; around the Reflection Lks; onPlummer Peak at 6000 ft; on the Kotsuck Cr trail at 5000 ft; at Chinook Pass; in the meadows at Sunrise; and on the trail to Burroughs Mtn.

Polygonaceae Juss.—BUCKWHEAT FAMILY

A large family of mostly perennial herbs (a few are annuals) including some distinctive wildflowers as well as many weeds. The leaves are alternate and not toothed; the stipules, except in *Eriogonum*, are united into a sheath that surrounds the stem. The inflorescence is a small to large, terminal or axillary cluster. The flowers are small and petals are absent; instead, the perianth of 5 or 6 sepals may be colored or petal-like. There are 3–9 stamens and a superior ovary with 3 styles. The fruit is a 3-angled achene, usually enfolded by the persistent perianth.

1	Lvs lacking stipules; fls clustered, in a "cup" formed of the fused involucral bracts	*Eriogonum*
1	Lvs with sheathing stipules; fls, if clustered, lacking fused involucral bracts	2
2(1)	Lvs kidney-shaped; lobes of the perianth 4	*Oxyria*
2	Lvs various but not as above; lobes of the perianth 5 or 6	3
3(2)	Lobes of the perianth 5	*Polygonum*
3	Lobes of the perianth 6	*Rumex*

Eriogonum Michaux—WILD BUCKWHEAT

Native perennials from woody rootstocks, with tufted, basal, stalked leaves; stipules are absent. The flowers are in headlike clusters on leafless stems and are carried on stalks in a more or less cuplike involucre; the involucre itself may be bracted. The perianth is generally 6-lobed and narrow at the base, with 9 stamens.

1   Stems erect and 15–40 cm tall; lvs more or less triangular ....
.... *E. compositum* var. *compositum*
1   Stems tufted and less than 15 cm tall; lvs lanceolate or oval 2

2(1) Lvs lanceolate to ovate, densely short-hairy below and green above; inflorescence with two linear bracts .... *E. pyrolifolium* var. *coryphaeum*
2   Lvs ovate to round, densely short-hairy above and below; bracts absent ..
.... *E. ovalifolium* var. *nivale*

Eriogonum compositum Douglas ex Benth. var. *compositum*—ARROW-LEAF BUCKWHEAT

The lvs of this characteristic subalpine plant are 3–10 cm long, with ovate lvs, heart-shaped or arrowhead-shaped where the blade meets the stalk, deep green and becoming hairless above but densely woolly beneath. The stems are unbranched, long-hairy, and 15–40 cm tall. The inflorescence is a compound umbel with leaflike bracts. The involucre is more or less woolly. The fls are creamy white and 4–6 mm long, on very long stalks.

Common on open rocky slopes above 5000 ft. Known from Mount Wow; at 5500 ft on Klapatche Ridge; at St. Andrews Park; at Goat Pass on Kotsuck Cr trail; and on the slopes above Tipsoo Lk.

Hitchcock and Cronquist (1973) state that *E. compositum* var. *leianthum* Hook. ranges east of the Cascades from Chelan County "to Mount Rainier where common" and into Idaho and Oregon. Variety *leianthum* differs from var. *compositum* in having hairless involucres. No collection examined in the course of this study showed this feature, and the early writers—Piper, Jones, and St. John and Warren—do not list both varieties in the Park.

Eriogonum ovalifolium Nutt. var. *nivale* (Canby) M. E. Jones—CUSHION BUCKWHEAT

A matted plant, with stems 3–15 cm tall. The lf blade is ovate to round, 2–8 mm long, and densely white-hairy on both sides. The fls are in a single small umbel, creamy yellow, and 3–5 mm long.

Uncommon, on rocky slopes, 5000-8000 ft. Only two collections were located, one from 7000 ft on the Cowlitz Rocks and one from an unspecified location in the Park.

Eriogonum pyrolifolium Hook.
var. *coryphaeum* Torr. & A. Gray
ALPINE BUCKWHEAT

Tufted and growing to form small cushions 20–30 cm across. The lvs are lanceolate to ovate and 10–40 mm long, green and perhaps sparsely hairy above, white-woolly below. The whitish fls are 4–6 mm long, in a single umbel on a stem 5–10 cm tall, just above 2 narrow bracts.

Very common on rocky alpine slopes, 5500-8000 ft on the south side of the Mtn and lower on the north side. Collected at Mildred Point at 5550 ft; in upper Paradise Park and north of McClure Rock; at 7500 ft on Wapowety Cleaver; on the slopes of Pinnacle Peak; on Goat Island Mtn at 6900 ft; at the lookout on Shriner Peak; at Sunrise and along the trail to Burroughs Mtn; at Frozen Lk; and at Eunice Lk. An early colonizer of disturbed ground.

Oxyria Hill—MOUNTAIN-SORREL

Oxyria digyna (L.) Hill—MOUNTAIN-SORREL

A hairless perennial with long-stalked basal, kidney-shaped lvs, 1.5–5 cm broad, with sheathing stipules; the lvs often turn reddish late in the season. The inflorescence is a compact panicle, held on a leafless stem 10–30 cm tall. The fls are about 2 mm long, 4-lobed, reddish brown, and nodding on slender stalks. The achenes are about 4 mm long, surrounded by a thin wing.

Very common in alpine and subalpine meadows, in wet places, 6000-8500 ft. Collected at Mtn Meadows and Spray Park; at Indian Henrys Hunting Ground; across Paradise Park; and on Burroughs Mtn at 7100 ft. Also known from 9000 ft on Wapowety Cleaver.

Polygonum L.—KNOTWEED

Annual or perennial herbs (*P. convolvulus* twines like a vine). Several are weeds and even the native species often look and behave as weeds. Indeed, *P. newberryi* is valuable for its ability to rapidly colonize areas of disturbed soil at higher elevations. American bistort is the only prominent wildflower. The leaves are alternate, with sheathing stipules. The inflorescence is made up of a few to many small clusters arranged in heads or panicles. The perianth is 5-lobed, with 3–9 stamens. The fruit is egg-shaped and the achenes brown or black.

1   Lvs heart-shaped; plant an annual twining vine ....
     .... *P. convolvulus* var. *convolvulus*
1   Lvs various, but not heart-shaped; plants not twining 2

2(1) Stems 1–2 m tall; lvs oval, about 20 cm long .... *P. cuspidatum*
2   Stems rarely more than 0.5 m tall; lvs linear to lanceolate or ovate, less than 20 cm long 3

3(2) Plants annual, typically of dry places 4
3   Plants perennial, typically of moist (or seasonally moist) places 8

4(3) Stems prostrate; weed of dry waste places and roadsides .... *P. aviculare*
4   Stems erect 5

5(4) Stipules sheathing the stem and fringed with long hairs on the upper margin; lvs lanceolate, to 20 cm long; weed in moist places ....
     .... *P. lapathifolium* var. *lapathifolium*
5   Stipules not sheathing, the upper margin various but not long-hairy; native plants of dry places 6

6(5) Lvs oval, tending to conceal the fls that are in the lf axils .... *P. minimum*
6   Lvs linear to narrowly oblanceolate, not overlapping 7

7(6) Fls in a slender spike, reflexed, 1–3 in the axils of each leaflike bract ....
     .... *P. douglasii* ssp. *douglasii*
7   Fls erect in crowded spikes .... *P. polygaloides* ssp. *kelloggii*

8(3) Fls white, in a dense spikelike raceme .... *P. bistortoides*
8   Fls greenish with some red, in small axillary racemes .... *P. newberryi*

Polygonum aviculare L.—YARD KNOTWEED

A mat-forming annual, with prostrate to erect stems 10–20 cm long. The lvs are 5–30 mm long, lanceolate or slightly wider, stalked, and pointed at the tip. The 2–8 white or pink fls are borne in the axils of the lvs; each is 2–3 mm long and has 8 stamens.

Very common weed at low elevations, usually seen on open roadsides; most frequent in the Park around the Nisqually entrance, at Longmire, and at Ohanapecosh. Also known from a collection made in the lower Carbon R V. Some, or perhaps all, of the plants in the Park may actually be *P. arenastrum* Jord. ex Boreau, another weed which is said to differ very slightly in the measurements of certain parts of the fls.

Polygonum bistortoides Pursh
AMERICAN-BISTORT

A perennial, from a short and thick rootstock, with mostly basal lvs and an unbranched flowering stem 30–60 cm tall. The lvs are long-stalked, narrowly oblong, 10–20 cm long and prominently marked by a light-colored midrib; the stem lvs, if any, are smaller, narrower, and not stalked. The fls are in a compact, oblong spikelike raceme. The perianth is 4–5 mm long, white, the lobes united only at the base and with 8 stamens.

Very common in subalpine meadows and one of the earliest fls to bloom at Paradise, 5000-6500 ft. Found at Mowich Lk and Mountain Meadows; at Spray Park; at Indian Henrys Hunting Ground; in Van Trump Park; across Paradise Park; in the Butter Cr Research Natural Area; at Tipsoo Lk; at Sunrise and on the trail to Burroughs Mtn; in upper Palisades meadow; and in Berkeley Park.

Polygonum convolvulus L. var. *convolvulus*
BLACK BINDWEED

An odd annual, with twining, vinelike stems that reach 20–100 cm long. The lvs are stalked and the blades, shaped like arrowheads, are 2–6 cm long. The inflorescence is slender and racemelike, in the axils of the lvs. The 5-lobed fls are greenish white and about 4 mm long.

A weed of disturbed places and at open roadsides, known in the Park from two old collections, made at Longmire and at Ohanapecosh.

Polygonum cuspidatum Siebold & Zucc.
JAPANESE KNOTWEED

A perennial, spreading by heavy rootstocks to form dense colonies. The stems can reach 1–2 m tall. The lvs are stalked, with large, flat, oval blades 15–20 cm long. Male and female fls are on separate plants, borne in tassel-like racemes grouped in the axils of the upper lvs. The fls are greenish white, about 4 mm long, and scarcely open.

A very troublesome weed, able to crowd out other vegetation on the low, wet ground it prefers. It is found in the Park on the Carbon R Rd, near Chenuis Falls, where Park Service have been attempting to eradicate it. It is also abundant around Ashford, just west of the Park.

Polygonum douglasii Greene ssp. *douglasii*
DOUGLAS'S KNOTWEED

An erect, branched annual, 20–40 cm tall. The lvs are oblanceolate to elliptic, to 3–4 cm long, reduced toward the top of the stem. The fls are in loose spikes in the axils, about 3 mm long, white to pinkish with a green midrib.

Uncommon, found in a few places above 5000 ft on dry, gravelly soil, as in the Silver Forest area and near the foot of the Cowlitz Glacier. This species is uncommon west of the Cascade crest.

Polygonum lapathifolium L. var. *lapathifolium*—WILLOW-WEED

A branched annual, with erect stems to 80 cm high. The lvs are stalked, 5–20 cm long, lanceolate, and hairy on the underside. The inflorescence is spikelike, 2–10 cm long, in the axils of the upper lvs as well as terminal. The fls are about 2 mm long, whitish, with 6 stamens.

Occasional weed on wet ground below 2000 ft, in the Ohanapecosh area.

Polygonum minimum
S. Watson
BROADLEAF KNOTWEED

A small, branched annual, with reddish stems 5–25 cm tall. The lvs are oval to ovate, stalkless or nearly so, and 5–10 mm long. The pinkish fls are mostly solitary in the axils, on erect stalks, and about 2 mm long, with 5–8 stamens.

C. V. Piper, writing in 1902, describes the plant as "common at 5000 to 6000 ft altitude," perhaps based on information from O. D. Allen, who found the plant on rocky ground in the Goat Mountains in the 1890s. It is known only from two collections: Allen's, plus one made in the early 1950s on the Reflection Lk trail, at an unspecified elevation and presumably on sandy, open ground.

Polygonum newberryi Small
FLEECEFLOWER, NEWBERRY'S KNOTWEED

A perennial with heavy roots and somewhat brittle stems, allowing the plant to spread when portions of the stem break off and take root where they've fallen. Mostly short-hairy, but sometimes hairless, the stems are typically unbranched and reach 30 cm long. The lvs are pale green, ovate-lanceolate to about oval, and 1–5 cm long. The inflorescence is a small, axillary raceme 1–2 cm long. The fls are about 3 mm long, greenish, with 8 stamens.

Common on the eastern side of the Park, in dry meadows and on stony flats, 6000-7500 ft. Collected at Sunrise; from Sunrise to Frozen Lk; in Moraine Park; and on the trail to Burroughs Mtn. Also found on the slopes of Pinnacle Peak. One of the first plants to colonize disturbed soils at higher elevations, it can often be found on kicked-up mounds of soil at trailsides.

Kartesz synonymizes *P. newberryi* with *P. davisae*, but the two seem sufficiently distinct, both morphologically and in range, to merit recognition as separate species, as is done in *The Jepson Manual* (Hickman, 1993).

Polygonum polygaloides Meisn. ssp. *kelloggii* (Greene) J.C. Hickman
KELLOGG'S KNOTWEED

A slender plant, sometimes unbranched, 2–7 cm tall, with stalkless, linear lvs 5–10 mm long. The greenish to whitish fls are in small clusters in the axils of the upper lvs and are 1.5–2 mm long, generally with 3 stamens.

Listed as "review – group 1" in *The Endangered, Threatened, and Sensitive Vascular Plants of Washington* (Washington Natural Heritage Program, 1997).

Uncommon native, found on open, seasonally moist ground, 4000-5000 ft. Listed by Jones in 1938 but not verified in the course of this study.

Rumex L.—DOCK

Coarse perennial herbs; most are weeds in the Park. The leaves may be quite large and often have wavy or crisped margins. The flowering stems are leafy and the inflorescence is an open to dense panicle. The 6-lobed flowers are greenish to reddish brown and arranged in whorls on the branches of the inflorescence. There are 6 stamens and 3 styles. The 3 inner perianth lobes become enlarged and hard, and enclose the maturing achene as "valves." The midrib between the perianth lobes may swell, becoming a bumplike "tubercule."

1	Lvs arrowhead-shaped, with a sour taste	*R. acetosella*
1	Lvs various, but not arrowhead-shaped, without a sour taste	2
2(1)	Native plant; stems branched below the inflorescence; lvs flat, light green	*R. salicifolius* var. *angustivalvis*
2	Introduced plants; stems generally unbranched; lvs dark green	3
3(2)	Lvs flat; the valves of the fruit strongly spiny	*R. obtusifolius*
3	Lvs wavy or crisped; the valves of the fruit smooth	4

4(3) Fls crowded in a dense panicle; lvs crisped on the margins .... *R. crispus*
4   Fls in whorls in loose panicles; lvs wavy but not crisp-margined ....
      .... *R. conglomeratus*

Rumex acetosella L.—SHEEP-SORREL

A perennial, able to colonize patches of ground by slender, wide-ranging rootstocks. The slender stems are 10–30 cm tall. The stalked lvs have the appearance of an arrowhead and are 3–10 cm long; the upper lvs are narrow to linear, shorter, and not stalked. Male and female fls are borne on separate plants and the inflorescence is a narrow, branched panicle. The fls are reddish and about 2 mm long. The valves are smooth and untoothed; tubercules are absent.

A common weed of wet ground at roadsides and around buildings at low elevations, especially at Sunshine Point, Longmire, Ohanapecosh, and in the lower White R Valley.

Rumex conglomeratus Murray
SHARP DOCK

A slender, erect perennial, with stems to nearly 1 m tall. The lanceolate-ovate lvs are 10–30 cm long, dark green, and crisped on the margins. The inflorescence is a loose panicle, leafy-bracted throughout. The fls are 2–3 mm long and reddish brown, in well-separated whorls. The valves are smooth and untoothed, with a large tubercule.

Common in shady places on roadsides and disturbed ground, as at the Nisqually entrance, Longmire, and in the White R V.

Rumex crispus L.—CURLY DOCK

"Curly" for the markedly distorted margins of the lvs, especially at their bases. This is a stout perennial, 50–150 cm tall. The lvs are dark green, narrowly oblong, and 15–30 cm long. The stem lvs are smaller and narrower. The inflorescence is a tall, dense, leafy panicle and the reddish brown perianth is about 5 mm long. The valves are untoothed and smooth, with tubercules.

The most frequently seen of the weedy docks in the Park, along most roads, reaching about 3000 ft.

Rumex obtusifolius L.—BITTER DOCK

A stout perennial, the stems reaching 40–60 cm tall. The lvs are oblong to ovate, cordate at the base, and 15–30 cm long, with flat margins. The inflorescence is of loose whorls; not leafy. The fls are 3–5 mm long and greenish. The valves are toothed; only one of the three bears a tubercule.

Occasional weed, noted at the roadside about 1 mi up Westside Rd and reported from Green Lk.

Rumex salicifolius Weinm. var. *angustivalvis* Danser
WILLOW-LEAVED DOCK

A perennial with spreading stems, less than 100 cm tall. The lvs are light green, flat-margined, and lanceolate or narrowly lanceolate. The fls are in crowded whorls; each is reddish and 3–4 mm long. The valves have small teeth and there are 1–3 tubercules.

Infrequent, in open gravelly places around buildings, as at Longmire. Wa*rren 1747*, from Longmire, was evidently the basis for *R. mexicanus* Meisn. in G. N. Jones's flora. *Warren 1747* actually is *R. salicifolius* var. *angustivalvis* rather than *R. salicifolius* var. *mexicanus*.

Portulacaceae Juss.
PURSLANE FAMILY

Annual or perennial, more or less succulent plants. The leaves are untoothed and alternate or opposite. The flowers have 2 sepals, 3–16 petals, and 2 to many stamens. The ovary is superior, with 2–8 styles, and develops into a dry capsule. *Lewisia* and *Claytonia* include some attractive wildflowers.

1   Fls in headlike clusters; petals 4, stigmas 2 .... *Cistanthe*
1   Inflorescence various, but not headlike; petals 5 to many, stigmas 3–8 .. 2

2(1)   Capsule circumsessile at the base, the top popping off like a lid; sepals 2–8 .... *Lewisia*
2   Capsule opening from the top down by 2 or 3 valves; sepals 2 3

3(2)   Lvs mostly basal; the stem lvs only 2, often fused together at their bases .... *Claytonia*
3   Stems leafy along their lengths .... *Montia*

Cistanthe Spach—CISTANTHE

A genus that has enjoyed quite a number of names, including S*praguea* and *Calyptridium,* by which it goes in *The Jepson Manual* (Hickman, 1993). A famous rock plant, *Cistanthe (Lewisia) tweedyi,* has been included in the National Park Service flora, but the basis for this is not known and Peter Dunwiddie did not include it (see "Doubtful and Excluded Species," p. 469).

Cistanthe umbellata (Torr.) Hershkovitz var. *caudicifera* (A. Gray) Kartesz & Gandhi
PUSSYPAWS

A distinctive alpine plant, growing as a prostrate mat from a deep taproot. The hairless lvs are in basal rosettes, spoon-shaped, and 1–3 cm long. The flowering stems are prostrate to erect, 2–6 cm long, and the inflorescence is a dense, rounded head. The sepals are papery, rounded, and 4–6 mm long; they remain pressed up to the 4 white or pink petals, which are about the same length but narrower. There are 3 stamens and a single style with a 2-lobed stigma. Altogether, the fls look rather like little pompoms.

Very common around the Mountain on dry, stony flats and slopes, 6000-9000 ft. Found in upper Paradise Park; at Sunrise and up to Burroughs Mtn.

Claytonia L.—SPRING BEAUTY

A large group of species that has been subject to much revision as species have been switched back and forth with *Montia*. The treatment used here follows Miller (1993).

These are perennial or annual herbs. The "spring beauty," *C. lanceolata,* rises from a tuber whereas the other perennials have short rootstocks. The plants are hairless and fleshy, with mostly basal leaves ; the stem leaves are limited to a single pair beneath the inflorescence. The inflorescence is a 1-sided raceme at the top of the stem. There are 5 petals, white or pink, and 5 stamens; the ovary is 1-chambered, with a single style and 3 stigmas.

1   Plants perennial, from a round tuber; stem lvs lanceolate to ovate 2
1   Plants annual, or if perennial then from a short rootstock; stem lvs more or less rounded 3

2(1) Fls yellow; stem lvs ovate .... *C. lanceolata* var. *chrysantha*
2   Fls white, marked with pink; stem lvs lanceolate ....
       .... *C. lanceolata* var. *lanceolata*

3(1) Stem lvs partly or fully united at their bases, forming a disk on the stem 4
3   Stem lvs not united 5

4(3) Stem lvs united on both sides of the stem, the disk 2–5 cm across; widest basal lf blades 1–4 cm across .... *C. perfoliata* ssp. *perfoliata*
4 Stem lvs united only on one side, the disk less than 2 cm across; widest basal lf blades to 1 cm across .... *C. rubra* ssp. *rubra*

5(3) Petals 8–13 mm long; bracts absent from the inflorescence .. *C. cordifolia*
5 Petals 6–8 mm long; each fl with 1 or 2 bracts .... *C. sibirica* var. *sibirica*

Claytonia cordifolia S. Watson
BROAD-LEAVED SPRING BEAUTY

A perennial from a short rootstock. The basal lvs are broadly ovate or cordate, 5–20 cm long overall, with the blade 2–6 cm wide. The stem is 10–30 cm tall and bears 2 lvs just above the middle; these are not stalked and not united, 2–5 cm long, and broadly lanceolate. The raceme is stalked and open, 3–10 cm long; bracts are not present. The petals are white and 10–13 mm long.

Infrequent, in wet places on stream banks and open slopes, 4000-5000 ft. Known from Mowich Lk and Spray Park; on Mount Wow at 5000 ft; at Cayuse Pass; along the trail from White R to Summerland at 4300 ft; and at Berkeley Park.

Claytonia lanceolata Pursh var. *chrysantha*
(Greene) C.L. Hitchc.—YELLOW SPRING BEAUTY

Differing from var. *lanceolata*, below, in the yellow color of the petals and smaller stature, 5–10 cm tall.

Rare in the Park. According to Hitchcock and Cronquist (1973), this subspecies is found in the Cascades from Mount Baker to Glacier Peak, well to the north of Mount Rainier. A collection was made on the Kotsuck trail at 5500 ft and a collection in the Park herbarium was made by A. S. Pope at an unspecified location in the Park.

Claytonia lanceolata Pursh var. *lanceolata*
WESTERN SPRING BEAUTY

A perennial from a round, deeply buried tuber. Each tuber usually produces several stems, these 5–20 cm tall, and 1 or 2 basal lvs. The latter can be 6–8 cm long and are narrowly oblanceolate. The stem bears 2 opposite lvs, these are not stalked and not united, lanceolate, and 2–5 cm long. The inflorescence is a rather compact raceme, above a small bract. The petals are white with pink veins, 5–12 mm long.

A common and early-blooming fl of subalpine meadows, above 5000 ft. Collected on Mount Wow; at Indian Henrys Hunting Ground; in the Butter Cr Research Natural Area; near the summit of Shriner Peak; at Tipsoo Lk; around Berkeley Park; and at Moraine Park.

Claytonia perfoliata Donn. ex Willd. ssp. *perfoliata*—MINER'S LETTUCE

An annual, with numerous basal lvs that are spoon-shaped (or sometimes long-stalked with an elliptical blade) and 3–15 cm long overall. The flowering stem is 10–20 cm tall, with 2 opposite lvs, fused at their bases on each side of the stem forming a disk 2–5 cm broad. The raceme is open and apparently 1-sided, with a single bract at the base. The petals are white with pink veins and 2–4 mm long.

Seen occasionally on damp ground among other low herbaceous plants in open woods and along trails, below about 2500 ft.

Claytonia rubra (T. Howell) Tidestr. ssp. *rubra* RED MINER'S LETTUCE

An annual, somewhat like a miniature version of *C. perfoliata*. "Red miner's lettuce" suggests the overall reddish cast of the stems and lvs. The plant is seldom more than 6–8 cm tall, with the largest basal lf blades less than 1 cm broad. The pair of stem lvs are fused on only 1 side of the stem and the resulting disk is less than 2 cm broad.

First observed in the Park in the course of this study. Typically seen in moss on rocks in openings in forests, in the area of the Nisqually entrance and on the lower part of Westside Rd.

Claytonia sibirica L. var. *sibirica*—CANDYFLOWER

The most frequently seen *Claytonia* in the Park, this is a perennial, growing from short rootstocks. The basal lvs are ovate and long-stalked, up to 25 cm long. The pair of stem lvs are not stalked but also not fused, ovate, and 1–6 cm long. The inflorescence is often branched, with a bract beneath each fl. The petals are white with pink veins, or pinkish, and 6–8 mm long.

Common and often abundant in moist forests up to 5000 ft, especially on shaded rocks and banks; occasionally seen at roadsides on gravelly slopes. Found at Mowich Lk; at the Nisqually entrance; along Westside Rd; at Longmire; at the Glacier View Bridge; in the Butter Cr Research Natural Area; along the Pinnacle Peak trail near Reflection Lks; along the road on the east side of the Cowlitz Divide; at 2500 ft on Chinook Cr; at Ohanapecosh; in the White R V below the campground; and along the Carbon R to about 3000 ft.

In some places where *C. sibirica* and *C. perfoliata* grow together, a fertile hybrid called *Claytonia washingtoniana* (Suksd.) Suksd. is found. Such plants have not been recorded for the state of Washington, but they should be looked for in the Park where the two parent species occur together. It is an annual, usually with 2–4 fls per bract, stem lvs that are fused on one side, and petals 4–5 mm long.

Lewisia Pursh—LEWISIA

A interesting trio of seldom-seen plants, quite dissimilar in appearance but each having 2 sepals that are not united at their bases, 5–11 petals, and capsules that open lidlike from near the base. All are perennials. See "Doubtful and Excluded Species," p. 469, for a discussion of *Cistanthe tweedyi*, formerly known as *Lewisia tweedyi*.

1   Lvs 2 or 3, on a slender stem that rises from a deeply buried tuber .... *L. triphylla*

1   Lvs many, from a heavy rootstock 2

2(1) Stems to 20 cm tall, with numerous fls in a spreading panicle .... *L. columbiana* var. *rupicola*

2   Stems less than 5 cm tall, 1-flowered .... *L. pygmaea*

Lewisia columbiana (Howell) B. L. Rob. var. *rupicola* (English) C.L. Hitchc.—ROCK LEWISIA

Growing from a short, thick, partly woody rootstock, with a basal rosette of lvs. The lvs are fleshy, narrowly oblanceolate, and 2–5 cm long. The inflorescence is an open panicle, with numerous fls, 10–30 cm high. The 2 sepals are roundish in outline and gland-toothed (as are the bracts on the stem and in the inflorescence). The petals are 8–10 mm long, are white to pink or rose, with darker veins.

Found in just a few places, mostly on the southwest side of the Park: in Klapatche Park and St. Andrews Park; on Mount Wow; on the southwest slope of Pyramid Peak; and on Eagle Peak. Also reported from Eunice Lk.

Lewisia columbiana var. *columbiana* has been claimed for the Park, on the basis of a misidentification of *Warren 1682* from Gobblers Knob, which is clearly var. *rupicola*.

Lewisia pygmaea (A. Gray) B. L. Rob.
DWARF LEWISIA

Also from a thick, short rootstock, with a basal tuft of fleshy, linear to linear-oblanceolate lvs 2–5 cm long. Each plant bears several 1-flowered stems, 1–5 cm tall, often shorter than the lvs and sometimes prostrate; rarely plants are seen with several fls per stem. The petals are 6–7 mm long, rose-purple, or less frequently white with rose veins. The sepals are toothed, the teeth generally gland-tipped.

Rare in the Park, cited by Jones in Paradise Park ("above Camp of the Clouds, 6300 ft," according to J. B. Flett) and at Panorama Point. A collection in the Park herbarium was made at 6000 ft in upper Berkeley Park.

Lewisia triphylla (S. Watson) B. L. Rob.
THREE-LEAF LEWISIA

A tuberous perennial with a threadlike stem 2–10 cm tall, bearing 2 or 3 whorled, linear lvs 1–5 cm long. A slender basal lf may be present in younger plants. One to several fls are borne on a short stalk above the stem lvs in an open cluster. The petals are white and about 4 mm long. The sepals are not toothed.

Uncommon and rather easily overlooked, on seasonally moist, stony subalpine slopes. F. W. Warren collected the plant on the lower slope of Pyramid Peak and O. D. Allen found it on "wet ledges" on Mount Wow.

Montia L.—MONTIA

Montia is separated from *Claytonia* by having more than 2 leaves on the stems. In other respects, plants of the two genera are very similar. The petals are often of unequal lengths.

1   Stem lvs opposite .... *M. fontana* ssp. *fontana*
1   Stem lvs alternate 2

2(1) Lvs linear, without a definite stalk or blade .... *M. linearis*
2   Lvs with a stalk and a blade, the blade more or less ovate 3

3(2) Basal lvs 10–20 mm wide; petals 10–15 mm long ....
.... *M. parvifolia* ssp. *flagellaris*
3   Basal lvs 5 mm wide or less; petals 7–12 mm long ....
.... *M. parvifolia* ssp. *parvifolia*

Montia fontana L. ssp. fontana
WATER-CHICKWEED

A sprawling annual, with stems reaching 30 cm long. The lvs are opposite, 5–20 mm long, narrowly oblanceolate, and not stalked. Flowers are borne in small axillary and terminal clusters, with 1 or 2 bracts below the lowest fls. The petals are white and 1–2 mm long; there are 3 stamens.

Known from a collection made in the 1930s by W. J. Eyerdam, on a stream bank below Chinook Pass, and not previously listed for the Park. Kartesz does not recognize varieties for *M. fontana*, but in Hitchcock and Cronquist (1973) the Eyerdam collection keys to var. *lamprosperma* (Cham.) Fenzl: the shiny black seeds are 1 mm long, with tiny, rounded bumps on the surface.

Montia linearis (Douglas ex Hook.) Greene
NARROW-LEAVED MONTIA

An erect, branched annual up to 25 cm tall. The lvs are linear and alternate, 1–10 cm long. Several fls are carried in open, 1-sided, terminal racemes, with a bract below the lowest fl. The petals are white and 4–6 mm long. There are 3 stamens.

Uncommon annual, growing on wet open ground below about 4000 ft. The species is more commonly found east of the Cascade Mountains.

Montia parvifolia (Moc. ex DC.) Greene ssp. flagellaris (Bong.) Ferris
"LARGER" LITTLE-LEAF MONTIA

Differing from ssp. *parvifolia* in the wider basal lvs, 5–10 mm wide, and larger fls borne in a definite raceme; the petals are 10–15 mm long and 5 stamens are present.

Evidently rare in the Park and previously not reported. Collected by Irene Creso in the early 1970s on the Carbon R near Fall Cr. This subspecies is common near the coast and in the Puget lowlands.

Montia parvifolia (Moc. ex DC.) Greene ssp. parvifolia
LITTLE-LEAF MONTIA

A perennial, often forming a mat as the rootstock branches as well as spreading by runners. The thick basal lvs are clustered and oblanceolate to narrowly spoon-shaped, 2–6 cm long and less than 5 mm wide; the stem lvs are ovate, thinner, alternate, and 5–10 mm long. The inflorescence is bracted and umbel-like or short raceme. The fls are pinkish, with 5 stamens. The petals are 7–12 mm long, of equal length.

Common, widespread, and occasionally abundant, on mossy rocks, cliffs, and banks, at streamsides, and sometimes on shaded roadsides, up to about 5000 ft. Found at Sunshine Point and around the Nisqually entrance; at Mowich Lk; along Westside Rd; at Longmire; in the Butter Cr Research Natural Area; through Stevens Canyon, at Box Canyon and on the east side of the Cowlitz Divide; around Ohanapecosh; along the road below Tipsoo Lk and on Naches Peak; and in riparian woods along the lower White R.

Primulaceae Vent.—PRIMROSE FAMILY

Perennial herbs with very attractive flowers; the three species found in the Park vary greatly in appearance. The flowers have 5 petals (although *Trientalis* usually has 6 or 7), and a superior ovary with 1 pistil and style which develops into a dry capsule. What unites the species as a family and separates it from other dicots with united petals is the fact that the stamens are placed opposite the petals.

1   Lvs basal; petals turned back .... *Dodecatheon*
1   Stems leafy; petals spreading 2

2(1) Lvs crowded at the top of a herbaceous stem; stamens exserted from the corolla .... *Trientalis*
2   Lvs arranged along the perennial stems; stamens hidden within the corolla .... *Douglasia*

Dodecatheon L.—SHOOTING STAR

Dodecatheon jeffreyi Van Houtte ssp. *jeffreyi*
JEFFREY'S SHOOTING STAR

An herb from a short rootstock and usually growing in clumps. The lvs are entirely basal, 15–40 cm long, and oblanceolate, narrowed to the base and sometimes lacking a stalk. The fls are in an umbel on a leafless stem 15–50 cm tall. The 5 (or rarely 4) petals are rose-purple and sharply reflexed, turned back away from the stamens and style. The lower portion of the petals are united into a short tube, which is white or yellow with a reddish ring at the throat. The 5 (or rarely 4) dark purple stamens are pressed together, surrounding the threadlike style, creating a spirelike structure that seems to spring from the downward-pointing fl. The capsule opens by teeth at the top.

A very common and conspicuous plant of wet subalpine meadows, along small streams, and even at roadsides; best seen above 5000 ft, but occasionally found as low as 4000 ft. Found at Mountain Meadows; at Mowich Lk and Spray Park (where white-flowered plants have been reported); at Indian Henrys Hunting Ground; throughout Paradise Park; in the meadow above Bench Lk; on the Olallie Cr trail at 4000 ft; at Moraine Park at 5800 ft; at upper Palisades Lk; and at Berkeley Park.

Douglasia Lindl.—DOUGLASIA

Douglasia laevigata A. Gray var. *ciliolata* Constance
CLIFF-PRIMROSE

A tufted perennial, much-branched, with the lvs tending
to be clustered toward the ends of the branches. The lvs
are spoon-shaped to oblanceolate, fringed with short
hairs on the margins, and 6–12 mm long. A bracted
umbel of pink fls is borne on a leafless hairy stem 2–8
cm high. The 5 free lobes of the corolla spread at right
angles from the narrow tube and the fl about 10 mm broad. Decorating
the throat at the top of the tube are 5 small crests.

Uncommon, on rocky slopes and cliffs above 6000 ft. Known from
ridges surrounding Paradise Park; the lower slope of Iron Mtn; and on
Mount Wow. Reported from the trail between Sunrise and Berkeley
Park.

Trientalis L.—STARFLOWER

Perennials from small tuberous roots. The stem may bear a few small
leaves, but the large leaves are whorled at the top of the stem. One or a
few flowers are carried on threadlike stalks that rise from the axils of the
uppermost leaves. The petals are united for a very short distance, with
5–7 lobes, and the fls are more or less flat and starlike. The fruit is a
small, round capsule.

1   Lvs whorled at the top of the stems; fls pink .... *T. borealis* ssp. *latifolia*
1   Lvs scattered on the stems below those whorled at the top; fls white ......
.... *T. europaea* ssp. *arctica*

Trientalis borealis Raf. ssp. *latifolia* (Hook.) Hultén
STARFLOWER

Often in small colonies, with slender stems 10–30 cm tall. There
are 4–7 major lvs at the top of the stem, which are broadly
lanceolate to oblanceolate, untoothed, and 3–8 cm long. The fl is
10–15 mm wide and light to dark pink.

Very common in open forests below 4500 ft, in all of the Park's
river valleys.

Trientalis europaea L. ssp. *arctica* (Fisch. ex Hook.) Hultén—NORTHERN STARFLOWER

Plants 5–15 cm tall, with mostly oblanceolate lvs 1–4
cm long that are scattered on the stem as well as
whorled at the top. The white fls are 12–16 mm wide,
generally with 7 lobes.

Infrequent on boggy ground and at the edges of
meadows, 2500-4000 ft. Known from Mtn Meadows;
most easily seen at Longmire.

Ranunculaceae Juss.—BUTTERCUP FAMILY

A large family in the Park. All are herbs with alternate leaves and often with colorful petals and sepals. In larkspur, monkshood, and columbine the flowers are irregular and showy; in many genera petals are absent. Sepals and petals, when present, number 3–15, and there are numerous stamens. There are 2 to many pistils (1 in *Actaea*), that develop into achenes, follicles, or berries.

1   Fls irregular, the upper sepal spurred or hoodlike; fls blue or purplish 2
1   Fls regular, the sepals all alike; fls yellow, white, or red, or petals absent 3

2(1) Upper sepal spurred; lvs deeply divided .... *Delphinium*
2   Upper sepal hoodlike; lvs lobed and toothed, but not deeply divided ......
      .... *Aconitum*

3(1) Fls red and yellow, spurred, nodding on the stems .... *Aquilegia*
3   Fls not colored red and yellow, erect and lacking spurs 4

4(3) Pistil 1, developing into a berrylike fruit; tall, coarse plants with compound lvs .... *Actaea*
4   Pistils 2 to several; fruit developing into an achene or follicle 5

5(4) Fruit a 1-seeded achene 6
5   Fruit a several-seeded follicle 9

6(5) Petals and sepals present, the petals yellow (if white, then the plant growing in water) .... *Ranunculus*
6   Petals absent; sepals present, green or colored, but not yellow 7

7(6) Sepals colored, white to bluish, persistent; lvs whorled at the top of low stems (or both basal and whorled) .... *Anemone*
7   Sepals greenish, soon falling; lvs various 8

8(7) Lvs alternate and compound on tall stems; inflorescence a raceme; stamens purplish .... *Thalictrum*
8   Lvs simple, palmately lobed, chiefly basal; inflorescence a flattish cluster; stamens greenish .... *Trautvetteria*

9(5) Lvs simple and toothed or scalloped .... *Caltha*
9   Lvs deeply divided, the lobes with rounded teeth .... *Trollius*

Aconitum L.—MONKSHOOD

Aconitum columbianum Nutt. var. *columbianum*—COLUMBIAN MONKSHOOD

A tall perennial, to 50–100 cm, with dark green, toothed, palmately lobed lvs 5–15 cm broad. The inflorescence is a tall, open raceme, with upcurved stalks. The dark blue to purple fls are irregular, 1.5–2 cm long, and helmetlike: the uppermost of the 5 sepals is the largest and arches over the lateral, enfolding sepals. Beneath the "hood" are 2 small, spurred petals, the stamens, and the 3–5 pistils. 3 additional, rudimentary petals may be present. Each pistil develops into a follicle 1–2 cm long.

Rare in the Park, and most often found east of the Cascades, but known from a few places on the east side of the Park: around Ohanapecosh; at 3800 ft on the trail to Three Lks; at an unspecified elevation on Chinook Cr; on the trail to Owyhigh Lks; and along the road below Cayuse Pass at 3500 ft.

Actaea L.—BANEBERRY

Actaea rubra (Aiton) Willd. ssp. *arguta* (Nutt.) Hultén—RED BANEBERRY

A perennial, typically with several branched stems, each 30–90 cm tall. The lvs are on long stalks, 20–70 cm long overall, ovate in outline, and divided into 3 major lobes, each lobe pinnately divided into 3–5 smaller, coarsely toothed leaflets. The inflorescence is a dense terminal raceme 2–3 cm long. The white fls are about 5 mm broad, with 4–10 very small petals; the numerous stamens are also white. The fruit is a bright red berry, 5–10 mm broad; white fruits are occasionally seen.

Frequent in openings in moist forests, to about 4000 ft. Found at Mowich Lk; on the roadside above Kautz Cr; at Longmire; along the lower part of the trail to Bench Lk; at Box Canyon; at Ohanapecosh; along the trail to the Grove of the Patriarchs; and at 2500 ft on Chinook Cr.

Anemone L.—WINDFLOWER

Although they lack petals, these species are nevertheless showy
wildflowers—the sepals are petal-like and white or brightly colored. The
perennial plants generally produce 1 stem, with basal leaves and
whorled stem leaves that form an involucre beneath the flower. These
species are usually 1-flowered and each flower has 5 sepals and
numerous stamens. The pistils are numerous and clustered in heads; the
fruit is a more or less hairy achene. *Pulsatilla occidentalis* has been
segregated from *Anemone,* by the nature of its style, but the *Flora of North
America* does not follow this.

1   Stem lvs simple, toothed but not divided; plants of lowland forests ....
      .... *A. deltoidea*
1   Stem lvs compound or very deeply divided; habitats various 2

2(1) Achenes short-hairy but not wooly; involucral lvs divided into three broad l
      obes; plant of lowland forests .... *A. lyallii*
2   Achenes wooly; lvs divided into many narrow segments; plants found near
      and above timberline 3

3(2) Styles featherlike, more than 20 mm long .... *A. occidentalis*
3   Styles not featherlike from long hairs, less than 5 mm long 4

4(3) Lvs nearly smooth; styles 2–4 mm long .. *A. drummondii* ssp. *drummondii*
4   Lvs long-hairy; styles less than 2 mm long .... *A. multifida* var. *saxicola*

Anemone deltoidea Hook.—COLUMBIAN WINDFLOWER, THREE-LEAVED ANEMONE

A delicate plant, from a slender, creeping rootstock, with
mostly hairless stems 10–25 cm tall. There is 1 3-parted basal
lf, with ovate, toothed leaflets, shorter than the flowering
stem; the 3 stem lvs are stalkless, in a whorl, toothed but
undivided, and 3–8 cm long. The fl is on a slender stalk,
with 5 white sepals, each 1–2 cm long. The
achenes are few in number, egg-shaped, and
hairy on the lower half.

Quite common in low elevation forests, to
about 3000 ft, mostly on the west side of the
Park, as at the Nisqually entrance, along
Westside Rd, and at Longmire.

Anemone drummondii S. Watson ssp. *drummondii*—DRUMMOND'S ANEMONE

A slender, softly hairy plant from a branched, woody
rootstock, reaching from 5 to 25 cm tall. There are several
basal lvs, 5–15 cm long overall, much divided into
relatively long, slender segments. The stem lvs are short-
stalked and also much divided. The 5 sepals are white,
tinged bluish on the outside, and 8–12 mm long. The
achene is roundish and wooly.

Uncommon, growing in crevices and on rocky slopes, 5000-7000 ft. Collections have been made at Spray Park, Frozen Lk, and on the Burroughs Mtn trail.

Anemone lyallii Britton—LYALL'S ANEMONE

Similar in stature and appearance to *A. deltoidea*, but the stem lvs have 3–5 leaflets that are on short stalks. The rootstock is short and the plant doesn't form the sort of colony typical of *A. deltoidea*. The sepals are whitish to bluish white, and 5–18 mm long. The achene is elliptical and hairy throughout.

Rare in the Park: this is chiefly a plant of forests close to the coast. [Inexplicably, it is said by St. John and Warren (1937) to be common in the Park.] Found in the Ohanapecosh R valley, from the campground area to Deer Cr along Hwy 123.

Anemone multifida Poir. var. *saxicola* B. Boivin
GLOBE ANEMONE

A sturdy anemone, with stems 15–40 cm tall from a woody rootstock. The stem and lvs are silky-hairy, and both the long-stalked basal and stalkless stem lvs are finely 2–3 times divided into narrowly lanceolate segments. There is generally 1 fl per stem, but stems with 2 or 3 fls are occasionally seen. The upper side of the petals is whitish to cream color and the underside usually is bluish. The achene is roundish and densely clothed with long hairs.

The rarest subalpine anemone in the Park, growing on wet, stony soil mostly on the east side of the Park. Found around Cayuse Pass. Collected in 1916 by J. B. Tarleton at an unspecified location and collected by O. D. Allen on Mount Wow at 6500 ft.

The treatment by Brian Dutton and others, in volume 2 of the *Flora of North America*, is followed here; they recognize fewer varieties than others have. Not many collections have been made in the Park and it is not clear that these collections are even representative. The National Park Service flora lists both var. *multifida* and var. *hirsuta*. The placement of the Park plants is complicated by the fact that, as Dutton notes, "Early-season plants of *A. multifida* var. *multifida* have solitary fls and will key to var. *saxicola*. *A. multifida* var. *tetonensis* and especially var. *saxicola* might be based on characteristics that are influenced primarily by environment; further study is warranted." At the same time, Kartesz considers that the "presence [of var. *hirsuta*] in the North American flora is not certain."

Anemone occidentalis S. Watson
WESTERN ANEMONE, WESTERN
PASQUE-FLOWER

Distinguished from other anemones by a
feature of the styles: when the achenes
are mature, these are very long,
featherlike, and drooping: "mop top" and
"mouse on a stick" are other common
names for the plant. The entire plant is
clothed with long, shining hairs. The lvs are
basal, from a thick, woody rootstock, deeply
divided 3 or 4 times into linear segments, and 10–
20 cm long overall. The flowering stem is 10–30 cm
tall (doubling in length as the seed-head matures). 2–
3 small, stalkless leaflike bracts are in a whorl beneath
the single fl on each stem. There are no petals, but the 6–
7 sepals are 2–3 cm long, white, and surround a center
dense with yellow stamens; the sepals may be bluish on the
outside. The achenes are in a rounded or elongated head.

Common and in places quite abundant; growing in alpine
and subalpine meadows, 4500-6500 ft. Collected at Indian Henrys
Hunting Ground; throughout Paradise Park; on Mazama Ridge; in Van
Horne Canyon at 4500 ft; at 5500 ft on Kotsuck Cr trail; between Tipsoo
Lk and Chinook Pass; around Sunrise and on the Burroughs Mtn trail.
Plants at Sunrise are notable for the sparsely short-hairy stems.

Aquilegia L.—COLUMBINE

Aquilegia formosa Fisch. ex DC. var. *formosa*
WESTERN COLUMBINE

One of the Park's most striking and unusual fls. Several
red and yellow blossoms nod in an open raceme on
branched stems that may reach 90 cm in favorable
locations. 5 red sepals spread away from the
tubular petals and are 10–20 mm long; each of the
5 petals consists of a downward-projecting
yellow blade beneath a long red spur, overall
15–20 mm long. The spreading sepals and
vertically oriented petals give the fl a "blocky"
appearance. Each petal spur is tipped with a
small sack of nectar, which is sought out by
hummingbirds. Numerous yellow stamens and
styles project downward from the "face" of the
fl. The lvs are mostly basal, bluish green, and
divided into 3 main lobes, each of which is again
3-times divided, overall to 30 cm long. After the
petals and sepals drop, the stalk straightens,
holding erect the 5 slender follicles.

Quite common, throughout the Park, chiefly above 2500 ft, along streams, on wet banks, and at the edges of meadows. Found at Mowich Lk; on Westside Rd above 2600 ft; along the trail between Klapatche and St. Andrews Parks; along the road between the Nisqually R and Ricksecker Point; on the lower slopes of Paradise Park; in the Butter Cr Research Natural Area; in Stevens Canyon; along the road on the east of Cowlitz Divide; between 3000 and 4000 ft along Hwy 123; at Tipsoo Lk and on the west slope of Naches Peak at 5500 ft; at Fish Cr; and at Eunice Lk.

Caltha L.—MARSH-MARIGOLD

Both subspecies of this hairless perennial have somewhat succulent, dark green leaves and stems. The broadly toothed leaves are basal (with a small leaf usually found on the flowering stem) and vary in shape between the subspecies. Petals are absent, but the 5–10 sepals are large and showy, white, and sometimes tinged blue on the backside. There are 1 or 2 flowers per stem, each 2–4 cm broad. The numerous stamens give the flower a bright yellow center. The fruit is a short follicle, set in small clusters.

1   Lvs round, nearly as broad as long .... *C. leptosepala* ssp. *howellii*
1   Lvs oval, longer than broad ....
.... *C. leptosepala* ssp. *leptosepala* var. *leptosepala*

Caltha leptosepala DC. ssp. *howellii* (Greene) Smith
WHITE MARSH-MARIGOLD

More apt than the next subspecies to have 2 fls per stem. The lvs are distinctly round in outline, 5–10 cm long and wide, with a heart-shaped base where the stalk joins the blade. Each follicle is borne on a very short stem in the cluster.

The less common subspecies in the Park, growing in wet places in meadows and bogs, above about 4000 ft. Found at Mtn Meadows and Spray Park; at Indian Henrys Hunting Ground; along Hwy 410 at about 4000 ft; and on Kotsuck Cr trail at 4500 ft. Also seen below Narada Falls, perhaps as a waif, occurring well below its usual elevation.

Caltha leptosepala DC. ssp. *leptosepala*
var. *leptosepala*
ELKSLIP MARSH-MARIGOLD

This subspecies almost always has 1 fl per stem and the lvs are elongated, more or less oval in outline, up to about 6 cm long, heart-shaped or even arrowhead-shaped at the base. The follicles are stalkless, in a cluster.

281

More common and typically at higher elevations than the above subspecies. Frequently seen on wet ground close to melting snow and along small streams throughout the Park; this is the common marsh-marigold of the Paradise and Sunrise meadows.

Delphinium L.—LARKSPUR

As intricate, in their own way, as columbine and monkshood. In larkspurs, the 5 blue to blue-purple sepals are enlarged and spreading; the uppermost sepal extends backwards as a narrow spur. The 4 petals are much smaller than the sepals and are whitish or light blue, forming an eye at the throat of the flower. The leaves are palmately divided and the inflorescence is a terminal raceme. The seeds develop in 1–5 follicles.

1   Plants large, 1–2 m tall; fls longer than the fl stalks; sepals blue-purple ... . *D. glaucum*
1   Plants less than 1 m tall, often shorter than 0.5 m; fl stalks longer than the fls; sepals bright blue .... *D. glareosum* ssp. *glareosum*

Delphinium glareosum Greene ssp. *glareosum* ROCKSLIDE LARKSPUR

The vivid blue of the fls makes it easy to spot this species at a distance. The stems are 10–80 cm tall, short-hairy and sticky towards the top, with the fls on long stalks in an open, branched raceme. The lvs are rounded in outline and up to about 4 cm across, divided into 3 major lobes, these further divided into narrow segments. The sepals are 15–20 mm long and spread widely; the petals are white. The slender follicles are 1.5–2 cm long.

Widespread in the Park and fairly common along stream banks and in meadows, 2500-5500 ft. Known from Lk Allen; at Longmire; at Box Canyon; at 5500 ft on the west slope of Naches Peak; at Sunrise; and along Ipsut Cr between the falls and the Pass. C. V. Piper found it at "Crater Lk," an early name for Mowich Lk.

Delphinium glaucum S. Watson PALE-LEAVED LARKSPUR

A plant with tall and slender stems, 1–2 m tall, mostly unbranched, glaucous and nearly hairless. The lvs have 5–7 irregularly toothed lobes and the basal lvs may be 20 cm wide. The inflorescence is very tall and narrow, the fls numerous and light to dark blue-purple, the sepals 10–12 mm long and not too widely spreading. The follicles are short and stout.

Uncommon, growing on moist, rocky slopes above about 6000 ft. Collected near the summit of Mount Wow; at 3750 ft on the South Puyallup R; around Tipsoo Lk; and at the lower falls of Ipsut Cr and at Ipsut Pass. Also reported from Mowich Lk.

Ranunculus L.—BUTTERCUP

An often-seen group of mostly yellow wildflowers, with an affinity for damp places. Petals and sepals number 5, with numerous yellow stamens and several pistils, which are arranged in small heads and which develop into achenes. Most are perennials, with basal leaves and tall, branched, leafy or bracted stems.

1   Plants aquatic; lvs dissected into threadlike segments; fls white .... *R. aquatilis*
1   Plants never truly aquatic; lvs various but not threadlike; fls yellow 2

2(1) Lvs broadly lanceolate, not toothed or divided; plants of wet ground at the shores of subalpine lakes .... *R. alismifolius* var. *alismifolius*
2   Lvs divided and usually toothed; habitat various 3

3(2) Lvs and stems hairless; lvs chiefly divided into three rounded, toothed lobes 4
3   Lvs and stems more or less hairy; lvs variously divided, the lobes sharp-tipped 5

4(3) Petals 8–12 mm long, dark yellow to orange-yellow .... *R. suksdorfii*
4   Petals 4–5 mm long, bright yellow .... *R. verecundus*

5(3) Petals 2–3 mm long, shorter than the sepals; achenes hairy .... *R. uncinatus* var. *parviflorus*
5   Petals 4–15 mm long, longer than the sepals; achenes smooth 6

6(5) Stems creeping, rooting at the lower lf nodes .... *R. repens* var. *repens*
6   Stems erect, not rooting at the lf nodes .... *R. acris* var. *acris*

Ranunculus acris L. var. *acris*—TALL BUTTERCUP
A perennial with tufted basal lvs and tall, leafy stems, lacking runners, and roughly hairy on the stems and lvs. The lower lvs are pentagonal in outline, deeply cut into sharp-pointed segments, and 10–15 cm long overall. The stems may reach 100 cm tall, each branch with several fls. The petals are 7–15 mm long, and the achene is 2–2.5 mm long, with a hooked beak less than 0.5 mm long.

An occasional weed along roads in open forests as on Westside Rd and near the Comet Falls trailhead. Also reported from Green Lk.

Ranunculus alismifolius Geyer ex Benth. var. *alismifolius*—WATER-PLANTAIN BUTTERCUP

A tufted, mostly hairless plant. The basal lvs are not toothed or minutely toothed, lanceolate, tapered to the stalk, and 5–8 cm long. The flowering stems are 15–20 cm tall; they may be single-flowered and leafless, or may have 2 or 3 fls on stalks above a pair of small lvs. The yellow petals are 8–10 mm long. The achene is 1.5–2.5 mm long, with a straight beak 0.5–1 mm long.

Rare in the Park, on wet ground around lakes and ponds and in wet meadows. Known from above 5000 ft, as at upper Palisades Lk. Reported from Clover Lk by Warren Tanaka. First reported for the Park by Marcia Hamann in her 1972 study.

Ranunculus aquatilis L.
WHITE WATER-BUTTERCUP

Rooted in mud in shallow water, with slender stems to 50 cm long. The submerged lvs are 1–2 cm long, round in outline and cut into numerous, threadlike segments. Floating lvs are sometimes present, and these are roundish to kidney-shaped in outline with 3 broad, toothed lobes. The fls are held above the water and have white petals 5–8 mm long, each with a yellow spot at the base. The achene 1–2 mm long, thick and wrinkled on the surface, with a straight beak less than 0.5 mm long.

Fairly common in shallow ponds and slow-moving streams, below about 6000 ft. Found in the pond at the foot of Tumtum Peak; at Lk George; at Longmire; and at Reflection Lks.

Kartesz recognizes no varieties. Most plants in the Park lack floating lvs and could be called var. *capillaceus* (Thuill.) DC. Some do have lobed (but not dissected) floating lvs and could be called var. *hispidulus* Drew.

An interesting filamentous alga, *Nitella gracilis* (Sm.) J. G. Agardh., or brittlewort, can be found forming mats in shallow ponds at low elevations. It could confused with the slender submerged lvs of *R. aquatilis*. The plant consists of a threadlike stem with regularly placed whorls of 6–8 slender branchlets, each of which is about 20 mm long. Male and female reproductive structures are borne in tiny reddish bodies near the bases of the branchlets; these become dark and more visible as they age.

Ranunculus repens L. var. *repens*
CREEPING BUTTERCUP

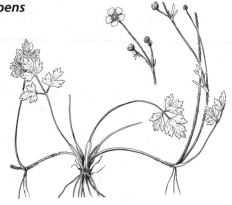

A weedy perennial, spreading
aggressively on long runners, with
short-hairy stems and lvs. The lvs are
triangular in outline, 2–8 cm long, and
deeply cut into 3 major lobes, these
also toothed; the stem lvs are smaller.
The lvs are usually marked with
irregular, lighter patches. The fls are
several on erect stems, with dark
yellow petals 10–13 mm long. The
achene is 2–3 mm long, with a curved
beak about 1 mm long.

Not uncommon, as a weed along roads and trails below about 3000 ft,
especially near the Park entrances. Known from Sunshine Point; along
Westside Rd; along the road on the east side of the Cowlitz Divide; at
Ohanapecosh, including the hot springs area; along the Carbon R near
Alice Falls; and at the Ipsut Cr campground. Some plants along trail to
the Grove of the Patriarchs have fls with 10 petals.

Ranunculus suksdorfii A. Gray
SUBALPINE BUTTERCUP, SPRUCE-FIR BUTTERCUP

Growing as low tufts, sometimes forming small clumps, with
flowering stems up to 25 cm tall, but often just 5–10 cm. The
plants are hairless, and the basal lvs are divided into 3 major
lobes, the lobes deeply toothed. Each flowering stem bears
1–3 dark yellow or yellow-orange fls. The petals are 8–12
mm long, and the achenes are 1.5–2 mm long with a
straight beak 0.5–1 mm long, in a dense and elongated
rather than spherical head.

Common and frequently abundant in subalpine
meadows and along open streams, 4500-7000 ft. An
early bloomer, often flowering at the snow's edge.
Found at Indian Henrys Hunting Ground; through
Paradise Park; at Goat Pass on the Kotsuck Cr trail;
at Chinook Pass; at Sunrise and on the Burroughs
Mtn trail; and at Berkeley Park.

Ranunculus uncinatus D. Don var.
parviflorus (Torr.) L.O. Benson
WOODLAND BUTTERCUP

A tall native perennial, somewhat similar to *R. acris*,
but with much shorter petals. The plant reaches 60
cm tall, with branching stems, and is lightly to
moderately short-hairy. The lvs are roundish to
broadly ovate, deeply 3–5-lobed. The lobes are
toothed, about 10 cm long including the long and

slender stalk. The lvs are sometimes marked with purple. The petals are pale yellow and 2–3 mm long, and shorter than the sepals. The achene is 2–2.5 mm long, hairy on the upper part, and has a stout, curved beak 1–2 mm long.

Locally common in the area of the Nisqually entrance and on lower Westside Rd, in open woods, reaching 4000 ft; also at Ohanapecosh.

See the list of "Doubtful and Excluded Species," p. 469, for a discussion of a similar species, *Ranunculus occidentalis*.

Ranunculus verecundus B.L. Rob. ex Piper—MODEST BUTTERCUP

A hairless perennial (the sepals are pubescent but quickly fall from the fl), 10–20 cm tall, with a few basal lvs and usually 1 flowering stem. The basal lvs are roundish, with a heart-shaped base, with 3 major lobes and rather regular teeth. 1–3 fls are borne on the bracted stem; the petals are 4–5 mm long, yellow, and somewhat longer than the sepals. The achenes are 1–5 mm long, with a short, hooked beak.

Uncommon, on rocky ridges and talus slopes, on the east side of the Park. Collected at Interglacier at 7500 ft.

Thalictrum L.—MEADOWRUE

Thalictrum occidentale Gray var. *occidentale* WESTERN MEADOWRUE

Until it fls, the plant looks very much like a columbine, with bluish green, mostly 3 times divided lvs, the segments 1–2 cm long with rounded teeth or lobes. The stems rise from a stout rootstock and reach 30–90 cm tall. The inflorescence is an open raceme and male and female fls are borne on separate plants; petals are absent. In the male fls, the stamens hang down and have large purplish anthers; in the female fls short, pale green sepals surround the numerous styles. The fruit is a rounded achene with prominent veins on the sides.

Widely scattered around the Mountain, 2500-5000 ft, in forest openings and at high elevation roadsides. Collected at Klapatche Park; below Van Trump Glacier; on Chinook Cr at 2500 ft; on Kotsuck Cr trail at 5000 ft; and along the highway below Tipsoo Lk at 5100 ft.

Trautvetteria Fisch. & C. A. Mey.—TASSEL-RUE

Trautvetteria caroliniensis (Walter) Vail var. *occidentalis* (Gray) C. L. Hitchc.—TASSEL-RUE

A perennial, generally hairless plant, with stems 50–90 cm tall, growing from spreading rootstocks. The lvs are kidney-shaped, on long stalks, deeply lobed and toothed on the margins, up to 10 or 12 cm broad; the few stem lvs are smaller and may lack stalks. The fls are numerous in a flat-topped cluster. Petals are absent and the 4–8 sepals soon fall, leaving a ball of stamens and styles. The fruit is a rounded achene.

Common along streams and in moist forests, up to about 5000 ft, chiefly on the west and south sides of the Park. Found on upper Westside Rd; at Mowich Lk and Spray Park; in lower Paradise V in the Butter Cr Research Natural Area; and on Chinook Cr at 2500 ft. The species, also called False Bugbane, is more often seen in the mountains north of Mount Rainier and in eastern Washington. It also grows in the Olympic Mountains.

Trollius L.—GLOBEFLOWER

Trollius laxus Salisb. ssp. *albiflorus* (Gray) A. Löve, D. Löve & S.L. Kapoor—AMERICAN GLOBEFLOWER

A very attractive plant, with the appearance of a buttercup, but with white fls and follicles rather than achenes. The plants are hairless, with mostly basal lvs and flowering stems 10–30 cm tall. The basal lvs are rounded in outline, with 5 broad lobes, which are bluntly toothed, overall 5–8 cm broad; the stalks of the basal lvs are slender, with membranelike sheaths at the base. 1 fl is borne on each leafy stem; the 5 white petals are 1–2 cm long and surround numerous yellow stamens and greenish styles.

Uncommon and first discovered in the Park by Marcia Hamann in 1972 at Huckleberry Basin and Hidden Lk, around 5400 ft, in the northeast corner of the Park. It has been reported at Clover Lk by Warren Tanaka.

Rhamnaceae Juss.—BUCKTHORN FAMILY

Shrubs or small trees, with small flowers in dense clusters and alternate, simple, toothed leaves. In our species, there are 5 sepals, petals and stamens; the ovary is partly inferior—that is, partly covered by the disk upon which the fl parts are borne.

1   Fls showy, white, in large clusters; fruit a dry capsule .... *Ceanothus*
1   Fls inconspicuous, greenish, in small clusters; fruit berrylike, black ....
     .... *Frangula*

Ceanothus L.—BUCKBRUSH, CEANOTHUS

Two very attractive, densely branched shrubs, with white, sweet-scented flowers in dense, narrowly pyramidal panicles at the ends of the side branches. In both, the leaves are glossy green and palmately 3-veined while the flowers are just 3–5 mm broad. The fruit is a rounded, dry, 3-chambered capsule.

1   Lvs evergreen, sticky and sweet-smelling .... *C. velutinus* var. *hookeri*
1   Lvs deciduous, neither sticky nor sweet-smelling .... *C. sanguineus*

Ceanothus sanguineus Pursh
REDSTEM CEANOTHUS, OREGON TEA TREE

A spreading shrub, 1–3 m tall, with mostly oval, coarsely toothed lvs that are 3–8 cm long. "Redstem" refers to the reddish color of the short branches that carry the 6–10 cm long fl clusters.

Uncommon, scattered in dry, open woods, to about 3000 ft in the Nisqually V, and at the south end of the road across the Cowlitz Divide.

Ceanothus velutinus Douglas ex Hook.
var. *hookeri* M. C. Johnst.
SNOWBRUSH, TOBACCO BRUSH

A shrub reaching 2 m tall, easily distinguished by the sweet and resinous smell of the lvs, especially on a warm day (hence one of the common names). The lvs are 6–8 mm long, oval, with fine glandular teeth on the margins; the upper surface is dark green and sticky to gummy. The fls are white to creamy white, in clusters 5–10 cm long.

Locally common on dry, rocky slopes and in openings in forests, to about 4000 ft, on the east side of the Park. Found in lower Stevens Canyon; at the Ohanapecosh entrance; at 3000 ft on the Shriner burn; and in the White R Valley.

Frangula P. Mill.—BUCKTHORN

Frangula purshiana (DC.) Cooper—CASCARA

In the Park, growing as a small tree, 3–12 m tall, with smooth gray bark. The lvs are oval to slightly obovate, finely toothed and glossy green on the upper side and short-hairy beneath, 5–15 cm long. The lvs are also prominently veined, in a "washboard" fashion. The greenish fls are about 3 mm long, in small umbels in the axils of the lvs. The seed is a blackish, berrylike drupe.

Rarely seen in the Park, although said by G. N. Jones to be common on moist ground below about 3000 ft, typically growing with red alder. One collection was located: made by O. D. Allen in the "upper valley of the Nisqually."

The Jepson Manual (Hickman, 1993) divides California species of *Rhamnus* into two subgenera. Subgenus *Rhamnus* has no petals, an exserted style, and scales covering the terminal bud. Subgenus *Frangula* has petals, an included style, and lacks scales on the terminal bud. Kartesz elevates these to full genera, and that use is followed here.

Rosaceae Juss.—ROSE FAMILY

A family of vast diversity of form and habit, including small lowland trees and creeping alpine perennials; only a couple of weedy potentillas are annuals. All have alternate, often pinnately divided leaves and most have prominent stipules. There are typically 5 or 10 sepals that are partly united in most species, forming bowl-like structure called a hypanthium. The hypanthium may partly or wholly surround the ovary and upon which the petals and stamens are borne. There are typically as many petals as sepals (petals are absent in *Sanguisorba*). The fls are often showy, in shades of white, yellow, or pink. Each fl has one or more pistils and numerous stamens. Many types of fruits are seen.

1	Shrubs or trees	2
1	Herbaceous plants, annuals or perennials (or slightly woody at the base of the stems)	14
2(1)	Ovary inferior; fruit an applelike pome	3
2	Ovary superior; fruit a drupe, follicle, hip, or achene	5
3(2)	Lvs pinnate; fls numerous in flat-topped clusters	*Sorbus*
3	Lvs simple or sometimes shallowly lobed; fls not in flat-topped clusters	4
4(3)	Fls in racemes; fruits blue-black	*Amelanchier*
4	Fls in small umbels; fruits greenish to reddish	*Malus*

5(2)   Lvs pinnately compound; fls showy, pink to deep rose; fruit of woody achenes enclosed in a hip (an expanded hypanthium .... *Rosa*

5   Lvs simple or compound (if pinnately compound, then the fls not pink); fruit various 6

6(5)   Lvs simple, not toothed to toothed 7

6   Lvs compound 12

7(6)   Lvs not toothed; fls in racemes, male and female fls on separate plants ...
       .... *Oemleria*

7   Lvs serrate to toothed or lobed; fls always perfect, with stamens and pistils in each fl 8

8(7)   Lvs palmately lobed 9

8   Lvs simple or pinnately lobed 10

9(8)   Fruit an aggregate of fleshy drupelets (raspberries and blackberries) ....
       .... *Rubus*

9   Fruit a dry, several-seeded follicle .... *Physocarpus*

10(8) Lvs pinnately lobed and toothed; fruit a dry, 1-seeded achene ....
       .... *Holodiscus*

10   Lvs simple, serrate; fruit a drupe or follicle 11

11(10) Pistil 1; fruit a fleshy drupe .... *Prunus*

11   Pistils 5; fruit a dry follicle .... *Spiraea*

12(6) Erect shrub; fls yellow .... *Pentaphylloides*

12   Plants prostrate or matlike, with trailing or creeping stems; fls white ..
       13

13(12) Petals 8; pistils many; fruit an achene .... *Dryas*

13   Petals 5; pistils 5; fruit a follicle .... *Luetkea*

14(1) Coarse plants, 1–2 m tall; lvs 2 or 3 times pinnately compound; fruit a follicle .... *Aruncus*

14   Plants smaller, mostly less than 0.5 m tall; lvs various, at most once-compound; fruit an achene 15

15(14) Petals absent, sepals 4; fls in dense terminal spikes .... *Sanguisorba*

15   Petals present, sepals 5; inflorescence not a spike 16

16(15) Stamens 5; petals yellow, very much shorter than the sepals . *Sibbaldia*

16   Stamens 10 or more; if petals yellow, then equal to or longer than the sepals 17

17(16) Fls white; leaflets 3; achenes set on the surface of a fleshy receptacle (strawberries) .... *Fragaria*

17   Fls yellow or purple; leaflets 3 to many; fruit a dry achene or follicle ...
       18

18(17) Styles persistent on the achenes at maturity; lvs pinnately compound with a large, rounded terminal leaflet .... *Geum*

18   Styles soon deciduous from the achenes; lvs various, if compound then the terminal leaflet not markedly larger than the others 19

19(18) Stamens 5; lvs pinnately dissected into narrow segments .... *Ivesia*
19   Stamens 10 or more; lvs compound but the leaflets not dissected .. 20

20(19) Fls purple; plants of swamps and shallow water .... *Comarum*
20   Fls yellow; plants of dry to moist places, but not of swamps or shallow water .... *Potentilla*

Amelanchier Medik.—SERVICEBERRY

Amelanchier alnifolia (Nutt.) Nutt. var. *semiintegrifolia* (Hook.) C.L. Hitchc.
WESTERN SERVICEBERRY, SASKATOON

An upright shrub 1–5 m tall. The lvs are deciduous, more or less oval but bluntish at the tip, toothed along the top half of the lf, and 2–4 cm long. The fls are in loose racemes at the ends of the branches. The 5 white petals are slender and 10–15 mm long while the sepals are just 3–4 mm long, triangular, and remain on the fruit. The ovary is inferior, with 5 styles, and develops into a bluish black, glaucous, applelike pome about 1 cm broad.

Fairly common, and most often seen along streams and at forest edges, as high as 5500 ft. Collections known from Westside Rd near the beaver dams; at Box Canyon; along the road on the east side of the Cowlitz Divide; at Ohanapecosh; and at 5500 ft on Lodi Cr below Berkeley Park.

Although sometimes called "saskatoon," the fruits of the variety *semiintegrifolia*, found in and west of the Cascades, are much less flavorful than the saskatoon of the Canadian prairies.

Aruncus L.—GOAT'S BEARD

Aruncus dioicus (Walter) Fernald var. *acuminatus* (Rydb.) H. Hara—GOAT'S BEARD

A herbaceous perennial, but so robust as to appear to be a shrub, with stems 1–2 m tall. The lvs are 2 or 3 times pinnately divided, the leaflets lanceolate to ovate with a long and tapered point, toothed on the margins, and 5–10 cm long. Male and female fls are borne on separate plants, each with 5 petals about 1 mm long. The fls are crowded on the slender, drooping branches of a 10–40 cm tall panicle. The fruit is a short, brownish follicle.

Common, but sometimes hidden among brush, reaching about 4000 ft on damp cliffs and banks in open woods. Found at Mowich Lk; along Westside Rd; around Longmire; along the lower trail to Bench Lk; at the head of Stevens Canyon near the road; at Ohanapecosh; in the White R campground; and along the Carbon R.

Comarum L. (formerly in *Potentilla*)—MARSHLOCKS

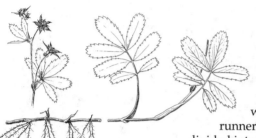

Comarum palustre L.
MARSH CINQUEFOIL

Very similar to *Potentilla* species, but easily separated by its purple fls. The plants are perennial and mat-forming, with creeping or even floating runners. The lvs are pinnately divided into 5–7 oblong, toothed leaflets (but may appear almost palmate), short-hairy on the underside, and 5–8 cm long. Several fls are carried on a leafy stem that reaches 30–90 cm tall. The petals are purple and 2–6 mm long, much shorter than the sepals. There are about 20 stamens; the fruit is an achene.

Evidently rare in the Park. G. N. Jones describes this as a plant of swamps and shallow water, while St. John and Warren (1937) cite *Warren 1616* from Longmire. Herbarium material was not found during the present study, but Warren Tanaka has also reported the plant at Longmire in the course of his rare plant surveys.

Dryas L.—MOUNTAIN-AVENS

Dryas octopetala L. ssp. *hookeriana* (Juz.) Hultén
WHITE MOUNTAIN-AVENS

A prostrate, mat-forming, evergreen shrub. The lvs are oblong to somewhat ovate, with rounded teeth on the margins, glandular and short-hairy on the underside, and 1–3 cm long. The fls are solitary on slender, leafless stalks in the axils of the lvs. The sepals are beset with stalked glands, and the fls are about 2 cm broad, white, with 7–10 white petals. There are numerous stamens and styles; the latter are long and feathery when the achene is mature.

Widespread around the mountain, forming mats on dry, open ridges, 6500-8500 ft .

Fragaria L.—STRAWBERRY

Strawberries are quite common in the Park, and easily recognized. These are low, tufted plants, spreading by runners to form small patches. The leaves are divided into 3 leaflets and 1 to several white flowers are borne on short stems. Each flower has 5 slender sepals and 5 broader bractlets. The fruit, much smaller than market strawberries, nevertheless is identical in form: a red, fleshy receptacle upon the surface of which the tiny achenes are borne.

In a note in *Flora of the Olympic Peninsula* (Buckingham et al., 1995), Nelsa Buckingham remarks that "there is an excess of names for western *Fragaria*," and at least seven names have been applied to the strawberries of Mount Rainier. Clearly, there are only three species in the Park: two with green leaves (and which are not always easily distinguished) and one with blue-green leaves . The characteristics by which the strawberries are keyed in Hitchcock and Cronquist (1973) do not hold up terribly well for plants from the Park and some hybridization is suspected, as seen in plants growing at the Nickel Cr bridge. Names here follow Kartesz, while the key is based upon that of Barbara Ertter in *The Jepson Manual* (Hickman, 1993).

Beginning in mid-July most years, one can start collecting the berries. The "greenleaf" strawberries bear several fruits on a cyme that reaches about to or above the lvs. The berries, which are 6 or 7 mm broad, are bright red when ripe, juicy and tart. *Fragaria virginiana* ssp. *platypetala*, the "blueleaf" strawberry, provides the tastier fruit, sweeter with a classic strawberry taste, if not quite as juicy; the berries are pink when ripe, about 10 or 12 mm broad, and usually solitary, borne near the crown of the plant, well beneath the lvs. Along the road for a mile or so south of the White R entrance, one can sometimes find greenleaf and blueleaf strawberries fruiting together.

Fragaria chiloensis has been listed for the Park. For a discussion, see "Doubtful and Excluded Species," p. 469.

1	Lvs bluish green, the upper surface typically somewhat glaucous; the central tooth at the tip of each leaflet much smaller than the two teeth flanking it .... *F. virginiana* ssp. *platypetala*
1	Lvs yellow-green to medium green; the central tooth at the tip of each leaflet about the same size as the two teeth flanking it 2
2(1)	Fls borne above the lvs; lower surface of the lvs green .... *F. vesca* ssp. *bracteata*
2	Fls borne at or below the lvs; lower surface of the lvs silvery-hairy .... *F. crinita*

Fragaria crinita Rydb.
WOODLAND STRAWBERRY, GREENLEAF STRAWBERRY

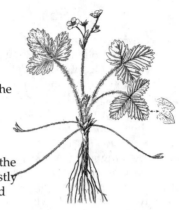

Quite similar to *F. vesca*, but the lower surface of the leaflet is silvery, from the longer hairs that lie flat on the surface; the upper side of leaflet is hairless. The flowering stem reaches 10 cm tall, usually no more than the height of the lvs.

Apparently much less common than the other "greenleaf" strawberry, found mostly in somewhat drier habitats in woods and along trails, to about 4000 ft.

Fragaria vesca L. ssp. *bracteata* (A. Heller) Staudt
MOUNTAIN STRAWBERRY, GREENLEAF STRAWBERRY

In this species, the lvs are short-hairy but nevertheless green on both surfaces. The leaflets are 15–50 mm long, obovate, coarsely toothed on the margins from the base of the leaflet to the tip, with 12–21 teeth; the central tooth at the tip of each leaflet is about the same size as the two teeth flanking it. The flowering stem is 10–30 cm tall, usually taller than the lvs, with 2–5 fls. The petals are 5–8 mm long. The fruit is bright red, less than 8 mm long.

At the roadside at the entrance to the Ohanapecosh campground, many of the plants have fls with toothed petals— 5–7 small teeth at the apex of each petal—a lovely effect.

Very common, often covering patches of ground, to about 4000 ft but usually lower, in open woods, at roadsides, and along trails. Generally on more shaded ground and in moister places than the *F. virginiana*, but the two do grow side by side, as along Westside Rd, at Ohanapecosh, and in the White R area. Found at the Nisqually entrance, up to almost 4500 ft on Westside Rd and on the slopes of Mount Wow; around Longmire; at the Glacier View Bridge; in lower Stevens Canyon; at Ohanapecosh, where abundant; and in the White R Valley.

Fragaria virginiana Duchesne ssp. *platypetala* (Rydb.) Staudt
VIRGINIA STRAWBERRY, BLUELEAF STRAWBERRY

The leaflets of this species are glaucous, with a distinctive bluish green color, hairless above and short-hairy below, 15–50 mm long, and obovate. The margins are toothed from the middle to the tip, with 7–11 teeth; the central tooth at the tip of each leaflet is much smaller than the two teeth flanking it. There is generally just 1 fl per stem, which may rise no more than a few centimeters above the crown of the plant. The petals are 5–9 mm long. The ripe fruit is pinkish, or splotched white and pink, 10–12 mm long.

As common as *F. vesca* in similar places and also carpeting the ground in favored locations; reaching almost 5000 ft on Sunrise Ridge, but more often at lower elevations; on dryish, shaded ground and at roadsides.

Geum L.—AVENS

Geum macrophyllum Willd. var. *macrophyllum*
BIGLEAF AVENS

A tufted perennial, with hairy lvs and stems. The lvs are pinnately divided into roundish leaflets; the 2–4 side leaflets are much smaller than the terminal leaflet, which itself is bluntly 3-lobed, kidney-shaped, and about 8 cm broad. The stem lvs are smaller, sharply toothed, and lack stalks. The fls are in an open, flat-topped cluster. Each is 1–2 cm broad, with 5 bright yellow petals; the sepals are turned back beneath the fl. The style persists as a long, hooked tip on the achene.

Very common, below about 3500 ft, often on disturbed ground along roads and trails; also in meadows. Collected along Westside Rd; at Longmire; at Ohanapecosh; in the White R V below the campground; and along the Carbon R below the glacier.

Holodiscus K. Koch
OCEAN-SPRAY, CREAMBUSH

Holodiscus discolor (Pursh) Maxim.
OCEAN-SPRAY

A shrub 2–5 m tall, with spreading, arching branches; the young branches are short-hairy. The lvs are ovate, blunt at the tip, with teeth on the margin above the middle, short-hairy on the underside and with perhaps a few hairs on the upper, and 2–5 cm long. The lvs of young shoots are wider and may be up to 10 cm long. The fls are creamy white, with 5 petals and 5 sepals, about 4 mm broad, in dense, drooping panicles. The 5 pistils develop into short follicles.

Common and very noticeable when in flower, in open woods on slopes and at roadsides. Well-displayed along the main road from the Nisqually entrance to Ricksecker Point and from Ohanapecosh through Stevens Canyon; also in the lower White R Valley.

Ivesia Torr. & A. Gray—IVESIA

Ivesia tweedyi Rydb.—TWEEDY'S IVESIA

Perennial, growing as a low clump. The lvs are mostly basal, short-hairy and glandular, once-pinnate with lobed leaflets, and less than 10 cm long overall. The fls are in a flat-topped cyme; the petals are yellow, with petals about 3 mm long and equaling the sepals. There are numerous stamens and 2–6 pistils that develop into achenes; unlike *Geum*, the style soon falls from the achene.

Listed as "watch" in *The Endangered, Threatened, and Sensitive Vascular Plants of Washington* (Washington Natural Heritage Program, 1997). First noted for the Park by Ola Edwards (1980).

A plant from chiefly east of the Cascades; known in the Park from a collection made north of Brown Peak, just above Bear Park.

Luetkea Bong.—PARTRIDGEFOOT

Luetkea pectinata (Pursh) Kuntze
PARTRIDGEFOOT

A prostrate, short-hairy, evergreen shrub, forming extensive mats, rooting along the creeping, woody stems. The lvs are set in tufts on the ends of the branches, and are palmately divided into narrow, sharp segments, 1–2 cm long. The flowering stems are 10–15 cm tall, leafy, and with dense, terminal racemes of white fls. The petals are about 3 mm long, as long as the 5 sepals, and surround 20 yellow stamens. The 5 pistils develop into short follicles.

Very common in dryish meadows and on open talus slopes, 6000-8500 ft; occasionally on favorable ground much lower, as at about 4500 ft at the Canyon Rim Overlook and at Mowich Lk. Abundant in Paradise Park; also known from Indian Henrys Hunting Ground; the Butter Cr Research Natural Area; at the summit of Shriner Peak; around Tipsoo Lk and on Naches Peak; at Sunrise and along the trail to Burroughs Mtn; at Berkeley Park; and at Eunice Lk.

Malus P. Mill.
APPLE

Malus fusca (Raf.)
C. K. Schneid.
OREGON CRABAPPLE

A small tree, 5–10 m tall, with mostly upright, often thorny, branches. The deciduous lvs are generally lanceolate-ovate, sharply toothed and may be lobed, but are not divided. The fls are in small umbels on short, lateral spurs along the branches. The 5 petals are white, 10–13 mm long, with numerous yellow stamens. The ovary is inferior, with 2–5 styles, and develops into a roundish to egg-shaped, reddish fruit up to 1.5 cm long; the floral parts often remain at the bottom of the fruit.

Common in swampy areas at low elevations, especially around Ohanapecosh.

Oemleria Rchb.—OSOBERRY

Oemleria cerasiformis (Hook. & Arn.) Landon—OSOBERRY

A multistemmed shrub 2–4 m tall. The deciduous lvs are oblanceolate, not toothed, and 5–10 cm long. Male and female fls are on separate plants, in pendulous racemes. The fls are shaped like long bells, white, 4–6 mm long, and have a rank smell. There are 15 stamens and 5 pistils; the fruit is a berrylike, purplish black drupe, glaucous and about 1.5 cm long.

Infrequent at low elevations in the major river valleys, on wet ground, especially under cottonwood and alder, sometimes forming thickets. An early bloomer, usually flowering by mid-April around Sunshine Point. Also called Indian Plum.

Pentaphylloides Duham. (formerly in *Potentilla*)
GOLDEN-HARDHACK

Pentaphylloides floribunda (Pursh) A. Löve
SHRUBBY CINQUEFOIL

A low shrub with woody stems 30–90 cm tall, otherwise similar to the species of *Potentilla*. (Another key difference is that the achene is densely hairy.) The lvs are

297

pinnate, with 5–7 narrowly oval leaflets; each leaflet is silky-hairy, with incurved margins, 8–18 mm long, and not toothed. There are 1 or 2 fls at the ends of the branches, with bright yellow, nearly round petals 5–10 mm long (about twice as long as the sepals).

Uncommon shrub on open rocky slopes, above about 5000 ft, reaching 8000 ft. Collections known from Ipsut Pass; near Cowlitz Glacier; on a ridge between the Winthrop Glacier and the White R; and on Burroughs Mtn; also reported from high ridges above Spray Park. On Second Burroughs Mtn, the plants are nearly prostrate.

Physocarpus Maxim.—NINEBARK

Physocarpus capitatus (Pursh) Kuntze
PACIFIC NINEBARK

The bark of this 2–5 m tall shrub is thin, brown, and tends to shred. The lvs are 3–5-lobed, toothed, mostly hairless on the upper side and with starlike hairs beneath, 3–10 cm long including the short stalk. The fls are in longish, headlike clusters at the ends of the branches. The 5 petals are roundish, white, and about 4 mm long. There are numerous stamens and 1–5 pistils, which develop into reddish, inflated follicles, about 1 cm long.
Uncommon, in open woods at low elevations. Known from the Ohanapecosh entrance area and at Green Lk.

Potentilla L.—CINQUEFOIL

With *Comarum* and *Pentaphylloides* segregated, the remaining potentillas may be characterized as annual or perennial herbs, with compound leaves and yellow flowers in cymes. There are 5 sepals that alternate with 5 bractlets, 5 petals, and 10 or more stamens (*Sibbaldia*, with 5 stamens, was also once included in *Potentilla*). There are several to many pistils and the fruit is an achene.

1	Lvs pinnately divided	2
1	Lvs trifoliate or palmately divided	3
2(1)	Lvs long and narrow, the leaflets toothed; petals about equal to the sepals .... *P. glandulosa* ssp. *glandulosa*	
2	Lvs about half as wide as long, the leaflets deeply cut; petals about twice as long as the sepals ...:............... *P. drummondii* ssp. *drummondii*	
3(1)	Leaflets 5	4
3	Leaflets 3 (the basal lvs rarely with 5 leaflets)	5
4(3)	Tufted alpine plants, with stems to 20 cm tall .... *P. diversifolia* var. *diversifolia*	
4	Erect plants, 30–60 cm tall, of low to middle elevations .... *P. gracilis* var. *gracilis*	

Potentilla diversifolia Lehm. var. *diversifolia*
VARIED-LEAF CINQUEFOIL

A tufted, mostly hairless plant, with a few branches from a heavy rootstock. The lvs are chiefly basal and palmately divided into 5–7 leaflets. Each leaflet is oblanceolate, with a blunt tip, and toothed above the middle; the middle leaflet is the largest, 10–20 mm long. The erect to ascending stems are 10–20 cm tall, with 5–20 fls. The petals are 6–7 mm long.

Fairly common on dryish subalpine meadows, talus slopes, and rocky ridges, 6000 to 8500 ft. Collections have been made at the summit of Mount Wow; at McClure Rock at 8500 ft; and on Sarvents Ridge at 6640 ft; reported from Berkeley Park and Burroughs Mtn.

Jones, in 1938, treated this taxon as two separate species: *P. diversifolia*, with lvs silky beneath and growing in meadows and on slopes in the subalpine zone; and *P. glaucophylla*, with lvs greenish on both sides and growing on alpine ridges. Hitchcock (1959-67) considered Jones's *P. glaucophylla* to be merely a montane ecotype, not meriting taxonomic recognition.

Potentilla drummondii Lehm. ssp. *drummondii*
DRUMMOND'S CINQUEFOIL

A tufted, short-hairy plant, with a few branches from a heavy rootstock. The lvs are pinnately divided (but, because the midrib can be very short, appear to be almost palmate), on stalks longer than the blades. The 5–11 leaflets are obovate, coarsely toothed, and 10–50 mm long. There are 10–20 fls in an open cyme, with petals 5–10 mm long, twice as long as the sepals.

An uncommon plant of subalpine meadows. Herbarium vouchers were not found in the course of this study. It was included in the Park flora by Jones but not by St. John and Warren.

Potentilla flabellifolia Hook. ex Torr. & A. Gray
FAN-LEAF CINQUEFOIL

A low, clustered perennial, mostly hairless. The basal lvs have 3 leaflets and are 10–12 cm long; the leaflets are broadly obovate to fan-shaped, coarsely toothed, and 1–3 cm long. The flowering stems reach 10–30 cm tall and carry 1–5 orangish yellow fls, with petals 6–10 mm long, which are broadly notched at the tip.

Common and often abundant in subalpine meadows and higher, 4500-7000 ft. Collected at Mowich Lk; at Indian Henrys Hunting Ground; through Paradise Park; in the Butter Cr Research Natural Area; at Reflection Lks; in the meadow near the summit of Shriner Peak; at Cayuse Pass; at Tipsoo Lk and Chinook Pass; at Sunrise, where the petals are often more yellow than the typical orange-yellow, and on the trail to Burroughs Mtn; by upper Palisades Lk; at Berkeley Park and at Eunice Lk.

Potentilla glandulosa Lindl. ssp. *glandulosa*
STICKY CINQUEFOIL

A tall perennial, glandular-hairy overall, with stems 30–100 cm tall. The lvs are pinnately divided into 5 or 7 lobes, the leaflets roundish to obovate and deeply toothed and hairy. There are usually numerous pale yellow fls in the open inflorescence; the ovate petals are 5–8 mm long, about equal to the sepals.

Plants on Westside Rd below 3000 ft differ from those at higher elevations, having pale yellow petals that are mostly shorter than the sepals.

Common through the Park below about 5000 ft, especially along roads and trails, often on dryish or stony ground. Found along Westside Rd; along the road on the eastside of the Cowlitz Divide; and on the roadside between Cayuse Pass and Tipsoo Lk.

Potentilla gracilis Douglas ex Hook. var.
gracilis—GRACEFUL CINQUEFOIL

A tufted perennial, hairy but not glandular. The lvs are palmately divided into 5–7 lobes; the leaflets are oblanceolate, coarsely toothed, white-hairy on the underside and dark green on the upper. The longest leaflet is 3–6 cm long. The flowering stems are 30–60 cm tall, with a few fls; the bright yellow petals are 8–10 mm long, widely notched at the tip, and much longer than the sepals.

Uncommon, found in a few places in dry meadows mostly at low elevations, as at Longmire, although a collection from 6000 ft in Glacier Basin is known and the plant can be found at 5600 ft on Naches Peak.

Potentilla norvegica L. ssp. *monspeliensis* (L.) Asch. & Graebn.—NORWEGIAN CINQUEFOIL

A coarse plant, with thickish, leafy stems 30–80 cm tall, clothed with stiff hairs. The lvs are divided into 3 coarsely toothed, obovate leaflets, 3–10 cm long. The petals are light yellow, about 6 mm long and somewhat shorter than the sepals.

An annual weed, occasionally found in dry meadows and on disturbed ground. Known from a collection made at Longmire.

Potentilla rivalis Nutt. var. *millegrana* (Engelm. ex Lehm.) S. Watson BROOK CINQUEFOIL

Annual or biennial, softly short-hairy in the upper stems and lacking glands. The lf is nearly palmate, with 3 oblanceolate leaflets that are 2–5 cm long, toothed, and hairy. There are numerous pale yellow fls in an open inflorescence, on a stem that reaches 60 cm. The petals are just 1.5–2 mm long, much shorter than the sepals, which are united into a bowl-shaped hypanthium.

Known from a 1888 collection made by C. V. Piper, at an unspecified location within the present confines of the Park. Cited by St. John and Warren (1937), from Longmire. The species is usually found east of the Cascades.

Potentilla villosa Pall. ex Pursh NORTHERN CINQUEFOIL

A short and tufted plant, from a woody rootstock, with flowering stems 10–20 cm tall. The stems and lvs are densely short-hairy, appearing silvery. The lvs are divided into 3 leaflets, each 2–4 cm long, broadly obovate, and deeply toothed. There are only a few fls on each stem, with widely notched petals 6–8 mm long.

Uncommon, on rocky flats and slopes, 5000-8500 ft, especially in the northeast of the Park, as at Owyhigh Lks and at the foot of Little Tahoma Peak. Collections were made by F. A. Warren on Mount Wow and on Iron Mtn above Indian Henrys Hunting Ground.

Prunus L.—PLUM, CHERRY

Prunus emarginata (Douglas) Walp. var. *mollis* (Douglas) Brewer in Brewer & Wats.
BITTER CHERRY

A tree 10–20 m tall, with erect branches and shiny, reddish brown bark. The deciduous lvs are often clustered on the branches; they are oval to obovate, toothed, and 3–8 cm long, on short stalks. The 3–10 fls are borne in flat-topped clusters on lateral spurs. The 5 petals are white and 4–8 mm long, surrounding numerous stamens. The fruit is an egg-shaped, red drupe, looking much like a cultivated cherry but very bitter.

Infrequent, in low-elevation mixed woods and often forming thickets. Found on Westside Rd, on the lower slopes of Mount Wow, and in the Butter Cr Research Natural Area.

Rosa L.—ROSE

Thorny or prickly shrubs, with alternate, pinnate lvs on arching stems. The leaf margins may be singly or doubly toothed. The flowers are large and attractive, followed by eye-catching red fruits, called hips, which contain several woody achenes. There are 5 sepals, 5 petals, and numerous stamens.

1   Sepals falling from the hip as the fruit matures; petals about 10 mm long .... *R. gymnocarpa* var. *gymnocarpa*
1   Sepals persistent on the maturing hip; petals 10–25 mm long 2

2(1) Leaflets sparsely glandular; fls generally solitary ..... *R. nutkana* var. *hispida*
2   Leaflets glandless; fls several in a cyme .... *R. pisocarpa*

Rosa gymnocarpa Nutt. var. *gymnocarpa*
WOOD ROSE

An open shrub; the stems are 0.5–2 m tall and bear slender, straight bristles. The lvs are elliptical in outline, hairless (but glandular-hairy on the stalk and mid-rib), with 5 or 7 leaflets. The margins are doubly toothed, the teeth tipped with glands. The largest leaflet is 1–3 cm long. There are 1–3 fls per cluster, on glandular stalks. The petals are about 10 mm long; the sepals are ovate, shorter than the petals, sometimes glandular on the back and united at their bases, falling together as a unit as the fruit matures. The hip is elliptical and a brilliant red-orange color, 4–6 mm in diameter.

Common in more or less open, dryish woods, frequently on slopes. Known from along the road at the south end of Backbone Ridge; around Reflection Lks; at Box Canyon; and in the White R V, as at Deadwood Cr; occasionally seen on open, dryish slopes on Mount Wow.

Rosa nutkana C. Presl var. *hispida* Fern.
NOOTKA ROSE

A loose shrub, with straight, somewhat flattened prickles on stems less than 1 m tall. The lvs are divided into 5 or 7 sparsely hairy and glandular leaflets, mostly singly toothed without glands on the tips of the teeth. The fls are usually solitary, on hairless stalks, with petals 1–2 cm long, which may be broadly notched at the tip. The sepals are as long as the petals, lanceolate, tapered to a slender tip, and mostly hairy on the back. The fruit is about 1 cm in diameter, with the withered sepals remaining at the bottom.

Rosa rainierensis was first described by G. N. Jones in his 1938 flora, for subalpine plants in the Park. The species was supposedly distinguished by features of the lvs, nature of the thorns, and size of the fls, but these features are highly variable in *Rosa*.

The least common of the roses in the Park, growing on rocky slopes and hillsides in open forests. Found along Westside Rd and on the slopes of Mount Wow; a collection was also made by Warren on a "hillside at Tipsoo Lk."

Rosa pisocarpa A. Gray—CLUSTERED ROSE

A stout shrub, 1–2 m tall, with relatively few but strong, broad-based prickles. There are typically 7 leaflets, the longest 1.5–3 cm long; the leaflets are hairless, with singly toothed, nonglandular margins. There are 2–8 fls per cluster, on hairless and nonglandular stalks 1–2 cm long. The petals are about 1.5 cm long; the sepals are about as long and finely glandular-hairy on the back. The hip is dark red, roundish, and about 1 cm in diameter.

Frequent in swampy places below about 2500 ft, as at the Nisqually and Ohanapecosh entrances.

Rubus L.—BLACKBERRY, RASPBERRY

An interesting and diverse genus, ranging from delicate forest creepers to rampaging roadside weeds. The stems usually trail on the ground and are prickly or thorny. Salmonberry and thimbleberry are exceptions—their erect stems can form dense thickets. The leaves are pinnately or palmately divided or lobed, usually with toothed margins. The flowers are mostly in small clusters and have 5 sepals and 5 petals, with numerous stamens and pistils. The fruit is the familiar blackberry or raspberry: technically, it's an aggregate of many small 1-seeded drupelets.

Although not used in this key, there is an easy way to tell the raspberries from the blackberries. In raspberries, the ripe fruit slips easily from the receptacle upon which the drupelets are borne; in the blackberries, the drupelets and the receptacle remain together when the ripe fruit is picked.

The three small, trailing species, *R. lasiococcus*, *R. nivalis*, and *R. pedatus*, grow together on a remarkable patch of ground along the Wonderland Trail on Rampart Ridge, at about 3200 ft.

1   Lvs lobed but not divided into leaflets 2
1   Lvs palmately divided into 3–5 leaflets.. 4

2(1) Upright shrub; fls white, 15–30 mm long ..... *R. parviflorus* var. *parviflorus*
2   Trailing plants; fls white or pink, less than 10 mm long 3

3(2) Fls pink; lvs dark, glossy green above .... *R. nivalis*
3   Fls white; lvs dull, light green above .... *R. lasiococcus*

4(1) Stems strongly angled, with heavy and curved prickles; larger lvs with 5 leaflets; introduced shrubs, with arching stems .... *R. laciniatus*
4   Stems slender and round, the prickles absent or slender; leaflets mostly 3 (5 in *R. pedatus*); native shrubs with upright or trailing stems 5

5(4) Tall, upright shrubs; fls magenta; berries yellow or red ....
         .... *R. spectabilis* var. *spectabilis*
5   Shrubs with trailing or weakly ascending stems; berries red, purplish, or black.. 6

6(5) Leaflets 5; stems mat-forming, prickles absent .... *R. pedatus*
6   Leaflets 3; stems weakly ascending or trailing, but not matlike; prickles present 7

7(6) Stems conspicuously glaucous, with hooked prickles .... *R. leucodermis*
7   Stems not glaucous, prickles straight or hooked 8

8(7) Prickles of the stems straight, slender; fruit reddish ....
         .... *R. idaeus* ssp. *strigosus*
8   Prickles of the stems hooked, stout and somewhat flattened; fruit black .
         .... *R. ursinus* ssp. *macropetalus*

Rubus idaeus L. ssp. *strigosus* (Michaux) Focke—RED RASPBERRY

A raspberry, with trailing stems that may reach 1–2 m long and straight, weak prickles. The lvs are 3-foliate, the leaflets are toothed or somewhat lobed, grayish with wooly hairs on the underside. The 3–5 fls are borne in loose clusters on lateral branchlets; the petals are white and 6–8 mm long. The fruit is red, roundish and about 1 cm broad; the drupelets are finely velvety on the surface.

Uncommon, in open woods below about 3000 ft, as at Longmire. St. John and Warren (1937) suggest that this species was introduced into the area by the Longmire family.

Rubus laciniatus Willd.—CUT-LEAF BLACKBERRY

A vigorous blackberry, with heavy, arching stems armed with strong, curved thorns. The lvs are 5-foliate and essentially evergreen; each leaflet is ovate in outline but very deeply cut into sharp teeth. The lvs are hairy on the underside but nevertheless appear green on both surfaces and the largest leaflet is 2–6 cm long. The inflorescence is a many-flowered, glandular panicle. The round petals are 7–15 mm long and white or pinkish; the egg-shaped fruit is black and hairless.

Known to be established in the Ohanapecosh and Nisqually entrance areas of the Park and at Longmire; also found along the lower Carbon R. Observed in 1998 on the road shoulder between Dewey Cr and Cayuse Pass. Another aggressive introduced blackberry, *Rubus armeniacus* Focke (see Czeska, 1999), the Himalayan blackberry, has spread into the upper Nisqually V around Ashford and could easily move into the Park. It has 5-foliate lvs that are toothed but not deeply cut and that are white-hairy on the underside; it has generally gone by the name *R. discolor* Weihe & Nees.

Rubus lasiococcus A. Gray DWARF BRAMBLE

An attractive ground covering raspberry: the trailing stems root at the lf nodes as they creep along the forest floor and can reach 2 m. Prickles are absent. The lvs are on very short lateral stems, 3–5-lobed, finely toothed, rounded in outline but somewhat wider than long, and 2–4 cm across.

The fls are mostly solitary in the axils of the lvs, with round petals about 5 mm long. The berry is red, of just 2–5 drupelets (small but tasty).

Very common in forests to about 6000 ft throughout the Park; also occasionally in brushy patches. Plants near Bench Lk have lvs with more pronounced teeth than typical.

Rubus leucodermis Douglas ex Torr. & A. Gray var. *leucodermis* BLACKCAP RASPBERRY

A shrub with trailing, clambering, or weakly ascending stems, armed with short, hooked prickles. The bark of the branches is strongly glaucous, appearing pale green. First-year stems have larger, 5-foliate lvs while the lvs of the fruiting branches are 3-foliate. The leaflets are ovate, sharply toothed, green above but white-hairy beneath; the largest are 2–6 cm long. The inflorescence is a few-flowered cluster; the petals are oblanceolate and 4–5 mm long, shorter than the sepals. The fruits, not often seen but quite delicious, are purplish black, rounded, about 1 cm broad, with a covering of thin, velvety hairs.

Common in open forests, along trails, and even at roadsides, reaching about 5000 ft.

Rubus nivalis Douglas ex Hook.—SNOW DEWBERRY

A creeping blackberry, the stems reaching about 1 m long, rooting at the lf nodes and with a scattering of slender, hooked prickles. The evergreen lvs are dark green and glossy on the upper surface, toothed and sometimes obscurely lobed, reaching 4–8 cm long and longer than wide. The pink to purplish fls are solitary or in pairs, with lanceolate petals 4–6 mm long. The fruit is red, velvety, and about 5 mm broad.

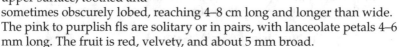

The plant only rarely flowers when growing in places in forests; in exposed places, it will occasionally produce fls, followed by small, sour fruits.

Uncommon, in dense forests, 2500-4000 ft on the south and west sides of the Park. Collected along Westside Rd on the lower slope of Mount Wow; along lower Tahoma Cr; on the Twin Firs loop trail; on Rampart Ridge and around Longmire; and at Ohanapecosh.

Rubus parviflorus Nutt. var. *parviflorus*
THIMBLEBERRY

An upright, unarmed raspberry, 1–2 m tall, with reddish brown bark that shreds. The ovate lvs are palmately 3–5-lobed, sharply toothed, softly hairy, with a blade 5–15 cm long. The 3–7 fls are borne in a flattish cluster, the stalks densely glandular. The white petals are oval and 1.5–3 cm long, textured like crepe paper. The fruit is red, 1.5–2 cm broad, and shaped like a flattened hemisphere.

Plants at Deadwood Cr on Hwy 410 have lvs that are sharply and coarsely lobed, appearing to be "pleated," quite different from the soft appearance typical of thimbleberry lvs.

Very common in open places in moist woods, along trails, and at roadsides, to about 4000 ft. Found at Mowich Lk; on Westside Rd; through Stevens Canyon; in the Butter Cr Research Natural Area; at Ohanapecosh; and in the White R Valley.

Rubus pedatus Sm.—FIVE-LEAVED BRAMBLE

Similar in growth habit to *R. lasiococcus*, but the lvs are fully divided into 5 obovate leaflets (rarely 3) that are sharply toothed and 5–15 mm long. The fls are solitary, with white petals 5–7 mm long. The red fruit is composed of 1–6 smooth drupelets.

Perhaps less common than *R. lasiococcus*, but found in similar habitats and often growing with that species, reaching about 5000 ft. Collected at Mtn Meadows and Mowich Lk; along the road between Kautz Cr and Longmire; in the woods at Ricksecker Point; at Box Canyon; along the trail to the Grove of the Patriarchs; in Berkeley Park; and at Eunice Lk.

Rubus spectabilis Pursh var. *spectabilis*
SALMONBERRY

A vigorous shrub, with erect stems, sprouting from the roots and able to form dense thickets. The stems are 1–3 m tall, with short, straight prickles. The lvs are divided into 3 ovate, toothed leaflets that may themselves be lobed, and 7–15 cm long. The fls are solitary or in pairs, with purplish red petals 10–15 mm long. The raspberrylike fruit is egg-shaped, smooth, and usually dark red; some plants have berries that ripen yellow.

Floyd Schmoe writes in *Our Greatest Mountain* (1925) :

> *Two forms of the species occur, one bearing yellow berries and the other bearing wine-colored berries that are somewhat better flavored than the yellow variety.*

Many people find the flavor, regardless of color, rather flat.

Common and often forming thickets in wet places along streams and at the margins of meadows and swamps, in each of the major river valleys, reaching nearly 5000 ft in Paradise Park and at Mowich Lk.

Rubus ursinus Cham. & Schltdl. ssp. *macropetalus* (Douglas ex Hook.) Roy L. Taylor & MacBryde—PACIFIC BLACKBERRY

A vigorous and widely spreading shrub with stems up to 1 m long, armed with slender, curved prickles. There are typically 3 leaflets (occasionally 5), each ovate, toothed, and tapered at the tip; the terminal leaflet is the largest, 3–4 cm long, and sometimes lobed. The 2–5 fls are borne in clusters at the tops of lateral stems. The petals are white, slender, and about 2 cm long; male and female fls are on separate plants. The fruit is egg-shaped, black, and 1.5–2 cm long.

Very common, especially on open ground at roadsides and in burned areas, to about 3500 ft. Known from the Nisqually entrance area; up Westside Rd; and through Stevens Canyon.

Sanguisorba L.—BURNET

Sanguisorba canadensis L.—SITKA BURNET

A perennial herb that, before it fls, somewhat resembles a small potentilla. The lvs are pinnately divided with 11–21 lobes; each lobe is roundish to oval, sharply toothed, and 1–2 cm long. The stems are 30–90 cm tall, with the fls in a dense spike 3–8 cm long. Petals are absent and the 4 whitish sepals are 2–3 mm long. There are 4 stamens, on long, flattened filaments. There are 1–3 pistils and the fruit is an achene.

Known form just two collections, made in wet meadows on Mount Wow above Lk George and in Berkeley Park. Typically found in the Pacific Northwest east of the Cascades.

Sibbaldia L.—SIBBALDIA

Sibbaldia procumbens L.
CREEPING SIBBALDIA

Formerly treated as a *Potentilla*. A mat-forming
plant with creeping, somewhat woody
stems. The basal lvs are on long, slender
stalks and divided into 3 leaflets of
roughly equal size, wedge-shaped with 3
teeth at the blunt apex, and 1–2 cm long. The
flowering stems are 5–15 cm tall, with 1 or 2
reduced lvs beneath a dense cluster of fls. The
petals are yellow and 2–4 mm long, much shorter
than the sepals. There are 5 stamens (or sometimes 6)
and the fruit is an achene.

Rather rare, in mountain meadows, 6000-8500 ft. Collections are
known from near McClure Rock at 7100 ft; on the ridge above Sluiskin
Falls; on Theosophy Ridge; in the Elysian Fields; and on the Burroughs
Mtn trail (at Frozen Lk, some plants are just 2 cm tall). Also reported
from Spray Park.

Sorbus L.—MOUNTAIN-ASH

Shrubs with several erect stems and large, alternate, deciduous,
pinnately divided leaves. Characteristics of the lvs are important in
separating the species, but are not always definitive. The white flowers
are borne in broad, flattish clusters. The flowers have 5 sepals and 5
petals, with 15–20 stamens and 2–5 styles. The ovary is inferior and
develops into a small, applelike pome, bright red at maturity, carried in
large numbers in clusters.

Sorbus aucuparia L., the rowan, is an introduced tree, spread by birds
far and wide in woods in the Puget lowlands. It has pointed leaflets that
are coarsely toothed, 13–17 in number, wider than those of *S. scopulina*. It
is not known from the Park, but a shrub found at Ricksecker Point does
not key convincingly to one of the three taxa listed below and seems to
show the leaf characteristics of *S. aucuparia*.

1   Leaflets sharply pointed and finely toothed most of their length; leaflets
      mostly 9 or 11 .... *S. scopulina* var. *cascadensis*
1   Leaflets blunt at the end and toothed on the upper half or three-quarters;
      leaflets 7 or 9 2

2(1)   Leaflets toothed above the midpoint, typically 9 ...... *S. sitchensis* var. *grayi*
2   Leaflets toothed from below the midpoint, typically 7 ....
      .... *S. sitchensis* var. *sitchensis*

Sorbus scopulina Greene var. *cascadensis* (G. N. Jones) C. L. Hitchc.—CASCADE MOUNTAIN-ASH

Shrubs 2–5 m tall, with many stems. The winter buds and new twigs are sticky, with white hairs. The lvs are divided into 9 or 11 oblong leaflets that are pointed at the tip and toothed along the margin nearly from the base to the tip. The fls are numerous, in somewhat rounded to flattish clusters 5–10 cm across. The petals are 5–6 mm long and the calyx is smooth on the outside. The fruit is scarlet, 8–10 mm in diameter.

Known from a few places in the Park, growing in thickets at the edges of woods and meadows. Collected at Van Trump Park; near Louise Lk; at Sunrise, and along the trail to Burroughs Mtn.

Sorbus sitchensis M. Roem. var. *grayi* (Wenzig) C.L. Hitchc. SITKA MOUNTAIN-ASH

A shrub reaching about 2 m tall. The buds and young twigs are covered with reddish brown hairs and are not sticky. The leaflets number 7 or 9, each oblong, rounded at the tip, and toothed on the margin only above the midpoint between the base and the tip; they're 3–6 cm long. The fl cluster is generally smaller and rounder than *S. scopulina*. The petals are 3–4 mm long and the calyx is short-hairy on the outside. The fruit is red and 7–10 mm in diameter.

A collection from Paradise made by Ken Miller, at the Creso Herbarium at Pacific Lutheran University, appears to represent a genetic dwarf, with very short internodes and leaflets all less than 3 cm long.

The common mountain-ash in the Park, widespread above about 4500 ft, growing in open, moist woods and in thickets. Found at Mowich Lk and Spray Park; at Paradise; at Reflection Lks and Louise Lk; at Bench Lk; in the Butter Cr Research Natural Area; at Tipsoo Lk; on the trail to Burroughs Mtn; and at Eunice Lk.

Sorbus sitchensis M. Roem. var. *sitchensis* SITKA MOUNTAIN-ASH

Very similar to var. *grayi*, but with lvs that are toothed on the margin above a point between the base and the midpoint. There is also a tendency for the leaflets to be more widely separated on the midrib.

Not previously reported for the Park, and known from two herbarium collections: from 1.25 mi below the White R campground and at the trailhead to Lk George at 4000 ft.

Spiraea L.—SPIREA

Shrubs with toothed but undivided, deciduous leaves. The flowers are small, in dense clusters, with 5 petals, 5 sepals, and 15–60 stamens. There are 5 pistils that develop into follicles. Except for the white-flowered *S. betulifolia*, they prefer wet or swampy ground, and often spread to form thickets.

1   Fls in flat-topped clusters 2
1   Fls in elongated clusters 3

2(1) Fls white .... *S. betulifolia* var. *lucida*
2   Fls pink to rose... *S. splendens* var. *splendens*

3(1) Lvs densely white-hairy on the underside .... *S. douglasii* var. *douglasii*
3   Lvs smooth beneath, or at most with hairs on the veins ....
           .... *S. douglasii* var. *menziesii*

Spiraea betulifolia Pall. var. *lucida* (Douglas ex Greene) C.L. Hitchc.—SHINY-LEAF SPIREA

Seldom reaching even 1 m tall, this species has oval to ovate, hairless, lvs that are 2–6 cm long and toothed on the margins above the midpoint. The white fls are in flat clusters 5–12 cm broad.

Rare in the Park, and much more common east of the Cascades. Found on the west slope of the ridge below Gobblers Knob; at 3000 ft along the trail on the Cougar Cr burn; along the road south of Cayuse Pass; and at Berkeley Park.

Spiraea douglasii Hook. var. *douglasii*
DOUGLAS'S SPIREA

A shrub with erect, strictly branched, stems 1–2 m tall; the young twigs are short-hairy. The lvs are oval to oblong, 4–10 cm long, short-hairy above and wooly beneath, and toothed on the margin near the tip. The rose-pink fls are in tall, pyramidal clusters 8–20 cm long at the ends of the branches.

Common on swampy ground and at the margins of lakes and meadows, to about 2500 ft, in the Ohanapecosh area and from the Nisqually entrance to Kautz Cr; also in the Butter Cr Research Natural Area.

Spiraea douglasii Hook. var. *menziesii* (Hook.) C. Presl
MENZIES'S SPIREA

Much like var. *douglasii*, except the lvs are smooth (or may have a few hairs on the veins beneath); the underside is glaucous.

Much less common, but found in habitats similar to the variety above, and reaching about 4000 ft. Collected at Longmire; around the Ohanapecosh entrance; and at Adelaide Lk.

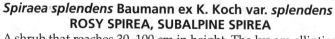

Spiraea splendens Baumann ex K. Koch var. *splendens*
ROSY SPIREA, SUBALPINE SPIREA

A shrub that reaches 30–100 cm in height. The lvs are elliptical to ovate, 1–7 cm long and toothed on the margin near the blunt tip. The fls are rose-pink, in dense clusters 2–4 cm broad that are wider than long and more or less flat-topped.

Listed as "review – group 2" in *The Endangered, Threatened, and Sensitive Vascular Plants of Washington* (Washington Natural Heritage Program, 1997).

Common and widespread, on moist hillsides and at the margins of lakes and meadows, 4500-6500 ft. Known from Mowich Lk and Spray Park; in Van Trump Park; at the saddle of Eagle Peak; in Paradise Park; along the Stevens trail on Mazama Ridge; along upper Butter Cr; at Box Canyon; at the Owyhigh Lks; above Lk James; at Seattle Park; at Tipsoo Lk and Chinook Pass; and at Eunice Lk.

Rubiaceae Juss.—MADDER FAMILY

Galium L.—BEDSTRAW

A prominent part of the herbaceous ground cover in middle-elevation forests. Two are weeds of only minor importance, while one native, fragrant bedstraw, contributes a scent of new-mown hay. The plants are usually weak and clamber over the ground or other vegetation, aided by hooked hairs on the slender, 4-angled stems. The leaves are in whorls along the stem and the small, greenish to whitish flowers (yellow, in one rare weed) are in clusters in the axils of the leaves or at the ends of the stems. There is no calyx, and the corolla is 4-lobed, united above the inferior ovary and spreading at the tips. There are 4 stamens and 2 styles; the ovary is 2-lobed and develops into a pair of nutlets.

1	Lvs in whorls of 4	2
1	Lvs in whorls of 5 or more	3
2(1)	Lvs 3-veined, broadly ovate; nutlets with hooked hairs	*G. oreganum*
2	Lvs 1-veined, linear to oblanceolate; nutlets smooth	*G. trifidum* ssp. *columbianum*
3(1)	Fls yellow; lvs linear; fls numerous in cymes	*G. verum*
3	Fls whitish; lvs wider; fls mostly in threes	4
4(3)	Plants annual; lvs oblanceolate; odor mild	*G. aparine*
4	Plants perennial; lvs broadly ovate; odor of crushed lvs sweet	*G. triflorum*

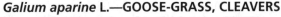

Galium aparine L.—GOOSE-GRASS, CLEAVERS

An annual with weak and trailing stems, 30–150 cm long, with small, hooked hairs. The lvs are in whorls of 6–8, linear to oblanceolate, and 2–5 cm long. The whitish fls are in few-flowered clusters in the axils, 2–3 mm broad and the nutlets are 2–3 mm in diameter, clothed with short, hooked hairs.

Common, often weedlike, native plant in open places in thickets and forests, and in low grasses at roadsides, below about 2000 ft. It is abundant some years at Sunshine Point, on Westside Rd, in the Ohanapecosh area, and in the lower Carbon R Valley.

Galium oreganum Britton
OREGON BEDSTRAW

A perennial, with numerous short, erect stems from a slender, creeping rootstock. The stems reach 10–30 cm tall, with lvs in whorls of 4; they are ovate to elliptical and 2–4 cm long, with 3 main veins (the other bedstraws in the Park have lvs with 1 main vein). The plant is mostly hairless, except for short hairs on the lf margins. The greenish-white fls are in terminal clusters and are about 2 mm broad. The nutlet is covered with long, hooked hairs.

Very common in forests where the shade is not too deep, reaching 5000 ft; sometimes forming dense patches. Found on Mount Wow; at the Nisqually entrance and along the main road to Ricksecker Point; at 4000 ft in Van Trump Park; at Ipsut Pass; below Indian Henrys Hunting Ground; at Dewey Cr along the road; at Ohanapecosh; on the trail to Eunice Lk; and near Tipsoo Lk.

Galium trifidum L. ssp. *columbianum* (Rydb.) Hultén
SMALL BEDSTRAW

A perennial with slender, sprawling stems 10–70 cm long; the stems are minutely rough-hairy. There are in whorls of 4–6 and are linear to narrowly oblanceolate, 5–15 mm long. The fls are in small clusters on the side branches and at the ends of the stems; each is white and 1–2 mm broad. The nutlet is smooth and 1–1.5 mm in diameter.

Several collections made by Charles Landes in the 1930s at Longmire key in Jones (1938) as well as in Hitchcock and Cronquist (1973) to *Galium cymosum* rather than *G. trifidum*. Kartesz synonymizes the two taxa.

Uncommon in the Park, usually on wet ground at the edges of meadows, as at Longmire.

Galium triflorum Michx.
FRAGRANT BEDSTRAW

A perennial with spreading stems to about 50 cm long and tending to form loose mats. There are 6 lvs per whorl, ovate or obovate and 1–5 cm long. The bruised lvs smell strongly of coumarin, the "new-mown hay" aroma. The cream-colored or greenish-white fls are in clusters of 3 on long stalks in the axils of the lvs. The nutlets have hooked hairs.

Common in moist, open woods to about 4000 ft in the major river valleys of the Park. Found at Sunshine Point and along Westside Rd; at Longmire; at the Canyon Rim overlook; in the Butter Cr Research Natural Area; at Ohanapecosh; and in the riparian woods along the lower White R.

Galium verum L.
LADY'S BEDSTRAW

A perennial with smooth, mostly erect stems 10–25 cm tall. The lvs are in whorls of 6–12, 10–25 mm long, and needlelike to linear. The yellow fls are in a dense panicle at the top of the stem. The fruit is smooth and about 1 mm in diameter.

A weedy perennial growing at Longmire in lawns and planted areas.

Salicaceae Mirb.—WILLOW FAMILY

Very common trees and shrubs, preferring streamsides and swampy ground. All have alternate, deciduous leaves with more or less prominent stipules. Male and female flowers are borne on separate plants, in dense, spikelike structures called catkins. Each flower has at its base a small scalelike bract. Sepals and petals are absent. There are 1 to many stamens in the male flowers and the female flowers have 1 pistil. The superior ovary has a short style and 2–4 stigma lobes. The fruit is a capsule. Each seed has a tuft of hairs, aiding dispersal by the wind.

1   Catkins pendulous, each with a bract cut into slender segments; tall trees with broadly ovate lvs (cottonwoods) .... *Populus*
1   Catkins ascending to erect, the bracts not divided; trees or shrubs, with narrow lvs (willows) .... *Salix*

Populus L.—COTTONWOOD

Aspen, *Populus tremuloides*, is not known in the Park although Jones listed it in his 1938 flora; see a discussion in "Doubtful and Excluded Species," p. 469.

Populus balsamifera L. ssp. *trichocarpa* (Torr. & A. Gray ex Hook.) T.C. Brayshaw—BLACK COTTONWOOD

Tall, robust trees, 20–50 m tall, with deeply fissured gray bark, with a crown broadest at the top. The winter buds are covered with a sticky resin, fragrant as they expand in the spring. The lvs are broadly ovate with a pointed tip, on long stalks, broadly toothed, and 2–5 cm long; they are dark green above and reddish brown below. The catkins appear early in the spring, before the lvs. The male catkins are 3.5–6 cm long, each fl with 30–60 stamens. The female catkins are 5–10 cm long; the fl has 3 stigmas. The capsule is round and the seeds are enveloped by long, white, cottony hairs.

Abundant tree on the banks of streams and rivers, reaching 4000 ft, and most frequent in the Nisqually, Ohanapecosh, and White R Valleys. Also known from the Ipsut campground.

Salix L.—WILLOW

A diverse group of trees and shrubs, ranging from tall trees in lowland river valleys to prostrate alpine shrubs. The buds are covered by a single scale, the color of which is important in keying some of the species. The leaves are usually long and narrow, with short stalks. Male and female flowers are borne on separate plants. The catkins may be pendulous or erect. There is 1 style (which may be so short as to appear to be absent) with 2 stigmas and 1–9 stamens (2 in most of the species below). The capsule may be smooth or hairy and is usually pear-shaped.

Difficult to key, partly because critical features—such as the stipules or capsule—may not be available at the time the identification is being attempted. Herbarium collections of the willows of the Park are very few, and precise locations are not available for some species.

1	Prostrate alpine shrubs	2
1	Erect trees or shrubs	3
2(1)	Catkins developing after the lvs; bud scales pale; lvs rounded at the tip .. .. *S. reticulata* ssp. *rivularis*	
2	Catkins developing at the same time as the lvs; bud scales dark; lvs pointed .... *S. cascadensis*	
3(1)	Tall tree, 10 m or more in height; lvs lanceolate and finely toothed; lf stalks bearing glands at the base of the lf blades; stamens 3–8........................ .... *S. lucida* ssp. *lasiandra*	
3	Shrubs or small trees less than 10 m tall; stalks lacking glands, lvs various; stamens 2 (1 in *S. sitchensis*)	4
4(3)	Ovary and capsule short- or long-hairy	5
4	Ovary and capsule smooth	8
5(4)	Stamens 1; lvs not glaucous	*S. sitchensis*
5	Stamens 2; lvs glaucous below	6
6(5)	Subalpine plants; lvs obovate; capsule appearing silvery from abundant short hairs .... *S. planifolia* ssp. *planifolia*	
6	Plants of low elevations; lvs various; capsule with short hairs but green . 7	
7(6)	Lvs linear to oblanceolate, to 1 cm wide	*S. geyeriana*
7	Lvs obovate, more than 1 cm wide	*S. scouleriana*
8(4)	Lvs green on both sides, not glaucous below	9
8	Lvs glaucous below	10
9(8)	Lvs lanceolate; catkins mostly not stalked	*S. boothii*
9	Lvs elliptical; anthers on short, leafy shoots	*S. commutata*
10(8)	Subalpine plants	*S. barclayi*
10	Plants of low to middle elevations	11
11(10)	Catkins appearing before the lvs; small tree . *S. lasiolepis* var. *lasiolepis* (see "Doubtful and Excluded Species," p. 469)	

11	Catkins appearing with the lvs; shrubs 12
12(11)	Lvs elliptic to obovate, about 3 times as long as broad or less, shiny green above .... *S. piperi*
12	Lvs lanceolate, 3–5 times as long as broad, dull green above 13
13(12)	Young twigs yellowish; lvs not toothed .... *S. lutea* (*see* "Doubtful and Excluded Species," p. 469)
13	Young twigs reddish brown; lvs finely toothed .... *S. prolixa*

Salix barclayi Andersson
BARCLAY'S WILLOW

A shrub with many spreading branches 1–3 m tall. The lvs are oval to obovate, silky-hairy when young (along with the new twigs) but becoming smooth, glaucous on the underside, and 2–5 cm long, with fine gland-tipped teeth on the margins. The 2–3 cm long catkins appear after the lvs have begun to unfurl. There are 2 stamens and the style is 1–1.5 mm long. The capsule is 6–8 cm long and smooth.

A fairly common willow of subalpine meadows and open, moist, rocky slopes, 5000-7000 ft. Good examples can be seen at Mowich Lk. (St. John and Warren listed *S. monticola* Bebb, a species of middle to high elevations east of the Cascades, probably based on a misidentification of *S. barclayi*; see "Doubtful and Excluded Species," p. 469.)

Salix boothii Dorn—BLUEBERRY-WILLOW

A shrub to about 2 m tall, with brownish, mostly smooth twigs. The lanceolate lvs are long-hairy when young, but soon become smooth and grow to 3–8 cm long, with mostly untoothed margins and green on both surfaces. The catkins are 2–6 cm long, appearing shortly before the lvs. There are 2 stamens and the style is very short, only about 0.5 mm long. The capsule is smooth and 3–4 mm long.

Infrequent; recently discovered in the Park (Adams, in an unpublished checklist from a 1980 field trip, available from the Washington Native Plant Society) and growing at the edges of drier meadows on the north side of the Park, as at Berkeley Park).

Salix cascadensis Cockerell
CASCADE WILLOW

Creeping shrub, with prostrate, much-branched stems to only about 50 cm long, rising about 5 cm high. The lanceolate lvs are less than 2 cm long. The catkins appear with the lvs, on the previous year's twigs, and are 5–20 mm long, with 5–20 fls. There are 2 stamens and the style is 1–1.5 mm long. The capsule is 4–5 mm long and grayish with dense, short hairs.

Uncommon on moist, rocky flats and meadows above about 6500 ft on the east side of the Park. Found along the Burroughs Mtn trail.

Salix commutata Bebb—VARIABLE WILLOW, UNDERGREEN WILLOW

Very similar to *S. barclayi*, differing chiefly in features of the lvs. They are green on both sides and not at all glaucous, about equally hairy on both sides when young, becoming more or less smooth late in the summer. The style is 0.5–1 mm long and the capsule is 5–6 mm long.

The most common of the subalpine willows in the Park, 5000-6500 ft. It grows along streams, at the margins of meadows, and in moist places on rocky slopes. T. C. Brayshaw (1996) recognizes three varieties of this widespread species; he would call the Mount Rainier plants var. *commutata*.

Salix geyeriana Andersson
GEYER'S WILLOW

A shrub about 3 m tall, with smooth, glaucous twigs. The lvs are narrowly oblanceolate, tapered at the tip, not toothed, dark green above and glaucous beneath, with silky white hairs, and 5–10 cm long; the stipules are small and obscure. The catkins appear with the lvs and are roundish, about 1 cm long. There are 2 stamens and the style is only about 0.2 mm long. The capsule is 5–6 mm long and short-hairy.

Rare in the Park and known only from herbarium collections made in the lower reach of the Nisqually R Valley.

Salix lucida Muhl. ssp. *lasiandra* (Benth.) E. Murray—SHINING WILLOW

A tree, 10 m or more tall, with a slender crown and fissured, yellowish brown bark. The lvs are lanceolate, with a long and slender tip, green and shiny above and glaucous beneath, finely toothed on the margins, with prominent glands on the lf stalk. The catkins appear with the lvs and are 2–9 cm long. There are 5–9 stamens and the style is only about 0.1 mm long. The capsule is smooth and 5–7 mm long.

The only true tree willow in the Park, common along low-elevation rivers and streams.

Salix piperi Bebb—PIPER'S WILLOW

A thicket-forming shrub 3–6 m tall. The lvs are oblanceolate to obovate, green and shiny above and glaucous beneath, coarsely toothed and 6–12 cm long, appearing after the catkins. The male catkins are 3–5 mm long and the fl has 2 stamens. The female catkins are 4–10 cm long and the style is about 1 mm long. The capsule is smooth or sparsely hairy, and 6–7 mm long.

A common shrubby willow along streams and rivers below about 2500 ft. Kartesz synonymizes *S. piperi* with *S. hookeriana* Barrett ex Hook., and G. R. Argus, who wrote the treatment of *Salix* in *The Jepson Manual* (Hickman, 1993), agrees, but notes that the "higher elevation glabrous form (Humboldt County, CA) warrants study." However, an examination of plants under both labels in the University of Washington herbarium shows that the coastal dune plant *S. hookeriana* is easily distinguished from the upland *S. piperi*. *Salix piperi* (as used here) once went by the name *S. hookeriana* var. *laurifolia*, and the two are obviously closely related. Nelsa Buckingham (1995), in her *Flora of the Olympic Peninsula*, reaches the same conclusion and also distinguishes the two species.

Salix planifolia Pursh ssp. *planifolia* TEA-LEAVED WILLOW

A shrub less than 1 m tall, with erect branches; the twigs are smooth. The lvs are 3–6 cm long, elliptical to narrowly obovate, not toothed, somewhat silky when young, at maturity dark green above and glaucous beneath; the stipules are very small and may sometimes be

319

absent. The catkins appear before the lvs and are 3 cm long. There are 2 stamens and the style is 1–1.5 mm long. The capsule is silvery with dense, short hairs.

A rare low shrub of rocky subalpine slopes. Known from a collection made on the Interfork of the White R. A collection from 4500 ft on the east side of Chinook Pass, not far outside the Park, is also known.

Salix prolixa Andersson
MACKENZIE'S WILLOW

A shrub 2–5 m tall, with mostly smooth twigs. The lvs are lanceolate to ovate and 4–8 cm long, the margins with small, gland-tipped teeth. The catkins are 2–8 cm long and appear at the same time as the lvs. There are 2 stamens and the style is about 0.5 mm long. The capsule smooth and 3–5 mm long.

Uncommon, growing at low elevations in river valleys. Collected at Ohanapecosh, on wet ground next to the hot springs.

Salix reticulata L. ssp. *nivalis* (Hook.) A. Löve, D. Löve & S.L. Kapoor—SNOW WILLOW

A creeping shrub, with stems to about 30 cm long and perhaps 5 cm high. The lvs are roundish, smooth, shiny, and dark green above and glaucous beneath, 7–15 mm long, with a distinctive netlike pattern of veins. The catkins appear after the lvs, at the tips of the season's new growth, and are just 5–10 mm long, with a few fls. There are 2 stamens and the style is less than 0.5 mm long. The capsule is smooth and 2–3 mm long.

Frequent on rocky alpine slopes on the north and east sides of the Park, 6500-7500 ft. Collected on the Fremont Peak trail at 7000 ft; at 6400 ft on Mineral Mtn; and on the trail to Burroughs Mtn.

Salix scouleriana Barratt ex Hook.
SCOULER'S WILLOW

Most often a shrub, but occasionally growing as a single-trunked small tree, 4–10 m tall, with gray bark and hairy twigs. The lvs are variable in shape: elliptical to oblanceolate, rounded or pointed at the tip, not toothed or with irregular teeth on the margin, and 3–8 cm long; the upper surface is smooth and dark green while the lower surface is short-hairy and glaucous. The catkins are 2–5 cm long and appear before the lvs. There are 2 stamens and the style is less than 0.5 mm long. The capsule is 7–9 mm long, covered with silky hairs.

Very common in wet places in open woods throughout the Park below about 3500 ft. Less commonly found along streams than is *S. sitchensis*.

Salix sitchensis Sanson ex
Bong.—SITKA WILLOW

A shrub 2–5 m tall, with smooth and glaucous twigs. The lvs are hairy when young and remain densely silvery-hairy on the lower surface, but become dark green and smooth on the upper; obovate to more or less spoon-shaped, and usually finely toothed, 5–10 cm long. The catkins are 2–8 cm long and appear at the same time as the lvs. There is 1 stamen and the style is less than 0.5 mm long. The capsule is silky and 4–6 mm long.

Very common along streams and rivers, reaching about 5000 ft, but more frequent below 3000 ft. Well-displayed around the Nisqually entrance and at Longmire.

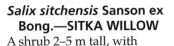

Saxifragaceae Juss.—SAXIFRAGE FAMILY

One of the largest families of plants in the Park. All are perennial herbs (shrubby relatives have been removed to separate families, the Grossulariaceae and the Hydrangeaceae). Many have neat rosettes of basal leaves, or else the leaves are alternate. There are typically 5 petals and sepals (occasionally 4 or 6), and typically 5 or 10 stamens (3 or 8 in some genera). The ovary is superior to partly inferior, surrounded by the partly fused sepals. The ovary usually has 2 (but ranging from 1 to 5) more or less fused chambers and develops into a capsule.

1	Fl 1 per stem; sterile filaments 5, alternating with the 5 stamens ....
	.... *Parnassia*
1	Fls several to many per stem; sterile filaments absent; number of stamens various 2
2(1)	Petals absent; sepals 4, greenish; stamens 8 .... *Chrysosplenium*
2	Petals present; sepals 5; stamens various, but not 8 3
3(2)	Petals 4, brownish purple; stamens 3; lf sometimes bearing a plantlet at the base of the blade .... *Tolmiea*
3	Petals 5; stamens 5 or 10; lvs not as above 4
4(3)	Stamens 5 5
4	Stamens 10 9
5(4)	Stems leafy, the lvs of the stem only a little smaller than the basal lvs 6
5	Stems leafless or with 1 or 2 much-reduced lvs or a few bracts 7
6(5)	Lvs kidney-shaped, deeply divided into rounded segments; stipule margin smooth .... *Suksdorfia*
6	Lvs cordate, sharply toothed; stipules bearing long slender bristles ....
	.... *Boykinia*
7(5)	Fls in panicles; petals not toothed, longer than the sepals .... *Heuchera*
7	Fls in racemes; petals toothed 8
8(7)	Petals with several short teeth at the apex; petals as long as the sepals ...
	.... *Elmera*
8	Petals divided into pinnate, threadlike lobes; petals longer than the sepals
	.... *Mitella*
9(4)	Ovary 2-chambered; stipules absent 10
9	Ovary 1-chambered; stipules present 11
10(9)	Pistils united only at the base; lvs hairless, thick, evergreen ....
	.... *Leptarrhena*
10	Pistils united at least half their lengths; lvs without the above combination of characteristics .... *Saxifraga*
11(9)	Valves of the fruit very unequal in size; fls in panicles; petals white, linear, not divided .... *Tiarella*
11	Valves of the fruit equal; fls in racemes; petals reddish, divided at the apex (fringelike) .... *Tellima*

Boykinia Nutt.

Boykinia occidentalis Torr. & A. Gray
WESTERN BOYKINIA

A tufted plant, with slender stems and mostly basal lvs, sparsely hairy with reddish bristles. The lvs are palmately lobed, toothed, and about 5 cm broad. The stems are 40–60 cm tall, with a few much-reduced lvs. The white fls are in a loose, widely branched panicle; the petals are 4–5 mm long.

Rare in the Park and found below about 2000 ft on the rocky banks of streams in the Nisqually V.

Chrysosplenium L.—GOLDEN-CARPET

Chrysosplenium glechomifolium Nutt.
PACIFIC GOLDEN-CARPET

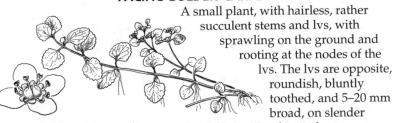

A small plant, with hairless, rather succulent stems and lvs, with sprawling on the ground and rooting at the nodes of the lvs. The lvs are opposite, roundish, bluntly toothed, and 5–20 mm broad, on slender stalks. The fls are solitary in the axils of the lvs. Petals are absent; instead, the 4 or 5 greenish lobes of the calyx spread widely, to 4–5 mm across. There are 8 stamens.

Rare in the Park and known from the Nisqually V, growing in swamps and on wet, shaded ground, to about 3000 ft.

Elmera Rydb.—ELMERA

Elmera racemosa (S. Watson) Rydb. var.
racemosa—YELLOW CORALBELLS

With a creeping rootstock, from which arise a few long-stalked basal lvs. The lf blades are rounded to kidney-shaped, 3–6 cm broad, unlobed but with shallow, rounded teeth, and glandular-hairy. The fls are in racemes on slender, glandular-hairy stems that are 10–25 cm tall and bear 1 or 2 small lvs. The sepals are fused more than half their length, forming a yellowish green, barrel-shaped calyx. The petals are white, with 3–7 very small teeth at the tip, 3–4 mm long and about equalling the lobes of the calyx.

Fairly common, but never abundant, on talus slopes, in rocky crevices, and on moraines, 5000-9000 ft. Known from above 5000 ft on Mount Wow; at Spray Park; across the upper reaches of Paradise Park; at the base of Little Tahoma Peak; at Sunrise and on Burroughs Mtn; at Frozen Lk; at Summerland; and on the west slope of Naches Peak at 5800 ft. Along the lower part of the trail to Frozen Lk, the plant has colonized a low retaining wall, growing in crevices.

Heuchera L.—ALUMROOT

Mostly coarse plants from short, vertical rootstocks. The leaves are roundish to kidney-shaped, and palmately lobed, the lobes shallow and with small, sharp teeth. The small flowers are on leafless stems, in open panicles. There are 5 slender, untoothed, white petals that are up to twice as long as the 5 sepals.

1	Flowering stem hairless at the base; If stalks hairless	*H. glabra*
1	Flowering stem and If stalks hairy throughout	2
2(1)	Lvs about as long as broad, the lobes of the If margin shallow and rounded	*H. micrantha* var. *micrantha*
2	Lvs longer than broad, the lobes deep and sharply toothed	*H. micrantha* var. *diversifolia*

Heuchera glabra Willd. ex Roem. & Schult.
SMOOTH ALUMROOT

Several stems rise from a tuft of basal lvs; stems and lvs are mostly hairless (although the branches of the panicle are glandular). The lvs are heart-shaped, palmately lobed, the lobes shallow and sharply toothed, and about 10 cm across. The stems are 15–60 cm tall, slender and clustered toward the tips of spreading branches. The petals are white, not toothed, and 2.5–3 mm long.

Common on moist, shaded cliffs and banks, to 5000 ft. Found along Westside Rd; on the Rampart Ridge trail; at the Glacier View Bridge; at Paradise Park; at Nickel Cr; at the Ohanapecosh entrance; on the banks of Fryingpan Cr at the Summerland trail crossing; on Eagle Peak; along the trail above Lk James at 5000 ft; and at Berkeley Park.

Heuchera micrantha Douglas ex Lindl. var. *diversifolia* (Rydb.) Rosend., Butters & Lakela
CREVICE ALUMROOT

This variety differs from var. *micrantha* in the shape of the lvs. In addition, the flowering stems and the stalks of the lvs are densely long-hairy.

Known from a collection made "near Longmire" but evidently much less common in the Park than the following variety.

Heuchera micrantha Douglas ex Lindl. var. *micrantha*—CREVICE ALUMROOT

Growing from a heavy rootstock, with sparsely hairy stems and lf stalks. The lvs are more shallowly lobed than those of var. *diversifolia*, and the teeth on the margin are rounded. The flowering stems may be numerous, with open, plumelike panicles of small fls; the branches of the inflorescence are glandular. The petals are 2–2.5 mm long.

Common on west side of the Park below 5000 ft, on rocky banks and cliffs, preferring drier situations than *H. glabra*. Found at Sunshine Point; at 3500 ft on the trail to Lk George; and at Longmire. Also reported for Mowich and Eunice Lks.

In a foray list compiled by Grace Patrick for the Washington Native Plant Society, *H. cylindrica* is listed as occurring on the trail between Chinook Pass and Sheep Lk, just beyond the Park boundary on the east side of Crystal Mtn. (The exact variety is not known.) Quite similar to *H. micrantha* var. *micrantha*, in bloom this plant may be distinguished by its congested, spikelike panicle and fls that have stamens shorter than the sepals.

Leptarrhena R. Br.—FALSE SAXIFRAGE

Leptarrhena pyrolifolia (D. Don) R. Br. ex Ser. in DC.—LEATHERLEAF-SAXIFRAGE

Distinctive for its tough evergreen lvs, which form neat rosettes. The lvs are oblong, 2–10 cm long, nearly stalkless, glossy green on the upper side, and have rounded teeth on the margins. The reddish stems are 20–40 cm tall, with 1 or 2 small lvs and a small cluster of fls at the top. The 5 petals are 2–3 mm long, overshadowed by the reddish calyx. The fruit is a pair of purplish red follicles.

Common on stream banks and springy ground, 4000-6500 ft. Collected between Mountain

Meadows and Mowich Lk; in the lower part of Paradise Park; at Reflection Lk; at Goat Pass on the Kotsuck Cr trail; at Tipsoo Lk; at 5000 ft in Seattle Park; along Sunrise Rd at Squaw Cr; below Frozen Lk; and in Berkeley Park. An interesting low-elevation colony can be found on a damp road bank just inside the White R entrance station.

Mitella L.—MITREWORT

Delicate plants, with intricately divided petals. The plants grow from rootstocks and have basal, long-stalked leaves, which are heart-shaped where the blade joins the stalk. The stems are leafless or may have 1 or 2 small leaves near the midpoint. The flowers are in a raceme. The calyx is broad and shallow, the 5 lobes widely spreading, with the 5 stamens borne on very short filaments on a disk surrounding the ovary (observation of their placement on the disk is important). The petals (except in *M. trifida*) are threadlike and pinnately divided, looking something like the teeth of a comb. The fruit is a capsule.

1   Flowering stems leafy; fls opening from the top of the raceme downward .... *M. caulescens*

1   Flowering stems leafless; fls opening from the bottom of the raceme upward 2

2(1) Stamens alternate with the sepals and opposite the petals . *M. pentandra*

2   Stamens opposite the sepals and alternate with the petals 3

3(2) Petals white, with 3 teeth at the tip .... *M. trifida* var. *trifida*

3   Petals greenish, pinnately divided into 5–9 threadlike segments .... *M. breweri*

Mitella breweri A. Gray
BREWER'S MITREWORT

A plant with wide, rather wrinkled lvs that are 3–8 cm across, round to kidney-shaped, and finely scalloped on the margin. The leafless flowering stem is 10–20 cm tall and glandular. The petals are greenish and divided into 3–6 segments.

The most common mitella in the Park, in some places carpeting the forest floor and more open places under trees, reaching about 6000 ft. Found at Mountain Meadows and Mowich Lk; at Klapatche Park; along the road between Longmire and Paradise; with trees in the Paradise meadows; in the Butter Cr Research Natural Area; along the trail to Bench Lk; along the Kotsuck Cr trail at 4500 ft; at Cayuse Pass, Tipsoo Lk, and Chinook Pass; below 5000 ft on Sunrise Rd; at Berkeley Park; and at Eunice Lk.

Mitella caulescens Nutt.
LEAFY MITREWORT

Unique for its leafy stems and sequence of bloom: the fls at the top of the raceme open first. The lvs are 5-lobed and more deeply lobed than the other species, with small, pointed teeth, sparsely hairy, and 3–5 cm across; the stem lvs are the same shape but smaller. The inflorescence reaches 15–30 cm tall. The fls are 5–7 mm broad and the white petals are divided into 7–9 segments; the stamens are alternate with the petals.

Found occasionally in mossy woods and on stream banks. Collections have been made along the Nisqually R below Longmire and at the Grove of Patriarchs.

Mitella pentandra Hook.—ALPINE MITREWORT,
FIVE-STAMEN MITREWORT

The lvs of this species are broadly ovate in outline, definitely longer than broad, up to 8 cm long, obscurely lobed, and shallowly toothed. The fls are 3–5 mm broad, with greenish petals that are divided into 7 or 9 segments, on leafless stems 10–30 cm tall.

Despite the common name, this is not an alpine plant at all, but, rather, is found between about 3000 and 5500 ft in moist, open woods and on stream banks. Collected at the Ipsut campground; Mowich Lk; at about 3000 ft on Westside Rd; near Longmire; at Narada Falls; in the lower part of Paradise Park; at the White R campground; and at Berkeley Park.

Mitella trifida Graham var. *trifida*
PACIFIC MITREWORT

A smaller plant, with lvs 2–4 cm broad that are about as long as broad and obscurely lobed, the lobes more or less rounded with blunt teeth. The petals are white, about the same length as the whitish sepals, perhaps 2 mm long, and divided into 3 blunt teeth at the tips. The flowering stem is 15–30 cm tall.

The least common mitrewort in the Park, known from a few collections in the valley of the White R, reaching about 6000 ft on rocky slopes.

Parnassia L.
GRASS-OF-PARNASSUS

Parnassia fimbriata Koenig var. *fimbriata*
FRINGED GRASS-OF-PARNASSUS

Hairless plants, with broad, kidney-shaped, untoothed lvs 2–6 cm broad that rise on slender stalks from a short rootstock. The flowering stem has a single small lf at its midpoint and is 20–40 cm tall, with a single fl at the top. The 5 petals are white, broader toward the tip and 8–12 mm long. The lower half of each petal is heavily fringed on the margin. There are 5 fertile stamens and 5 sterile staminodia, each of which is short and fleshy, with 5–9 blunt teeth.

Very common in subalpine bogs and meadows and along streams, mostly above 4000 ft, but reaching lower elevations along cool, shaded streams. Found at Mowich Lk; at about 3000 ft on Westside Rd; in Paradise Park; along the road between Cayuse Pass and Tipsoo Lk; on Shaw Cr at the bridge; at 4000 ft on the trail from White R to the Owyhigh Lks; and at Summerland at 5300 ft.

Saxifraga L.—SAXIFRAGE

A genus of great variety in the Park, including many fine wildflowers. All are perennial herbs, with basal lvs, although the flowering stems may bear a few small leaves or leaflike bracts. The flowers are solitary or in cymes or panicles, and in some species are slightly irregular. The 5 petals are conspicuous and the ovary is superior to slightly inferior, of 2 nearly distinct chambers; there are 10 stamens, the filaments of which are sometimes expanded. The fruit is a capsule or follicle.

1　Lvs opposite; fls 1 per stem, with purple petals .... *S. oppositifolia*
1　Lvs alternate or basal; fls several per stem, with white to yellowish petals
　.. 2

2(1)　Lvs rounded in outline, coarsely toothed and on slender stalks 3
2　Lvs longer than broad, stalked or not, even to shallowly toothed 7

3(2)　Teeth of the lvs doubly (or irregularly) toothed; lf stalks hairy; some of the fls usually replaced by plantlets .... *S. mertensiana*
3　Teeth of the lvs simple (once-toothed); lf stalks hairless; no fls replaced by plantlets 4

4(3)　Fls 2–4 per stem; basal lvs to 15 mm long and broad; flowering stems leafy
　.. *S. rivularis*
4　Fls numerous on the stems; basal lvs greater than 20 mm broad; flowering stems leafless or with a few small bracts 5

5(4)   Lf blades tapered to the stalk; filaments of the stamens somewhat broadened below the anthers .... *S. lyallii*

5   Base of the lf blades heart- or kidney-shaped; filaments broadened, winglike 6

6(5)   Petals nearly round; inflorescence glandular-hairy .... *S. odontoloma*

6   Petals elliptical; inflorescence with long hairs, not glandular .... *S. nelsoniana* ssp. *cascadensis*

7(6)   Lvs with 3 lobes at the tip; fls about 5 or 6 per stem .... *S. cespitosa* ssp. *cespitosa* var. *emarginata*

7   Lvs toothed or not, but not 3-lobed at the tip; fls usually more numerous on each stem 8

8(7)   Lvs not toothed, evergreen, to 10 mm long 9

8   Lvs toothed, deciduous, mostly more than 20 mm long 11

9(8)   Lvs hairless or with a few hairs on the margin toward the base; petals white .... *S. tolmiei*

9   Lvs with hairs along the margins; petals white with purplish to orangish spots 10

10(9) Lvs lanceolate, pointed; hairs on the lf margin coarse, relatively few ... . *S. bronchialis* ssp. *austromontana*

10   Lvs spoon-shaped, rounded at the tip; marginal hairs many, slender ... . *S. bronchialis* ssp. *vespertina*

11(8) Lvs oval to ovate, finely toothed; petals oval, all equal in size; fls never replaced by plantlets .... *S. occidentalis*

11   Lvs oblanceolate and coarsely toothed; petals lanceolate, unequal in size; fls often replaced by plantlets 12

12(11) Many fls replaced by plantlets .... *S. ferruginea* var. *vreelandii*

12   Fls not replaced by plants .... *S. ferruginea* var. *ferruginea*

Saxifraga bronchialis L. ssp. *austromontana* (Wiegand) Piper
YELLOW-DOT SAXIFRAGE

Tufted, with the evergreen lvs crowded on the short spreading stems; old brown lvs clothe the lower stems. In this subspecies, the lvs are 1–1.5 cm long, lanceolate and tapered to a sharp tip, fringed on the margins with very short (less than 0.3 mm) coarse hairs, but not toothed. The flowering stems are 10–20 cm tall, with several narrow lvs and spreading branches. The petals are 5–6 mm long and spotted yellow or orange.

Not common, but more widespread than var. *vespertina*, growing in dryish, rocky places, usually with some shade, 3000-8000 ft (the upper limit according to Ola Edwards in her study of the alpine meadows on Mount Rainier). Found mostly on the west and south sides of the Park: at Ipsut Pass; on the trail between Klapatche and St. Andrews Parks; at 3000 ft along the Nisqually R above Longmire. Also known from Mazama Ridge overlooking Louise Lk, from Mount Fremont, at Berkeley Park, and at Eunice Lk.

Saxifraga bronchialis L. ssp. *vespertina* (Small) Piper
YELLOW-DOT SAXIFRAGE

In this subspecies the lvs are less than 1 cm long, broadest at the blunt tip, and fringed with longer hairs, 0.3–0.5 mm long. The petals are somewhat shorter, 4–5 mm long.

Known from a few collections made in the southwest corner of the Park, in rocky places on Mount Wow and Iron Mtn.

Saxifraga cespitosa L. ssp. *cespitosa* var. *emarginata* (Small)
Rosend.—TUFTED SAXIFRAGE

A tightly tufted plant, with short branches densely clothed by the lvs. The lvs are 5–15 mm long, short-hairy chiefly on the margins, and broadest toward the 3-lobed tip. The flowering stem is about 15 cm tall, with only a few fls. The showy, white petals are nearly round and 5–6 mm long.

Infrequent, growing on cliffs and rocky ridges, 7000-8500 ft. Found at Paradise and at Sunrise.

Saxifraga ferruginea Graham var.
ferruginea—RUSTY SAXIFRAGE

In this variety the fls are not replaced by plantlets.

Known with certainty in the Park from an Irene Creso collection made in a "wet place between Longmire and Paradise," and from observations at Box Canyon.

Saxifraga ferruginea Graham var. *vreelandii*
(Small) Engl. & Irmsch—RUSTY SAXIFRAGE

The basal lvs of this species taper only gradually to the lf stalk, and are wedge- to spoon-shaped, 3–6 cm long, with a few broad teeth above the middle. The flowering stems are 10–30 cm tall and are glandular-hairy. The fls are frequently replaced by small, bulblike plantlets. The fls, when they occur, are somewhat irregular, with 3 larger upper petals 4–6 mm long and with 2 yellow spots at the base of each, and 2 lower, more slender white petals.

Common on seasonally wet cliffs and rocks, often in moss, and along rocky stream banks, 2500-6000 ft. Known from around Mowich Lk and Spray Park; at Van Trump and Klapatche Parks; at Longmire; along the road at Christine Falls; at the Glacier View Bridge; at a few places in Paradise Park; at 4500 ft on the Kotsuck Cr trail; along the road on the east side of the Cowlitz Divide; at Tipsoo Lk; on the White R Rd near Shaw Cr; and at Eunice Lk.

Saxifraga lyallii Engl. ssp. *hultenii* (Calder & Savile) Calder & Savile—RED-STEMMED SAXIFRAGE, LYALL'S SAXIFRAGE

An attractive little plant, with fan-shaped, toothed lvs that are squared off at the point of attachment to the slender stalk, overall less than 3 cm long. The 1–3 fls are borne on the short branches of the inflorescence that reaches up to 10 cm tall. The petals are white and the filaments of the stamens are broadened at the middle.

Listed as "watch" in *The Endangered, Threatened, and Sensitive Vascular Plants of Washington* (Washington Natural Heritage Program, 1997).

Rare and known from just one location, on a seasonally wet cliff along the road on the east side of the Cowlitz Divide at about 3000 ft. Not previously noted in the Park and perhaps a new southernmost station for the species in the state of Washington. Hitchcock and Cronquist (1973) say that *S. lyallii* ranges south from Alaska to the Cascades of north-central Washington, and Hultén (1968) places the range of ssp. *lyallii* on the east slope of the Rocky Mountains.

Saxifraga mertensiana Bong. MERTEN'S SAXIFRAGE

A low-growing plant, with lvs that are nearly round in outline and kidney-shaped at the base. The lvs are 3–10 cm broad, thickish, sparsely hairy, divided along the margin into fairly regular, shallow, blunt lobes. The lobes themselves have small teeth and the lf stalks are hairy. The fls are in a loose panicle, with slender, glandular branches. Usually a few of the fls are replaced by tiny plantlets. The white petals are 3–4 mm long.

Uncommon, found in crevices on moist cliffs and rocky banks. Unlike the other two round-leaf saxifrages—*S. nelsoniana* and *S. odontoloma*—with which this species is keyed, *S. mertensiana* is not typically found along streams. Collections are known from Mount Wow and at 6000 ft above Chinook Pass;

reported by Piper from the north side of Cowlitz Glacier and by Peter Dunwiddie from Mowich Lk.

Saxifraga nelsoniana D. Don ssp. *cascadensis* (Calder & Savile) Hultén
NELSON'S BROOK SAXIFRAGE

Quite similar to *S. mertensiana*. Here, the lvs are singly toothed on the margin and 3–7 cm broad. The lf stalks are hairless but the branches of the inflorescence are long-hairy. No fls are replaced by plantlets. The petals are 3–4 mm long, oval, and white, lacking colored spots.

Common in crevices in rocks along streams, up to 6000 ft. Collected at Mtn Meadows and along Mowich Lk Rd; in Spray Park; at 3000 ft on the Nisqually R; at 6000 ft in Van Trump Park; at Paradise; at Silver Falls; along the trail to the Owyhigh Lks; on Naches Peak; at 5500 near Summerland; at Berkeley Park; and at Elysian Fields.

Saxifraga occidentalis S. Watson
WESTERN SAXIFRAGE

Of coarser appearance than the other saxifrages, this species has ovate or oval lvs 4–8 cm long, shallowly toothed, and tapering gradually to a broad stalk. The young lvs are reddish-hairy on the underside. The stems are 5–20 cm tall and glandular-hairy. The inflorescence ranges from small and rounded to rather expanded and pyramid-shaped. The petals are white, oval, and 3–4 mm long; the filaments of the stamens are somewhat expanded.

Uncommon, in seasonally moist places on cliffs and rocky slopes, 4000-6000 ft. Collected at Mowich Lk; above Indian Henrys Hunting Ground; at 4000 ft on Tahoma Cr; and at 6000 ft in Paradise Park near Sluiskin Falls. Kartesz recognizes no varieties of this variable species. Plants from the Park key, in Hitchcock and Cronquist (1973), mostly to var. *occidentalis*, but some also to var. *allenii*.

Saxifraga odontoloma Piper—BROOK SAXIFRAGE

Another species similar to *S. mertensiana*, but sharing with *S. nelsoniana* a lf margin that is once-toothed. The lvs tend to be somewhat broader than the latter species, up to 8 cm broad, while the branches of the inflorescence are minutely glandular. The petals are 2–3 mm long, round, on a short stalk, with 2 yellow spots at the base.

Less common than *S. nelsoniana*, but in the same sorts of habitats, reaching about 5500 ft. Found at Mowich Lk; along the road above the Glacier View Bridge; along the Ohanapecosh R at 2500 ft; on the trail to Summerland from the White R; and in Berkeley Park.

Saxifraga oppositifolia L.
PURPLE SAXIFRAGE

Unique in the genus for having opposite lvs, this species forms spreading mats of stems crowded with lvs. The lvs are 3–5 mm long, evergreen, 4-ranked, mostly ovate, and fringed with hairs on the margins. The fls are solitary on slender stalks, from the axils of the lvs, with purple or pinkish purple petals 8–9 mm long.

A rare and distinctive saxifrage, known in the Park only from the northeast corner. It has been collected at 5500 ft on Fryingpan Cr and at an elevation of 7000 ft on Burroughs Mtn.

Saxifraga rivularis L.—PYGMY SAXIFRAGE

A weakly tufted plant with kidney-shaped lvs and stems bearing 2–4 fls. The lvs have 3–6 broad teeth and are 5–15 mm long, on very slender stalks. The stems are less than 10 cm tall, with 1–3 small lvs. The white petals are 3–6 mm long.

Described by early writers as a very rare plant, found in crevices of rocks. St. John and Warren (1937) say the plant was found on a rock wall at Sluiskin Falls in 1895 by O. D. Allen. A sheet in the Park herbarium, of *Mitella trifida*, collected on Mount Wow by Allen, includes a single, unlabeled plant of *S. rivularis*. It is not certain that the saxifrage itself came from that location.

Listed as "sensitive" in *The Endangered, Threatened, and Sensitive Vascular Plants of Washington* (Washington Natural Heritage Program, 1997).

Saxifraga tolmiei Torr. & A. Gray
TOLMIE'S SAXIFRAGE

A mat-forming plant, with short branches crowded with lvs. The dark green lvs are alternate on the stems, 2–10 mm long, cigar- to spoon-shaped, thick and inrolled on the margins, and hairless or with a few long hairs near the base. The flowering stems are 2–8 cm tall, usually leafless, with 1–3 fls. The petals are 5–6 mm long, white, and oblanceolate; the filaments of the stamens are expanded and appear petal-like (and at first glance the fl might appear to have 10 petals).

Common, 5500-9000 ft, along small streams, on pumice flats, and on talus slopes, often where snow lingers into the summer. It tends to favor disturbed ground and sometimes even colonizes the edges of trails. Found at St. Andrews Park; in upper Paradise Park; at 6500 ft in Seattle Park; at Sunrise and near Frozen Lk; and at 7100 ft on Burroughs Mtn. It's also abundant on the east side of Naches Peak. This plant is named for Dr. William Tolmie, surgeon to the Hudson's Bay Company, who visited the northeast corner of what is presently the National Park in 1833.

Suksdorfia A. Gray—SUKSDORFIA

Suksdorfia ranunculifolia (Hook.) Engl.
BUTTERCUP SUKSDORFIA

Similar in appearance to some of the saxifrages, this genus is distinguished by having 5 stamens. Bulblets are often found along the rootstock and in the axils of some of the lvs. The lvs are on long stalks, fan-shaped and divided into 3 major lobes, each lobe with 3 or 4 broad teeth. The flowering stem is leafy, 10–30 cm tall and forms a flat-topped panicle. The petals are white and 6–7 mm long and the sepals are conspicuously glandular.

Fairly widespread in the Park, but nowhere abundant. In crevices on damp cliffs and among rocks, 3000-5000 ft. Found at Paradise; at Box Canyon; along the road on the east side of the Cowlitz Divide; at Dewey Cr on Hwy 410; and along the road between Cayuse Pass and Tipsoo Lk.

Tellima R. Br.—FRINGECUP

Tellima grandiflora (Pursh) Douglas ex Lindl.
FRAGRANT FRINGECUP

Named for the spreading, reddish petals that are notably fringed at the tips. The plant has mostly basal, heart-shaped lvs that are 3–6 cm broad, short-hairy, 3–5-lobed and coarsely toothed. The flowering stem is 50–80 cm tall, with 2–3 small lvs and a tall raceme. The petals are 7–10 mm long and rather broad; there are 10 stamens.

Common in moist forests below 4500 ft on the south side of the Park. Found along Westside Rd; at Longmire on the Trail of the Shadows; at the Glacier View Bridge; in Stevens Canyon; around Ohanapecosh and at the Grove of the Patriarchs; and along Hwy 410 to about 4000 ft.

Tiarella L.—FOAMFLOWER, COOLWORT

Delicate woodland plants, from creeping rootstocks, with a few basal lvs and leafy stems. The leaves and stems are short-hairy. The leaves are palmately lobed, variously toothed, and 3–8 cm broad. The stems are 15–40 cm tall, with 1–3 smaller lvs. The flowers are in a narrow, loose panicle. The white petals are 2–3 mm long and the sepals are also whitish. There are 10 stamens. The capsule develops into two halves very unequal in size.

Mixed populations of the two varieties are frequently found, as at Sunshine Point, around Longmire, and along the trail to Silver Falls. In such places, hybrids can sometimes be seen. At the roadside not far beyond Sunshine Point, for example, on a patch of ground no more than 1 m across, were found plants with (1) basal leaves and stem leaves both 3-lobed, (2) basal and stem leaves simple but deeply divided, and (3) 3-lobed basal leaves and simple stem leaves; one plant here even had finely cut, laciniate leaves. [See Ganders and Ganders, 1983, for a discussion of leaf polymorphism in *Tiarella*; they argue that *T. trifoliata* var. *laciniata*, recognized in Hitchcock and Cronquist (1973), is merely a hybrid, not meriting varietal status.]

1   Lvs shallowly to deeply lobed but not fully divided ....
.... *T. trifoliata* var. *unifoliata*
1   Lvs fully divided into 3 lobes .... *T. trifoliata* var. *trifoliata*

Tiarella trifoliata L. var. *trifoliata*—FOAMFLOWER

Very common in moist forests below 3,500 ft, often forming dense carpets. Found chiefly on the west and south sides of the Park, as at the Nisqually entrance; at Longmire; and at 3,500 ft along the Ohanapecosh R.

Tiarella trifoliata L. var. *unifoliata* (Hook.) Kurtz.—FOAMFLOWER

As common but more widespread than the above variety, reaching 5500 ft, and the only variety found in the northeast corner of the Park. Found on Ipsut Cr at 2300 ft; at Mowich Lk; between the Nisqually entrance and the Glacier View Bridge; in the Butter Cr Research Natural Area; at Box Canyon; around Ohanapecosh; at 4000 on Dewey Cr; near Tipsoo Lk; in the White R V; at Berkeley Park; and at Eunice Lk.

Tolmiea Torr. & A. Gray
PIGGYBACK PLANT, YOUTH-ON-AGE

Tolmiea menziesii (Pursh) Torr. & A. Gray—PIGGYBACK PLANT

Familiar as a houseplant, although the wild version is only occasionally seen to produce plantlets on the lf blades. The lvs and stems are light green and rather coarsely hairy. The basal lvs are ovate, heart-shaped at the base, shallowly lobed, toothed, and 5–15 cm long. The stems are 40–60 cm tall, with 1 or 2 smaller lvs and a tall raceme. The petals are purplish to almost brown, usually 4 in number, and 6–8 mm long; there are 3 stamens.

Common in moist woods, mostly on the south side of the Park, reaching about 4000 ft. Found at the Nisqually entrance; along Westside Rd; at Longmire; at the Glacier View Bridge; around Ohanapecosh; and in the lower Carbon R Valley.

Scrophulariaceae Juss.—FIGWORT FAMILY

Second only to the Asteraceae in terms of the number of members in the Park's flora, the figwort family also contributes a large number of the most-noticed wildflowers, including penstemons, paintbrushes, and louseworts. Most are perennial herbs, although a few penstemons have woody stems and a small number (chiefly weeds) are annuals or biennials. The flowers are weakly to strongly irregular and 2-lipped, but unlike the mint family, the figworts have neither square stems nor distinctive odors. There are typically 5 sepals, which are partly fused, and the petals are united to form a short to long tube, with 5 free lobes. There are generally 4 or 5 petals that are fused most of their length to form a tubular corolla. There are 4 stamens in most of the genera; a 5th stamen may be present as a sterile filament or gland. The ovary is superior, of 2 chambers, and develops into a capsule with numerous, tiny seeds.

Synthyris schizantha Piper, fringed synthyris, is said by Hitchcock (1959–1967) to occur in the "Cascade Range near Mount Rainier." This is an attractive, tufted perennial with heart-shaped, lobed leaves and flowers in a small cluster, with 4 blue-violet, deeply fringed petals. "Near Mount Rainier" may refer to a Flett collection made near Elbe, some distance west of the Park; it is not known from within the Park.

1	Stamens with anthers 5; corolla nearly regular	*Verbascum*
1	Stamens with anthers 2 or 4 (a 5th sterile filament may be present); corolla 2-lipped	2
2(1)	Lvs alternate	3
2	Lvs opposite	6
3(2)	Lower lip of the corolla long-spurred; fls yellow	*Linaria*
3	Lower lip of the corolla not spurred	4
4(3)	Corolla 30–50 mm long, tubular, the anthers not enclosed in the upper corolla lip; weedy biennial more than 1 m tall	*Digitalis*
4	Corolla to 30 mm long, strongly 2-lipped, the upper lip (galea) enclosing the anthers; seldom more than 70 cm tall	5
5(4)	Galea straight and slender, beaklike; bracts brightly colored	*Castilleja*
5	Galea hoodlike, curved, or resembling an elephant's trunk; bracts greenish	*Pedicularis*
6(2)	Stamens 2; the 2 lobes of the upper lip fused to form a single wide, flat lobe (the corolla thus appearing to be 4-lobed)	*Veronica*
6	Stamens 4; upper lip of the corolla with 2 lobes	7
7(6)	Lobes of the calyx united into a tube	8
7	Lobes of the calyx free or united only at the base	11
8(7)	Corolla bright yellow or rose-pink	9
8	Corolla white, blue, blue-violet, or blue and white (*Penstemon confertus* is cream-colored to pale yellowish)	10

9(8)   Calyx tube inflated; upper corolla lip forming a galea .... *Rhinanthus*
9   Calyx tube strongly angled; corolla 2-lobed, galea absent .... *Mimulus*

10(9) Fls attached directly to the stem; corolla white with purple marks; lower
       lvs toothed .... *Euphrasia*
10   Fls on slender stalks; upper lip of corolla whitish, the lower blue; lower
       lvs not toothed .... *Collinsia*

11(7) Filaments of the anthers hairless at the base .... *Penstemon*
11   Filaments of the anthers short-hairy at the base .... *Nothochelone*

Castilleja Mutis ex L. f.—PAINTBRUSH

The scarlet and magenta paintbrushes are signature plants of subalpine meadows, and one species, *C. cryptantha*, was long thought to be strictly endemic to the Park. All are unbranched (although especially vigorous specimens of *C. miniata* may be branched), the leafy stems rising from singly or in tufts from short rootstocks. The leaves are alternate, undivided or lobed, and attached directly to the stems. The flowers are in racemes or spikes, each flower with a colorful bract (in fact, in most species more of the color of the "flowers" is actually to be found in the bracts). The calyx is 4-lobed and forms a tube; the corolla is 2-lipped: the upper 2 lobes form an elongated structure called a galea, which projects forward from the corolla tube to cover the stamens while the 3 lower lobes are somewhat petal-like or mere teeth. There are 4 stamens. Paintbrushes are described as "hemiparasites," for they use specialized roots to obtain at least some part of their nutrients from the roots of other plants. This treatment follows that of Mark Egger for the *Flora of North America.*

1   Corolla and calyx yellowish, the corolla hidden or nearly hidden by the dull-
       colored bracts; stems slender, to 15 cm tall; lvs glandular-hairy and grayish
       green .... *C. cryptantha*
1   Corolla red to rose-purple (rarely orange-red or yellowish or whitish); bracts
       brightly colored and not hiding the corolla; plants usually more than 20 cm
       tall; lvs various 2

2(1)   Lvs not divided; bracts not divided or with 2 short lateral teeth at the apex
       .... *C. miniata* var. *miniata*
2   Lvs and bracts deeply divided, with 2–6 slender lateral lobes 3

3(2)   Corolla and bracts rose-purple .... *C. parviflora* var. *oreopola*
3   Corolla and bracts red (rarely orange-red or yellowish) 4

4(3)   Tufted alpine plants, to 20 cm tall; lvs linear, with short, curled hairs ....
       .... *C. rupicola*
4   Erect plants of low to middle elevations, more than 30 cm tall; lvs lanceolate
       to ovate, with long, soft, more or less straight hairs ....
       .... *C. hispida* var. *hispida*

Castilleja cryptantha Pennell & G. N. Jones
MOUNT RAINIER PAINTBRUSH

Also known as "obscure paintbrush" for the way its fls are hidden by the bracts, this is an unimpressive plant compared to the other paintbrushes. It reaches 10–15 cm and is long-hairy throughout. The lvs are mostly undivided and 2–3 cm long. The bracts (and sometimes the upper lvs) have 3 pointed lobes and are greenish to somewhat purplish. The corolla is dull yellow and 12–16 mm, somewhat shorter than the bracts. The galea is 4–5 mm long.

Currently listed as a "species of concern" by the U.S. Fish and Wildlife Service and listed as "sensitive" in *The Endangered, Threatened, and Sensitive Vascular Plants of Washington* (Washington Natural Heritage Program, 1997). Long believed to be endemic to Mount Rainier National Park, this species has been found not far east of the Park boundary in the Wenatchee National Forest, on the American Ridge trail (the collection, made in 1989, is in the herbarium at the University of Washington). Mark Egger, who is writing about *Castilleja* for the *Flora of North America*, points out that the American Ridge site is "part of the greater Mount Rainier ecosystem."

Found on a small number of sites in dryish subalpine meadows on the north side of the Park: at Sunrise; Frozen Lk; at Interglacier Basin; in Grand Park; near Mystic Lk; in Moraine Park; above the Owyhigh Lks; in Berkeley Park; near Clover Lk; and on a lateral moraine of the Carbon Glacier. It generally occurs on ground where the soil derives from pumice, which the prevailing winds have carried to the north and east sides of Mount Rainier. Earlier collectors misidentified this species as *C. chrysantha* and *C. levisecta*.

Castilleja hispida Benth. var. hispida
HARSH PAINTBRUSH

The common lowland species, with roughly hairy stems 30–50 cm tall. The lvs are 2.5–5 cm long, broadly lanceolate and mostly 3–5-lobed, with long, soft, more or less straight hairs. The bracts are also 3–5-lobed, and are usually scarlet but are also occasionally yellow or orange. The corolla is 25–30 mm long and the galea 10–15 mm.

Common on dry, gravelly soil below about 4000 ft; frequently seen at roadsides. Known from Westside Rd; along the road on the east side of the Cowlitz Divide; in the Ohanapecosh entrance area; to about 4500 ft on Sunrise Rd; and reported from Berkeley Park. At about 2500 ft on Westside Rd, plants are found that range from nearly yellow to pale orange to orange-red to bright red. Kartesz considers *C. hispida* to have subspecies; Mark Egger, who is preparing an account of the genus for the Flora of North America, uses varieties and that use is followed here.

Castilleja miniata Douglas ex Hook. var. *miniata*
SCARLET PAINTBRUSH

Thinly hairy, with tufted stems 20–40 cm tall, and lanceolate, undivided lvs. The ovate bracts have a pair of small, sharp teeth near the tip and are scarlet. The corolla is about 30 mm long, the galea not quite 15 mm long.

Common in meadows and occasionally in openings in woods, mostly above 5000 ft but know to occur as low as 2800 ft on Westside Rd. Found at Mowich Lk; on the trail between St. Andrews and Klapatche Parks; Indian Henrys Hunting Ground; on the west side of the Nisqually R above the bridge; at Paradise; at 5500 ft at Goat Pass; along the road from Cayuse Pass to Chinook Pass; at Sunrise and along the trail to Burroughs Mtn.

Castilleja parviflora Bong. var. *oreopola* (Greenm.) Ownbey
MAGENTA PAINTBRUSH

A low plant, with stems just 15–30 cm tall, with some hairs on the upper stems and lvs. The lvs are 2–4 cm long, with 1 or 2 pairs of slender lobes. The bracts of the fls are rose-purple to magenta, with a pair of lobes. The galea is up to about half the length of the 20–30 mm long corolla.

Common and abundant in subalpine meadows throughout the Park. Collected at Mowich Lk and Spray Park; in Klapatche Park; at Indian Henrys Hunting Ground and Van Trump Park; across Paradise meadows; at Reflection Lks; on Kotsuck Cr trail at 5000 ft; in the meadow at the summit of Shriner Peak; at Chinook Pass and Tipsoo Lk; at Sunrise and on the trail to Burroughs Mtn; in Seattle Park; in Moraine Park; at Berkeley Park; and at Eunice Lk.

Reports of *C. parviflora* var. *albida* from Berkeley Park are almost certainly based on misidentifications of albino variants of var. *oreopola*. (See "Doubtful and Excluded Species," p. 469.)

Castilleja rupicola Piper ex Fernald—CLIFF PAINTBRUSH

A small species with tufted stems less than 20 cm tall, hairy on both the lvs and stems. The lvs are linear, with 1–3 pairs of slender lobes, overall 2–3 cm long. There are relatively few fls in the inflorescence. The bracts are tipped scarlet; the calyx and corolla are scarlet as well. The corolla is 20 mm long, and the galea is much longer than the tube.

Common on cliffs and rocky slopes above about 5500 ft, chiefly on the north and east sides of the Park. Found between Cayuse Pass and Chinook Pass; at Sunrise; around Frozen Lk; on the trail to Burroughs Mtn; in Seattle Park; on Mineral Mtn near Mystic Lk; and reported from ridges around Spray Park. A Warren collection in the Park herbarium is labeled "cliffs between Iron and Crystal Mountains," perhaps meaning between Iron and Copper Mountains, on the southwest side of Mount Rainier. The species was discovered on Mount Rainier by C. V. Piper.

Collinsia Nutt.—BLUE-EYED MARY

Collinsia parviflora Lindl. var. *parviflora*
SMALL-FLOWERED BLUE-EYED MARY

A small plant, easy to overlook when it grows with grasses and other low herbs. An annual, with slender stems to 20 cm tall, but often less, and opposite lvs. The lower lvs are stalked and roundish, the upper more linear and lacking stalks, and 1–2 cm long. The blue fls are on long and slender stalks in the axils of the upper lvs; the corolla is cocked at an angle from the stem, strongly 2-lipped, and 6–7 mm long. There are 4 fertile stamens; a 5th is represented by a small gland inside the corolla.

Common below about 6500 ft in dryish, open woods, on banks, and at roadsides. Found along Westside Rd; in Stevens Canyon; at Box Canyon; on the road along the Cowlitz Divide; in the Ohanapecosh area; at the White R campground; and on the moraine along the Carbon Glacier.

Digitalis L.—FOXGLOVE

Digitalis purpurea L. var. *purpurea*—FOXGLOVE

A conspicuous weedy biennial. The first year's growth is a rosette of large, felty lvs; not until the second year is the tall flowering stem, which may reach 1.5 m, produced. The basal lvs are more or less ovate and 15–25 cm long; the stem lvs are smaller, narrower, and alternate. The fls nod in a 1-sided raceme and are usually purplish, but sometimes nearly white, and spotted inside, overall 3–5 cm long.

Common at low elevations along roadsides and on open ground along rivers, as along Westside Rd; at Longmire; at the Ohanapecosh entrance; and in the White R Valley.

Euphrasia L.—EYEBRIGHT

Euphrasia nemorosa (Pers.) Wettst.
COMMON EYEBRIGHT

A small annual weed, usually hidden among grasses at roadsides. It reaches 15 cm tall on good soil and has opposite, mostly hairless, toothed lvs less than 1 cm long. The fls are in a leafy-bracted raceme. The bracts are leaflike, with long teeth tipped with bristles. The fls are 5–10 mm long, and strongly 2-lipped: the upper lip, of 2 lobes, forms a blunt hood and is usually light blue, while the lower lip has 3 small, spreading lobes that are white with purple lines. There are 4 stamens, hidden beneath the upper corolla lip.

A low-growing weed, first noted in the Park in this study. Found about 1 mi south of the White R entrance, in low grass at the side of the road, blooming late into the fall.

Linaria Mill.—TOADFLAX

Two weeds, both seldom encountered in the Park. The tall stems carry spikes of vividly-colored flowers: each is strongly 2-lipped, the upper bannerlike and the lower spreading, with a bright orange swelling (called a palate) that nearly closes the throat of the flower. The tube extends as a long, downward-projecting spur. There are 4 stamens and the fruit is a spherical capsule.

1   Lvs ovate; fls 3-5 cm long .... *L. dalmatica*
1   Lvs linear; fls 2-3 cm long .... *L. vulgaris*

Linaria dalmatica Mill.
DALMATIAN TOADFLAX

Weedy perennial, forming clumps of stems to 100 cm tall. The stems and lvs are hairless and glaucous. The lvs are 2 to 5 cm long, ovate or widest at the middle, and more or less clasp the stem. The fls are in a loose spike; each is 3-5 cm long.

Found in 1999 on the roadside north of Cayuse Pass at about 4300 ft.

Linaria vulgaris Mill.—BUTTER-AND-EGGS

A perennial weed, spreading by rootstocks and forming small patches, with upright, slender stems that may reach 60–80 cm tall. The plant is hairless and the alternate, stalkless, linear lvs are somewhat glaucous and 2–5 cm long. The bright yellow fls are in a long spike; each is strongly 2-lipped and 2–3 cm long. The palate is darker than that of *L. dalmatica.*

Occasional weed of roadsides and waste ground. Collected once at the roadside at Laughingwater Cr and reported from 3000 ft along the road across Cowlitz Divide.

Mimulus L.—MONKEYFLOWER

Cheerful flowers, bright yellow or pink, often decorated with spots and markings, common in a variety of moist habitats, along streams and in meadows. These are annual or perennial herbs with opposite leaves. The flower is weakly to strongly 2-lipped, usually with an angled calyx and 4 stamens; they are borne on slender, frequently long, stalks in the axils of the upper leaves.

1	Corolla pink or rose 2
1	Corolla yellow 3
2(1)	Slender annuals, to 10 cm tall; corolla to 1 cm long .... *M. breweri*
2	Vigorous perennials, more than 30 cm tall; corolla 3.5–5 cm long .... *M. lewisii*
3(1)	Slender annual, usually less than 15 cm tall; corolla about 10 mm long ... . *M. alsinoides*
3	Perennials, often taller; corolla 15–30 mm long 4
4(3)	Lvs pinnately veined, slimy-sticky; calyx teeth about equal in size 5
4	Lvs palmately veined, hairy or not, but not slimy or sticky; upper calyx tooth longer than the others 6
5(4)	Plants more than 50 cm tall; lvs mostly not stalked .... *M. moschatus* var. *sessilifolius*
5	Plants to 30 cm tall; lvs stalked .... *M. moschatus* var. *moschatus*

6(4)   Fls in a leafy-bracted raceme; lvs toothed, 1–5 cm long .... *M. guttatus*
6   Fls mostly solitary, occasionally in pairs; lvs not toothed, 1–2 cm long 7

7(6)   Plants branched, the branches generally creeping on the ground; lvs less than 1 cm long .... *M. tilingii* var. *cespitosus*
7   Plants more erect, the branches not creeping; lvs 1–2.5 cm long ....*M. tilingii* var. *tilingii*

Mimulus alsinoides Douglas ex Benth.
CHICKWEED MONKEYFLOWER

Often only a few centimeters tall when growing on shallow soils, this annual plant can reach 25 cm on favorable ground; 15 cm is perhaps an average height. The lvs are 5–15 mm long, on short stalks, ovate to roundish, and toothed. The fls are 8–14 mm long and moderately 2-lipped, with a large red blotch or smaller red spots on the lower lip.

Plants in the Park usually have hairs and small red spots on the palate of the lower lip. These are features by which Hitchcock and Cronquist separate *M. alsinoides* from *M. washingtonensis*. However, Park plants have lower calyx teeth that are long and blunt, diagnostic for *M. alsinoides*. *Mimulus washingtonensis* is a rather rare plant, found in a few places in eastern Washington and Oregon.

Fairly common on wet cliffs and rocky banks, typically on very shallow soil, reaching about 4500 ft; blooming early and easily overlooked. Found along the main road on the south side of Tumtum Peak; about 1.5 mi up Westside Rd; at about 3500 ft on the Rampart Ridge trail; and on mossy rocks where the road crosses the Cowlitz Divide.

Mimulus breweri (Greene) Coville
BREWER'S MONKEYFLOWER

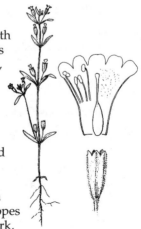

A small annual, just 3–10 cm tall, generally with a single, slender, glandular-hairy stem. The lvs are narrowly lanceolate, stalkless, not toothed, and 1–2 cm long. The fls are light rose, 5–10 mm long, and weakly 2-lipped.

Evidently infrequent in the Park, with just two herbarium collections found. One was made by J. B. Flett at "Snipe Lk, Mt Tacoma 5700 feet," a place name he seems to have used for a meltwater pond below the Interfork Glacier. Piper collected it on "dry cliffs near Camp of the Clouds" (perhaps on or near Alta Vista). The plant should be sought on open slopes and cliffs on the south and east sides of the Park, above 5000 ft.

Mimulus guttatus DC.—YELLOW MONKEYFLOWER

A common and variable perennial, spreading by runners and rootstocks, with mostly hairless stems 20–60 cm tall. The lvs are round to ovate, toothed, palmately veined, and 1–5 cm long; the lower lvs are on short stalks. The fls are yellow, 2–4 cm long, and strongly 2-lipped; the lower lip is usually red-spotted.

Very common along streams in all parts of the Park, around springs, and on wet ground in meadows, reaching about 4500 ft.

Mimulus lewisii Pursh
PURPLE MONKEYFLOWER, LEWIS'S MONKEYFLOWER

A vigorous perennial, forming clumps of stems 30–60 cm tall. The stems and lvs are sticky and long-hairy. The lvs are 3–5 cm long, lanceolate to ovate, toothed, and not stalked. The fls are bright rose-red to pale pink, strongly 2-lipped with spreading lobes and marked with yellow in the throat; white-flowered plants are occasionally found.

Common above 5000 ft, occasionally reaching lower elevations; along streams, around springs, and in wet meadows throughout the Park. It frequently grows with *M. guttatus* at middle elevations. (The infraspecific designation "forma *alba*" has been used for white-flowered plants collected in the Park.)

Mimulus moschatus Douglas ex Lindl. var. *moschatus*
MUSK MONKEYFLOWER

A lowland perennial that reaches the Park in a few places. The plants are sticky-slimy and densely hairy, and spread by rootstocks. The stems mostly sprawl across the ground and are 10–30 cm long. The lvs are lanceolate to ovate, toothed, and 2–5 cm long; at least the lower lvs are short-stalked. The fls are bright yellow, weakly 2-lipped, and 2–3 cm long.

Uncommon, growing on wet, usually sandy soil to almost 5500 ft. Found at Longmire; in Stevens Canyon; around the Ohanapecosh entrance; and at Chinook Pass.

Mimulus moschatus Douglas ex Lindl. var. *sessilifolius* A. Gray
LARGER MUSK MONKEYFLOWER

A more upright version, with stems 50 cm or more tall. The lvs are not stalked.

A single herbarium collection of this variety was found, made at "Schmitz Park," a location now unknown. According to Hitchcock and Cronquist (1973), this is the common variety west of the Cascades.

Mimulus tilingii Regel var. *caespitosus* (Greene) A. L. Grant
LARGE MOUNTAIN MONKEYFLOWER

A low, mostly hairless perennial, growing as tufts or small mats and spreading by rootstocks and runners. The branched stems are 5–20 cm tall, with ovate, stalkless lvs 2–10 mm long. The fls are 2–3 cm long,' strongly 2-lipped, yellow and spotted with red at the throat.

The common variety of this monkeyflower in the Park, found in subalpine meadows and along streams above 5000 ft. Known from Spray Park; Paradise meadows as high as 7400 ft; at Summerland; and at Berkeley Park. A collection was also made at Longmire.

Mimulus tilingii Regel var. *tilingii*
LARGE MOUNTAIN MONKEYFLOWER

Differing in stature from the variety above: the stems are seldom branched and the lvs are 10–25 mm long.

This robust variety is known only from two collections made by J. R. Slater: at Lee Cr near Mowich Lk, and at Cayuse Pass.

Nothochelone Straw—NOTHOCHELONE

Nothochelone nemorosa
(Douglas ex Lindl.) Straw
WOODLAND-PENSTEMON

A penstemon look-alike, and formerly included in that genus. The distinguishing features can be seen by dissecting the fl: in *Nothochelone* there is a specialized disk with nectar-producing glands at the base of the ovary and the filaments of the stamens are hairy at the base, whereas in *Penstemon* such a disk is absent and instead nectar is produced at the bases of the hairless filaments. The stems are erect or leaning, unbranched and reach 30–60 cm tall. The lvs are opposite, lanceolate, sharply toothed, and 5–10 cm long. The inflorescence is a panicle, with glandular-hairy branches. The corolla is blue-purple, 2.5–3 cm long, tubular and gently curved, and 2-lipped. The anthers are notably woolly (those true *Penstemon* species with woolly anthers are low, mound- or mat-forming plants).

Very common, 3000-6000 ft in somewhat shaded places on cliffs and rocky banks, often at roadsides; occasionally on rocks above streams. Found at Ipsut Pass; along the road to Mowich Lk at 4500 ft; at Ricksecker Point; at Bench Lk; at Box Canyon; along the road from Stevens Canyon to the Grove of the Patriarchs; at 4000 ft on the Shriner Peak trail; on rocky outcrops along Silver Falls trail; to 4500 ft on Sunrise Rd; on the Owyhigh Lks trail; along the road south of the White R entrance; and at Eunice Lk. Also reported from below Spray Park. This plant has been confused by some collectors with *Penstemon serrulatus*, largely, it would seem, on the basis of similarities in the lvs.

Pedicularis L.—LOUSEWORT

A genus much like *Castilleja*, the paintbrushes, in some respects, and, like them, conspicuous members of meadow communities at higher elevations. All are perennials, with both basal lvs and more or less leafy stems. The leaves are variously toothed or lobed, often fernlike, while the flowers are in racemes or spikes. The corolla is strongly 2-lipped; the upper lip forms a galea (an elongated structure that projects forward from the corolla tube to cover the stamens), which in several species is curved into unique shapes. In some species there is a projection called a "beak" at the tip of the galea. There are 4 stamens.

1	Lvs toothed but not divided; upper lip of the corolla slender and curved, reaching the lower lip	2
1	Lvs pinnately divided, the lobes toothed or again divided; upper lip of the corolla various	3
2(1)	Fls pink to purplish	*P. racemosa* ssp. *racemosa*
2	Fls white	*P. racemosa* ssp. *alba*
3(1)	Upper lip of the corolla hoodlike, with no beak or with a blunt or sharp point less than 1 mm long	4
3	Upper lip of the corolla not hoodlike, the beak long and slender	6
4(3)	Lvs chiefly basal, those of the stem smaller; calyx 7–8 mm long	*P. rainierensis*
4	Lvs chiefly on the upper stem, the basal lvs if any of the same size; calyx 10–12 mm long	5
5(4)	Inflorescence more or less hairy; common	*P. bracteosa* var. *latifolia*
5	Inflorescence hairless; fairly rare in the Park	*P. bracteosa* var. *flavida*
6(3)	Beak straight, at a right angle to the corolla tube	*P. ornithorhyncha*
6	Beak curved	7

7(6) Corolla reddish purple; the beak curved upward and outward, resembling an elephant's trunk .... *P. groenlandica* ssp. *surrecta*

7 Corolla whitish; beak rounded and partially tucked into the lower lip ......
.... *P. contorta* var. *contorta*

Pedicularis bracteosa Benth. var. *flavida* (Pennell) Cronquist—BRACTED LOUSEWORT

Variety *flavida* is the one commonly found in the Cascade Range south of Mount Rainier. It differs from var. *latifolia* in its yellowish fls and in technical characteristics of the calyx. In var. *flavida*, the free tips of the lateral sepals are very slender and glandular; in var. *latifolia*, they are shorter and broader and not at all glandular. In var. *flavida*, hairs in the inflorescence are limited to the calyx lobes and margins of the bracts.

This is the only variety found on the Naches Peak loop, where it is abundant on open slopes. Two herbarium collections were located, neither of which carries locality information: C. V. Piper collected it "in meadows 6500 ft" and O. D. Allen found it "on Mount Rainier, 7000-8000 ft."

Pedicularis bracteosa Benth. var. *latifolia* (Pennell) Cronquist—BRACTED LOUSEWORT

The tallest of the lousewort species in the Park, reaching 100 cm. The stems and lvs are hairless, but the inflorescence is typically hairy. The stem lvs are pinnately divided into slender, toothed lobes and are 5–20 cm long; the basal lvs are similar. The inflorescence may be branched and the bracts are lanceolate to ovate and undivided. The corolla is purplish, but sometimes mostly yellowish, and 15–22 mm long; the galea is about 10–15 mm long, straight at the base and hooded at the tip.

Common in meadows and on moist slopes, 5000-7000 ft. Known from Mowich Lk; along the trail between Klapatche and St. Andrews Parks; at Indian Henrys Hunting Ground; in Paradise meadows; in the Butter Cr Research Natural Area; at Goat Pass; northeast of Hidden Lk at 5850 ft; in meadows around Tipsoo Lk; at Sunrise; at Grand Park; at Berkeley Park; and at Ipsut Pass.

Pedicularis contorta Benth. var. *contorta* COILED-BEAK LOUSEWORT

A species 15–40 cm tall, with chiefly basal lvs, which are pinnately divided into slender, toothed lobes, hairless or sparsely hairy; the stem lvs are notably smaller. The bracts are broad, with numerous narrow teeth. The corolla is white and 10–13 mm long; the upper lip is curved into a half-circle and tucks into the very broad lower lip.

Common in meadows, 5000-7000 ft, and especially abundant on the east side of the Park. Collected around Mowich Lk; at Indian Henrys Hunting Ground; at 6000 ft in Van Trump Park; in Paradise Park; at Sunrise and around Frozen Lk; in Moraine Park; at the foot of Goat Island Mtn; at Berkeley Park; and at Eunice Lk. This species was first collected in 1833 by William Tolmie in the northwest corner of the present Park.

Pedicularis groenlandica Retz. ssp. *surrecta* (Benth.) Piper
ELEPHANT'S HEAD

20–60 cm tall, with hairless stems and lvs. The numerous basal lvs are 5–20 cm long, lanceolate, and pinnately divided into slender, toothed lobes, and are often reddish colored; the stems are leafy, the size of the lvs reduced upwards. The corolla is pinkish purple to reddish purple, about 10 mm long, with a long and narrow beak from the upper lip that is 10–15 mm long, curved downward and outward, like the trunk of an elephant.

Common in wet meadows, 4000-7000 ft. Found at Mtn Meadows; in the lower part of Paradise meadows; at Reflection Lks; in the meadow above Bench Lk; at 5000 ft at Three Lks; at Shriner Lk; at Sunrise; at Moraine Park; and in Berkeley Park.

Pedicularis ornithorhyncha Benth.
BIRD'S-BEAK LOUSEWORT

A small plant, 10–20 cm tall, with narrowly lanceolate, pinnately divided into short, toothed lobes. The lvs are 3–12 cm long and usually hairless; there are 1 or 2 stem lvs. The fls are in short, dense clusters. The corolla is rose-purple and 10–15 mm long; the beak of the galea is 2–4 mm long and bent at a right angle to the tube.

Common in meadows and on moist slopes, 6000-7000 ft across the Park. Known from Spray Park and Mowich Lk; in Klapatche Park; at Van Trump Park; at Indian Henrys

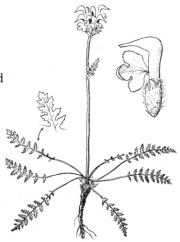

Hunting Ground; in Paradise meadows; around Reflection Lks; in the Butter Cr Research Natural Area; at Goat Pass; on Naches Peak at 5600 ft; at Sunrise and at 7000 ft on Burroughs Mtn; at Moraine Park; and at Eunice Lk. This species was first discovered by Dr. Tolmie on his 1833 Mount Rainier collecting expedition.

Pedicularis racemosa Douglas ex Benth. ssp. *alba* Pennell WHITE-FLOWERED SICKLETOP LOUSEWORT

Much like ssp. *racemosa*, but with white fls.

Not verified with herbarium collections, but this subspecies can be found along the Naches Peak loop trail. It is the common subspecies in eastern Washington.

Pedicularis racemosa Douglas ex Benth. ssp. *racemosa*—SICKLETOP LOUSEWORT, RAM'S-HORN LOUSEWORT

Easily distinguished by the lvs, which are evenly toothed but not divided into separate lobes. The lvs are long-lanceolate and 3–8 cm long, on short stalks; young plants are usually shaded reddish purple. The corolla is 1–1.5 cm long and pink to purplish; the upper lip is strongly curved so that its slender tip touches the twisted, broad lower lip.

Common in meadows and open woods at lower elevations than the other *Pedicularis* species, 3000-6000 ft. Found at Mowich Lk; on Mount Wow; at Indian Henrys Hunting Ground; in the lower part of the Paradise meadows; around Reflection Lks; on Eagle Peak; in the Butter Cr Research Natural Area; at Goat Pass; at Tipsoo Lk and Chinook Pass; at Frozen Lk; and in Berkeley Park.

Pedicularis rainierensis Pennell & Warren MOUNT RAINIER LOUSEWORT

A hairless species with stems 10–40 cm tall and mostly basal lvs that are pinnately divided into slender, narrow lvs 3–10 cm long. The stem lvs are few and much smaller. The inflorescence is fairly short and densely flowered. The corolla is 1–1.5 cm long and yellow; the upper lip is gently curved and hooded at the tip while the lower lobes are small, narrow, and spread very little from the throat of the fl.

Listed as "sensitive" in *The Endangered, Threatened, and Sensitive Vascular Plants of Washington* (Washington Natural Heritage Program, 1997). This species was once considered to be endemic to

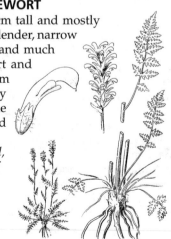

Mount Rainier National Park, but Laura Potash (1991) has reported that the plant is known from a moist subalpine hillside at the southern end of the Mount Baker – Snoqualmie National Forest" outside the Park boundary (possibly, that is, in the Crystal Mtn ski area?).

Fairly common but nowhere abundant, 5000-7000 ft, growing chiefly on north-facing slopes; it seems to be absent from the Paradise area. Found at Mowich Lk and Spray Park; at Klapatche Park; at Indian Henrys Hunting Ground; at Van Trump Park; at Chinook Pass; in Seattle Park; at Sunrise; at 5850 ft by Hidden Lk; at Berkeley Park; along the trail to Clover Lk; and at Eunice Lk.

Penstemon Schmidel—PENSTEMON

Some of the finest shrubby wildflowers in the Park are penstemons, including the brilliant red-pink cliff penstemon, *P. rupicola*, which is well-displayed on the highway approaches to Paradise, and the bright blue cascade penstemon, *P. serrulatus*. A 1997 book by Dee Strickler, *Northwest Penstemons*, is an excellent guide and reference to the genus.

Penstemons are common in the Park and are found at all elevations, growing as mid-sized shrubs or spreading mats. All have opposite, usually toothed, leaves; the lower leaves are usually short-stalked. The flowers are in racemes, panicles, or in whorled clusters and are strongly 2-lipped and tubular, open at the throat. There are 4 stamens and a 5th, sterile filament. A close examination of the anthers is important in this genus: key features are whether or not they are woolly and whether the pollen sacs spread apart and the way in which they open.

1	Anthers densely wooly	2
1	Anthers hairless or with, at most, a few straight hairs	6
2(1)	Erect plants, 30–60 cm tall; lvs 5–10 cm long .... see *Nothochelone*, above	
2	Plants with prostrate or short erect stems, never more than 30 cm tall; lvs to 6 cm long	3
3(2)	Stems erect, not matlike, to 30 cm tall; lvs lanceolate, 3–6 cm long .... *P. fruticosus*	
3	Stems prostrate, forming mats, the flowering stems less than 15 cm tall; lvs rounded to obovate, less than 2 cm long	4
4(3)	Lvs glue-green and glaucous; corolla bright pink to rose-red .... *P. rupicola*	
4	Lvs green, not glaucous; corolla blue-purple to violet	5
5(4)	Lvs not toothed, rounded at the tip; corolla 1.5–2.5 cm long .... *P. davidsonii* var. *davidsonii*	
5	Lvs toothed, sharp at the tip; corolla 2.5–3.5 cm long .... *P. davidsonii* var. *menziesii*	
6(1)	Anther horseshoe-shaped at maturity, the two anther sacs opening only at the top .... *P. serrulatus*	
6	Anther sacs opening along their lengths and spreading apart at maturity, thus the anther not horseshoe-shaped	7

7(6) Corolla pale yellow or cream-colored .... *P. confertus*
7   Corolla blue to blue-purple 8

8(7) Lvs toothed, ovate; fls in 4 or 5 whorls per stem .... *P. ovatus*
8   Lvs not toothed, elliptical-lanceolate; fls in 1 or 2 whorls per stem ....
      .... *P. procerus* var. *tolmiei*

Penstemon confertus Douglas ex Lindl.
YELLOW PENSTEMON
Somewhat mat-forming, with only a few flowering stems 20–50 cm tall. The plant is hairless and the lvs are lanceolate to oblanceolate, not toothed, and 3–15 cm long. The fls are in narrow whorls, widely spaced at the top of the stem. The corollas are creamy white to yellowish and usually tipped downward, and 8–12 mm long, with a fairly slender tube; the throat of the fl is usually purple-spotted or -streaked. The anthers spread widely and open their full length.

Infrequent on talus slopes and moraines, 4000-7000 ft. Collected along the trail between Klapatche and St. Andrews Parks; on the southwest side of Mount Ararat; near the terminus of the Nisqually Glacier; at 5500 ft on Eagle Peak; on drier parts of the Paradise meadow; and Sunrise. Jones, in his 1938 flora, considered *P. confertus* and *P. procerus* to constitute a single species.

Penstemon davidsonii Greene var. *davidsonii*
DAVIDSON'S PENSTEMON
Differing from the more common var. *menziesii* in having lvs that are shorter, 5–10 mm long, and smaller fls, the corollas 1.5–2.5 cm long.

Uncommon; found in a few places among rocks above about 6500 ft. Found at 7000 ft between Panorama Point and McClure Rock; in the upper part of the Butter Cr Research Natural Area; on Burroughs Mtn; and on Mount Fremont. The variety reaches its northern limit of distribution at Mount Rainier.

Penstemon davidsonii Greene var. *menziesii* (D.D. Keck)
Cronquist—MENZIES'S PENSTEMON
A mat-forming plant with prostrate stems, able to form patches up to 50 cm across. The lvs are short-stalked, 8–20 mm long, oval to obovate, and toothed. The flowering stems are leafy and 3–10 cm tall, with several blue-violet fls.

Common in crevices of cliffs and among rocks on talus slopes, 5000-8000 ft; scarce on Mount Rainier itself. Found on Mount Pleasant at 6400 ft; on the ridge above Spray Park; in the Tatoosh Range; on Mazama Ridge and at Louise Lk; along the trail between Tipsoo and Dewey Lks; on the upper part of the west slope of Crystal Mtn; near Frozen Lk; at Berkeley Park; at Moraine Park; and at Eunice Lk.

Penstemon fruticosus (Pursh) Greene var. *fruticosus*—SHRUBBY PENSTEMON

A low, rounded shrub that reaches about 30 cm tall. The lvs are lanceolate to elliptical, usually toothed, but sometimes without teeth, pointed at the tip, hairless, and 2–4 cm long. There are typically 3–10 blue-violet fls in a simple raceme. The corolla tube is relatively wide and straight, with 2 prominent ridges on the lower lip. The anther sacs are woolly and spread widely.

A common species east of the Cascade summit; the occurrence here west of the Cascades is unusual.

Found in a few places on seasonally moist cliffs and banks, 4500-7000 ft. Found at Panorama Point; on Sunrise Ridge; and along the trail to Burroughs Mtn. At a low elevation, on the roadside below the White R campground, a patch of normal-colored purple-flowered plants surrounds a lovely pink-flowered plant.

Penstemon ovatus Douglas ex Hook. BROAD-LEAVED PENSTEMON

A tufted plant with stems 60–120 cm tall, minutely hairy overall. The lvs are lanceolate to ovate, sharply toothed, stalked below and stalkless higher on the stem. The fls are in several whorled clusters, on short stalks. The corolla is blue and 14–18 mm long, expanded at the throat. The anther sacs spread widely and are hairless. Superficially similar to the common *P. serrulatus*, but easily distinguished by an examination of the anthers.

This species is known for the Park only from an Allen collection made about 1895 in the "mountains near the upper valley of the Nisqually." Jones describes it as a plant of "banks and cliffs" at low elevations. It may not be well-established in the Park.

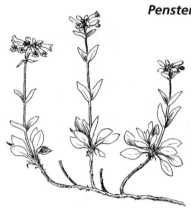

Penstemon procerus Douglas ex Graham var. *tolmiei* (Hook.) Cronquist
SMALL-FLOWERED PENSTEMON

Tufted, with several stems to about 15 cm tall. The lvs are lanceolate to narrowly elliptical, not toothed, not stalked, and hairless. The fls are in several tightly whorled clusters, blue or blue-violet, but occasionally creamy white. The corolla is 8–12 mm long, with a narrow tube, and whitish at the throat. The anther sacs are hairless and spread widely.

The variety *tolmiei* was discovered by Dr. William F. Tolmie on his 1833 visit the area now on the west side of the Park.

Common on moraines, pumice flats, and in meadows, 5000-8000 ft, mainly on the north and east sides of the Park. Found at Spray Park; at Indian Henrys Hunting Ground; in Paradise Park, up to 7000 ft between Panorama Point and McClure Rock; in the Butter Cr Research Natural Area; at 5900 ft at Interglacier; at Upper Palisades Lk; around Sunrise and at Sunrise; on Mount Fremont; and on the Burroughs Mtn trail. White-flowered plants are occasionally seen.

The variety *procerus* has been listed for the Park, but no basis is known for this. *Penstemon rydbergii* has also been included on lists for the Park, based on a citation in St. John and Warren (1937) of *Penstemon hesperius*; the collection, *Warren 1674*, is actually *P. procerus* var. *tolmiei*. For both of these, see the list of "Doubtful and Excluded Species," p. 469.

Penstemon rupicola (Piper) Howell
CLIFF PENSTEMON

A striking plant that grows as a mat or low mound, with flowering stems that are 5–10 cm tall. The lvs are oval to ovate, thick and leathery, toothed, finely hairy, glaucous, gray-green, and 5–7 mm long. There are 1–4 fls in a raceme on each stem. The corolla is deep pink to rose-red, 3–4 cm long, somewhat curved, with 2 sharp ridges on the lower lip. The anther sacs are densely woolly and spread widely; often the anthers protrude from the throat of the fl.

Common on cliffs and among rocks on talus slopes, 5000-8000 ft but occasionally lower, as near Narada Falls (where it makes an effective contrast to the more numerous *Penstemon serrulatus*) and at Box Canyon. Found at the Canyon Rim overlook; around Paradise Park; on Mazama Ridge; through Stevens Canyon and along the road across Cowlitz Divide; on the road between Cayuse and Chinook Passes; and at Sunrise. Plants found around Sunrise are strongly short-hairy.

Penstemon serrulatus Menzies ex Rees
CASCADE PENSTEMON

A tall but not shrubby plant, with unbranched stems 15–60 cm tall. The lvs are lanceolate to ovate, sharply and irregularly toothed, bright green, short-stalked on the lower part of the stem, and 2–10 cm long. The fls are in a single, somewhat headlike cluster. The corolla is bright blue, 1.5–2.5 cm long, with a short and relatively broad tube, although plants on the Kautz Cr mud flow are typically rose-purple. The anthers are horseshoe-shaped and usually have a few short hairs at the top.

Very common through the Park, reaching almost 6000 ft, generally along streams and in moist places among rocks. Found between Mountain Meadows and Mowich Lk; at Spray Park; at Ipsut Pass; along Westside Rd to 4000 ft; at Longmire; at the Glacier View Bridge; in the lower part of Paradise Park; in the Butter Cr Research Natural Area; in Stevens Canyon and at Box Canyon; along the main road on the east side of Cowlitz Divide; at the Ohanapecosh entrance; at Goat Pass; along the road between Cayuse Pass and Tipsoo Lk; along the road between the White R "Y" and Sunrise; at Eunice Lk; and in the Carbon R V up to the glacier snout. Plants on shaded wet rocks above Dewey Cr on Hwy 123 have light blue fls.

Rhinanthus L.—RATTLEBOX

Rhinanthus minor L. ssp. *minor*
LITTLE RATTLEBOX

An annual, with erect, usually unbranched stems 10–40 cm tall. The lvs are opposite and lanceolate, not stalked, 1–3 cm long and coarsely toothed; the upper lvs are ovate-lanceolate. The fls are in a leafy-bracted spike. The calyx lobes are united, with four small teeth at the tips, pear-shaped in side view, and about 1 cm long, becoming remarkably inflated and flattened as the fl matures. The corolla is 2-lipped, with the upper lip forming a humplike galea reaching about 6–9 mm beyond the calyx. The corolla is bright yellow (the tiny upper lobes of the galea are bluish, a nice study with a hand lens). "Rattlebox" is named for the sound made by the seeds when the mature capsule is shaken.

A small colony is found at the shop area at Kautz Cr, partly shaded by alder on a bank above the Nisqually R, growing among low grasses and herbs. Not previously known from the Park.

Verbascum L.—MULLEIN

Verbascum thapsus L.—COMMON MULLEIN

A biennial weed of impressive stature, with thick stems that can reach 2 m. The lvs and stems are feltlike, with dense, short hairs. The lvs are thickish, oblong to ovate, not toothed, and 10–40 cm long. The fls are yellow, in a dense spike. The corolla is nearly regular, with a very short tube and broad, spreading lobes, 16–20 mm broad. There are 5 stamens.

Occasional weed on low-elevation roadsides, especially along the White R.

Veronica L.—SPEEDWELL

A genus of annual and perennial herbs that can be found in most parts of the Park, both as weeds and as native plants. *Veronica cusickii* is prominent in subalpine meadows. The leaves are mostly opposite, at least on the lower stems; the stems are upright or sprawling. The flowers may be solitary in the axils of the leaves, or in racemes, which may terminate the stems or be axillary. The flowers are blue, or sometimes white, small, and nearly regular, with 4 lobes; the lower lobe is narrower. There are 2 stamens. The capsule is flattened and more or less heart-shaped.

1   Fls in racemes that arise from the axils of the lvs on the upper stem 2
1   Fls in racemes that terminate the stems 5

2(1)   Plant hairless 3
2   Plant short-hairy 4

3(2)   Lvs short-stalked, ovate, toothed .... *V. americana*
3   Lvs not stalked, linear-lanceolate, not toothed .... *V. scutellata*

4(2)   Lvs coarsely toothed, not stalked; corolla 10 mm broad .... *V. chamaedrys*
4   Lvs finely toothed, short-stalked; corolla 6–7 mm broad ....
   .... *V. officinalis* var. *officinalis*

5(1)   Style 6–9 mm long, as long as or longer than the capsule; lvs hairless, neither toothed nor stalked .... *V. cusickii*
5   Style less than 3 mm long, shorter than the capsule; lvs various 6

6(5)   Subalpine perennial; lvs all opposite .... *V. wormskioldii*
6   Low- to mid-elevation annuals or perennials; upper lvs and bracts alternate
   7

Veronica americana Schwein. ex Benth.
AMERICAN BROOKLIME

A sprawling perennial, with hairless stems 30–60 cm long, typically branched. The lvs are elliptical to ovate, broadly toothed, short-stalked, and 2–8 cm long. The fls are in slender, loose racemes in the axils of the upper lvs. The corolla is blue, marked with darker veins on the lobes, and 4–5 mm broad.

Common in swamps and along streams below 3000 ft. Found in ponds between the Nisqually entrance and Longmire and at Ohanapecosh.

Veronica arvensis L.
SMALL SPEEDWELL

An annual weed, short-hairy overall, with weak stems to 20 cm long. The lvs are ovate, broadly toothed, stalked on the lower stem and not stalked above, or broadly elliptical, 2–15 mm long. The fls are solitary in the axils of the upper lvs and leaflike bracts. The corolla is blue to purplish and 2–3 mm broad.

A common weed on roadsides and along trails. Frequently seen between the Nisqually entrance and Longmire.

Veronica chamaedrys L.
GERMANDER-SPEEDWELL

A perennial weed, from a creeping rootstock and short-hairy overall. The stems are mostly upright, 10–40 cm tall. The lvs are 1–2.5 cm long, ovate but blunt at the tip, coarsely toothed, and not stalked. The fls are in racemes from the axils of the upper lvs; the corolla is about 10 mm broad and bright blue with a white center.

A weed, first noted during fieldwork for this study at Sunshine Point, where it grows on open, moist ground among low grasses.

357

Veronica cusickii A. Gray
CUSICK'S SPEEDWELL

A finely hairy to hairless perennial, from slender rootstocks, tending to form loose mats. The lvs are dark green, not toothed, and 1–3 cm long. The fls are in an elongated terminal raceme, with small bracts. The dark blue (rarely white) corolla is 10–12 mm broad. The style is distinctive—as long as or longer than the mature capsule. The stamens are showy and longer than the corolla.

Common and widespread in meadows and along streams, above 5000 ft. Collected at Spray Park; in Klapatche Park; at Indian Henrys Hunting Ground; at Paradise; along Nickel Cr at 5000 ft; in the Butter Cr Research Natural Area; at Tipsoo Lk and Chinook Pass; at Upper Palisades Lk; above about 5000 ft on Sunrise Rd and around Sunrise; on the trail to Burroughs Mtn; and at Eunice Lk. White-flowered plants have been called *V. allenii*, for Professor O. D. Allen, the first resident botanist at Mount Rainier.

Veronica officinalis L. var. *officinalis*
COMMON SPEEDWELL, FOREST SPEEDWELL

A short-lived, finely hairy perennial, with 1–3 cm long, oval to obovate, finely toothed lvs that are short-stalked. The fls are in fairly dense, slender racemes in the axils of the lvs. The corolla is light blue and 4–8 mm broad.

A weed found in a few places near the Park entrances and at Longmire.

Veronica peregrina L. ssp. *xalapensis*
(Kunth) Pennell—PURSLANE SPEEDWELL

An upright annual, with linear to narrowly oblong lvs up to 2 cm long; the stems have short, glandular hairs. The fls are in a loose raceme, in the axils of short leaflike bracts. The corolla is white or pale blue, and 2–3 mm wide.

A native species, known from a collection made at Longmire.

Veronica scutellata L.—MARSH SPEEDWELL

Perennial, with hairless, lax stems from a rootstock, up to 50 cm long. The lvs are narrowly lanceolate, neither stalked nor toothed, and 2–4 cm long. The fls are in delicate racemes in the axils; the corolla is white to blue and 5–7 mm broad.

A native skullcap, growing in shallow ponds, that fls as the ponds dry in midsummer. Found between the Nisqually entrance and Tahoma Cr and around Ohanapecosh.

Veronica serpyllifolia L. ssp. *humifusa* (Dickson) Syme—THYME-LEAF SPEEDWELL

Native perennial, from a slender rootstock and hairy throughout and with sprawling stems up to 30 cm long. The lvs are elliptical to broadly ovate, blunt at the tip and broadly toothed. The fls are in a slender, terminal raceme, with at least some glandular hairs on the stem. The corolla is 6–7 mm broad and bright blue, with darker lines.

Uncommon native, found in open places in woods and along trails. Known from the Nisqually entrance area and Chinook Pass.

Veronica serpyllifolia L. ssp. *serpyllifolia*— THYME-LEAF SPEEDWELL

Differing from var. *humifusa* in several ways: the fls are white to pale blue, the plant is less hairy, and the upper stem is not glandular.

An occasional roadside weed around the Ohanapecosh area and on the trail to the Grove of the Patriarchs.

Veronica wormskioldii Roem. & Schult. AMERICAN ALPINE SPEEDWELL

Similar to *V. cusickii*, this is also a perennial, from a creeping rootstock. The stems are mostly erect, 5–30 cm tall, with long, wavy hairs. The lvs are broadly toothed or without teeth, oval to ovate, hairy, and 1–3 cm long. The fls are in a terminal raceme, on a glandular-hairy stem; the corolla is blue and 6–10 mm broad. The style is shorter than the mature capsule and the stamens are hidden within the corolla.

Common in moist mountain meadows and along streams, above 5000 ft, sometimes growing with *Veronica cusickii*. Found at Spray Park; through Paradise Park; on Dewey Cr where it crosses the road; between Cayuse Pass and Chinook Pass; at Sunrise and at Frozen Lk; at Berkeley Park; and above the snout of the Carbon Glacier.

Solanaceae Juss.—NIGHTSHADE FAMILY

Solanum L.—NIGHTSHADE

Represented by two weeds that are seldom seen in the Park. Both are annuals, with hairy, often glandular stems and lvs (occasionally only sparsely so). The lvs are alternate, stalked, and not toothed. The fls are in small panicles in the axils of the lvs. Fl parts are in 5s, with both the sepals and petals partly united. The fruit is a round berry.

1   Fruit black; stems mostly hairless .... *S. americanum*
1   Fruit yellowish; stems hairy .... *S. sarrachoides*

Solanum americanum Mill.
BLACK NIGHTSHADE

Native to North America but certainly not to the Park, this is an annual with spreading, branched stems that are mostly hairless. The lvs are ovate and not toothed, 2–15 cm long. The fls are in umbel-like clusters in the lf axils. The corolla is white, 3–6 mm broad, and deeply lobed. The berry is black.

A rare weed in the Park, collected a few times at Longmire in recent years.

Solanum sarrachoides Sendtn.
HAIRY NIGHTSHADE

A branched plant, with ovate, coarsely toothed lvs 1–4 cm long. The corolla is white and the free lobes are about equal in length to the tube and spread widely, about 1 cm broad. The fruit is a greenish yellow berry.

Listed by Jones, in his 1938 flora, as a roadside weed. It is infrequently seen and herbarium collections were not found in this study.

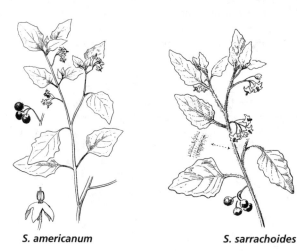

S. americanum   *S. sarrachoides*

Urticaceae Juss.
NETTLE FAMILY

Urtica L.—NETTLE

Urtica dioica L. ssp. *gracilis* (Aiton) Selander—STINGING NETTLE

A perennial, growing from spreading rootstocks and forming small stands of unbranched, 4-angled stems, reaching 1–2 m tall. The stems and lvs are clothed with stinging hairs. The lvs are opposite, ovate, coarsely toothed, and 5–15 cm long. Male and female fls are borne in separate clusters in the axils of the lvs and somewhat resemble catkins. Petals are absent and there are 4 or 5 sepals and stamens. The fruit is a small achene.

Common around the Nisqually, Ohanapecosh, and White R entrances to the Park, on wet ground along rivers, reaching about 2000 ft. Also found at the Grove of the Patriarchs.

Valerianaceae Batsch—VALERIAN FAMILY

Valeriana L.—VALERIAN

Two valerians are found in the Park, one of which, Sitka valerian, is a prominent meadow wildflower. They are mostly hairless perennials, with opposite leaves, which are pinnately divided into 3–7 leaflets. The flowers are borne on squarish stems in dense, rounded or flat-topped clusters. The corolla is 5-lobed and slightly irregular. The ovary is inferior and develops into an achene; the 5 lobes of the calyx develop into featherlike bristles at the top of the achene.

1　Lvs mostly basal, the leaflets untoothed or nearly so; known only from the lower Carbon R .... *V. scouleri*

1　Lvs found mostly on the flowering stems, the leaflets coarsely toothed; very common .... *V. sitchensis*

Valeriana scouleri Rydb.—SCOULER'S VALERIAN

A low plant, with mostly basal lvs and stems less than 60 cm tall. The leaflet margins are mostly smooth. The fls are white to pinkish, in rounded clusters. The achenes are 5–6 mm long.

A species of low elevations in western Washington, noted once by Mary Fries in the lower Carbon R V near Cataract Cr , where it grew on a low cliff. Not previously reported in the Park.

361

Valeriana sitchensis Bong.
SITKA VALERIAN

A taller plant with hairless lvs and square stems 60–120 cm tall. Each leaflet is 3–8 cm long, coarsely toothed and lanceolate to ovate or roundish; the basal lvs are very long-stalked. The fls are white and fragrant, in dense, flat or rounded clusters. Plants with pinkish fls are occasionally seen. The corolla is 6–8 mm long, with a long tube and fairly short lobes. The achenes are about 4 mm long.

Smaller plants are sometimes found on cool slopes at lower elevations [these lower-elevation plants were distinguished by Jones as var. *hookeri* (Shuttlew.) G. N. Jones]. Despite their smaller size, these plants are easily distinguished from *Valeriana scouleri* by the characteristics in the key.

Very common, 4500-6500 ft throughout the Park along streams, in meadows, and in wet places in open woods; also on dryish slopes on the north side of the mountain.

Violaceae Batsch—VIOLET FAMILY

Viola L.—VIOLET

Low-growing perennial herbs, from generally thick rootstocks and in some species spreading by runners. The Park's violets have alternate, undivided, often heart-shaped, long-stalked leaves. Flower parts are distinct and in 5s, with a superior ovary. The flowers are irregular, with 2 upper petals, 2 lateral petals, and 1 lower petal that is both larger and spurred. The lateral petals are often bearded with short, thick hairs. The anthers are crowded around the style and the fruit is a capsule.

1	Fls blue or white	2
1	Fls yellow	8
2(1)	Fls blue or blue-violet	3
2	Fls white (petals sometimes tinged bluish on the back)	5
3(2)	Lvs 2–5 cm broad; style head not bearded	*V. langsdorfii*
3	Lvs 1–3 (rarely to 4) cm broad; style head bearded with short hairs	4
4(3)	Fls 5–15 mm long, with a pouchlike spur; sepals fringed with hairs	*V. howellii*
	(see "Doubtful and Excluded Species," p. 469)	
4	Fls 15–20 mm long, with a slender spur; sepals hairless	*V. adunca* var. *adunca*

5(2) Fls and lvs on a slender, ascending stem .... *V. howellii*
(see "Doubtful and Excluded Species," p. 000)

5 Fls and lvs arising from the rootstock 6

6(5) Fl stalks 10–15 cm tall; upper petals tinged bluish on the back; capsule short-hairy .... *V. palustris* var. *palustris*

6 Fl stalks 7–10 cm tall; upper petals pure white; capsule hairless 7

7(6) Teeth of the lvs rounded, shallow; lvs to 2.5 cm broad ....
.... *V. macloskeyi* var. *macloskeyi*

7 Teeth of the lvs rounded, deep; lvs more than 2.5 cm broad ....
.... *V. macloskeyi* var. *pallens*

8(1) Stems bearing both lvs and fls only at the top; lvs heart-shaped at the base and tapered to the pointed tip .... *V. glabella*

8 Stems bearing lvs and fls almost along the full length; lvs not both heart-shaped at the base and long-tapered at the tip 9

9(8) Lvs squared-off or somewhat tapered at the base, coarsely toothed ....
.... *V. purpurea* ssp. *venosa*

9 Lvs heart- to kidney-shaped at the base, more or less even to shallowly toothed 10

10(9) Plants with long runners; lvs thickish and purple-spotted beneath ......
.... *V. sempervirens*

10 Plants with short or no runners; lvs thin and not spotted . *V. orbiculata*

Viola adunca Sm. var. *adunca*
BLUE VIOLET, WESTERN DOG VIOLET

A variable plant, a fact reflected by the many names that have been applied to it. It grows in small clumps from thin, upright, usually branched rootstocks. The lvs and stems are short-hairy. The lvs are both basal and on the flowering stems (obscurely so in dwarfed subalpine plants), ovate, sometimes heart-shaped at the base, broadly toothed, and 1–3 cm long, rather blunt at the tip. The fls are long-stalked, dark blue, white at the throat, and 10–15 mm long; the spur is slender and 4–7 mm long. There is a tuft of hairs at the top of the style.

Common in meadows and on flats around lakes, mostly above 4000 ft, reaching 6000 ft. Found along the trail between Klapatche and St. Andrews Parks; around Indian Henrys Hunting Ground; at Longmire; at Goat Pass at 5500 ft; around the Paradise V; at Reflection Lks; northeast of Hidden Lk at 5850 ft; at Shadow Lk; at Moraine Park; on the northeast slope of Palisades Peak at 5600 ft; and at Berkeley Park. Plants can also be found among the grass and weeds in front of the Ohanapecosh Visitor Center (perhaps brought in several decades ago when a botanical garden was maintained there).

Viola glabella Nutt.
STREAM VIOLET, PIONEER VIOLET

From a thick, horizontal rootstock, with deciduous, mostly hairless basal and stem lvs. The basal lvs are on long, slender stalks; the lf blades are heart-shaped, usually wider than long and tapered to a short, sharp tip, broadly toothed, and 3–8 cm wide. The stem lvs are usually 2, borne at the top of the 5–8 cm stem just below the fls, smaller and narrower than the basal lvs. There are 1–3 yellow fls per stem, each 9–13 mm long; the lower 3 petals have purple veins.

Very common in moist woods, ranging up to almost 5000 ft. Found at the Nisqually entrance and on Westside Rd; at Longmire; in lower Paradise Park; at Box Canyon; in Stevens Canyon; on Chinook Cr; at Ohanapecosh; at a few places along the road below Tipsoo Lk; to about 5000 ft on Sunrise Rd; and in Moraine Park.

Viola langsdorfii (Regel) Fisch. ex Ging.
ALEUTIAN VIOLET

A small and compact plant, from a thick, horizontal rootstock, less than 15 cm tall. The lvs are on rather short, erect stems, heart-shaped, broadly toothed, and 2–5 cm wide. The fls are blue to blue-violet with darker veins on the lower petals, on long stalks, and 15–25 mm long.

A species typically found in more or less open, moist areas in the Puget lowlands and near the coast. Known in the Park from a collection that is in the University of Washington herbarium made early this century by J. B. Flett at "Simple Lk, Mount Tacoma." The place name has not been identified. The sheet was labeled by Flett as "*Viola* sp." and was annotated in 1992 by Harvey E. Ballard Jr. as *V. langsdorfii*. The specimen shows petals almost fully faded but clearly once blue. Runners are absent but there is a short, erect stem. The stalks of the lvs and fls are hairless and the stipules are mostly narrowly ovate or broader.

Viola macloskeyi Lloyd ssp. *pallens* (Banks ex DC.) M. S. Baker—SMALL WHITE VIOLET

Small plants, usually not much more than 10 cm tall, hairless, and spreading by runners to form small patches. The lvs are all basal, mostly round in outline, kidney-shaped at the base, and broadly toothed; the blades are 1–4 cm wide. The fls are white, on tall, slender stems, and 5–8 mm long; the lower petals have dark veins and the lateral petals are usually not bearded.

Evidently rare in the Park, and a plant of bogs and swamps at middle elevations. St. John and Warren cite it for "wet places, Snipe Lk (White R District)," where it was collected by J. B. Flett. ("Snipe Lk" is a place name Flett used for a meltwater pond below the Interfork Glacier.)

An unpublished checklist of plants at Berkeley Park, made in 1978 by Virginia and A. B. Adams for the Washington Native Plant Society, names *V. macloskeyi* var. *macloskeyi*. The variety is listed by Dunwiddie (1983) and appears on the National Park Service flora. It differs in having smaller, more shallowly toothed lvs. It has not been verified with herbarium collections or recent discoveries in the field.

Viola orbiculata Geyer ex Holz.
ROUND-LEAVED VIOLET

Growing from a more or less vertical rootstock and lacking runners, with both basal lvs and leafy stems. The lvs are thinnish (and usually not persistent through the winter), medium green, round in outline, kidney-shaped at the base and blunt at the tip, with broad, shallow teeth, and 1–4 cm wide; the stem lvs are much smaller and distributed all along the stem. The stems are 4–8 cm tall, bearing a few fls. The fl is yellow, marked with darker veins, and about 12 mm long.

Common in moist to dryish conifer woods up to about 5500 ft. Found between Mtn Meadows and Mowich Lk; around the Nisqually entrance and up to 3600 ft on Westside Rd; at Longmire and on Rampart Ridge; on the west side of the Nisqually R above the Glacier View Bridge; in brushy places in lower Paradise Park; at Box Canyon; around Ohanapecosh; on the Kotsuck Cr trail at 5000 ft; to 5500 ft on Sunrise Rd; and at Berkeley Park.

Viola palustris L. var. *palustris*
MARSH VIOLET

A slender plant, from a long, horizontal rootstock, also bearing runners. The lvs are all basal, with stalks that can reach 15 cm; the lf blade 2–4 cm broad, heart-shaped with a blunt tip, and broadly and shallowly toothed. The fl stalks about equal those of the lvs; each bears a single white fl 8–12 mm long; the lower petals are marked with dark veins and the upper petals are often pale blue on the backsides.

Common on the banks of streams in open places, in wet meadows, and in swampy areas, reaching about 5000 ft. Found at Mtn Meadows and at Lee Cr between Mowich Lk and Spray Park; at the edge of the Longmire meadow; below Narada Falls; along the road on around seeps below Paradise; around Reflection Lks; at 5000 along the Kotsuck Cr trail; and at Ohanapecosh.

Viola purpurea Steven ssp. *venosa* (S. Watson) M. S. Baker & J. C. Clausen
PURPLE-BACKED VIOLET, GOOSEFOOT VIOLET

Growing as small tufts, from a short, ascending rootstock, reaching about 10 cm tall and short-hairy on the stems and lvs. The lvs are 1–2.5 cm long, blunt to tapered at the base (not at all heart-shaped) and more or less triangular in outline, deeply toothed, and purplish on the underside (altogether reminiscent of the footprint of a goose). The leafy stems are usually shorter than the basal lvs and bear 1 or 2 fls. The fls are 8–12 mm long and deep yellow, with dark brown veins on the lower petals; the upper petals are brownish or purplish on the backs.

Chiefly a species from east of the Cascade Mountains, and uncommon on the east side of the Park, on open ridges. Known from a herbarium collection made "near Tipsoo Lk."

Viola sempervirens Greene
EVERGREEN VIOLET

A low plant, with a vertical rootstock and long, slender leafy runners. The basal lvs are 1–4 cm wide, heart-shaped and somewhat pointed at the tips, rather thickish, evergreen, hairless or sparsely hairy, and marked on the underside by many small purple spots. The flowering stems bear smaller lvs and reach 5–10 cm tall. The fls are lemon yellow, 12–15 mm long, and marked with dark veins on the lower petals.

Often found with the far more common V. *orbiculata*, in moist woods below 5000 ft, chiefly on the west side of the Park. Collected near Mowich Lk and near Longmire. *Viola sarmentosa* Douglas was once listed for the Park, based on a misidentification of *V. sempervirens*.

Viscaceae Batsch (formerly Loranthaceae)
MISTLETOE FAMILY

Arceuthobium tsugense (Rosend.) G. N. Jones ssp. *tsugense*—HEMLOCK DWARF MISTLETOE

A yellowish green plant, growing only as a parasite on the branches of its host tree. The stems are 3–12 cm long and branched; the opposite lvs are reduced in size to short scales. Male and female fls are borne on separate plants. The fls are greenish and very small, in short spikes; petals are absent. The male fls have 3 sepals and 3 stamens, and the female fls have 2 sepals and 1 short style. The fruit is a small, sticky berry.

Fairly common on western hemlock, *Tsuga heterophylla*, forming small "witch's brooms" where the tangles of small, leafless mistletoe branches emerge from the host branch.

Only this species has been verified for the Park, but it would not be surprising if others were to be found, for almost every conifer species in the Northwest can be parasitized by a dwarf mistletoe. Because the host trees in the Park can be so tall, perhaps an efficient way to hunt for dwarf mistletoes would be to examine blown-down branches following a windstorm.

The best way, in the field, to distinguish the currently recognized species is by the host. Thus:

• *Arceuthobium abietinum* Engelm. ex Munz is found on *Abies* species, the true firs. Its stems are 6–20 cm long and branched.

• *Arceuthobium douglasii* Engelm. is found on *Pseudotsuga menziesii*, Douglas fir. It has very short, rarely branched stems to 2 cm long.

• *Arceuthobium tsugense* (Rosend.) M. E. Jones ssp. *mertensianae* Hawksw. & Nickr., on *Tsuga mertensiana*, mountain hemlock. It has somewhat shorter branches than ssp. *tsugense*.

367

Monocotyledonous Plants (Monocots)

The second largest group of plants in the Park, in terms of numbers of species, the monocots comprise one of the two major groups of angiosperms (plants in which the seeds are formed within a specialized container, or fruit, in contrast to the "naked" seeds of the gymnosperms, or conifers). Monocots include sedges, rushes, lilies, orchids, and grasses, plus a few others, including duckweed, skunk-cabbage, pondweed, bur-reed, and cattails. Most members of the family in the Park are perennials; none are shrublike or treelike.

Monocots
Cotyledon (seed lf) 1
Lvs mostly parallel-veined
Petals, sepals, stamens typically
  borne in sets of 3

Dicots
Cotyledons 2
Lvs mostly net-veined
Petals, sepals, stamens typically borne in
  sets of 4 or 5, rarely 3

Key to the Monocots

1   Plants aquatic, the stems and lvs growing underwater or floating on the surface, although the flowering stem may reach above the surface 2
1   Plants terrestrial, or if growing in the water, then the stems and all the lvs reaching above the surface 4

2(1) Plant a small, oval, leaflike, floating thallus with single, threadlike root in the water .... Araceae (*Lemna*)
2   Plants with normal stems and lvs; roots anchored in soil or mud 3

3(2) Fls unisexual (each bearing stamens or a pistil but not both), in rounded heads; lvs long, linear, and floating on the surface .... Sparganiaceae
3   Fls bisexual (each with stamens and a pistil), whorled in spikes; submerged lvs linear, floating lvs oval .... Potamogetonaceae

4(1) Perianth segments inconspicuous and scalelike or bristlelike, or absent; fls some shade of green or brown (in *Lysichiton* in the Araceae, the small fls are surrounded by a bright yellow spathe) 5
4   Perianth segments present and well-developed, petal-like, colored (greenish in some orchids) 9

5(4) Inflorescence a massive columnar spadix, partially enclosed by an ample, bright yellow, cloaklike spathe (the actual fls are small greenish) ....
.... Araceae (*Lysichiton*)
5   Inflorescence not a spadix, spathe absent 6

6(5) Fruit a capsule of fused chambers; perianth segments 6, scalelike, regular, and arranged in 2 series (rushes) .... Juncaceae
6   Fruit an achene, nutlet, or grain; perianth segments reduced in size and/or much modified 7

7(6) Fls unisexual (each bearing stamens or a pistil but not both) and numerous, packed into a single dense, erect, spikelike inflorescence, the male fls at the top; perianth segments absent (cattails) .... Typhaceae

7   Fls bisexual (each with stamens and a pistil) or unisexual (if unisexual, then not combining features as above); perianth segments present 8

8(7) Stems typically triangular, solid or stuffed with pith; fls spirally arranged on the axis of the spikelet; each fl subtended by 1 bract (sedges) ....
      .... Cyperaceae

8   Stems typically round, hollow; fls arranged in two ranks on the axis (sometimes just 1 fl per spikelet); each fl subtended by 2 bracts (grasses)
      .... Poaceae

9(4) Ovary superior .... Liliaceae

9   Ovary inferior 10

10(9) Fls regular .... Iridaceae

10   Fls irregular .... Orchidaceae

Araceae Juss.—ARUM FAMILY

(*Lemna* was formerly in a separate family, the Lemnaceae)

A large family of mostly tropical plants, the arums are united by features of the flower and inflorescence. Since *Lemna* rarely produces flowers, and since studying those rare blossoms calls for a dissecting microscope, suffice it here to say that our two native genera couldn't appear more different.

1   Tiny floating aquatic plant .... *Lemna*

1   Huge terrestrial plant .... *Lysichiton*

Lemna L.—DUCKWEED

Lemna minor L.—COMMON DUCKWEED

Usually noticed only when enough of the tiny plants are grouped together on the water's surface. Each plant consists of an oval, floating thallus about 2 mm across and a single, threadlike root that is 5–10 mm long. True lvs are absent. The upper surface of the thallus is smooth and glossy green. Fls are extremely rare and the plant reproduces mostly by budding; thus the thalli are usually found in pairs or threes.

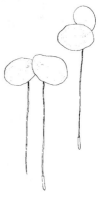

Not previously reported for the Park. It can be found in the quiet water of a small, sunny pond along the main road on the southwest side of Tumtum Peak, at 2200 ft.

Lysichiton Schott—SKUNK-CABBAGE

Lysichiton americanus Hultén & H. St. John
SKUNK-CABBAGE

One of the first plants to bloom in the spring.
The flowering stem appears first, soon
followed by the glossy green lvs that rise on
short stalks from the muck. The erect lvs are
elliptical to oblanceolate and can be 150 cm
long. A bright yellow, cloaklike spathe 10–20
cm long partially surrounds the thick spadix, a
columnar structure upon which the fls are borne.
The fls are small and greenish yellow, with a foul
smell.

Common and conspicuous on low, swampy
ground in forests, throughout the Park below
3500 ft; occasionally in more open places and
along roadside ditches. A nice stand grows
by the road at the downstream end of the Longmire meadow.

Most references use "*Lysichitum americanum*" or "*Lysichiton
americanum*" for this plant. The correct name of the genus, *Lysichiton*, is
masculine in gender and therefore requires the masculine *americanus*.

Cyperaceae Juss.—SEDGE FAMILY

A large family of grasslike plants, numerous in the Park and often
forming the dominant vegetation in wet meadows and streambanks. The
leaves are 3-ranked on triangular or rounded stems. The flowers are
small and greenish or brown. They can be perfect (each flower with both
stamens and a pistil) or unisexual (each flower bearing stamens or a
pistil but not both). They are borne in spikelets that are in turn arranged
in heads, spikes, or umbels. In *Carex*, each female flower is enclosed in a
special structure called a perigynium. In several genera, the petals and
sepals are represented by bristles ("perianth bristles") that rise from
beneath the ovary and correspond to the sepals and petals of showier fls.

Several of the sedges are quite striking plants, and the cotton-grasses,
of the genus *Eriophorum*, are attractive in fruit.

1   Fls unisexual (each bearing stamens or a pistil but not both); the ovary
     enclosed in a saclike perigynium .... *Carex*
1   Fls with both stamens and a pistil; the ovary not enclosed in a perigynium
     2

2(1) Perianth bristles very long and conspicuous, the spike appearing cottony
     .... *Eriophorum*
2   Perianth bristles short and inconspicuous 3

3(2)   Spikelet or spikelets set above 1 or more leaflike bracts; base of the style slender, not persistent, the achene merely with a short beak .... *Scirpus*

3   Spikelet solitary, no leaflike bract present; base of the style enlarged and persistent on the achene as a tubercule .... *Eleocharis*

Carex L.—SEDGE

"Sedges have edges" is as good a starting point as any in dealing with this large genus of grasslike plants. A cross section shows that the stems are triangular with sharp and sometimes raspy edges; the leaves, therefore, are 3-ranked on the stems. The blades of the lowest leaves are often reduced to sheaths, and the upper blades may be flat, folded, or inrolled. Leaf measurements, usually of the width, are made on the lower normal leaves of the stems.

Carex flowers are always unisexual; that is, the male and female organs are in separate flowers, which in turn are arranged in one or more spikes on the stem. In species with just one spike, the male flowers are usually fewer in number and are crowded at the top of the spike while the more numerous female flowers are below. Several possibilities exist in species with two or more spikes.

• The terminal spike may consist of male flowers only.

• It may be androgynous, with both male and female flowers, the male flowers placed above the female flowers.

• It may be gynaecandrous, with both male and female flowers, but female flowers at the top of the spike and the male flowers below.

The lower spikes are usually of mixed sexes, but may be wholly female.

The male flowers have 3 stamens (rarely 2) and the male spike, or portion of a spike, is typically much more slender than, and therefore easily distinguished from, the female portion. Even after flowering, the wispy anthers are usually visible under magnification. (As with the grasses, at least a 10x lens is necessary for the successful use of the key. A 20x lens and a dissecting microscope are often helpful.)

Each female flower is borne in a saclike structure, formed by a modified bract, called a perigynium (plural: perigynia). Details of the perigynium are essential in using this key: the shape, dimensions, color, and surface features all must be observed under magnification. The perigynium may be blunt at the tip, but is more often extended to form a beak. The surface of the perigynium is sometimes hairy and may bear distinctive veins or a winged margin. The perigynium encloses the ovary, which develops into an achene. The stigmas, which may number 2 or 3, protrude from a tiny mouth at the top of the perigynium to catch the wind-borne pollen. At the base of the perigynium is a scale, which more or less covers the perigynium and is largely responsible for the overall color of the spike.

Forty-one *Carex* species are found in the Park, and, unlike the grasses, characteristics of stature and habitat are of little use in separating them, making a close examination of features of the spike and perigynium essential. One field guide, *Plants of the Pacific Northwest Coast*, by Pojar and MacKinnon (1994), illustrates many of the species that are found in the Park, and provides useful technical descriptions as well.

The sedges of the Park are poorly represented in herbaria, and location information for many species is unfortunately sketchy. The most thorough collecting was done by Irene Creso during the 1970s. Her material is in herbaria at Pacific Lutheran University and the University of Puget Sound.

An important point must be made about this key to the sedges: it is constructed to work efficiently only for the *Carex* species that are found in the Park. It does not separate the species into natural subgroups, nor does it attempt to account for all the variation that can be seen in many species. It cannot, therefore, be used with confidence in other geographic areas in Washington.

1	Spikes solitary on the stems; stigmas 3 (rarely 2 in *C. nardina*)	Group 1
1	Spikes several per stem; stigmas 2 or 3	2

2(1)	Spikes androgynous or the terminal spike male	Group 2
2	Spikes gynaecandrous or the terminal spike female	Group 3

Group 1

1	Lvs flat or folded along the midvein, mostly 1 mm or more wide	2
1	Lvs inrolled at the edges and needlelike, rarely more than 1 mm wide ...	4

2(1)	Perigynia spreading from the axis of the spike at maturity	*C. pyrenaica*
2	Perigynia remaining flattened against the axis or merely ascending at maturity	3

3(2)	Rootstock absent, stems densely tufted; spike oblong, light brown	*C. nardina var. hepburnii*
3	Stems few from a rootstock; spike egg-shaped, nearly as long as wide, dark brown	*C. engelmannii*

4(1)	Perigynia spreading or reflexed from the axis at maturity; growing at 5000 ft or higher	*C. nigricans*
4	Perigynia flattened on the axis or ascending at maturity; seldom found above 5000 ft	5

5(4)	Lvs about 1 mm wide; perigynia 2–4 mm long	*C. leptalea* ssp. *leptalea*
5	Lvs 1.5–3 mm wide; perigynia 5–6 mm long	*C. geyeri*

Group 2

1	Stigmas 2 and the achenes lens-shaped	2
1	Stigmas 3 and the achenes 3-cornered...	11

2(1)	Spikes elongated and cylindrical, at least the lower spikes on stalks	3
2	Spikes egg-shaped to elliptical, not stalked	7

3(2) Bract of the lowest spike obviously sheathing the stem; perigynia yellow-brown to golden at maturity .... *C. aurea*

3 Bract of the lowest spike not sheathing the stem; perigynia green to straw-colored or brown 4

4(3) Sheaths of the lowest lvs shredding with age .... *C. nudata*

4 Sheaths of the lowest lvs not shredding, mostly remaining intact with age 5

5(4) Perigynia with evident veins on both faces ..... *C. lenticularis* var. *lipocarpa*

5 Perigynia without veins or the veins indistinct 6

6(5) Perigynia flat and compressed, about 3 mm long .... *C. scopulorum* var. *bracteosa*

6 Perigynia inflated, 2–3 mm long .... *C. aperta* (see "Doubtful and Excluded Species," p. 469)

7(2) Stems single or a few rising together from a slender, creeping rootstock . ... *C. praegracilis*

7 Stems tufted, the rootstock absent or short but not creeping 8

8(7) Spikes fewer than 10 per stem and so densely crowded into a head that they cannot be easily distinguished .... *C. hoodii*

8 Spikes more numerous and less crowded, so that they are more or less easily distinguished 9

9(8) Lf sheaths reddish-dotted, not cross-wrinkled on the side facing the stem .... *C. cusickii*

9 Lf sheaths cross-wrinkled on the side facing the stem, not reddish-dotted 10

10(9) Lvs 4–8 mm wide; perigynia 4–5 mm long .... *C. stipata*

10 Lvs 1.5–3 mm wide; perigynia 3–4 mm long .... *C. neurophora*

11(1) Perigynia short-hairy 12

11 Perigynia smooth 15

12(11) Plants with creeping rootstocks; perigynia velvety 13

12 Plants without creeping rootstocks; perigynia sparsely hairy 14

13(12) Lvs folded; the lowest lvs reduced to sheaths at the base of the stems .... *C. pellita*

13 Lvs flat; lowest lvs with well-developed blades .... *C. halliana* (see "Doubtful and Excluded Species," p. 469)

14(12) Plants to 30 cm tall; perigynia with a stalklike base and abruptly constricted at the top of the body to the beak .... *C. rossii*

14 Plants 40–90 cm tall; perigynia with a very short base and tapered from the body to the beak .... *C. luzulina* var. *ablata*

15(11) Style remaining attached to the achene at maturity (dissect the perigynia to see this); perigynia with a long, tapered 2-toothed beak 16

15 Style falling from the achene before maturity; perigynia various, but not at once long, tapered, and sharply 2-toothed 17

16(15)   Perigynia 4–6 mm long, spreading from the axis at maturity .... *C. utriculata*

16   Perigynia 7–9 mm long, ascending at maturity .... *C. exsiccata*

17(15)   Bracts of the spikes with well-developed sheaths; perigynia with a long-tapered beak 18

17   Bracts of the spikes without sheaths; perigynia lacking a long-tapered beak 19

18(17)   Lvs 5–12 mm wide; perigynia 4.5–6 mm long, beak blunt .... *C. hendersonii*

18   Lvs 2–6 mm wide; perigynia 3–4.5 mm long, beak 2-toothed .... *C. luzulina* var. *ablata*

19(17)   Spike cylindrical, 5–10 cm long; lvs 10–20 mm wide .... *C. amplifolia*

19   Spikes less than 5 cm long, more or less elliptical; lvs less than 10 mm wide 20

20(19)   Perigynia beakless, or at most with a blunt tip that is less than 0.2 mm long; lateral spikes on slender, nodding stalks; rootstocks long and conspicuously hairy .... *C. limosa*

20   Perigynia with a short, 2-toothed beak; lateral spikes not nodding (except occasionally in *C. spectabilis*, which has a short, smooth rootstocks) 21

21(20)   Lowest lvs with well-developed blades; perigynia somewhat inflated .. .. *C. raynoldsii*

21   Blades of the lowest lvs rudimentary and scalelike; perigynia flattened 22

22(21)   Female spikes 8–12 mm long; lowest spike typically nodding; perigynia ovate to nearly round .... *C. spectabilis*

22   Female spikes 10–20 mm long; lowest spike typically erect; perigynia elliptical .... *C. paysonis*

Group 3

1   Stigmas 3; spikes on short or long stalks 2

1   Stigmas 2; spikes not stalked 3

2(1)   Spikes clustered into a loose head, on short, erect stalks .... *C. albonigra*

2   Spikes on long, nodding stalks .... *C. mertensii*

3(1)   Margins of the perigynia winged 4

3   Margins of the perigynia not winged 10

4(3)   Perigynia planoconvex (rounded on one side, flat on the other), with distinct veins on the faces, 4-5 mm long; spikes crowded into a dense head .... *C. phaeocephala*

4   Perigynia planoconvex or flattened; other characteristics of the perigynia not combined as above; the spikes usually readily distinguishable 5

5(4)   Scales smaller and paler than the perigynia, exposing the edges of the perigynia to view and giving the spike a bicolored appearance; inflorescence a crowded head 6

5   Scales equal in size to the perigynia, concealing the latter, the spikes not notably bicolored; inflorescence open to crowded, but not head-like 8

6(5)   Perigynia flattened, about 1 mm wide with a long and slender beak .... *C. microptera*

6   Perigynia planoconvex, 1.5-2 mm wide, the beak relatively short 7

7(6)   Stems 10-30 cm tall; perigynia brown with a metallic luster .... *C. pachystachya*

7   Stems 20-60 cm tall; perigynia golden-brown .... *C. preslii*

8(5)   Perigynia broadly winged, 4-6 mm long, generally strongly veined on the faces .... *C. phaeocephala*

8   Perigynia narrowly winged, 3-4 mm long, obscurely veined on the faces 9

9(8)   Plant to 30 cm tall; perigynia 3-4 mm long .... *C. leporinella*

9   Plant to 70 cm tall; perigynia 3.5-5.5 mm long .... *C. praticola*

10(3)   Perigynia widely spreading or reflexed at maturity 11

10   Perigynia ascending or somewhat spreading at maturity 12

11(10)   Margin of the perigynia smooth; spikes crowded into a dense head ... . *C. illota*

11   Margins of the perigynia minutely toothed; spikes distinct on the axis .... *C. echinata* ssp. *echinata*

12(10)   Spikes crowded into a dense, elongated head 15-30 mm long .... *C. arcta*

12   Spikes distinct on the axis of the inflorescence 13

13(12)   Beak of the perigynia short or nearly obsolete (less than 0.5 mm long) 14

13   Beak of the perigynia well-developed (0.5 mm long or more) 15

14(13)   Lvs 1-2 mm wide; perigynia 2-4 mm long and 1-1.5 mm wide .... *C. laeviculmis*

14   Lvs 2.5-5 mm wide; perigynia 3.5-4.5 mm long and about 1 mm wide .... *C. deweyana* var. *deweyana*

15(13)   Perigynia 5-10 in each spike; lvs 1.5-2 mm wide .... *C. brunnescens*

15   Perigynia 10-30 in each spike; lvs 2-4 mm wide .... *C. canescens* ssp. *canescens*

Carex albonigra Mack.—BLACK-AND-WHITE SEDGE

Named for the appearance of the spikes, in which the scales are dark brown with a whitish band at the upper edge. It grows in tufts, with stems 10–30 cm tall and is gynaecandrous, with 3 stigmas. The perigynium is about 3 mm long, and widest above the midpoint, with a short beak. The 2–4 spikes, on very short and erect stalks, are crowded into a dense head.

Uncommon at and above timberline on dryish, rocky slopes. It has been found at about 8500 ft, above Paradise.

Carex amplifolia Boott—BIG-LEAF SEDGE

"Big-leaf" for the ample dimensions of the flat lvs that are 1–2 cm wide. The stems reach 50–100 cm tall and spread vigorously by strong runners. It is androgynous, with 3 stigmas. The perigynium is 3 mm long, egg-shaped, with a long and sometimes curved beak; the scales are brown. The spikes are long and slender, up to 10 cm long and just 1 cm wide.

Common and prominent in low-elevation swamps. It can be found in the Nisqually V, along Westside Rd, and in places in the valley of the Ohanapecosh R.

Carex arcta Boott
NORTHERN CLUSTER SEDGE

Tufted and medium-sized, with stems 20–60 cm tall. The flat lvs are 2–4 mm wide and glaucous. The 7–10 spikes are contracted into an oblong head. It is gynaecandrous, with 2 stigmas. The perigynium is 2–3.5 mm long, egg-shaped, with a long, 2-toothed beak. The scales are straw-colored to light brown.

Fairly common on wet soils in open woods, in meadows, and occasionally at roadsides, 3000-4000 ft.

Carex aurea Nutt.
GOLDEN SEDGE

An attractive species, for the distinctive golden color of the spikes on stems 10–30 cm tall. The lvs are light green and slender, 2–4 mm wide. The leaflike bract below the lowest spike has an obvious sheath. Golden sedge is androgynous, with a slender, inconspicuous male spike and several loose female spikes in a short, fairly open cluster. The perigynia are 2–3 mm long, nearly round or somewhat wider at the top, lacking a beak; stigmas 2.

Uncommon, on wet, often stony ground, in meadows and on gravelly stream banks. It has been collected in the meadow at Longmire.

Carex brunnescens (Pers.) Poir. ssp. *brunnescens*—BROWN SEDGE

A tufted plant with stems 20–50 cm tall and narrow lvs, 1.5–2 mm wide. The 4–8 spikes are in a loose inflorescence. The species is gynaecandrous, with 5–10 fls per spike. The perigynia are about 2 mm long, oval, with a short, 2-toothed beak; stigmas 2.

Rare in the Park and growing at and above timberline on seasonally wet ground in meadows and on open slopes. A collection was made by Huntley in the 1930s at Ipsut Pass, but the plants are small and poorly mounted.

Carex canescens L. ssp. *canescens*—GRAY SEDGE

A lower-elevation counterpart to *C. brunnescens*, gray sedge differs in its flat lvs that are 2–4 mm wide and heads with more fls. It is gynaecandrous; the beak is not prominently 2-toothed. Stigmas 2.

Common on swampy ground in meadows and in open places in forests, to about 4000 ft. Collections have been made at Mtn Meadows and at Longmire.

Carex cusickii Mack.—CUSICK'S SEDGE

A large sedge, with 50–100 cm stems and flat, 2–6 mm wide lvs. The front of the sheath of the lf is colored with reddish dots. Cusick's sedge is androgynous, with 2 stigmas. The spikes are rather crowded into a narrowly pyramidal head, but not so crowded that the individual spikes cannot be discerned. The perigynium is brown, 3–4 mm long, and widest at the

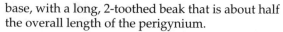

base, with a long, 2-toothed beak that is about half the overall length of the perigynium.

Rare, or at least seldom reported for the Park; it is known to grow in the meadow at Longmire.

Carex deweyana Schwein. var. *deweyana*
DEWEY'S SEDGE

Mostly growing as slender stems from creeping rootstocks, reaching 100 cm tall, but often just half that. The lvs are flat, pale green, and 2.5–5 mm wide. The 4–8 spikes per stem are carried in a loose head on weak stems. The species is gynaecandrous, with 2 stigmas. The perigynia are light green, 3.5–4.5 mm long, relatively narrow (about 1 mm), with a long, toothed beak; they remain erect to flattened against the axis at maturity.

Common on wet ground in rather open places in forests, up to about 4500 ft. It has been found at Mtn Meadows; in the meadow at Longmire; in places between Kautz Cr and Narada Falls; and on the Carbon R at Fall Cr.

Carex echinata Murray ssp. *echinata*
BRISTLY SEDGE, STAR SEDGE

A tufted plant, with stems reaching 50 cm and flat lvs 1.5–2 mm wide. The 3–4 spikes are arranged loosely along the erect stem. The spikes are gynaecandrous, with 2 stigmas. The perigynia are 3.5–4 mm long, widest at the base, and abruptly tapered to a slender, 2-toothed beak. The common names refer to the appearance of the mature spikes. The scales of the perigynia are long and sharply pointed while the scales and perigynia themselves spread widely at maturity, becoming "starlike."

Chiefly a species of areas near the coast in Washington, but collected in the Park at a place described as "near Ohanapecosh 3000 ft." It grows in swamps and sphagnum bogs.

Carex engelmannii L. H. Bailey
ENGELMANN'S SEDGE

A small plant, growing as tufts on a spreading rootstock, to 15 cm tall, with lvs about 1 mm wide that are folded along the midvein. A solitary, androgynous spike about 1 cm long tops the stem. The perigynia and scales are dark brown and 4 mm long; stigmas 3.

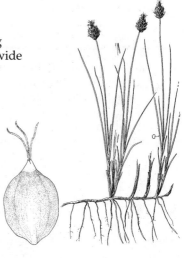

Very common, 5000-6000 ft, reaching 8000 ft on a few ridges on the south side of the mountain, and collected by O. D. Allen at 9500 ft, probably above Paradise. Herbarium collections have also been made on a moraine of the Van Trump Glacier and in Berkeley Park. Called "Brewer's sedge" in older references.

Carex exsiccata L. H. Bailey—INFLATED SEDGE

A vigorous, clump-forming species, from spreading rootstocks, reaching 100 cm tall. The lvs are 3–6 mm wide; the midvein is thickened and forms a keel. The species is androgynous, with 2–4 male spikes and 1–3 female spikes in a long, narrow inflorescence; stigmas 3. The perigynia are tinged reddish, very large, 7–9 mm long, widest at the base, and tapered to a deeply toothed, slender beak. The perigynia are ascending at maturity, a key difference from C. *utriculata*.

Common in shallow ponds and on boggy ground below 3000 ft in the valleys of the major rivers, as in the pond at the Stevens Canyon entrance station and in the lower Nisqually V.

Carex geyeri Boott—ELK SEDGE

Mid-sized, reaching about 40 cm tall, from a heavy rootstock and flat, stiff lvs 2–4 mm wide. The 3–8 spikes are crowded into a single straw-colored head on the stem. The spike is androgynous, with 1–3 perigynia each 5–7 mm long, narrow and wider toward the top, with a very short beak; stigmas 3.

Uncommon, growing on dryish, open or lightly wooded slopes, 5000-7000 ft on the east side of the Park. It has been found at 6760 ft at Panhandle Gap.

Carex hendersonii L. H. Bailey—HENDERSON'S SEDGE

A tall, tufted plant, to 100 cm tall, with wide, flat lvs, 5–12 mm broad. The spikes are held on short, erect stalks in an open, androgynous inflorescence. The bracts of the spikes, especially the lowest, sheath the stem at their bases. The perigynia are 4.5–6 mm long, green, with many conspicuous veins, widest at the middle and tapered to a long beak; stigmas 3.

Fairly common on swampy ground in meadows and forests in lowland western Washington. It is known from the Park in the Nisqually V, and has been collected at Longmire.

Carex hoodii Boott—HOOD'S SEDGE

Reaching 40–80 cm tall and growing in clumps, with narrow lvs 1.5–3.5 mm wide. The 5–10 androgynous spikes are so closely crowded into a terminal head that it gives the appearance of being one large spike; pulling away a few spikes shows the true nature of the inflorescence. The perigynia are an attractive copper-brown color with green edges, 3.5–5 mm long and spreading at maturity, with a long, tapered, 2-toothed beak; stigmas 2.

Known for the Park only from a single collection in the Park herbarium, one that was made at Berkeley Park. It was not known to earlier writers. The species is common and widespread in a variety of habitats in the Pacific Northwest and may well be more frequent in the Park than the record indicates.

Carex illota L. H. Bailey
SHEEP SEDGE, SMALL-HEAD SEDGE

Growing in tufts with stems to 30 cm tall, with lvs 1.5–3 mm wide. The 3–5 gynaecandrous spikes are crowded into a dense head that is up to 15 mm long, but frequently just 5 mm long. The perigynia are 2.5–3 mm long, widest at the middle, smooth and dark brown to nearly black, with 2 stigmas; the margins are thin-edged and not toothed.

Common on moist, rocky slopes, in meadows, and on stream banks above 5000 ft. It has been collected at Spray Park, at Paradise, and below Chinook Pass.

Cyperaceae

Carex laeviculmis Meinsh.
SMOOTH-STEM SEDGE

Densely tufted and reaching 30–60 cm tall, but also spreading by rootstocks. The lvs are 1–2 mm wide, pale green, and lax. The spikes are well-separated on the stem, except for the top 2 or 3, and gynaecandrous. The greenish to brown perigynia are 2–4 mm long and 1–1.5 mm wide, prominently veined, and tapered rather abruptly to a long beak; stigmas 2.

Common and found in moist meadows, along streams, and occasionally in wet places in open woods, up to nearly timberline. Collections are known from between Mowich Lk and Spray Park; at Longmire; and at a place described as "near Sunrise entrance."

Carex lenticularis Michx. var. *lipocarpa* (T. Holm)
L. A. Standl.—LAKESHORE SEDGE

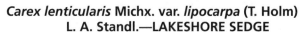

Tufted, with short rootstocks, and reaching 15–40 cm tall; the longest lvs may be much longer than the flowering stem and 1–3.5 mm wide. The uppermost spikelet is male, and long and slender. The 3–5 female spikes are erect and in a loose cluster; each is densely flowered and nearly cylindrical. The perigynia are about 2.5 mm long, widest at the middle, veined, and greenish. The scales are nearly black and smaller than the perigynia, giving the spikes a two-toned appearance; stigmas 2.

One of the most common species of *Carex* in the Park, in wet meadows, at lakeshores, and on stream banks, 2500-6000 ft. It can be found at Mountain Meadows; around Mowich Lk and Spray Park; on the shore of Lk George; at Longmire; at Paradise; by Louise Lk; at Three Lks; at Tipsoo Lk; and in Berkeley Park.

Carex leporinella Mack.—LITTLE HARE SEDGE

A small, tufted plant to 30 cm tall, with inrolled lvs 0.5–2 mm wide. The spikes are gynaecandrous and arranged in a loose cluster. The perigynia are golden-colored, 3–4 mm long, slender, obscurely veined on both faces and notably planoconvex. The perigynia are also narrowly winged, with the wings continuing the curve of the back of the perigynia; stigmas 2.

Rare in the Park, or at least seldom collected. Found at the edges of lakes and in wet meadows near timberline. It can be found at Louise Lk.

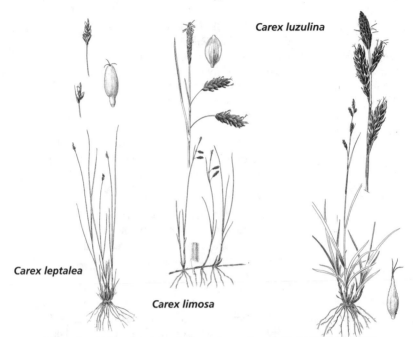

Carex luzulina

Carex leptalea

Carex limosa

Carex leptalea Wahlenb. ssp. leptalea—FLACCID SEDGE

"Flaccid," for the slender, weak, often nodding stems and lvs, this sedge is 15–30 cm tall with lvs 1–1.5 mm wide. A single, greenish, androgynous spike is borne on each stem. The perigynia are few, 2–4 mm long, beakless, and slightly wider near the base; stigmas 3.

Common up to about 4000 ft, in swamps, wet meadows, along streams, and on lake shores. It can be seen in the Longmire meadow.

Carex limosa L.—MUD SEDGE

A slender plant, reaching 40 cm tall, from a long rootstock; the roots are notably hairy, the yellowish hairs feltlike. The lvs are 2–3 mm wide and pale green. The terminal male spike is slender and about twice the length of the elliptical female spikes; the latter nod on down-curved stalks. The perigynia are glaucous and green or brown, egg-shaped, 2.5–4 mm long, and minutely roughened on the upper half. The scales are brown; stigmas 3.

Primarily a low-elevation plant of sphagnum bogs, it is found in the Park along the Nisqually R, to an elevation of about 4000 ft.

Carex luzulina Olney var. *ablata* (L. H. Bailey) F. J. Herm. WOODRUSH SEDGE

Densely tufted, with short lvs 2–6 mm wide. The stems are 30–90 mm tall. Androgynous, with a slender male spike and 3–7 female spikes grouped rather closely. The perigynia are light green to purplish, 3.5–4 mm long and short-stalked, with a long and tapered, 2-toothed beak. The perigynium is sparsely short-hairy and often so only toward the beak; stigmas 3.

Very common in wet meadows, on lakeshores and along streams, 4000-6000 ft.

Carex mertensii Prescott ex Bong.—MERTEN'S SEDGE

Forming large clumps, with stems reaching 120 cm and flat lvs 3–7 mm wide. Gynaecandrous, with 5–8 spikes on long, nodding stalks. The spikes are 2 cm or more long, elliptical and up to 1 cm broad; stigmas 3. The perigynia are pale green, about 5 mm long, widest at the middle and tapered toward each end, with a very short beak. The perigynia are notable for being much larger than the achenes. The scales are dark brown and much smaller than the perigynia, giving the spikes an attractive two-colored, patterned appearance.

Common along streams and on moist ground in open places in woods, reaching about 6000 ft. Collections have been made along the Carbon R near the Park entrance; along the Lk George trail; at Narada Falls; at Ricksecker Point; in lower Paradise Park; and along the Emmons Glacier trail.

Carex microptera Mack.
SMALL-WING SEDGE

Plants 20–60 cm tall, with flat lvs 2–6 mm broad. The 4–7 gynaecandrous spikes are generally indistinct in a crowded head. The spikes appear bicolored: the brown perigynia are both darker and larger than the scales. The perigynia themselves are 3–5 mm long, about 1 mm wide including the thin wing, veined, and tapered to a long, slender beak; stigmas 2.

Fairly common along lakeshores, on stream banks, in meadows, and in wet places in open woods. Collections have been made near Narada Falls; near Ricksecker Point; at 4500 ft near the road in Stevens Canyon; at 6400 ft on the Rim Trail at Sunrise; and at Eunice Lk.

Carex nardina Fr. var. *hepburnii* (Boott) Kükenth.—SPIKENARD SEDGE

A densely tufted alpine plant, to 15 cm tall but often less, with long, needlelike lvs. Each stem carries a single spike that is brown and 5–12 mm long. The perigynia are 4 mm long, widest above the middle and tapered to a short beak. Stigmas usually 3, but sometimes 2.

Fairly common, on rocky ridges above timberline, chiefly on the south side of the Park. It has been found at about 8000 ft above Paradise and near Van Trump Glacier; and on the upper slopes of the south side of the Tatoosh Range in the Butter Cr Research Natural Area. An early collection made by O. D. Allen came from 10000 ft, according to the label on the specimen at the Park herbarium (probably from above Paradise).

Carex neurophora Mack.
ALPINE NERVED SEDGE

Tufted and sometimes forming large clumps, with stems about 50 cm tall. The lvs are flat and 1.5–3 mm wide, well-distributed along the stems. The lf sheaths are distinctive: lighter in color, of a different texture, and cross-wrinkled. The spikes short, fewer than 10, and crowded into an oblong head. The perigynia are 3–4 mm long, widest at the base and tapered to the tip, and prominently veined; stigmas 2.

Rare in the Park; the only collection located in the course of this study came from along Tatoosh Cr, at an unspecified elevation. More common southward in the Cascades, it typically grows at middle elevations on swamps and wet meadows.

Carex nigricans C. A. Mey.
BLACK ALPINE SEDGE

Tufted and spreading by thick rootstocks, often forming mats. The lvs are flat and 1.5–3 mm wide. The stems are 5–20 cm tall and bear 1 dark brown to nearly black androgynous spike. The perigynia are 4 mm long, widest at the middle, tapered to the tip and with a prominent stalk at the base; stigmas 3. At maturity, the perigynia spread from the axis of the spike.

One of the most common and abundant sedges in the Park, often the dominant plant in subalpine meadows, 5500-7500 ft and more frequent on the south side of the mountain than the north. Also found at Tipsoo Lk and in the upper part of the Butter Cr Research Natural Area. It often forms colonies on ground where the snow is late in melting and is often the dominant plant on the margins of subalpine ponds and on the shores of high-elevation lakes and ponds.

Carex nudata W. Boott—TORRENT SEDGE

Typically growing in large, raised clumps, with stems to about 100 cm tall and sturdy, flat lvs 2–4 mm wide. The sheaths of the lowest lvs become shredded and fibrous in age. The spikes are cylindrical and 2–5 cm long. The perigynia are greenish, 2.5–4 mm long, about 1.5 mm wide, elliptic in outline, and veined; stigmas 2. The scales are smaller and brown.

Listed for the Park for the first time in this study, based on a collection made by Irwin in 1951 at Chinook Pass and held at the Slater Museum at the University of Puget Sound. The listing is tentative, for the exact location of the collection is not recorded.

Elsewhere in its range from California to southern Washington, it grows among rocks at the edges of fast-flowing streams; the species would seem out of place at Chinook Pass, unless the collection was made some distance downhill. The collection may yet prove to have been made on the Yakima County side of the Pass.

Carex pachystachya Cham. ex Steud.
THICK-HEADED SEDGE

A slender, tufted plant 15–30 cm tall, with flat lvs 2–4 mm wide. The 2–5 spikes are crowded into a dense, roundish head about 1 cm across. The perigynia are 3–5 mm long, planoconvex and widest at the middle, brown with a metallic luster, and with a long, tapered, 2-toothed beak; stigmas 2.

Common in wet places, 4000-6000 ft, including meadows, at lakeshores, and along streams; also occasionally in wet ditches along roads. It has been collected in Klapatche Park; at Paradise; at Tipsoo Lk; and along the road to Sunrise "4 miles beyond White R."

Carex paysonis Clokey
PAYSON'S SEDGE

Slender and tufted, with stems 20–40 cm tall and with short, flat lvs 2–4 mm wide. Androgynous; the male and the female spikes elliptical, dark brown, and overlapping but not crowded on the stem; the lowest spike is held erect. The perigynia are 3–4 mm long, nearly round, and abruptly contracted to a short beak; stigmas 3.

Common in wet subalpine meadows, especially along small streams, mostly above about 6000 ft. Collected at several places in Paradise Park—at the base of the Edith Cr trail, near Sluiskin Falls, at 6500 ft on the Skyline trail, at Panorama Point, and on Mazama Ridge. Also found above Reflection Lks and at 6500 ft at Sunrise.

Carex pellita Muhl. ex Willd.—WOOLLY SEDGE

A clump-forming sedge, growing from stout rootstocks, with stems to 100 cm tall but often less. The lvs are flat and 2–5 mm wide. Androgynous, with the female spikes widely spaced on the stems. The spikes are cylindrical, with 20–50 fls. The perigynia are 2.5–3.5 mm long, obovate, 2-toothed with a short beak, and finely hairy. The reddish brown scales are narrower than the perigynia and have a long pointed tip; stigmas 3.

Fairly common in the Park, in swamps, wet meadows, and at lakeshores below about 5000 ft. It can be found at Longmire.

Carex phaeocephala Piper—DUNHEAD SEDGE

Forming patches of tufted plants, from branching rootstocks, with short stems, to 30 cm tall. The lvs are 1.5–2 mm wide, borne mostly at the bases of the stems. The spikes number 2–5, crowded into a dark brown, pyramidal head. The perigynia are 4–5 mm long, slender and widest at the middle, veined, with a short, slender beak; stigmas 2.

Common on dryish, rocky slopes and flats, 6000-7500 ft. Collections have been made at 7150 ft on the moraine of Van Trump Glacier and at 7385 ft on McClure Rock.

Carex praegracilis W. Boott
CLUSTERED FIELD SEDGE

Stems to about 50 cm tall, rising singly from a slender rootstock, with lvs 1.5–3 mm wide, flat or folded. The 5–10 spikes are crowded, but the lowest may be set somewhat apart. The perigynia are planoconvex and 3–4 mm long, and about 1 mm wide near the base, and tapered to a beak; stigmas 2.

Listed by St. John and Warren in 1937, citing a Warren collection made at Longmire, and said to be fairly common. The specimen was not found in the course of this study, nor was other herbarium material found. The species grows widely across western North America but is evidently rare west of the Cascades. It favors alkaline ground, so the Longmire location makes some sense.

Carex praticola Rydb.—MEADOW SEDGE

Densely tufted, with stems reaching 70 cm tall and with flat lvs 2–4 mm wide. The 4–7 short, green gynaecandrous spikelets are fairly well-spaced on the stem, which usually nods at the tip. The perigynia are 3.5–5.5 mm long and 1.5 mm wide, planoconvex and widest at the middle, with a long, tapered beak; stigmas 2.

Meadows, stream banks, and open forests, below 6000 ft. It is found at Paradise, generally growing in rock crevices.

Carex preslii Steud.—PRESL'S SEDGE

Quite similar to *C. pachystachya*. Densely tufted, with stems 20–60 cm tall and flat lvs 2–4 mm wide. The inflorescence is a somewhat open head, with the lowest 1 or 2 spikes set apart. The perigynia are golden-brown, ovate, about 4 mm long and 1.5–2 mm wide (including the narrow wing), tapered to a long, 2-toothed beak; stigmas 2. The scales are reddish brown.

Common in wet meadows, as well as in moist ditches along roads, 5000-6000 ft, chiefly in the Paradise area.

Carex pyrenaica Wahlenb. ssp. *pyrenaica*
PYRENAEAN SEDGE

Densely tufted, with stems reaching just 15 cm tall and with short, inrolled lvs about 1 mm wide. (It differs from the more abundant, mat-forming *C. nigricans* in lacking spreading rootstocks.) The spike is solitary, dark brown, and 1–2 cm long. The perigynia are 3–4 mm long and spreading at maturity; stigmas 3.

Common on open, rocky slopes at and above timberline. It has been collected at 8000 ft near the Van Trump Glacier, and at 6000 ft at the forks of the Paradise River. O. D. Allen found it at 9500 ft, according to the herbarium label (probably above Paradise).

Carex raynoldsii Dewey
RAYNOLDS'S SEDGE

Tufted and spreading by rootstocks, with stems 20–70 cm tall. The lvs are flat, 3–8 mm wide. Androgynous, with 4 or 5 spikes in a loose cluster. The perigynia are obovate, 3.5–4.5 mm long and 1.5–2 mm wide, contracted to a very short beak, green and veined; stigmas 3.

Listed for the Park on the basis of a find made in 1974 in upper Paradise Park. It was included in a list of endangered and threatened taxa by Ola M. Edwards (1980). This species is on the "monitor" list of the Washington Natural Heritage Program.

Carex rossii Boott—ROSS'S SEDGE

Loosely to densely tufted, from stout rootstocks, with stems 10–30 cm tall. The lvs are flat, 1–3 mm wide, and scattered along the stems. The terminal male spikelet is about 1 cm long, longer and more slender than the 1–3 female spikelets. Often 1 or 2 additional spikes are borne on very short stems near the base of the plant. The perigynia are egg-shaped and widest toward the tip, 2.5–5 mm long, abruptly contracted to a pronounced 2-toothed beak, and short-hairy.

Uncommon in the Park, and found from 4500 ft to about timberline, on dry, rocky slopes. It can be found along the Naches Loop trail. A collection from Ipsut Pass was also located during this study, and O. D. Allen found it at 6000 ft in the "upper valley of the Nisqually."

Carex scopulorum T. Holm var. *bracteosa* (L. H. Bailey) F. J. Herm.—ROCKY MOUNTAIN SEDGE

Loosely tufted, 30–50 cm tall, with flat lvs about 4 mm wide. Androgynous, with 2 or 3 elongated and cylindrical spikes. The perigynia are 2–4 mm long obovate and about half as wide as long, nearly beakless, and purplish on the upper half; stigmas 2. Jones treated plants lacking purplish shades in the perigynia and with smaller scales as *C. accedens*.

Widespread in the Park, 4000-6000 ft, on boggy ground and in wet meadows. Collections have been made at Chinook Pass; at Glacier Basin; and at Berkeley Park.

Carex spectabilis Dewey SHOWY SEDGE

Spreading on short rootstocks and forming clumps, with stems 25–70 cm tall that are often purplish at the base, and with flat lvs 2–5 mm wide. The terminal spike is male, with 3–4 female spikes loosely arranged below, the lowest on nodding stalks; stigmas 3. The perigynia are 3–5 mm long and oval, with a short beak. The scales and the upper half of the perigynia are purple-black; the long-elliptical spikes, therefore, are an attractive dark color.

Common, widespread, and abundant on open, wet ground, 5000-7500 ft. Collections have been made at Spray Park; at Lk George; along the trail between Klapatche and St. Andrews Parks; at Van Trump Park; throughout Paradise Park; in the Butter Cr Research Natural Area; at 6000 ft on Cowlitz Ridge; at Louise Lk; at Chinook Pass and Tipsoo Lk; around Dewey Lk; at Sunrise and on First Burroughs Mtn; and at Berkeley Park.

Carex stipata Muhl. ex Willd. var. *stipata* SAWBEAK SEDGE

Tufted perennial from stout rootstocks, with stems 30–90 cm tall and flat lvs 4–8 mm wide. Androgynous, with numerous, stalkless spikes densely clustered into elongated clusters. The perigynia are 4–5 mm long, strongly veined, with a tapered, 2-toothed, and minutely toothed beak about half the overall length of the perigynia. The perigynia are yellow-brown and the scales a darker brown; stigmas 2.

Uncommon, on moist or boggy ground in forests below about 4000 ft. It has been collected around the Carbon R entrance and near Longmire.

Carex utriculata Boott
BLADDER SEDGE

Forming large clumps from wide-spreading rootstocks, with stems to 120 cm tall. The lvs are 3–10 mm broad and are notable for the internal crosswalls between the veins. The 2–4 slender male spikes top the stem. The 2–4 female spikes are green and up to 10 cm long and about 1 cm thick, cylindrical, and with very many fls in lengthwise rows. The perigynia are veined, 4–6 mm long, wider near the middle, and tapered to a long, slender beak; the beak is deeply 2-toothed, the teeth spreading. The style dries and becomes hard with age, remaining attached to the achene; stigmas 3.

Common in bogs and on wet ground below about 4000 ft. It has been collected at Mountain Meadows and at Longmire.

Eleocharis R. Br.—SPIKE-RUSH

Eleocharis obtusa (Willd.) Schult.
OVATE SPIKE-RUSH

Sedgelike in overall appearance, with a single brownish, egg-shaped spike 4–10 mm long at the top of a 5–40 cm tall stem. Unlike species of *Carex* in the Park, *E. obtusa* is an annual, growing in dense tufts of stems, the lvs reduced to mere sheaths on the lower stems. The fls are bisexual (with both stamens and a pistil), with 2 stigmas. Each fl is borne on the axis of the solitary spike at the base of a scale; a perigynium-like structure is absent. Surrounding the ovary are 6–8 slender, barbed perianth bristles, and the base of the style is enlarged, remaining as a small projection on top of the mature achene.

Uncommon on wet, muddy ground in the lower valleys of the major rivers, as around Ohanapecosh and Longmire.

Eriophorum L.—COTTON-GRASS

Another sedgelike group of plants, but easily distinguished by the greatly lengthened, white, soft perianth bristles that give the maturing spike the appearance of a cotton boll. Three species are found in the Park, but only one is common.

1   Spikelet solitary on the stem; lvs folded their whole length *E. chamissonis*
1   Spikelets 2 or more per stem 2

2(1) Bracts beneath the inflorescence 2 or more ....
     .... *E. angustifolium* ssp. *scabriusculum*
2   Only 1 bract beneath the inflorescence .... *E. gracile* var. *gracile*

Eriophorum angustifolium Honck. ssp. *scabriusculum* Hultén—NARROW-LEAVED COTTON-GRASS

Stems slender and 30–60 cm tall, rising singly or a few together from a spreading rootstock. The lvs are represented by brown sheaths at the base of the stems and by green stem lvs only near the tip; the latter are 3–8 mm wide and channeled, or V-shaped. The several spikelets are in a loose cluster, on nodding stalks. The 2 leaflike bracts are at the base of, and longer than, the inflorescence. The bristles are white.

Collections are known from Mtn Meadows, Mowich Lk, and Mystic Lk; reported from the meadow above Bench Lk, Reflection Lks, and from Tipsoo Lk.

An unusual plant found at Mountain Meadows seemed to combine the overall look of *E. angustifolium* with the small-scale features of the achene found in *E. gracile*.

Eriophorum chamissonis C. A. Mey. CHAMISSO'S COTTON-GRASS

Typically found in dense stands of stems 20–60 cm tall, rising from spreading rootstocks. The lvs are narrow, to about 1.5 mm wide. A single spikelet is borne on each stem, just above 2 scalelike bracts. The bristles are reddish brown, at least near the base.

Evidently rare in the Park, on boggy ground at low elevations below about 3000 ft. It is known from Longmire.

Eriophorum gracile Koch ex Roth var. *gracile* SLENDER COTTON-GRASS

Somewhat intermediate in appearance between the other two species of the Park: 3 or 4 spikelets are above by a single leaflike bract. The stem lvs are channeled, or V-shaped, for at least half their length. The bristles are white.

Rare in the Park, and known from a plant list for Berkeley Park made by Virginia Adams and A. B. Adams in 1978 (a list printed and distributed by the Washington Native Plant Society). One collection, made in 1896 by J. B. Flett, was made at the "base of Mount Tacoma."

Scirpus L.—BULRUSH

Sedgelike plants, the Park's two species are easy to distinguish from *Carex* species by the lack of a perigynium. Instead, the perianth parts are present as long or short bristles. *Scirpus cespitosus*, with its single spikelet, might be mistaken for an *Eleocharis*; close examination of the nature of bracts, if any, beneath the spikelet is necessary. The loose, open inflorescence of *S. microcarpus* is distinctive among members of the Cyperaceae in the Park.

1   Inflorescence a single spike; bracts of the involucre very short, erect, and not leaflike .... *S. cespitosus*
1   Inflorescence of numerous spikelets on spreading branches; bracts of the involucre long, leaflike and spreading .... *S. microcarpus*

Scirpus cespitosus L.—TUFTED CLUBRUSH
This species is distinguished by having a single spike per stem and by characteristics of the bracts of the inflorescence. It grows in tufts, with nearly round stems 10–40 cm tall on short rootstocks. The lvs are reduced to sheaths at the base of the stems. The single spike is about 3–6 mm long, slender, few-flowered, and brown.
    Uncommon. Found in a few places on wet, boggy ground above about 5000 ft, as below Mowich Lk.

Scirpus microcarpus J. Presl & C. Presl
SMALL-FRUITED BULRUSH
A vigorous perennial from long rootstocks, with triangular stems 100–150 cm tall. The lvs are 10–20 mm wide and deeply channeled along the midvein. Numerous spikelets are borne in an umbel-like inflorescence just above several leaflike bracts. The spikelets are egg-shaped, green, and 3–5 mm long.
    Found in a few low-elevation swamps, as at Longmire and at the Stevens Canyon entrance station. It also grows in the small pond along the road on the south side of Tumtum Peak.

Iridaceae Juss.—IRIS FAMILY

A true iris, *Iris missouriensis* Nutt., appears to be established in the meadow at Longmire, growing in a large clump about 15 yards out beyond the visitor walk. The author first noted its presence in the Park in 1996 and it seemed to have grown well in 1997. A common species east of the Cascade Mountains, Hitchcock and Cronquist (1973) note its presence on the west side of the mountains only on a few of the San Juan Islands. It also is found on the dry, northeast corner of the Olympic Peninsula. It seems possible that seed of this iris was carried to the site by migrating waterfowl.

Sisyrinchium L.—SISYRINCHIUM

Sisyrinchium idahoense E.P. Bicknell var. *macounii* (E.P. Bicknell) Douglas M. Henderson— BLUE-EYED GRASS

A tufted perennial, with flowering stems reaching 40 cm tall and with slender, flattened, grasslike lvs. The fls are flattish and about 2 cm in diameter, blue-violet with a small yellow center. The sepals and petals are very nearly alike, with attractive dark veins and a short point at the tip.

Frequent in the meadow at Longmire, but not known elsewhere in the Park.

Juncaceae Juss.—RUSH FAMILY

A family of grasslike plants, with two genera: *Juncus*, the true rushes, and *Luzula*, the wood-rushes. Anatomically the flowers resemble those of some lilies, but in outward appearance they are quite different. The perianth segments, 3 sepals and 3 petals, are alike, and are sometimes collectively called "tepals." The flowers are greenish, brown, or purple-brown and small, generally no more than 7 mm long. The inflorescence can be congested and headlike, or the flowers may be arranged singly along the branches of an open and spreading inflorescence. In the Park, only one rush, *Juncus bufonius*, is an annual.

Rushes and wood-rushes are mostly plants of meadows and other moist places, although some rushes are found on open slopes and some of the wood-rushes prefer moist places in forests.

1   Capsule 1-celled, with 3 seeds; lvs lax, hairy on the margins .... *Luzula*
1   Capsule 3-celled, with many seeds; lvs stiff, not hairy on the margins .....
     .... *Juncus*

Juncus L.—RUSH

Tufted perennials, with hairless, rounded or flat leaves; one species is an annual. A basic distinction in the genus is based on the appearance of the lower bract of the inflorescence. In some species, the bract looks exactly like a vertical extension of the stem, so that the inflorescence appears to be borne along the side of the stem. In other species, the bract is relatively short and flat and is notably different from the stem. The usual number of stamens is 6; *J. effusus* has 3 and *J. xiphioides* may have either 3 or 6. Most rushes in the Park have dark brown or purple-brown fls. The shape of the mature capsule is significant in separating some of the species.

1   Inflorescence appearing to be lateral: the lowest bract of the inflorescence cylindric and resembling an extension of the stem 2
1   Inflorescence terminal, the lowest bract of the inflorescence not resembling an extension of the stem 6

2(1)   Fls 1–4 per stem 3
2   Fls usually more than 6 per stem 4

3(2)   Capsule pointed; blades of the inner lvs well-developed .... *J. parryi*
3   Capsule blunt; blades of the inner lvs reduced, bristlelike .... *J. drummondii* var. *subtriflorus*

4(2)   Stamens 3, anthers usually shorter than the filaments; stems densely tufted .... *J. effusus* var. *gracilis*
4   Stamens 6, the anthers about equal to or longer than the filaments; stems rising singly or in small tufts from the rootstock 5

5(4)   Anthers 3–5 times longer than the filaments; inflorescence open .... *J. balticus* var. *balticus*
5   Anthers about equal to the filaments; inflorescence congested .... *J. filiformis*

6(1)   Annual, mostly shorter than 20 cm tall and much-branched; fls arranged singly along the branches.................................... *J. bufonius* var. *bufonius*
6   Perennial, mostly taller than 15 cm; fls in headlike clusters 7

7(6)   Lf blade flattened and turned so that an edge faces the stem; internal crosswalls of the hollow lvs present .... *J. xiphioides*
7   Lf blade flattened, with the flat side facing the stem; crosswalls absent . 8

8(7)   Heads solitary (or rarely 2) on the stem .... *J. mertensianus*
8   Heads 3 or more per stem, in a panicle 9

9(8)   Perianth segments 2–3.5 mm long; plant 10–25 cm tall .... *J. covillei* var. *obtusatus*
9   Perianth segments 4–5 mm long; plant 15–40 cm tall .... *J. regelii*

Juncus balticus Willd. var. *balticus*
BALTIC RUSH

A tufted rush, to 60 cm tall. The lvs are represented by brown scales at the bases of the flowering stems. The many fls are arranged on the branches of an open, apparently lateral, panicle. The perianth segments are 4–5 mm long and purplish brown. The anthers are distinctive among the rushes in the Park: they are 3–5 times longer than the filaments.

Common on moist ground in meadows, occasionally in ditches and along streams, below 3500 ft. Frequent at Longmire.

Juncus bufonius L. var. *bufonius*
TOAD RUSH

The stems of this annual species may reach 20 cm in height, but are often less and sometimes even sprawl on the ground. The lvs are usually inrolled, less than 1 mm wide. The fls, 4–7 mm long, are arranged singly on the profusely branched inflorescence. The anthers are shorter than the filaments.

Very common on moist ground below about 3000 ft, preferring disturbed soil, as at Longmire.

Juncus covillei Piper var.
obtusatus C.L. Hitchc.
COVILLE'S RUSH

A small, tufted rush, about 20–25 cm tall, with slender lvs about 2 mm wide. The inflorescence has 2–4 erect branches with several small clusters of dark brown fls on each. The perianth segments are about 3 mm long; the anthers are slightly longer than the filaments.

Uncommon; found up to 5000 ft on river banks and around lake shores, especially on sandy soils. It has been collected along the lower reach of the White R. It is chiefly a plant from closer to the coast and south into California.

Juncus drummondii E. Mey. var. subtriflorus (E. Mey.) C.L. Hitchc.—DRUMMOND'S RUSH

A low-growing species, reaching 35 cm, with matlike tufts of short, leafless stems. The lower bract of the inflorescence appears to be a continuation of the stem, but is just 1–3 cm long. The 1–3 fls are borne in the small inflorescence. The perianth segments are brown and comparatively long, 5–7 mm.

Very common, 5000-6500 ft, on rocky ridges and open slopes, at and above timberline. Collections have been made at Klapatche Park; Paradise; Three Lks; at Tipsoo Lk and Chinook Pass; at Sunrise and on Burroughs Mtn. Also known from Green Lk.

Juncus effusus L. var. gracilis Hook.—COMMON RUSH

The tallest rush in the Park, reaching 100 cm. The lvs are reduced to brown sheaths at the bases of the flowering stems. The inflorescence appears to be lateral and is usually compact, with many fls on short branches. The perianth segments are 2–3 mm long. The anthers of the 3 stamens are shorter than the filaments.

Common throughout the Park on wet ground: in meadows, in ditches, on stream banks, and at lake shores, reaching about 5000 ft. Weedy around buildings at Longmire.

Juncus filiformis L.—THREAD RUSH

With the lvs reduced to brownish sheaths, the slender, tufted stems reach 20–60 cm tall. The lower bract of the involucre is remarkably long, at

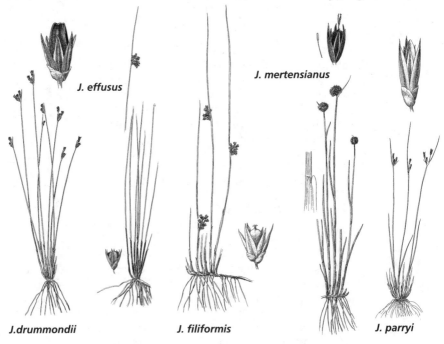

J. effusus

J. mertensianus

J.drummondii   J. filiformis   J. parryi

least half the length of the flowering stem. The apparently lateral inflorescence is rather loose, with 6–8 fls. The perianth segments are 2.5–3.5 mm long, and the anthers about equal the filaments.

Rare in the Park; known from herbarium collections made in wet meadows at Mystic Lk and Sunset Park.

Juncus mertensianus Bong.—MERTENS'S RUSH

Growing 15–45 cm tall, with inrolled lvs 1–2 mm wide. The stems are slender, with 1 or 2 lvs and a solitary head of dark brown fls. The perianth segments are 4 mm long and the anthers are about the same length as the filaments. Plants with 2 heads are sometimes seen, as at Narada Falls.

Very common and widespread in the Park, 4000-7500 ft, on wet ground in meadows and along streams. It can be found at Mowich Lk and Spray Park; on Mount Wow at Lk George; in upper Van Trump Park; at Narada Falls; at Paradise, up to Panorama Point; at Clover Lk; on Dewey Cr at 4000 ft, near the road; at Tipsoo Lk and Chinook Pass; and at Berkeley Park.

Juncus parryi Engelm.—PARRY'S RUSH

Very similar to *J. drummondii*, this species is best distinguished by its pointed capsule and lvs that are threadlike and 3–7 cm long.

Uncommon, on open, rocky slopes, 5000-7000 ft. (*Juncus drummondii* is more common in this habitat.) Collections have been made at Mowich Lk; on the slope below Panorama Point; and at Chinook Pass. Also reported from Berkeley Park.

Juncus regelii Buchenau—REGEL'S RUSH

Usually tufted, with slender stems 15–40 cm tall. The lvs are well-developed and 1–3 mm wide. Each stem bears 1–3 dense, many-flowered heads. The perianth segments are 4–5 mm long and brown. The filaments are about equal in length to the anthers.

J. xiphioides

Apparently uncommon in the Park, or at least seldom collected or reported. It grows on wet ground, about 3000-4500 ft.

Juncus xiphioides E. Mey. DAGGER-LEAF RUSH

A stout rush, 30–60 cm tall with flat lvs 3–10 mm wide; the lvs are borne on the stem so that one edge faces the stem. The panicle is loose and composed of numerous heads. The perianth segments are dark brown and about 3–6 mm long. Plants may have 3 or 6 stamens; 6 is the more common condition.

J. regelii

Very common, ranging up to about 5500 ft, in wet meadows, in marshes, and on stream banks.

Juncaceae

Luzula L.—WOOD-RUSH

Distinctly more grasslike than the true rushes, *Juncus*, with flat, lax leaves that are usually hairy on their lower margins. (Unless otherwise specified, leaf measurements refer to the basal leaves.) The appearance of the inflorescence can be variable: in headlike clusters or open, branched panicles. The flowers have greenish or brownish perianth segments. As with the true rushes, a close examination under magnification of the length of the perianth segments as well as the anthers and their filaments is often crucial in making identifications. The seed capsule holds just 3 seeds in the single chamber. Wood-rushes typically grow on drier ground than do true rushes.

1   Fls on rudimentary stalks in headlike clusters 2
1   Fls mostly solitary on slender stalks in open panicles 5

2(1)   Clusters of fls on very short stalks and crowded into a spikelike panicle; inflorescence nodding on the stem .... *L. spicata*
2   Clusters stalked, not crowded together, nodding or erect 3

3(2)   Stalk of the inflorescence very slender and nodding, 2–4 per stem; perianth segments about 2 mm long .... *L. arcuata* ssp. *unalaschcensis*
3   Stalk of the inflorescence erect, generally 5 or more per stem; perianth segments 2.5 mm or longer 4

4(3)   Heads 1–3 cm long; perianth segments 3–4.5 mm long; anthers twice as long as filaments .... *L. comosa*
4   Heads 0.5–1 cm long; perianth segments 2.5–3 mm long; anthers about equal to the filaments .... *L. multiflora* ssp. *multiflora* var. *multiflora*

5(1)   Lvs 8–12 mm wide; anthers much longer than the filaments ....
                   .... *L. glabrata* var. *hitchcockii*
5   Lvs 6–8 mm wide; anthers shorter than or about equal to the filaments ..
                   6

6(5)   All branches of the inflorescence stiff and widely spreading; perianth segments light brown .... *L. divaricata*
6   Branches of the inflorescence variously curved, nodding, or spreading, but not all stiff and wide-spreading; perianth segments greenish or purplish brown 7

7(6)   Lvs of the stem 5–8 mm broad; perianth segments usually greenish; plants of low-elevation forests, but reaching about 5000 ft in open woods ....
                   .... *L. parviflora*
7   Lvs of the stem 2–3 mm broad; perianth segments purplish brown; plants of open subalpine and alpine slopes and ridges .... *L. piperi*

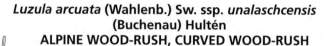

Luzula arcuata (Wahlenb.) Sw. ssp. *unalaschcensis* (Buchenau) Hultén
ALPINE WOOD-RUSH, CURVED WOOD-RUSH

A small plant, just 10–15 cm tall, with slender tufted stems. The lvs are narrow, 1–3 mm wide, and often folded rather than flat. The inflorescence is of 2–6 heads, each with several fls, on slender, drooping stalks.

Known for the Park from just one old collection, made by J. B. Flett in 1919. Flett found the plant "on moraines, [at] Spray Park, at the foot of the Flett Glacier." St. John and Warren (1937) reported that Jones gave them the information that Flett "collected the specimens on Mazama Ridge, near Sluiskin Falls," but there is no confirmation for the plant at this location.

The identification as *L. arcuata* ssp. *unalaschcensis* was made by C. V. Piper, to whom Flett sent much material. Piper wrote to Flett in 1920 to tell him that "this is the first time it has been found within United States territory."

Luzula arcuata is a common plant of high-elevation meadows in the Coast ranges of British Columbia and north through Alaska. Listed as "sensitive" in *The Endangered, Threatened, and Sensitive Vascular Plants of Washington* (Washington Natural Heritage Program, 1997).

Luzula comosa E. Mey.—FIELD WOOD-RUSH

Plants 10–40 cm tall, with lvs markedly hairy, at least when young, and 3–6 mm wide. The clusters of fls are crowded into an oblong-cylindric head to 30 mm long. The perianth segments are 3–4.5 mm long and light brown.

Common in open forests and along trails, below about 4000 ft, especially around the Nisqually entrance and Longmire. Also reported from Berkeley Park.

Luzula divaricata S. Watson
SPREADING WOOD-RUSH

A rush with a distinctive appearance due to the stiff, straight, widely spreading branches of the inflorescence. Otherwise, it's rather ordinary: about 30–40 cm tall, with hairless lvs that are 4–10 mm wide. The fls, arranged singly on slender stalks in an open panicle, are brown and about 2.5 mm long.

Evidently uncommon in the Park. A herbarium collection from Paradise is convincing, but collections in the Park herbarium made at Sunrise at 6400 ft may actually represent immature specimens of *L. piperi*. Another doubtful collection was made along the Reflection Lks trail. The species grows on gravelly soil on hillsides and in open woods, 4500-6500 ft.

Luzula glabrata (Hoppe ex Rostk.) Desv. var. *hitchcockii* (Hämet-Ahti) Dorn—HITCHCOCK'S WOOD-RUSH

A tall plant, to 60 cm, which grows in loose tufts, with broad, flat lvs that are 8–12 mm wide; plants near timberline are shorter. The fls are borne at the tips of a widely branched, mostly erect panicle. The branches sometimes droop, leading to possible confusion with *L. arcuata*, but the fls of *L. glabrata* are solitary rather than in small clusters, and the dark brown perianth segments are longer, 3–3.5 mm long.

Common in woods, 5000-6500 ft on moist ground. Found on Gobblers Knob; through Paradise Park; at Tipsoo Lk and Chinook Pass; at Sunrise and at Sunrise; at Moraine Park; and reported for Berkeley Park.

Luzula multiflora (Ehrh.) Lej. ssp. *multiflora* var. *multiflora*—MANY-FLOWERED WOOD-RUSH

Tufted, with slender stems to 50 cm tall and rather narrow lvs, 2–4 mm wide. Notably long, white hairs fringe the point where the lf joins the stem. The headlike inflorescence is much smaller than that of the similar *L. congesta*, less than 1 cm long and sometimes just half that. The fls are also darker brown and smaller, with perianth segments 2.5–3.5 mm long.

Uncommon, on moist, open slopes near timberline. It has been collected at 5300 ft in St. Andrews Park.

Luzula parviflora (Ehrh.) Desv. ssp. *parviflora* SMALL-FLOWERED WOOD-RUSH

One of the taller wood-rushes, to 60 cm tall, but with fls just 2–2.5 mm long. The lvs are thin and numerous. Those of the stem are 5–8 mm wide. The branches of the open panicle are long and nodding; the fls are usually solitary, but may also occur in 2s.

Plants from Paradise typically have floral bracts that are "lacerate-fimbriate," so finely cut as to appear fringed, a characteristic of *L. piperi*, but otherwise they conform to *L. parviflora* (in which the bracts are typically irregularly toothed but not fringed).

Very common, chiefly below 3000 ft, in openings in moist forests, as well as along trails. It can be found occasionally as high as 5000 ft, as at Paradise. Frequent at the Nisqually, Ohanapecosh, and White R entrances; at Klapatche Park; between Narada Falls and the lower meadows at Paradise; in Stevens Canyon where the road crosses Stevens Cr; and at Three Lks.

Luzula piperi (Coville) M. E. Jones
PIPER'S WOOD-RUSH

A densely tufted wood-rush, 10–40 cm tall, which varies in height according to the harshness and degree of exposure of its habitat. The lvs are thickish and relatively few. Those of the stem are 2–3 mm wide. The inflorescence is an open panicle with nodding branches; the perianth segments are purple-brown and 1.5–2 mm long.

Common on open, stony or sandy soils at and above timberline, 5000-7500 ft, chiefly on the north and east sides of the Mtn. It has been collected at Paradise on Pebble Cr; at 5000 ft on the Kotsuck Cr trail; on the Naches Peak loop; along the Burroughs Mtn trail; at Berkeley Park; and at 7000 ft at St. Elmo Pass.

Luzula spicata (L.) DC.—SPIKED WOOD-RUSH

To 40 cm tall, but usually less, and densely tufted. The lvs are narrow, 1–3 mm wide, and usually folded. The inflorescence is congested and spikelike, the fls are stalkless or on rudimentary stalks. The perianth segments are brown and 2–2.5 mm long.

Infrequent, on open, rocky alpine slopes, 7000-9000 ft. Known for the Park from two herbarium collections, made above Chinook Pass and at Paradise at 7300 ft between Pebble Cr and McClure Rock.

Liliaceae Juss.—LILY FAMILY

For the lilies, an exception is made to the use of J. L. Reveal's families (see Introduction, p. 5). Traditional classification systems have grouped the lilies into a large, artificial family that has satisfied no one. Arthur Cronquist, an important figure in Pacific Northwest botany and author of a widely accepted "integrated system of flowering plants," himself wrote for the *Flora of North America*:

I would be happy enough to divide this group into several families, if I could find a reasonable way to do it, but I have not found the way. . . . We still await a comprehensive reorganization of the lilies into several families more comparable to other recognized families of angiosperms.

Under the system advocated by J. L. Reveal, the Park's lilies would be divided into 8 families, the Alliaceae, Convallariaceae, Liliaceae, Melanthiaceae, Themidaceae, Tofieldiaceae, Trilliaceae, and Xerophyllaceae. Until a new system (and there are other alternatives, too) gains wide acceptance, it makes sense in the present study to maintain the traditional family concept.

Liliaceae

(The term "petals" is used in the keys below; actually, most of our lilies have three petals and three sepals, the so-called perianth segments, which are so alike in appearance that collectively they can be called "tepals.")

Convallaria majalis L., the lily-of-the-valley, is listed in the National Park Service flora, and there was a time, many years ago, when ornamental plantings persisted around Park buildings. No herbarium material was found in the course of this study, and the plant is not included below.

1   Fls in an umbel at the top of a leafless stem; the lvs basal and grasslike. 2
1   Fls not in umbels; the stems more or less leafy (if the stems leafless, then the basal lvs not grasslike) 3

2(1) Lvs onion-scented; petals separate ....*Allium*
2   Lvs not onion-scented; lower portion of the petals united into a tube .....
.... *Triteleia*

3(1) Lvs always 3, in a whorl at the top of a naked stem ....*Trillium*
3   Lvs various, basal or on the stems, if 3 then not in a whorl at the top of the stem 4

4(3) Lvs basal, grasslike but very tough; fls in a dense, headlike raceme on a tall, leafy stem .... *Xerophyllum*
4   Lvs if basal then not tough and grasslike; fls various 5

5(4) Petals reddish orange, curved back; fls on tall, leafy stems .... *Lilium*
5   Petals colored otherwise 6

6(5) Fls bell-shaped, brownish purple mottled with yellow; plants glaucous ....
.... *Fritillaria*
6   Fls of other shapes and colors; plants not glaucous 7

7(6) Lvs all basal; stems lvs, if any, reduced in size to small bracts 8
7   Lvs basal and on the flowering stems 10

8(7) Fl solitary on naked stem, white, erect; plants with slender rootstocks .....
.... *Clintonia*
8   Fls 1 to several, erect or nodding, white or colored; plants with bulbs ... 9

9(8) Fls purplish brown, numerous in a raceme or panicle .... *Stenanthium*
9   Fls white or yellow, 1 to several in a loose raceme .... *Erythronium*

10(7) Fls solitary on a slender stem; lvs mostly basal, linear; stem lvs reduced in size .... *Lloydia*
10   Fls in terminal racemes or panicles, or solitary in the axils of the lvs; stems often with well-developed lvs or sometimes with lvs reduced in size 11

11(10) Fls pendant, solitary or 2–3 at the ends of leafy branches or in the lvs of axils 12
11   Fls erect, numerous, in racemes or panicles at the ends of the branches 13

12(11) Fls solitary in the axils of the lvs .... *Streptopus*
12   Fls solitary or 2–3 at the ends of leafy branches .... *Disporum*
13(11) Lvs ovate to heart-shaped, arranged along the stems, little reduced in size upwards .... *Maianthemum*
13   Lvs mostly basal and slender, the stem lvs much reduced 14
14(13) Lvs equitant (arranged on opposite sides of the stem, with an edge turned toward the stem); fls arranged in groups of 3 in the fl cluster .. .. *Tofieldia*
14   Lvs not equitant, attached on all sides of the stem; fls arranged singly in the cluster .... *Zigadenus*

Allium L.—ONION

With a characteristic onion scent, the alliums of the Park are attractive spring flowers, with dense umbels of flowers at the tops of slender, naked stems. Neither species is often seen here.

1   Umbel on a nodding stem .... *A. cernuum* var. *obtusum*
1   Umbel erect on the stem .... *A. validum*

Allium cernuum Roth var. *obtusum* Cockerell
NODDING ONION

The distinguishing feature of this onion is the U-shaped bend in the top of the leafless stem that reaches 50 cm tall, so that the umbel of pink fls hangs downward. The lvs are grasslike and slender, up to 5 mm wide.

Uncommon, found chiefly on dry, wooded slopes in the southwest corner of the Park, as on Mount Wow at 5000-5700 ft.

Allium validum S. Watson—SWAMP ONION

The white to rose fls of the swamp onion are arranged in a dense umbel on an erect stem 30–60 cm tall and the lvs are flat, about 10 mm wide. The plants generally grow in clumps, expanding as the bulbs multiply.

Uncommon, in wet meadows in the northwest corner of the Park, as at 3500 ft at Mountain Meadows.

Clintonia Raf.—BEADLILY

Clintonia uniflora (Menzies) Kunth
QUEEN'S CUP, BRIDE'S BONNET

Often seen growing in deep forests, where its creeping rootstocks can form attractive patches of plants. Each plant has 2 or 3 oval, glossy green lvs up to 20 cm long, from the bases of which a single, 5–10 cm tall stem rises, bearing a single fl. The widely spreading white petals show well against the dark background and are followed by a single, dark blue, beadlike berry about 1 cm in diameter.

J. B. Flett, evidently unhappy with the Loyalist sound of "queen's cup," tried but failed to win support for "alpine lily" as a common name for this plant.

Very common throughout the Park to 5000 ft, but in greatest numbers in deep, low-elevation forests. Collections have been made at the Nisqually entrance; on Westside Rd and at Lk George; in the woods around Longmire; at 2500 ft on Chinook Cr; in the Butter Cr Research Natural Area; around Ohanapecosh; and at the White R entrance.

Disporum Salisb. ex D. Don—FAIRY-BELLS

Distinctive among the lilies of the Park for their branched stems, the two fairybells species might not be taken for members of the lily family at all, until the flowers are examined. The flowers, whitish bells that hang from the ends of the branches, are followed by red berries.

1	Petals 8–13 mm long and spreading, revealing the stamens; lvs hairless on the upper side .... *D. hookeri* var. *oreganum*
1	Petals 18–25 mm long, hiding the stamens; lvs short-hairy on the upper side .... *D. smithii*

Disporum hookeri (Torr.) Nichols. var. *oreganum* (S. Watson) Q. Jones
HOOKER'S FAIRY-BELLS

A plant with sparingly branched stems 30–80 cm tall. The ovate lvs are 5–12 cm long, with pointed tips and margins fringed with short hairs. The fls usually hang in pairs at the tips of the branches and the petals spread outward from about the midpoint, revealing the stamens.

Common in moist forests below about 3000 ft. Found on Westside Rd at 2200 ft; at 3000 ft on the east slope of Tumtum Peak; at Longmire; and around Ohanapecosh.

Disporum smithii (Hook.) Piper
LARGE-FLOWERED FAIRY-BELLS

About the same height as *D. hookeri*, but more freely branched. The lf margins are not fringed with short hairs. The fls are usually in groups of 3–5 at the tips of the branches. The petals spread slightly only at the tips, and the stamens remain hidden.

A plant common in the Puget lowlands but evidently rare in the Park. It is known only from an unnumbered collection made by O. D. Allen in the early 1890s, in the "upper valley of the Nisqually." It was not listed by Jones (1938) or by St. John and Warren (1937).

Erythronium L.—FAWN-LILY

In sheer numbers, the erythroniums are a dominant element of subalpine meadows throughout the Park. Among the earliest flowers to bloom at Paradise and Sunrise, these lilies can often be found pushing their leaves up through melting snow, blooming not long after the snow has receded. From between the paired, oval, bright green, basal leaves rises a naked stem bearing one to several nodding flowers. As in the Columbia lily, the petals are curved backward, displaying the downward-hanging stamens. The fruit is a fleshy capsule.

1   Fls white; lvs narrowed to a distinct stalk .... *E. montanum*
1   Fls yellow; lvs indistinctly narrowed to a broad stalk, or stalkless ....
.... *E. grandiflorum* ssp. *grandiflorum*

Erythronium grandiflorum Pursh ssp. *grandiflorum*
GLACIER LILY

The yellow, sweet-scented fls of this lily are most often found one per stem, but in mass the effect is fully as striking as that achieved by the multiflowered *E. montanum*. Where the two grow together, the glacier lily will be found to bloom somewhat earlier. The stems reach 15–30 cm. The petals are 2–3 cm long, with a pale zone at the base, and wider than those of the next species.

Common and abundant in wet subalpine meadows on the east side of the Park; less frequent elsewhere. Found between Mtn Meadows and Mowich Lk; on Mount Wow; at Indian Henrys Hunting Ground; in the Paradise meadows; at Reflection Lks; on Kotsuck Cr trail at 5500 ft; at Tipsoo Lk and Chinook Pass; at Sunrise and on the Burroughs Mtn trail; and at Berkeley Park. (The name used by Allen in 1891 is amusing: *E. grandiflorum* var. *parviflorum*, which translates to "small-flowered large-flowered erythronium.")

Erythronium montanum S. Watson—AVALANCHE LILY

There are usually 2 or 3 white fls on the 15–40 cm stems, although plants with 10–12 fls have been reported. The petals are 2–3 cm long and fairly narrow.

In a colony of plants, one will see many single lvs rising from the soil; plants take many years to grow from seed to the stage where they are ready to bloom.

Common and abundant in subalpine meadows and in open woods and thickets near timberline, often blooming at the edge of melting snow. Collected between Mountain Meadows and Mowich Lk; around Lk George and on Mount Wow; at Klapatche Park; at Indian Henrys Hunting Ground; along the road between Narada Falls and Paradise; throughout Paradise Park; at Reflection Lks; along the road between Bench Lk and Stevens Cr; on the Kotsuck Cr trail; at Tipsoo Lk; at Sunrise; at Berkeley Park; and at Windy Gap at 5800 ft.

Fritillaria L.—FRITILLARY

Fritillaria affinis (Schult.) Sealy var. *affinis*—CHOCOLATE-LILY

The chocolate-lily resembles the Columbia lily in stature but is smaller, with 2 or 3 whorls of lvs 10–15 cm long on an erect stem. The stems reach 30–60 cm tall, and it and the lanceolate lvs are notably glaucous. The brownish fls are distinctive: bell-shaped and nodding, each oblong petal 1–3 cm long. The fruit is a blocky, 6-sided capsule.

In other parts of its range, the chocolate-lily tends to grow in brushy places and is not always easy to see. In the seedling stage, its single lf mimics the appearance of the Columbia lily, which is common in the Park.

Collected by O. D. Allen on Mount Wow at 5500 ft, a remarkably high elevation for this plant, in open woods. It is also found along Hwy 123, at the foot of a cliff between Deer Cr and Cayuse Pass. The plant is fairly frequent in the Puget lowlands and would be at about its upper elevation limit in the Park. Kartesz uses the familiar name, *F. lanceolata*, although recent scholarship (Turill and Sealy, 1980) shows *F. affinis* to have precedence according to the rules for naming plants.

Lilium L.—LILY

Lilium columbianum Hanson in Baker—COLUMBIA LILY

One of the most striking of the Park's wildflowers, bearing jewel-like bright red-orange fls on tall stems, capable of reaching a height of almost 200 cm, although often shorter. The narrow, lower lvs are in whorls and up to about 20 cm long; the upper lvs are scattered on the stem. The fls are borne on slender, curved stalks, so that they hang downward. The petals are 3–4 cm long, curved backward and decorated with purple spots. The 6-sided capsule is about 2.5 cm high and the seeds are flattened. Sometimes called "tiger-lily," for its resemblance to the classic tiger-lilies of eastern North America.

Common in open woods on dryish ground and often seen at roadsides, reaching almost 5000 ft and avoiding the deeper forests. Found on Westside Rd at 2600 ft; at Hessong Rock near Spray Park; along the trail between St Andrews and Klapatche Parks; on Rampart Ridge; in the lower reaches of Paradise Park; on the Nisqually Vista trail; on Mazama Ridge; at Box Canyon; at Reflection Lks; on the east side of the Cowlitz Divide along the road; at Silver Falls; around Ohanapecosh; north of Ohanapecosh along Hwy 123 to about 4000 ft; along the road south of the White R entrance; at Berkeley Park; and at Clover Lk. Very nice plants can be seen along the road downhill from the point where it crosses onto the east side of Cowlitz Divide. A vigorous plant collected at 4600 ft on Shriner Peak had six blossoms.

Lloydia Salisb.—ALPINE LILY

Lloydia serotina (L.) Rchb. ssp. *serotina*
ALPINE LILY

Not much more than 15 cm tall, the delicate alpine lily has grasslike lvs, just 1–2 mm wide, that easily disappear among the lvs of the grasses and sedges with which it often associates. The single white blossom, about 2 cm across and marked with dark veins, however, stands out. The plant is more common farther north in the Cascades.

Listed as "watch" in *The Endangered, Threatened, and Sensitive Vascular Plants of Washington* (Washington Natural Heritage Program, 1997)

A rare lily of open rocky slopes above timberline, chiefly on the east side of the mountain. One herbarium collection was located during this study, made by F. A. Warren in the 1930s on Mount Wow, on a "shaded mossy cliff." Also reliably reported from the west slope of Crystal Mtn.

Maianthemum G. H. Weber ex Wiggers—MAIANTHEMUM

Maianthemum, as used here, incorporates two species that formerly were placed in the genus *Smilacina*. The three species are characterized by clusters of small, white flowers that are borne in racemes or panicles at the ends of leafy stems. Unlike *Disporum*, the stems are not branched. The leaves are alternate, heart-shaped to ovate, and not much reduced in size toward the top of the stem.

1   Petals and stamens 4; stem less than 30 cm tall, lvs generally 2–3 per stem .... *M. dilatatum*

1   Petals and stamens 6; stems 30–100 cm tall, lvs many 2

2(1) Stems 30–50 cm tall; fls few in a raceme .... *M. stellatum*

2   Stems 50–100 cm tall; fls many in a panicle .... *M. racemosum* ssp. *amplexicaule*

Maianthemum dilatatum (A.W. Wood) A. Nelson & J. F. Macbr.—WILD LILY-OF-THE-VALLEY

This species was formerly the sole Northwest member of the genus *Maianthemum*, and remains unique among our lilies, with its fl parts in 4s. The lvs are heart-shaped, on long stalks, and strongly resemble those of the cultivated lily-of-the-valley, to 20 cm long and 10 cm wide. The fl is about 2.5 mm long; the stamens are somewhat longer than the petals. The berries are about 6 mm in diameter and become red at maturity.

Common in moist woods below 4000 ft, often carpeting patches of ground in suitable places. Found at the Nisqually entrance; on Westside Rd; around Longmire; at Narada Falls; at Box Canyon; and around Ohanapecosh.

Maianthemum racemosum (L.) Link. ssp. *amplexicaule* (Nutt.) LaFrankie LARGE FALSE SOLOMON'S SEAL

Despite the name "*racemosum,*" the fragrant fls of this vigorous plant are arranged not in racemes but in plumelike panicles. The stems are tall, to 100 cm, and the lvs are ovate to lanceolate, 6–12 cm long, and clasp the stem at their bases. The fl is 1–2 mm long, shorter than the stamens. The berries are red and 5–7 mm in diameter.

Fairly common, reaching about 5000 ft, mainly on the west and south sides of the Park. Found at the Nisqually entrance; on Westside Rd at Fish Cr; on the Van Trump trail; around Longmire; at Narada Falls; around Ohanapecosh and at the Grove of the Patriarchs; and on Chinook Cr at 2500 ft. This species is often seen in thickets and in dense patches of other wildflowers on moist slopes.

Maianthemum stellatum (L.) Link—STAR-FLOWERED FALSE SOLOMON'S SEAL

A much smaller and more delicate plant than *M. racemosum*, this species bears just a few white, starlike fls in a zig-zag raceme at the end of each 20–50 cm tall stem. The lvs are 5–12 cm long and do not have clasping bases. The fls are 4–6 mm long, exceeding the length of the stamens. The berries are purplish and 7–10 mm in diameter.

More common than *M. racemosum*; reaching 4000 ft in open woods and in thickets. Found at the Nisqually entrance and on Westside Rd; at Longmire; at the Glacier View Bridge; in the Butter Cr Research Natural Area; at Box Canyon; at Ohanapecosh; and in the White R V about 1 mi below the campground. Exceptionally robust plants may be found at the Box Canyon parking area in a salmonberry thicket.

Stenanthium (A. Gray) Kunth STENANTHIUM

Stenanthium occidentale A. Gray BRONZE BELLS, MOUNTAINBELLS

A handsome plant, with purple-brown, bell-shaped fls about 1.5 cm long, arranged in an open raceme or panicle on stems that reach 15–40 cm tall. The tips of each petal are curved sharply backward. The 2 or 3 grasslike lvs are 10–20 cm long, from the base of the flowering stem. The fruit is an oblong capsule. The scent of the fls has been described as "tangy."

Fairly common in forests on moist slopes and along streams, often on cliffs and rocks. Found on a cliff along the Van Trump trail; at Longmire; at Silver Falls and up the Ohanapecosh R canyon to 3500 ft; in the White R campground; at Chinook Pass; at 5000 ft on a cliff on Naches Peak; and along the road south of the White R entrance to 3500 ft.

Streptopus Michx.—TWISTED STALK

This group of lilies takes its common name, "twisted stalk," from the unique manner in which the stalk of each flower is twisted, so that the flower is rotated beneath and hidden by the leaf. The flowers are in the axils of the upper leaves on the stems. The plants spread by rootstocks and differ in size, but the species are most easily distinguished by flower color and form. The flowers are followed by red berries.

1   Petals spreading widely, reddish at the base and yellow-green toward the tips .... *S. streptopoides* ssp. *streptopoides* var. *brevipes*

1   Fls bell-shaped, the petals spreading only at the tips if at all, whitish to rose-colored 2

2(1)   Fls greenish white, the petals spreading at the tips ....
.... *S. amplexifolius* var. *amplexifolius*

2   Fls rose-colored, the petals not spreading at the tips ....
.... *S. lanceolatus* var. *curvipes*

Streptopus amplexifolius (L.) DC. var. *amplexifolius*—CLASPING-LEAVED TWISTED STALK

"*Amplexifolius*" refers to the lvs, which at their bases surround and clasp the stem. The 5–10 cm long lvs are ovate, glaucous on the underside, and have fine teeth on the margins. This is the biggest of the three species, with stems 50–100 cm tall. The fls are bell-shaped, 9–15 mm long; the tips of the petals spread or curve backward. The berry is 10–15 mm in diameter.

Widespread but not common, in rich soil in forests below 4000 ft, especially along streams. Found at Mountain Meadows and Mowich Lk; along Tahoma Cr at 3600 ft; at Narada Falls; at Box Canyon; on the Kotsuck Cr trail at 4000 ft; at Silver Falls; and in the White R campground.

Streptopus roseus Michaux var. *curvipes* (Vail) Fassett—ROSY TWISTED STALK

In this species, the stems reach about 30 cm tall, with lvs that are stalkless but do not clasp the stem; the lvs are 3–8 cm long and not glaucous beneath. The fls are 10–12 mm long, bell-shaped; the petals do not spread at the tip. The berry is 7–10 mm in diameter.

Fairly common, on moist soil in forests, 3000-5000 ft. Collected at Mountain Meadows; at Lk George; at Narada Falls; on the trail to the Grove of the Patriarchs; and on Dewey Cr at 4000 ft near the road.

Streptopus streptopoides (Ledeb.) Frye & Rigg ssp. *streptopoides* var. *brevipes* (Baker) Fassett—SMALL TWISTED STALK

The smallest of the twisted stalks, growing 15–20 cm tall. Like the rosy twisted stalk, the lvs are shiny beneath and do not clasp the stem, and are 3–5 cm long. The fls are saucer-shaped and only 2–4 per stem, about 6–7 mm across. The berry is 6–10 mm across.

Uncommon, in forests, 4000-5000 ft. Collected between Mowich Lk and Spray Falls; on Klapatche Ridge; along Tahoma Cr on the trail to Indian Henrys Hunting Ground; up the road from White R on the way to Sunrise; at 4000 ft near Alice Falls; and along the trail up Ipsut Pass from the north. Also reported from Longmire.

Tofieldia Huds.—TOFIELDIA

Tofieldia glutinosa (Michx.) Pers. ssp. *brevistyla* C.L. Hitchc.—TOFIELDIA

Growing to about 30–40 cm tall, the epithet "*glutinosa*" refers to the abundant, sticky glands on the upper part of the plant (which make an interesting study with a hand lens). The fls are about 4 mm long, whitish or greenish white and crowded in a raceme at the top of the stem. The grasslike lvs are 3–5 mm broad and are shorter than the flowering stem. The plant is quite striking at the end of the summer, when the fls are replaced by bright reddish purple seed capsules.

Found on boggy ground and wet subalpine meadows, above about 5000 ft, but also reported in the Longmire meadow. Common but rather easily overlooked. Found between Mountain Meadows and Mowich Lk; at Spray Park; at Chinook Pass; in Glacier Basin at 5500 ft; and at Berkeley Park. An early collection by O. D. Allen was made at "4000 ft, shore of lake at base of Tatoosh Mountains" (possibly Bench Lk?).

Trillium L.—TRILLIUM

Trillium ovatum Pursh ssp. *ovatum* WESTERN TRILLIUM

One of the few fls in the Park that is as likely to be found at the roadside entrance stations as at the fringes of subalpine meadows. "*Trillium*" refers to the profound three-ness of the plants: 3 lvs at the top of the stem, surmounted by a single flower of 3 sepals and 3 petals. Plants can reach 40–50 cm tall, but are often just half that, and the broadly ovate leaves are 5–20 cm long. The fragrant fls are set atop a 3–5 cm long stalk. This is one of the few lilies in the Park in which the sepals are markedly different from the petals. The sepals are greenish and 1–3 cm long while the 2–5 cm long petals are snow white, becoming pink and eventually purplish as they age. The stigma is 3-lobed and the fruit is a fleshy, 3-faced capsule.

Liliaceae

Very common throughout the Park to almost 5000 ft, in deep to open forests in moist places, flowering early in the season and sometimes blooming near melting snow. Found at the Nisqually entrance and along Westside Rd; along Tahoma creek between 2000 and 4000 ft; at the Twin Firs trail; on Rampart Ridge; at Narada Falls; at 4900 ft in lower Paradise Park; in the Butter Cr Research Natural Area; at Box Canyon; along the road between Bench Lk and Stevens Cr; at 4500 ft on the Kotsuck Cr trail; around Ohanapecosh; and in the lower White R V. Especially robust plants are found along the Twin Firs trail.

Triteleia Douglas ex Lindl.—BRODIAEA

Triteleia hyacinthina (Lindl.) Greene—WHITE BRODIAEA

Resembling the wild onions, with grasslike basal lvs and an umbel of fls at the top of a naked stem, but lacking the onion scent. Brodiaeas are common through the Puget lowlands and on the plains east of the Cascades, although none has been previously reported in the Park.

The stems of the white brodiaea reach about 50 cm tall, each with a tight umbel of white or bluish white, bell-shaped fls. The lvs are 10–30 cm long and about 1 cm wide. Each fl is about 1 cm long and each petal has a bluish or greenish midvein. There are 6 fertile stamens, the filaments of which are widened at the base.

Known from a collection made in 1951 on the "Reflection Lk trail" (*Huntley 738*, at the Slater Museum, University of Puget Sound). Huntley identified the plant as *Brodiaea coronaria* (Salisb.) Engl., but the fls are white and crowded in the umbel, on very short stalks, with six fertile stamens. The white brodiaea is known to occur at middle elevations in the mountains of north-central Washington, but its occurrence in the Park is still surprising.

In 1903, J. B. Flett of Tacoma wrote to C. V. Piper, of the faculty at Washington State College in Pullman, about finding "*Brodiaea grandiflora* above where the old trail leading into Paradise V crosses the swamp." He wrote that it was not very abundant here. The specimen he sent to Piper has not been found. In a later letter, Flett describes the plant as growing "near Paradise R Falls on the old trail above the swamp."

Veratrum L.—FALSE HELLEBORE

Tall and distinctive plants of subalpine meadows and along open streams, only the green-flowered *V. viride* is common in the Park. The plants are sometimes called "corn lilies," for the putative resemblance to cultivated corn: tall, to 2 m, with leafy stems and flowers in panicles suggestive of corn tassels. The stalkless leaves are oval to lanceolate, 15–30 cm long, and "pleated" along the lengthwise veins.

1   Fls white, on ascending branches .... *V. californicum* var. *caudatum*
1   Fls greenish, on branches that droop .... *V. viride*

Veratrum californicum Dur. var. *caudatum* (Heller) C.L. Hitchc.—WHITE FALSE HELLEBORE

Difficult to separate from the much more common *V. viride* when not in fl, this species carries its numerous fls on the branches of an erect panicle. The petals are white and greenish at the base, with a green midvein; they measure about 15 mm long and are widest at the base.

Common in the mountains of California and Oregon, but rare in the Park and known from just two locations: at Longmire and at Chinook Pass.

Veratrum viride Aiton GREEN FALSE HELLEBORE

In this species, the lower branches of the open panicle droop, as if burdened by the large number of fls each carries. The fls are green, marked with darker green veins and dark green at the center. The petals are 8–10 mm long.

Common and occasionally abundant in subalpine meadows and on boggy ground, above about 4000 ft, although a few plants can be seen near the White R entrance station in a cold ditch. Found at Mowich Lk; at Indian Henrys Hunting Ground; in a swamp on Backbone Ridge at 4000 ft; in the Paradise meadows; in the Butter Cr Research Natural Area; at Reflection Lks; above Marie Falls at 5600 ft; at Three Lks; on the trail to Burroughs Mtn; at Tipsoo Lk; at Berkeley Park; and along the Ipsut Cr trail.

Xerophyllum Michaux—BEARGRASS

Xerophyllum tenax (Pursh) Nutt.—BEARGRASS

An important plant in native economies; its tough, wiry, grasslike basal lvs, which may reach nearly 1 m in length, were woven into a variety of articles, including mats, baskets, and capes. The stems reach 1.5 m tall and bear a few, short, stiff lvs. The fl clusters are striking: the dense racemes bloom from the bottom upwards, producing a bulbous cluster of white, airy, starlike blossoms beneath a "nipple" of unopened fl buds. Each fl is on a stalk 2–5 cm long; the petals are oblong and about 1 cm long and spread widely. The fruit is a dry, 3-lobed capsule about 5 mm long.

Very common in meadows and on slopes in open, sometimes dry, woods; most common 4500-5500 ft, but seen as low as 2000 ft in places, and reaching 6000 ft. Found at Spray Park; at Lk George; on Rampart Ridge and around

Longmire; at Narada Falls; at the Canyon Rim overlook; in the Paradise meadows; in Stevens Canyon; at the Ohanapecosh entrance; on the Laughingwater Cr trail at 3000 ft; in the Butter Cr Research Natural Area; along the road between Cayuse Pass and Tipsoo Lk and at Chinook Pass; on the Emmons Glacier trail; in the lower White R V; and at Berkeley Park. Plants in open forests seldom bloom or produce stems with reduced numbers of fls.

Zigadenus Michx.—DEATH-CAMAS

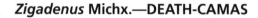

Zigadenus venenosus S. Watson var. *venenosus*—DEATH-CAMAS

The common name, as well as the species epithet, refers to the poisonous bulb of the plant. Somewhat similar to *Tofieldia*, with grasslike basal lvs 10–40 cm long and a dense cluster of cream-colored fls at the top of an erect stem, and growing in similar habitats. The stems reach 50 cm tall and bear a few short lvs. The fls are on short stalks up to 1 cm long and each 8–10 mm long petal has a green spot at the base.

Evidently rare in the Park and most likely reaching the area from slopes east of the Cascade Crest in recent years. Found only along Hwy 123 uphill from Dewey Cr, where it grows on seasonally damp ground below cliffs and along wet roadside ditches. No herbarium collections from the Park are known and the species was not listed by Jones or by St. John and Warren.

Orchidaceae Juss.—ORCHID FAMILY

The forests of Mount Rainier provide ideal habitats for the 15 orchid species found in the Park. Most are common enough and easy to find, although a few are rare, and one, *Corallorhiza trifida*, has not previously been reported in the Park. Most species grow in deep, organically rich soils in forests where snow cover protects them from severe winter weather. Except for the calypso, our orchids are not especially "orchidlike," until, that is, one examines the flowers closely with low-power magnification, when the intricate, exotic forms of the flowers become evident.

The flowers are irregular and bilaterally symmetrical. The 3 sepals are petal-like in appearance, the upper one usually erect and the lateral two spreading. Two of the 3 petals are similar, while the third is broadened or otherwise differentiated into a "lip." The inferior ovary twists, as the flower bud develops, through a 180° turn so that the "lip," which was initially the upper petal, appears then to be the lowest petal. The single stamen is fused with the style into a structure called a column. The fruit is a oblong capsule, stuffed with tiny, dustlike, seeds.

Two genera lack chlorophyll; without green leaves, they gain nutrients through a complex symbiotic relationship with soil fungi.

1	Green lvs absent	2
1	Plants with green lvs, at least up until flowering time	3
2(1)	Plant white	*Cephalanthera*
2	Plant colored, pale yellow to reddish or purplish	*Corallorhiza*
3(1)	Lip with a spur	4
3	Lip lacking a spur	5
4(3)	Lvs basal, withering by flowering time	*Piperia*
4	Lvs basal and on the stem, green at flowering time	*Platanthera*
5(4)	Plant with 1 lf and 1 fl	*Calypso*
5	Plants with more than 1 lf; fls several to many in racemes or spikes	6
6(5)	Lvs 2, opposite, at the top of a short stem	*Listera*
6	Lvs several, basal or arranged along the stem	7
7(6)	Lvs in a flat basal rosette, dark green and often marked with white	*Goodyera*
7	Lvs arranged along the stem, light green and not marked with white	*Spiranthes*

Calypso Salisb.—CALYPSO, FAIRY-SLIPPER

Calypso bulbosa (L.) Oakes var. *occidentalis* (Holz.) Boivin—CALYPSO

A small plant with a single ovate lf 3–5 cm long arising from a bulblike corm. A spicy-fragrant single fl is borne on a slender stem reaching 15 cm tall, and is the most classically formed of our orchids, with flaring pink petals and a white to pinkish, slipper-shaped lip that is mottled with purple-brown spots and streaks; yellow hairs ornament the center of the lip. The sepals and petals are lanceolate and 1.5–2.5 cm long. Infrequent, but locally abundant in a few places, reaching about 4500 ft, often growing on well-decayed logs. Found in the forest at the Nisqually entrance; on Westside Rd at 2300 ft; at Longmire; on the slope above the Glacier View Bridge; and around Ohanapecosh, especially at the start of the hot springs loop trail, where plants with very light-colored blossoms can be found. Scarce in the White R Valley.

Cephalanthera L. C. Rich.—PHANTOM ORCHID

Cephalanthera austiniae (A. Gray) A. Heller
PHANTOM ORCHID

This plant lacks any chlorophyll and obtains its nutrients from a complex symbiotic relationship with soil fungi. The stems reach 20 cm tall and bear 10–20 fls in a loose raceme; the fls are 12–15 mm long. The plant is pure white, with a waxy luster, except for a yellow spot decorating the base of the lip. The upper petals and sepal form something of a "hood" over the column.

Listed as "watch" in *The Endangered, Threatened, and Sensitive Vascular Plants of Washington* (Washington Natural Heritage Program, 1997).

Very rare; known in the Park only from a collection made by O. D. Allen in 1892 in the "upper Nisqually valley." This orchid was not located in the field in the course of this study. A nearly white variant of the western coralroot, *Corallorhiza mertensiana*, is occasionally seen; it can be separated easily from the phantom orchid by the short spur formed by its two lower sepals, while the phantom orchid lacks a spur, has a yellow spot on the lip, and smells of vanilla. (The incorrect spelling *"austinae"* is used in many references.)

Corallorhiza (Haller) Chatel.—CORALROOT

This group of orchids takes its name from the coral-like root structure of the plant. Most species are reddish in color and reach about 30–40 cm in height; one is shorter and pale yellow-green. The flowers are in racemes and the leaves are represented by sheathlike scales. Coralroots are abundant in the Park, in moist to dryish forests, and usually grow in loose clumps of several to many stems.

1   Lip not lobed or toothed; sepals, petals, and lip heavily striped .... *C. striata* var. *striata*
1   Lip with lateral lobes or teeth near its base; sepals, petals, and lip often spotted or with thin veins, but not heavily striped 2

2(1) Plants pale yellow or greenish yellow; sepals 1-veined, 4–6 mm long; lip white and unmarked... *C. trifida*
2   Plants typically reddish purple or reddish brown (rarely yellowish); sepals 3-veined, 8–12 mm long; lip whitish to pinkish, spotted or lightly veined . 3

3(2) Lip spotted; spur absent or represented by a small bulge; column to 5 mm long .... *C. maculata*
3   Lip lightly veined; spur present; column 6–8 mm long ...... *C. mertensiana*

415

Corallorhiza maculata (Raf.) Raf.
SPOTTED CORALROOT

Distinctive for the magenta spots on a white lip, the spotted coralroot is far less frequently seen and favors somewhat drier ground than *C. mertensiana*. Overall the plant is usually reddish brown, but can also be yellow-brown. The stems are 20–30 cm tall and the sepals and petals are 6–10 mm long. The lip is shorter and broader than the lateral sepals, roughly toothed at the tip, and with 2 very short lobes on either side at its base. The spur formed by the sepals is less than 0.5 mm long, or absent.

Uncommon, growing in dense forests and reaching almost 4000 ft. Found at Sunshine Point and on Westside Rd; on Tahoma Cr above the road; on the slope above the Ohanapecosh entrance; at 3000 ft on Laughingwater Cr trail; at 3500 ft on the Shriner Peak trail; and at the White R entrance.

Corallorhiza mertensiana Bong.
WESTERN CORALROOT

The most frequently seen coralroot in the Park, usually found in groups of several to many stems. The stems are 15–40 cm tall. The sepals are 7–10 mm long. The lip is shorter and broader, with 3 dark reddish veins and 2 small teeth at either side of its base. The spur is 0.5–2.5 mm long. The color is typically reddish purple, including the lip, but plants can also be found in shades of pink and yellow. Nearly white plants can be found in the lower White R V. Another variant, seen on the lower slope of Tumtum Peak and at Cougar Rock, shows "broken" stripes on the lip; that is, the lip is striped near the throat and spotted near the tip.

Common and abundant in lower-elevation forests throughout the Park, up to 5000 ft; often the only plant in the understory of dense forests.

Corallorhiza striata Lindl. var. *striata*
HOODED CORALROOT,
STRIPED CORALROOT

An easily distinguished species: each petal and sepal shows 3–5 bold purple stripes; the fls also lack spurs. The stems are 15–40 cm tall and reddish brown to purple. The sepals and petals, 12–14 mm long, are relatively broad and spread more widely than the other species, giving this orchid the most nearly regular fl of any coralroot in the Park. The lower lip lacks basal teeth or lobes; instead, 2 short, lengthwise ridges can be seen.

Evidently rare in the Park ("not common," according to G. N. Jones), and said to prefer drier forests than the other coralroots. (The only herbarium collection found, made by C. F. Brockman, was misidentified and is actually *C. mertensiana*.)

Corallorhiza trifida Chatel.—EARLY CORALROOT

This species has been reported from only a few places in the state of Washington, although it is said to be the most common leafless orchid in North America. The plants are smaller, 20–25 cm tall, and bear fewer fls than is typical for the other three species. The color is a pale yellowish green, indicating that some chlorophyll is present. The lateral sepals are 4–6 mm long, while the lip is about 5 mm long and pure white. The upper petals curve such that their tips touch above the column.

Rare. Two herbarium collections were located in this study which have before this time not been noted by other writers. One, in the herbarium at Pacific Lutheran University, was made by J. R. Slater at Mountain Meadows in the 1950s. The other, at the Park herbarium, was made on the "West R trail" at 2700 ft (that is, around Longmire). The latter was called *C. mertensiana* by C. F. Brockman in the 1930s, but bears a pencil note "?" made by H. W. Smith in the 1940s. These plants have light-colored sepals just 4 mm long that appear to be 1-nerved and that are not striped or spotted. In 1996, one group of 6 stems was found at 2300 ft on Westside Rd, 1.8 mi north of the main Park road, growing in a pocket of deep, rich soil under *Abies amabilis* and *Pseudotsuga menziesii*.

Goodyera R. Br. ex Aiton f.—RATTLESNAKE-PLANTAIN

Goodyera oblongifolia Raf.—RATTLESNAKE-PLANTAIN

A most attractive plant, with neat, flat rosettes of ovate, dark green lvs that are streaked with white, netlike markings. A great deal of variation can be seen in the amount and complexity of the lf veining; rarely, the lvs are entirely green. They are lanceolate to elliptical and 3–6 cm long. The small (6–8 mm long), greenish white fls are unremarkable, in a dense raceme on a stem 20–30 tall. The upper petals and sepal are partly fused and form a hood over the column.

Very common in dryish forests below about 4500 ft. Found at Mowich Lk; at the Nisqually entrance; at 3800 ft on Tahoma Cr; on the Kautz mudflow; on Rampart Ridge and around Longmire; at Ohanapecosh and Silver Falls; in the White R V up to 4500 ft; and in the Carbon R V up to the Windy Gap trailhead.

Listera R. Br.—TWAYBLADE

The twayblades take their name from the two opposite leaves at the middle of each stem; the leaves are ovate to rounded and generally 2–5 cm long. The plants are small, rarely reaching much more than 20 cm tall. Each slender stem bears a few green (rarely reddish brown), spurless flowers, with wide-spreading sepals and petals that are 2–4 mm long. The three species are best distinguished by characteristics of the lip.

1   Lip forked half its length; lvs heart-shaped at the base .... *L. cordata*
1   Lip blunt, or shallowly notched at the tip but not deeply forked; lvs rounded at the base 2

2(1)   Lip shallowly notched at the tip, narrowed at the base; ovary glandular .. .. *L. convallarioides*
2   Lip blunt or rounded at the tip, not narrowed at the base; ovary smooth .... *L. caurina*

Listera caurina Piper—WESTERN TWAYBLADE

In addition to the points noted in the key, this species can be distinguished from the less-common *L. convallarioides* by two short, flaring hornlike structures at the base of the lip; the lip declines at an angle of about 45° from the ovary. The column is about 2 mm long. The plant is often found growing on mossy patches of ground and is the tallest, sometimes reaching 30 cm.

Common in moist forests, reaching about 4000 ft; often found growing with *L. convallarioides*. Found at Mowich Lk; on Westside Rd; along Tahoma Cr on the trail to Indian Henrys Hunting Ground; on Rampart Ridge and around Longmire; at Ohanapecosh; in the lower White R V; and at Ipsut Pass.

Listera convallarioides (Sw.) Nutt. ex Elliott
BROAD-LIPPED TWAYBLADE

Usually less than 20 cm tall, this species has a column about 3 mm long and a lip that is more or less in line with the ovary.

The least common of the three twayblades in the Park, preferring wetter ground. Found at about 3000 ft on the Nisqually R above Longmire; at about 3000 ft on the Muddy Fork of the Cowlitz R above Box Canyon; near the Ohanapecosh entrance (where hundreds grow on a patch of muddy ground in a flood channel); in the White R campground; and in the springs area along the road south of the White R entrance.

Listera cordata (L.) R. Br.—HEART-LEAF TWAYBLADE

Besides the long-forked lip, the plant has attractive, heart-shaped lvs. Most plants have greenish fls, but plants with reddish or reddish brown fls are also seen. Kartesz recognizes two varieties, var. *cordata* and var. *nephrophylla* (Rydb.) Hultén, said to differ in the color of the fls and lf size. Variety *nephrophylla* has green fls, while var. *cordata* is said to have fls that may be green but more often show some red or reddish brown, often in the lip. Hitchcock (1959-67) says that the supposed varieties grow in mixed populations in the Northwest while Coleman (1995) says that the varieties cannot be reliably separated in California. In the Park, most plants bear green fls, but at 2300 ft along Westside Rd, a large colony can be found in which fl color ranges from green to reddish brown. The same thing can be seen in a small colony on the Twin Firs trail, and the recognition of two varieties in the Park does not appear to be justified.

Common in moist forests, reaching about 4500 ft. Found between Mountain Meadows and Mowich Lk; at the Nisqually R; along Tahoma Cr on the trail to Indian Henrys Hunting Ground; at Longmire; at Box Canyon; around Silver Falls; at the White R entrance; and at Ipsut Falls.

Piperia Rydb.—REIN-ORCHID

Formerly grouped with *Platanthera* in the genus *Habenaria*, the piperias native to the Park grow in drier forests and have basal leaves that wither by flowering time. The flowers, while small, are unusually attractive. The sepals have 1 prominent vein.

1 Upper sepals whitish with a green midvein; fls 10–12 mm across at the mouth .... *P. elegans* ssp. *elegans*
1 Upper sepals green; fls 5–6 mm across .... *P. unalascensis*

Piperia elegans (Lindl.) Rydb. ssp. *elegans*
ELEGANT REIN-ORCHID

The fls are borne in a dense, slender spike on stems about 30–50 cm tall. There are 2–4 oblanceolate lvs 5–25 cm long. The sepals and petals, including the lower lip, are not strongly differentiated (although the upper petals are narrower) and spread widely. The fl spurs are long, reaching about 10 mm, slender, and gracefully curved downward.

Uncommon, at low elevations in somewhat dry, open forests, as on the Silver Falls trail.

Piperia unalascensis (Spreng.) Rydb.
ALASKAN REIN-ORCHID

In this species, the fls are fewer and widely spaced on the stems, which are typically less than 40 cm tall. The 2 or 3 basal lvs are oblanceolate and 7–15 cm long. The lower sepals are notably wider than the upper sepal or the petals, and do not spread as widely as those of *P. elegans*. The spur is 2–3 mm long and only slightly curved.

Also uncommon, known from a few collections made in the southwest corner of the Park, 3000-5000 ft: on the north fork of the Puyallup R, on Mount Wow, and on the lower slope of Mount Ararat above Indian Henrys Hunting Ground.

Platanthera L. C. Rich.—BOG ORCHID

Also segregated from *Habenaria*, the bog orchids grow in moist to wet, open places and retain green leaves on the stem through the flowering period. The sepals have 3 prominent veins. The upper sepal and 2 petals meet at their tips to form a hood over the column, while the lateral sepals spread widely. A spur is present, and beneath each flower is a small, leaflike bract.

1   Fls green; spur shorter than the lip and sacklike .... *P. stricta*
1   Fls white; spur equal to or longer than the lip, more or less slender 2

2(1) Spur curved, very slender, much longer than the lip .... *P. leucostachys*
2   Spur almost straight, about as long as the lip and not especially slender ..
     .... *P. dilatata* var. *dilatata*

Platanthera dilatata (Pursh) Lindl. ex Beck var. *dilatata*
DILATED BOG ORCHID

Quite similar in overall appearance to *P. leucostachys*, differing in the nature of the spur, which is about as long as the lip, more or less straight, and not especially slender.

Evidently rare in the Park, and known with certainty only from collections made in wet meadows along the road to Mowich Lk. F. A. Warren made a collection at Longmire in the 1930s, which was annotated by an unknown person as "var. *dilatata*." It has a broad spur less than the length of the lip and may represent a hybrid with *P. stricta*.

Platanthera leucostachys Lindl.—WHITE BOG ORCHID

Variable in height, depending on elevation, 20–80 cm tall. The stems are very leafy, the larger lvs lanceolate and 20 cm long. The inflorescence is a dense spike. The sepals and upper petals are 4-mm long. The lip is 6–8 mm long, abruptly wider below and narrower toward the tip. The spur is about 10 mm long, gracefully curved downward from the ovary.

Called "bog candles" by some, this orchid is often conspicuous at roadsides: the tall plants bear spikes crowded with white fls and often occur in dense colonies. A good candidate for the magnifying glass, with a graceful spur and wide-spreading sepals; the scent of cloves is another reason to take a close look.

Common in wet meadows and around springs, 3000-4500 ft. Found at Mountain Meadows and Mowich Lk and at Longmire. A few can be found in wet ditches along the main road below Longmire as low as 2400 ft.

Platanthera stricta Lindl.
SLENDER BOG ORCHID

When viewed at a distance, the greenish stems and fls tend to disappear among the grasses and sedges with which it often grows. The stems are 20–60 cm tall and the lanceolate lvs up to 15 cm long. The fls are greenish, in a loose to dense spike. The sepals are 3–6 mm long and the lip is slender and 5–7 mm long. The spur is about 3 mm long and shaped like a sack.

A number of collectors at Mount Rainier have misidentified *P. stricta*, calling collections instead *Habenaria sparsiflora* or *H. hyperborea*. The confusion is most often seen with pressed specimens of densely flowered plants, where the short spur is evidently overlooked and the lip is viewed as the spur.

Very common and in some places abundant, in wet meadows, along streams, in ditches, and around springs, 2500-5000 ft. Found between Mountain Meadows and Spray Park; in the Longmire meadow; along the road between Longmire and Paradise; between Cayuse Pass and Chinook Pass; along the road between the White R and Sunrise, to about 5000 ft; and at Berkeley Park.

Spiranthes L. C. Rich.—LADIES' TRESSES

Spiranthes romanzoffiana Cham.
HOODED LADIES' TRESSES

Beautiful for the geometrically precise arrangement of the fragrant fls, set in 3 ranks that twist in spirals along the stems. The stems reach 30–40 cm tall, above the mostly basal, 7–15 cm long, slender lvs; the stem lvs are much shorter. The cream-colored fls are about 1 cm long, narrow, with the lip extended tonguelike, out and down, from a hoodlike cover formed by the other petals and sepals. Under bright light and magnification, they take on a crystalline appearance. There is no spur.

421

Fairly common, typically in wet meadows, reaching 6000 ft, but also colonizing open, disturbed ground and other unusual places. The latter include patches of moss on bare rock under alder at Box Canon; under cottonwood on the Kautz mudflow; and on the sunny shoulder of the road at the Grove of the Patriarchs parking lot. More-conventional locations include the Longmire meadow; along the Nisqually R near the Glacier View Bridge; in lower Paradise Park; and in the meadow above Bench Lk at 4700 ft.

Poaceae (R. Br.) Barnh.—GRASS FAMILY

With 33 genera, the grasses make up the second largest family of plants in the Park, following the sunflower family. As far as possible, the keys in this section rely on characteristics of the plants that are easily observed or measured, but the fact is that identification of the grasses is difficult, and many of the junctions in the keys call for an examination of features of the plants, especially the spikelet; using a 10x hand lens is recommended. Serious study of grasses requires a dissecting microscope, but a 20x hand lens will solve many problems.

Grasses bear their flowers in a small structure called a spikelet. The spikelet itself is an aggregate of a number of smaller organs, which are combined and arranged in a multitude of variations. It is on these variations (which may include the absence of particular organs) that the classification of the grasses is based.

A pair of bractlike glumes sits at the base of each spikelet. At the heart of the spikelet are one or more flowers, each called a floret. The floret is placed immediately above a pair of tiny lodicules, which correspond to the petals and sepals of our showier wildflowers. The essential organs of the floret are stamens, typically two or three, and a pistil, made up of an ovary, two styles, and two stigmas. In some grasses, the florets may bear only stamens or only pistils. These features are only rarely used in the following keys.

Each floret is borne in a lemma, the axil of a bract. The size of a lemma, the degree to which it is hairy or smooth, the veins upon its surface, and the presence or absence of an awn (an elongation of a vein of the lemma) are all of critical importance in identifying the grasses. In most grasses, the lemma is paired with a second bract, the palea, which usually partially enfolds the lemma. In some species the palea is much reduced in size.

Much variation is seen with respect to the arrangement of the spikelets on the branch or branches of the inflorescence. A spikelet-bearing branch is called an axis (plural: axes). The inflorescence may consist of multiple axes forming a panicle. The spikelets may be borne on stalks on a single axis in a raceme, or the spikelets may be attached directly on the axis, an arrangement called a spike.

Characteristics of the leaves are also useful in identifying the grasses. Grass leaves are always two-ranked on the hollow stems and are attached at nodes, solid and usually swollen joints on the stem. The lower part of a leaf, the sheath, surrounds the stem like a tube and is split partly or nearly all of its length. What is commonly viewed as the "leaf" is called the blade, the free portion that rises from the sheath. The blade may be flat, folded, or inrolled along the margins. Where the sheath meets the blade, an evident to obscure structure called the ligule is found; the ligule is often collarlike and represents a short extension of the sheath. In some species, the ligule is nearly absent or is represented by a ring of hairs. In a few genera, two earlike lobes called auricles may be found projecting from the base of the blade.

The First Book of Grasses, by Agnes Chase, is recommended as an excellent introduction to the intricacies of grass anatomy. Mary Barkworth, of Utah State University, is compiling a major new reference on grasses, to be published in 2000, entitled "Manual of Grasses for the Continental United States and Canada," which can be expected to introduce revisions of some of the names listed below. Work by Robert Soreng, of the Smithsonian Institution, on a catalog of New World grasses (the "Tropicos" database) should also be consulted—<http://mobot.mobot.org/Pick/Search/pick.html>

The following key, with its emphasis on characteristics of habitat and stature, is designed to work efficiently for the grasses that occur within the Park. It should not be relied on to yield accurate results in regions outside the Park. Another consequence of the design of this key is that many of the grasses are keyed more than one way: species of *Poa*, for example, are found in each of the five groups.

"Annual" grasses sprout from seed, grow to maturity, flower, and set seed in one growing season. They will generally be found to have fibrous roots—a spreading net of thin, branched roots, all of about equal diameter. "Perennial" grasses live more than one season. They have roots that are thickened at the crown of the plant or take the form of thin to thick, more or less horizontal rootstocks that creep below the soil surface. It is often possible with the perennial grasses to see the remnants of the previous season's stalks.

As used in this key, "disturbed" places are those where the soil has been scraped, turned over, driven or walked on, or deliberately landscaped. The disturbing influence is almost always attributable to humans, although natural processes such as landslides and fire may give nonnative grasses a foothold.

1	Annuals, including weeds and native species	Group 1
1	Perennials, including weeds and native species	2
2(1)	Weedy, introduced species, usually of roadsides and waste places, only rarely in undisturbed places	Group 2
2	Native species, usually of woods, meadows, or rocky slopes (including a few weedy species also keyed in Group 2)	3
3(2)	Dwarf, mostly tufted plants, mostly less than 15 cm tall, growing at subalpine or alpine elevations, above 5000 ft	Group 3
3	Medium-height to tall plants, 20–150 cm, at low to middle elevations, below 5000 ft	4

4(3)   Plants of wet meadows and along streams .... Group 4
4   Plants of drier meadows, open places, and woods .... Group 5

Group 1—Annual Grasses

1   Small, delicate plants, less than 20 cm tall 2
1   Plants 30 cm or more tall 5

2(1)   Lemmas not awned 3
2   Lemmas awned.. 4

3(2)   Panicles spikelike; spikelets 1-flowered .... *Muhlenbergia*
3   Panicles open, spikelets on evident branches; spikelets of 2 or more florets
         .... *Poa*

4(2)   Spikelets 5–10 mm long.. *Vulpia*
4   Spikelets 3–4 mm long.. *Aira*

5(1)   Spikelets contracted into a dense, spikelike panicle 6
5   Spikelets in loose panicles, on evident branches 7

6(5)   Inflorescence plumelike, the spikelets obscured by the awns of the lemma
         .... *Polypogon*
6   Inflorescence brushlike, the spikelets visible between the bristles of the axis,
         lemmas not awned .... *Setaria*

7(5)   Awns of the lemmas straight .... *Bromus*
7   Awns of the lemmas bent 8

8(7)   Stems 20–50 cm tall; spikelets ascending to erect .... *Deschampsia*
8   Stems 40–100 cm tall; spikelets nodding... *Avena*

Group 2—Introduced and Weedy Perennial Grasses

1   Most or all of the spikelets replaced by plantlets .... *Poa*
1   All spikelets normal 2

2(1)   Lf blades and stems velvety, appearing whitish or grayish .... *Holcus*
2   Lf blades and stems hairless or short-hairy, but not velvety, appearing greenish
         3

3(2)   Inflorescence an open panicle with ascending branches 4
3   Inflorescence contracted and spikelike 5

4(3)   Spikelets compressed, somewhat longer than wide, crowded on one side
         of the axis; branches of the inflorescence erect .... *Dactylis*
4   Spikelets slender, much longer than wide; branches of the axis open and
         pyramidal (the lower branches longer than the upper) ....*Puccinellia*

5(3)   Spikelets loosely 2-ranked on opposite sides of the axis; the axis visible
         between the spikelets .... *Lolium*
5   Spikelets densely arranged around the axis, mostly obscuring it 6

6(5)   Inflorescence 10–35 cm long; lemmas not awned 7

6 Inflorescence less than 10 cm long, typically less than 7 cm; lemmas awne
.... 8

7(6) Auricles present; ligule less than 1 mm; panicle narrow, unbranched or
with 1 lower branch.. *Festuca*
7 Auricles absent; ligule 4–10 mm long; panicle narrow but evidently branched
.... *Phalaris*

8(6) Ligules 4–9 mm long; spikelets all fertile, alike .... *Phleum*
8 Ligules less than 1.5 mm long; each fertile spikelet paired with a sterile
spikelet of very different appearance .... *Cynosurus*

Group 3—Subalpine and Alpine Tufted Grasses

1 Spikelets 2-ranked on opposite sides of the axis in terminal spikes ....
.... *Elymus*
1 Spikelets in racemes or panicles (which may be dense and appear spikelike),
not 2-ranked.. 2

2(1) Spikelets 1-flowered 3
2 Spikelets of 2 or more florets 5

3(2) Panicle very dense and spikelike .... *Phleum*
3 Panicle open to contracted, but not spikelike 4

4(3) Loosely tufted annual with reclining stems; lf blades 1–2.5 mm wide ......
.... *Muhlenbergia*
4 Perennials, with densely tufted and erect stems; lf blades less than 1 mm
wide .... *Agrostis*

5(4) Edges of the lf joined at the tip in an upturned "prow"; lemmas not
awned .... *Poa*
5 Lf blades flat or the edges inrolled, but not prowlike; lemmas awned 6

6(5) Awns of the lemmas bent; lf blades 1–3 mm wide .... *Trisetum*
6 Awns of the lemmas straight; lf blades less than 1 mm wide .... *Festuca*

Group 4—Grasses of Wet Meadows and along Streams

1 Lemmas awned, the awns exserted at least 2 mm beyond the glumes .. 2
1 Lemmas not awned, or the awns scarcely if at all exserted beyond the glumes
.... 3

2(1) Spikelets slender, few, arranged like flags in a loose, 1-sided raceme ....
.... *Pleuropogon*
2 Spikelets many, in panicles .... *Calamagrostis*

3(1) Spikelets 1-flowered 4
3 Spikelets of 2 to many florets 5

4(3) Panicle drooping; lemmas often with a rudimentary awn at the tip; mature
spikelets disarticulating (breaking away) from the stalk as a unit, including
the glumes.. *Cinna*

4   Panicle not drooping; lemmas, if awned, bearing the awn at about the middle of the back; spikelets disarticulating above the glumes, leaving the glumes attached to the stalk .... *Agrostis*

5(3)   Lemmas keel-like on the back and purplish at the tips .... *Poa*
5   Lemmas rounded on the back, greenish 6

6(5)   Lemmas short-awned .... *Vahlodea*
6   Lemmas not awned 7

7(6)   Uncommon weeds, 50–200 cm tall; first glume with 1 major vein and second glume with 3, or both glumes with 3–5 veins 8
7   Common natives, 40 cm or more tall; glumes each with 1 major vein ... 9

8(7)   Plant to 50 cm tall; first glume with 1 major vein, second glume with 3 .. .. *Puccinellia*
8   Plant 75–200 cm tall; both glumes with 3–5 major veins .... *Phalaris*

9(8)   Lf sheaths closed to nearly the top .... *Glyceria*
9   Lf sheaths open to the base .... *Torreyochloa*

Group 5—Native Grasses of Other Habitats

1   Flowering stems enlarged and bulbous (onionlike) at the base (the rare weedy grass, *Poa bulbosa*, has stems that are somewhat bulbous; most of its florets are replaced by plantlets) .... *Melica*
1   Flowering stems not enlarged and bulbous at the base 2

2(1)   Spikelets 1-flowered 3
2   Spikelets of 2 or more florets 6

3(2)   Awns of the lemmas 30–40 mm long, bent .... *Achnatherum*
3   Lemmas not awned, or with awns less than 5 mm long 4

4(3)   Base of the lemmas heavily bearded with straight hairs; lemmas with a long or short awn .... *Calamagrostis*
4   Base of the lemmas not bearded; lemmas awned or not 5

5(4)   Panicle drooping; lemmas often with a rudimentary awn at the tip; mature spikelets disarticulating (breaking away) from the stalk as a unit, including the glumes .... *Cinna*
5   Panicle not drooping; lemmas, if awned, bearing the awn at about the middle of the back; spikelets disarticulating above the glumes, leaving the glumes attached to the stalk .... *Agrostis*

6(2)   Inflorescence a spike, the spikelets not stalked (or on a stalk less than 0.5 mm long) and arranged in two rows on opposite sides of the axis 7
6   Inflorescence a panicle, the spikelets evidently borne on stalks and not arranged in 2 rows 10

7(6)   Spikelets flattened, attached singly to the axis 8
7   Spikelets not markedly flattened, attached to the axis in groups of 2–4 . 9

8(7)   Introduced, weedy plant, with extensive creeping rootstocks; lf blades flat, about 10 mm broad, green .... *Elytrigia*

8   Native plant, bearing short runners; lf blades inrolled, about 6 mm broad, and becoming inrolled in age, glaucous .... *Pascopyrum*

9(7)   Mature inflorescence easily breaking apart, the axis disarticulating ....
.... *Hordeum*

9   Axis of the inflorescence continuous and not disarticulating as the spikelets mature and fall .... *Elymus*

10(6)   Lemmas not awned.. 11

10   Lemmas awned 12

11(10)   Spikelets 4–8 mm long; the tips of the lf blades like the prow of a canoe (the edges of the blade upturned and joined at the tip) ...... *Poa*

11   Spikelets 7–12 mm long; tips of the lf blades flat, merely pointed and not prowlike .... *Festuca*

12(11)   Lower margins of the lf blade (where the lf joins the stem) with earlike lobes; lemmas longer than the glumes 13

12   Lf blades not so lobed; lemmas shorter than the glumes 15

13(12)   Uncommon weedy grass with sweet-scented lf blades; florets 3 per spikelet, the lower 2 sterile and represented by empty lemmas ....
.... *Anthoxanthum*

13   Native or weedy grasses; lf blades with no special odor; spikelets with 2 to several florets, but not with 2 sterile florets beneath 1 fertile floret
.... 14

14(13)   Lemmas 2-toothed at the tip, with the awn arising from between the teeth .... *Bromus*

14   Lemmas with a single point at the tip, the awn arising from the tip or the tip merely long and slender .... *Festuca*

15(12)   Lemmas awned at or below the middle of the back; lemmas blunt at the tip, with 2–4 very short teeth 16

15   Lemmas awned from the tip or near the tip; lf blades divided or not at the tip, but not 2–4-toothed 17

16(15)   Lemmas awned at the middle; stem lf blades 4–6 mm wide ....
.... *Vahlodea*

16   Lemmas awned below the middle; stem lf blades 1–3 mm wide.........
.... *Deschampsia*

17(15)   Lemmas deeply 2-cleft at the tip, long-hairy over the rounded back and on the margins; axis of the spikelet hairless .... *Danthonia*

17   Lemmas undivided or with 2 slender bristles at the tip, hairless over the keeled back and on the margins; axis of the spikelet hairy ...... *Trisetum*

Achnatherum Beauv.—NEEDLEGRASS

Achnatherum occidentale (Thurb. ex S. Watson) Barkworth ssp. *occidentale*—WESTERN NEEDLEGRASS

A tufted grass reaching about 50 cm in height, with narrow branches in an ascending panicle. The awns are interesting and unique among the grasses of the Park: nearly 4 cm long, hairy, and bent twice

Rare in the Park, and found on dry, open slopes among rocks and in dry woods, 5000-6000 ft. Known from a collection made at 5600 ft in Glacier Basin.

In her treatment of species formerly grouped in *Stipa*, for *The Jepson Manual* (Hickman, 1993), Mary E. Barkworth uses three separate genera, including *Achnatherum*. That practice is followed here.

Agrostis L.—BENTGRASS

One of the largest and most abundant groups of grasses at Mount Rainier. Several occur as weeds, but most are natives, found in most habitats except deep forests. "Bent" refers to the awns of some species, which are bent at midlength.

Recent authorities have included plants formerly called *Agrostis thurberiana* with *A. humilis*. The following key presents a path to *A. humilis* by two routes, for the plants of the two formerly separate species are easily distinguished, at least as they occur at Mount Rainier.

1	Plants tufted, less than 20 cm tall, growing above 6,000 ft	2
1	Plants 30–120 cm tall, tufted or rhizomatous and spreading	4

2(1)	Lemmas awned from below the middle of the back	*A. geminata*
2	Lemmas not awned	3

3(2)	Spikelets crowded on the branches; lower lf blades 3–6 cm long	*A. variabilis*
3	Spikelets not crowded; lower lf blades 1–3 cm long	*A. humilis*

4(1)	Introduced species, typically weedy at roadsides and around buildings	5
4	Native species, growing in meadows, woods, and on open slopes	6

5(4)	Plants with runners or rootstocks, these less than 5 cm long, slender, and not scaly; ligules less than 2 mm long	*A. capillaris*
5	Plants with rootstocks, these to 25 cm long, stout and scaly; ligules 2–6 mm long	*A. gigantea*

6(4)	Plants with well-developed runners or rootstocks	7
6	Runners and rootstocks absent	9

7(6)	Plants more than 50 cm tall (reaching 120 cm), growing at elevations below 4000 ft; lemmas much shorter than the glumes	*A. diegoensis*
7	Plants to 50–60 cm tall, growing at 4000–6000 ft; lemmas about equal to the glumes	8

8(7)   Spikelets 2 mm long .... *A. aequivalis*

8   Spikelets 3 mm long .... *A. humilis*
           (for plants formerly known as *A. thurberiana*)

9(6)   Lemmas awned... 10

9   Lemmas not awned 11

10(9) Panicle loose, the spreading-ascending branches to 6 cm long; the spikelets borne at the ends of the branches ....*A. geminata*

10   Panicle contracted, the ascending-erect branches to 2.5 cm long; the spikelets borne from near the base of the branch to the tip................ ....*A. exarata* var. *minor*

11(10) Panicle very diffuse, the branches spreading-ascending and reaching 15 cm long .... *A. scabra* var. *scabra*

11   Panicle contracted to open, the longest branches less than 6 cm long 12

12(11) Panicle less than 15 cm long; lower lf blades to 5 cm long and 2 mm wide .... *A. idahoensis* var. *idahoensis*

12   Panicle 15–30 cm long; lower lf blades 10–30 cm long and 2–4 mm wide ....*A. oregonensis*

Agrostis aequivalvis (Trin.) Trin.—ARCTIC BENTGRASS

The plant grows to about 60 cm tall from strong rootstocks, with an open, rather sparse-flowered panicle. The lf blades are mostly flat and up to 3 mm wide. The spikelets are purplish.

    Generally found in wet meadows and bogs near the coast, but occasionally occurring in the Park in meadows at around 5000 ft.

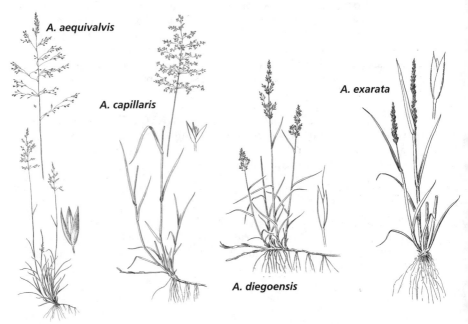

A. aequivalvis

A. capillaris

A. exarata

A. diegoensis

Agrostis capillaris L.—COLONIAL BENTGRASS

A vigorous grass, reaching almost 80 cm tall and spreading by rootstocks, this species is sometimes used as a lawn grass. The lf blades are flat and 3–5 mm wide. The panicle is open, with widely spreading branches that bear spikelets only at their tips.

The plant has spread in Paradise Park, reaching above 6600 ft on the Skyline Trail, and it has also been found in the Longmire meadow. Some herbarium collections of this species have been misidentified as *Agrostis alba*, a pasture grass that has not been verified as occurring in the Park.

Agrostis diegoensis Vasey—LEAFY BENTGRASS

Reaching 120 cm tall, this species is a prominent (if never really abundant) member of meadow communities. The panicle is narrow but not congested. The lf blades are flat and about 3 mm wide.

Rather common below about 4000 ft in meadows; also in open woods. Collections are known from along the Nisqually R at the Glacier View Bridge and along the lower reach of the road to Sunrise. A collection housed at the University of Washington was made at 6000 ft on the moraine of Emmons Glacier, an unusual location.

Agrostis diegoensis has recently been merged under the species concept of *A. pallens* Trin. In Washington, however, "*A. pallens*" is a plant of sand dunes on the coast, from Gray's Harbor southwards. Because the two species are easily distinguished, the earlier name is maintained here.

Agrostis exarata Trin. var. *minor* Hook.
SPIKE BENTGRASS

Named "spike" for the congested panicle, which, though not a true spike, is dense and narrow compared to the other *Agrostis* species in the Park; the branches are short and erect, bearing spikelets nearly to the base. The plant reaches 30–90 cm tall, with flat lf blades 7–8 mm wide.

One of the most common bentgrasses in the Park and often seen at roadsides. It also grows in open woods but is seldom found in meadows. It grows in the area of Cayuse Pass, but is typically found below 3000 ft.

Agrostis geminata Trin.—TWINNED BENTGRASS

Essentially a dwarfed, alpine version of *Agrostis scabra*, this species reaches about 30 cm tall and has a fairly open panicle, its branches often in pairs. The lf blades are inrolled, 5–10 cm long and just 0.5 mm wide. The two other alpine *Agrostis* species in the Park, *A. humilis* and *A. variabilis*, have slender, narrow panicles.

Apparently uncommon, growing on moist to dryish slopes at and above timberline; it has been found on Burroughs Mtn by J. B. Flett. More recent herbarium collections were not located in the course of this study. Hitchcock and Cronquist (1973) did not separate *A. geminata* and *A. scabra*.

Agrostis gigantea Roth—BLACK BENT, REDTOP

A vigorous, weedy bentgrass, found at roadsides and occasionally around buildings and developed areas, below about 3500 ft. Stems rise 50–120 cm from a thick, horizontal rootstock. The lf blades are flat and 5–8 mm wide. The panicle is open, with stiff, reddish purple branches in several whorls.

Collections have been made along the Carbon R near Fall Cr and at the old bridge over the South Puyallup R. G. W. Jones listed *Agrostis alba* L. for the Park, but herbarium research has not verified this species. That name was evidently misapplied to *A. gigantea*.

Agrostis humilis Vasey—ALPINE BENTGRASS

In its "classic" form, this species is a small, tufted plant of seasonally moist places at and above timberline, reaching an elevation of 10000 ft. It rarely grows to more than 10 cm in height and has a slender, constricted panicle, usually purplish in color. The folded lf blades are less than 4 cm long and 1 mm wide.

It is found at higher elevations than is the similar *A. variabilis*, from which it differs in having a palea nearly as long as the lemma.

This little version of alpine bentgrass is chiefly found on the east side of the Park, although it has also been collected in upper Paradise Park. Collections are also known from along the Tipsoo–Dewey Lk trail and from Burroughs Mtn.

More robust, taller plants, formerly called *Agrostis thurberiana*, that grow to about 50 cm tall and have lf blades that are flat and about 2 mm wide, are now also incorporated within *A. humilis*. The panicle is usually greenish and the plants are found most often on wet ground on the east side of the Park above 5000 ft. Herbarium collections are known from along Edith Cr at 5700 ft; at the head of Stevens Canyon; in a wet ditch between Cayuse Pass and Tipsoo Lk; and along Lodi Cr below Berkeley Park.

Agrostis idahoensis Nash var. *idahoensis*
IDAHO BENTGRASS

A small grass, to about 30 cm tall, with a loose, branched panicle that is up to 10 cm long. The lf blades are flat and about 2 mm wide. A few spikelets are borne at the end of each threadlike branch.

Uncommon, growing in wet meadows above about 4500 ft and best known from lower Paradise Park.

Agrostis oregonensis Vasey
OREGON BENTGRASS

Distinguished from *Agrostis idahoensis* by its larger size overall: The plant can reach 90 cm tall and the panicle can stretch to about 30 cm from top to bottom. The lvs are flat and 2–4 mm wide.

A rare grass, of wet meadows and streambanks, sometimes in open woods, below about 3,000 ft.

Agrostis scabra Willd. var. *scabra*
ROUGH BENTGRASS, TICKLEGRASS

Of larger stature, but otherwise similar overall to *A. geminata* and not separated from that species by Hitchcock and Cronquist (1973). It reaches 60 cm tall and has a very diffuse panicle, which can equal half the height of the plant. The flat lf blades are 2–3 mm wide. The panicle is purplish, with ascending branches.

Common in the Park, in moist to wet places in meadows, open woods, and at roadsides, 2500-6000 ft. Collections have been made at Longmire; along the stream below the Nisqually Glacier; in Paradise Park; on the Emmons Glacier trail at 4500 ft; and near the mouth of Fish Cr.

Agrostis variabilis Rydb.
MOUNTAIN BENTGRASS

A tufted plant that reaches about 30 cm tall, with a contracted panicle. The lf blades are inrolled and are less than 7 cm long and 1 mm wide. The lemmas are usually not awned, but occasionally have a short (less than 2 mm) straight awn. The palea is absent, a feature that helps to distinguish this species from the similar *A. humilis*.

Very common in dry meadows and on rocky ridges, 5000-7000 ft. Collections are known from Paradise Park and at 6000 ft in Glacier Basin.

Earlier floras of the Park included *Agrostis rossae* Vasey, a plant that grows only around the hot springs of Yellowstone National Park. Almost all herbarium collections that were once labeled "*A. rossiae*" have been annotated as *A. variabilis*; a few are *A. humilis*. The error seems to have originated with C. V. Piper in his 1906 *Flora of Washington*.

Aira L.—HAIRGRASS

Delicate annual, weedy grasses, with several slender stems reaching 20–40 cm tall. They can often be found on freshly turned ground at roadsides and are not known to occur higher in the Park than about 3000 ft. The awns are slender, bent once, and about as long as the lemmas.

1   Panicle narrow and spikelike, the spikelets longer than the stalks .... *A. praecox*
1   Panicle wide with divergent branches, the spikelets shorter than the stalks .... *A. caryophyllea*

Aira caryophyllea L.—SILVER-HAIRGRASS
A weed of dry ground in disturbed places, only rarely seen in the Park, and possibly only intermittently introduced (that is, failing to maintain ongoing populations). A collection is known from Longmire,

Aira praecox L.—EARLY HAIRGRASS
A weed of similar habitats as the above species and also rarely seen.

Anthoxanthum L.—VERNAL GRASS

Anthoxanthum odoratum L.—SWEET VERNAL GRASS
A tufted, weedy perennial grass, about 60 cm tall, with a tawny-colored, spikelike panicle. The spikelet is composed of 3 florets, only the uppermost of which is fertile; each of the two sterile florets bears an awn. The common name comes from the strong, sweet scent of the roots and stem bases.

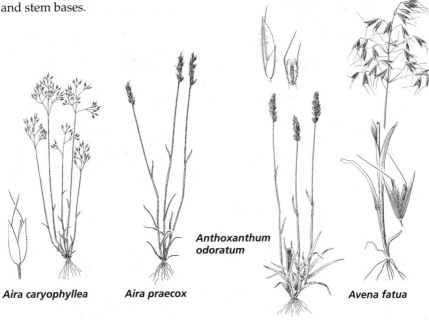

Aira caryophyllea   Aira praecox   Anthoxanthum odoratum   Avena fatua

Listed for Mount Rainier by Regina Rochefort, Park Botanist, but no herbarium material has been found. Sweet vernal grass should be expected in wet meadows and on open roadsides below about 3000 ft.

Avena L.—OAT

Avena fatua L.—WILD OATS

A weedy annual grass, especially noticeable at maturity, when the tall, dry, straw-colored stems stand out against the surrounding vegetation. The 3-flowered spikelets nod at the ends of slender branches in a loose panicle. The awns of the lemmas reach 30–40 mm.

Of sporadic occurrence and seen occasionally around Longmire and along trails where horses are permitted.

Bromus L.—BROME, CHEAT

Bromes are quite common in the Park, and several weedy species are prominent along trails and roads at lower elevations. In general, the bromes are tall, vigorous plants with broad, flat leaf blades that are variously roughened or hairy. The spikelets are large and in most species have 4–7 florets. The awns are generally prominent.

1   Spikelets flattened; lemmas and glumes keeled on the back.................. 2
1   Spikelets more or less cylindrical; lemmas and glumes rounded on the back 3

2(1) Auricles absent at the base of the lf blade (where the lf joins the stem); spikelets 1 or 2 per branch .... *B. sitchensis* var. *sitchensis*
2   Slender auricles present; spikelets several per branch .... *B. carinatus*

3(2) Native perennial, of moist places in open woods below about 3000 ft .... *B. vulgaris*
3   Weedy annuals, mostly of roadsides and disturbed ground 4

4(3) Awns of the lemmas 10–15 mm long .... *B. tectorum*
4   Awns less than 10 mm long 5

5(4) Panicle dense; spikelets ascending to erect .... *B. hordeaceus* ssp. *hordeaceus*
5   Panicle open; at least the lower spikelets spreading to nodding .... *B. commutatus*

Bromus carinatus Hook. & Arn.—CALIFORNIA BROME

The flattened spikelets distinguish this tall (to 100 cm) brome. Auricles at the base of the lf blade separate this species from the much less common *B. sitchensis*. The panicle is loose, with 2 or more spikelets borne at the end of each ascending branch.

Fairly common native grass, found both in open forests and on disturbed ground along trails and roads, 3000-5500 ft. Collections have been made along the trail between Klapatche and St. Andrews Parks; at Round Pass north of Mount Wow; and along the road between Cayuse Pass and Tipsoo Lk.

Bromus commutatus Schrad.
MEADOW BROME

An annual grass reaching about 60 cm tall, with short ascending to spreading branches. The sheaths, and to a lesser extent the lf blades, are long-hairy.

An uncommon roadside weed, known from around the Nisqually entrance.

Bromus hordeaceus L. ssp. *hordeaceus*
SOFT BROME

A weedy annual grass, characterized by soft-hairy lf blades and sheaths, and spikelets arranged in a contracted, upright panicle.

One collection of the plant was located, in the Park herbarium, described as coming from along Hwy 410 about 1 mi south of the White R entrance. It's evidently quite scarce in the Park and was not seen in the field in the course of this study.

Bromus sitchensis Trin. var. *sitchensis*
SITKA BROME

A graceful native grass, similar to *B. carinatus* but easily distinguished by the absence of auricles. The lf blades are mostly hairless. The panicle is quite large, with spreading or drooping branches.

This species is evidently rare in the Park, and just two collections are known: one made by Irene Creso at Sunshine Point and one by O. D. Allen in the "upper valley of the Nisqually." It was not listed in Jones (1938) or in St. John and Warren (1937), nor was it included in Dunwiddie's list (1983).

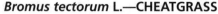

Bromus tectorum L.—CHEATGRASS

The spikelets of this weedy annual grass are smaller than those of other bromes in the Park, seldom reaching much more than 15 mm in length. The plant is also rather short, reaching about 50 cm tall, with an open panicle.

Known from around Longmire, cheatgrass also is found sporadically along roadsides. Plants in the Park are probably var. *glabratus* Shear.

Bromus vulgaris (Hook.) Shear
COLUMBIAN BROME

A tall perennial grass, reaching 120 cm, with a smallish panicle (to about 15 cm high). The branches droop but the spikelets are large, about 25 mm long.

The common low-elevation brome in the Park, chiefly below 3000 ft along roads and in open woods, on moist ground. Collections are known from Sunshine Point and in the valley of the White R near Shaw Cr.

Calamagrostis Adans.—REED-GRASS

Medium-height to tall perennial grasses, from spreading rootstocks. The spikelets, with awned lemmas, are 1-flowered and arranged in dense to open panicles. The base of each floret is beset with short to long straight hairs.

1	Awns straight, about equal in length to the glumes .... *C. canadensis* var. *langsdorfii*
1	Awns of the lemmas bent, exserted 2–5 mm beyond the glumes 2
2(1)	Stems 30–60 cm tall; panicle compact; of drier places .... *C. purpurascens* var. *purpurascens*
2	Stems 20–40 cm tall; panicle somewhat open and interrupted; of moister places .... *C. sesquiflora*

Calamagrostis canadensis (Michx.) P. Beauv. var. *langsdorfii* (Link) Inman—BLUEJOINT

A vigorous grass, to 120 cm tall with a pyramidal panicle that is about 20 cm high and purplish. The lf blades are flat and 5–10 mm wide. The glumes are 5 mm long and the awn of the lemma is short and straight. The hairs at the base of the lemma about equal the lemma in length.

Common at middle elevations, 2500-5500 ft, growing in wet meadows and on stream banks. Collections have been made at Longmire; at an unspecified elevation on Tatoosh Cr; near Sluiskin Falls; along the road between Cayuse Pass and Tipsoo Lk; and in Grand Park at 4800 ft.

A collection in the Park herbarium approaches var. *canadensis* (known as var. *acuminata* Vasey in Hitchcock and Cronquist, 1973): the glumes are 4 mm long and tapered to a slender tip, bearing a few short, stiff hairs; the hairs are about as long as the lemma; the awn is short and straight. It came from the meadow on the Bench at 4800 ft.

Calamagrostis purpurascens R. Br. var. *purpurascens*
PURPLE REED-GRASS

A more or less tufted plant, 30–60 cm tall, with short rootstocks. The lf blades are up to 5 mm wide and softly hairy on the upper surface. The panicle is purplish and the hairs at the base of the lemma are short.

Unlike the two other *Calamagrostis* species, this one grows in open places on dryish, rocky slopes, in a fairly narrow band between 5000 and 6000 ft, chiefly on the west side of the Park, although Piper (1901) reported it from "rocky ridges north of Cowlitz Glacier." Collections are known from Mount Wow and Ipsut Pass.

Calamagrostis sesquiflora (Trin.) Tzvelev
ONE-AND-A-HALF REED-GRASS

Besides the points mentioned in the key, this species may be distinguished from the similar *C. purpurascens* by its longer awn, which in *C. sesquiflora* is exserted from the glumes 3–5 mm (compared to about 1.5 mm in *C. purpurascens*). The common as well as the scientific names refer to the occasional presence of a sterile floret in the spikelet.

In open, moist places, 3000-6000 ft and known from Tolmie Peak, Ipsut Pass, Klapatche Park, and along the trail between Tipsoo Lk and Dewey Lk.

Cinna L.—WOODREED

Cinna latifolia (Trevir. ex Göpp.) Griseb.
SLENDER WOODREED

A tall, slender grass, reaching 120 cm, with an open, drooping panicle that can reach 30 cm long. The flat lf blades tend to stand out from the stem at right angles. The lemmas often have a rudimentary awn at the tip.

Occasional on swampy ground in open woods, up to about 5000 ft, and usually seen on the east side of the Park: collections have been made at the White R campground and at Three Lks. It also occurs on moist, disturbed ground.

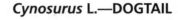

Cynosurus L.—DOGTAIL

Cynosurus cristatus L.—CRESTED DOGTAIL GRASS

A tufted perennial grass, with distinctive panicles: dense
and spikelike, curved toward the top. Each fertile spikelet
is paired with a sterile spikelet; the latter have flattened
glumes that are toothed at the top and empty lemmas. The
plant reaches 60 cm tall.

An occasional weedy grass of roadsides and open,
disturbed ground, below 3000 ft and chiefly in the Longmire
area.

Dactylis L.—ORCHARD-GRASS

Dactylis glomerata L. ssp. *glomerata*—ORCHARD-GRASS

A weedy perennial, cultivated for hay
and now established in many parts of
the Park. The panicle has a decidedly
interrupted appearance, with the spikelets
clustered in 1-sided, small bundles called
"glomerules." The plants have extensive
creeping rootstocks and the stems can reach 150
cm.

A very common roadside weed in open
places; also reaching nearly 5500 ft on disturbed
ground, as at Paradise and Longmire.

Danthonia DC.—OATGRASS

Danthonia intermedia Vasey
TIMBER OATGRASS

A tufted grass, reaching a height of 60 cm, bearing
slender stems and narrow, purplish, spikelike panicles.
The spikelets are relatively few per stem, between 6 and
10, and quite large: the glumes are about 15 mm long
and the lemmas 10 mm, making this grass a good
candidate for an introduction to the details of grass
anatomy. The bent and twisted awns reach 10 mm.

A common grass in moist meadows and on open
slopes, 4500-6000 ft; occasionally on boggy ground.
Collected at Paradise on Marmot Hill at 6000 ft; on the
Bench at 4800 ft; and at Mystic Lk.

Deschampsia P. Beauv.—HAIRGRASS

Little in the way of obvious features unites the three species in this genus: annual or perennial, with open to contracted panicles, of medium height. The glumes and lemmas are distinctively shiny, giving a unique look to the spikelet; each spikelet is composed of two florets.

1   Plants annual; stems 15–40 cm tall .... *D. danthonioides*
1   Plants perennial; stems 30–90 cm tall 2

2(1) Inflorescence contracted and spikelike; lower lf blades about 1 mm wide
....*D. elongata*
2   Inflorescence open, the branches spreading; lower lvs 1.5–3 mm wide ...
.... *D. cespitosa* ssp. *cespitosa*

Deschampsia cespitosa (L.) P. Beauv. ssp. *cespitosa*
TUFTED HAIRGRASS

The tallest of the hairgrasses in the Park, reaching 120 cm, but often half that at higher elevations. The lf blades are rather stiff and folded along the midvein, about 3 mm wide. The panicle is loose and can be 30 cm high; the branches are arranged in whorls.

    Quite common in a range of habitats: meadows, rocky slopes, and open forests, in wet to dryish places. Collections have been made on the trail between Mowich Lk and Spray Park; on the Lodi Canyon trail at 5700 ft; at Mystic Lk; and in Berkeley Park.

Deschampsia danthonioides (Trin.) Munro ex Benth.
ANNUAL HAIRGRASS

An annual grass with a few stems, reaching 60 cm tall, and inrolled lf blades 1–2 mm wide. The spikelets are borne at the ends of ascending branches.

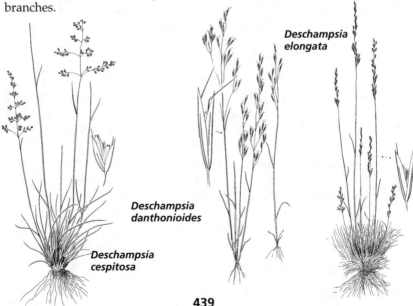

Deschampsia elongata

Deschampsia danthonioides

Deschampsia cespitosa

Found infrequently on dryish, often disturbed ground at middle elevations: one known site is along the highway at the Glacier View Bridge.

Deschampsia elongata (Hook.) Munro ex Benth.
SLENDER HAIRGRASS

Quite different in aspect from *D. cespitosa*: the spikelets are borne on very short, erect stalks, giving a spikelike appearance to the panicle. At a closer scale, the lemmas of slender hairgrass are shorter, 2.5 mm compared to 4 mm.

A common grass found at lower to middle elevations, reaching about 5000 ft, usually in dryish meadows but also in open places in woods. Collections from near Sunshine Point and from Mystic Lk were found.

Elymus L.—WILD-RYE

Differing from the true ryegrasses, *Lolium*, in the way the spikelets are set on the axis. In *Lolium*, the flattened spikelet is turned so that an edge of the spikelet faces the axis; in *Elymus*, the flatter side of the somewhat more cylindrical spikelet faces the axis. In addition, our species of *Lolium* are introduced and weedy, growing mostly at roadsides and in lower-elevation meadows and fields. *Elymus* species are native and found from middle elevations to above timberline. Auricles are present. In some species of *Elymus* both the glumes and the lemmas are awned.

Synonyms abound with this genus and its relatives: some *Sitanion* and *Agropyron* species have been folded into *Elymus*, while another genus, *Elytrigia*, has been segregated.

1 Glumes awnlike, more than 30 mm long, sometimes divided; lemmas very long, 20–90 mm, and widely divergent 2
1 Glumes pointed to short-awned; lemmas not awned or, more often, short-awned, the awn rarely more than 20 mm long 4

2(1) Awns of the lemmas to 35 mm long, spreading ....
.... *E. canadensis* var. *canadensis*
2 Awns of the lemmas mostly much longer, reaching 80–90 mm long 3

3(2) Inflorescence 2–8 cm long; glumes divided, with 2 awns; awns of the lemmas spreading .... *E. elymoides* ssp. *elymoides*
3 Inflorescence 8–18 cm long; glumes not divided; awns of the lemmas erect
.... *E.* x *hansenii*

4(1) 1 spikelet on each branch of the inflorescence; lemmas not awned or the awns less than 5 mm long .... *E. trachycaulus* ssp. *trachycaulus*
4 2–4 spikelets on each branch of the inflorescence 5

5(4) Lemmas with awns 10–30 mm long..................... *E. glaucus* ssp. *glaucus*
5 Lemmas not awned or with awns less than 5 mm long ....
.... *E. glaucus* ssp. *virescens*

Elymus canadensis L. var. *canadensis*
CANADIAN WILD-RYE

A robust grass, growing in clumps and reaching 150 cm in height. The lower lf blades are about 1 cm wide, with smooth sheaths and auricles 1–2 mm long. Both the glumes and the lemmas bear long awns.

Fairly common, 4000-5500 ft, in moist places in meadows and in open woods. It has been collected on the Carbon R at Ranger Cr; on the trail between Klapatche and St. Andrews Parks; and below Chinook Pass. First noted in the Park in this study; although the collections were made between about 1950 and 1970, by Marcus Huntley and Irene Creso, they evidently remained unexamined in the herbaria of Pacific Lutheran University and the University of Puget Sound.

Elymus elymoides (Raf.) Swezey ssp. *elymoides*
SQUIRREL-TAIL GRASS

The distinctive reddish, bushy ("squirrel-tail") appearance of the panicle is due to the long, sharply spreading awns of the glumes and lemmas, which can reach 4–7 cm long. The plants are usually less than 30 cm tall; the erect spike is 2–8 cm long.

Very common on rocky slopes and in dry meadows, from near timberline to about 8500 ft. Collections have been made at 8500 ft on the west side of the Muir snowfield; on Sarvent's Ridge; at 7000 ft at St. Elmo Pass; at 6000 along the trail to Dege Peak; and at 7000 ft in Glacier Basin.

Subspecies of *E. elymoides* are recognized by Mary E. Barkworth in *The Jepson Manual* (Hickman, 1993).

Elymus glaucus Buckley ssp. *glaucus*
BLUE WILD-RYE

"Blue" refers to the strongly glaucous, blue-green lf blades. This grass grows to about 100 cm tall. The lf blades are flat and about 10 mm wide. The awns of the lemmas reach 20–25 mm long.

Common on dry ground along roads and trails, and in open woods. Herbarium collections have been made at Sunshine Point; Longmire; along the highway between Cayuse Pass and Tipsoo Lk; on the slope of Chenuis Mtn at an unspecified elevation; and at 5000 ft below Windy Gap.

Elymus glaucus Buckley ssp. *virescens* (Piper) Gould
BLUE WILD-RYE

Found usually at higher elevations, ssp. *virescens* is shorter than the previous variety, reaching 60 cm. The best feature to distinguish the two are the lemmas, which in ssp. *virescens* are not awned or merely long-pointed.

Common in more or less open forests above about 5000 ft, but seldom collected: a collection made at Ipsut Pass is known.

Elymus x *hansenii* Scribn.—SQUIRREL-TAIL GRASS

A much larger alpine grass than *E. elymoides*, growing to 80 cm tall, with a longer, purplish panicle (8–18 cm) that sometimes nods under its own weight. The awns are often longer as well, to 8 cm.

Less common than the other squirrel-tail, *E. elymoides* ssp. *elymoides*, and found in similar habitats. Collections are known from 6000 ft "between the forks of the Paradise R;" and from cliffs above the Owyhigh Lks at 5800 ft. Said by Mary E. Barkworth (in Hickman, 1993) to be a hybrid between *Elymus multisetus* (formerly called *Sitanion jubatum*) and *Elymus glaucus*.

Elymus trachycaulus (Link) Gould ex Shinners ssp. *trachycaulus*
SLENDER WHEATGRASS

A tufted grass, growing 20–60 cm tall, with short, stout spikes. The lemmas are sharp-pointed or bear a short awn.

Said by Jones (1938) to be a rare grass of rocky slopes above timberline.

The species occasionally forms a hybrid with *Hordeum jubatum*, called *Elyhordeum* x *macounii* (Vasey) Barkworth & Dewey. A 1951 collection at the Slater Museum of the University of Puget Sound (*Huntley 767*, made at Ipsut Pass) is labeled *Elymus macounii* Vasey. The specimen is in poor condition and the identification is, at best, tentative. In any case, the hybrid is sterile and is not likely to be often seen.

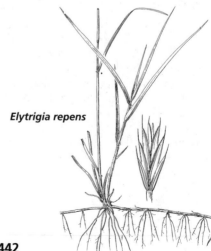

Elytrigia repens

Elymus trachycaulus

Elytrigia Desv.—QUACK-GRASS

Elytrigia repens (L.) Desv. ex W.D. Jacks. var. *repens*
QUACK-GRASS

A pernicious weed of lawns and pastures at lower elevations throughout western Washington, quackgrass spreads rapidly and widely by wiry rootstocks. The leaf blades are usually hairy, with well-developed auricles. The stems are about 100 cm tall, bearing 1 or 2 leaf blades, and the spike, 7–12 cm long, is slender and crowded.

Known in the Park from one collection from Paradise, at the site of the old horse barn, which was made by Wakefield in his 1965 exotic plant survey. Apparently it was not in the Park at the time Jones wrote his flora in 1938.

Festuca L.—FESCUE

A group of handsome perennial bunchgrasses, with mostly smooth stems and leaf blades; the weedy *F. pratensis* grows in loose clumps. The leaf blades are flat or inrolled and the panicle varies from loose and open to fairly compact. The glumes are shorter than the lowest floret and of unequal size; the lemmas of most species are awned (an awn is sometimes absent in *F. pratensis* and sometimes in *F. viridula*).

See the genus *Vulpia* for an annual species of fescue.

1	Lf blades flat, the lower lf blades 3–10 mm wide	2
1	Lf blades folded or inrolled, 0.5–3 mm wide	3
2(1)	Lemmas not awned or with a short point less than 0.5 mm long	*F. pratensis*
2	Awns of the lemmas 5–20 mm long	*F. subulata*
3(1)	Lemmas not awned or no more than 1 mm long; panicle open, the branches mostly erect	*F. viridula*
3	Lemmas awned, the awns at least 1 mm long; panicle open to narrow	4
4(3)	Awns equal to or exceeding the lemmas in length, about 7 mm long; panicle loose and open, the lower branches drooping	*F. occidentalis*
4	Awns less than the lemmas in length, 1.5–4 mm long; panicle narrow with erect branches	5
5(4)	High-elevation grass, growing on rocky slopes; plant to about 20 cm tall; inflorescence about 5 cm long	*F. brachyphylla* ssp. *brachyphylla*
5	Grass of low to middle elevations, growing in woods and meadows and on open slopes; plant 60–100 cm tall; inflorescence 10–15 cm long	*F. rubra* ssp. *rubra*

Festuca brachyphylla Schult. & Schult. f. ssp. *brachyphylla*—ALPINE FESCUE

Seldom reaching even 20 cm tall, this grass is often found tucked between small rocks. The lf blades are up to 5 cm long and tightly folded, almost needlelike. The 4 mm long spikelets are arranged in a rather contracted, spikelike panicle.

Also called "sheep fescue," this is a very common grass at high elevations on open slopes, usually above timberline. It has been collected in Paradise Park and at the base of McClure Rock; and at 7000 ft on Burroughs Mtn.

Festuca occidentalis Hook. WESTERN FESCUE

Vigorous and loosely clumped, with stems reaching nearly 100 cm. The lf blades are inrolled and about 10–25 cm long. The panicle is open and loose, and the lower branches usually droop. The spikelet is 7–8 mm long and sometimes purplish; the awns of the lemmas reach 5–6 mm.

Common in open forests at middle elevations and occasionally on open slopes, below about 4000 ft. Collections have been made near the Glacier View Bridge at 3900 ft and on a ridge close to the Ohanapecosh entrance.

Festuca pratensis Huds.—MEADOW-FESCUE

The well-developed auricles set this perennial species apart. In addition, the young lf blades are flat and the lemmas are not awned, or bear awnlike points less than 0.5 mm long. The stems reach 50–120 cm tall.

Occasional weed along roadsides, known from the Carbon R entrance and Longmire. A plant collected by Irene Creso at Paradise was identified as *Festuca arundinacea*—the auricles of the specimen are not well-preserved, but do not seem to be fringed with short hairs, and the plant is almost certainly *Festuca pratense*.

Kartesz (1998) places this species in *Lolium, L. pratense* (Huds.) S. J. Darbyshire. Susan Aiken, who wrote the treatment of *Festuca* in *The Jepson Manual* (Hickman, 1993), maintains it in *Festuca*, and that practice is followed here.

444

Festuca rubra L. ssp. *rubra*—RED FESCUE

A tall, loosely tufted grass, reaching 100 cm but generally half that at higher elevations. The stem is reddish at the base and the sheaths of the basal lf blades tend to shred in age. The lf blades are smooth, inrolled, and lack auricles. The spikelets are usually reddish purple, occasionally pale green. The awns of the lemmas are short, about 3 mm long.

Common in a wide variety of habitats, from open forests to meadows and dry, rocky slopes, to 7000 ft, although usually below 5500 ft. Found at Sunshine Point; near Narada Falls; at Paradise; and on Burroughs Mtn.

Festuca subulata Trin.—BEARDED FESCUE

With stems 50–120 cm tall, this impressive grass differs from other fescues in the Park in having stems that are leafy most of the way to the panicle. The lf blades are flat and the lower ones about 1 cm wide; auricles are absent. The panicles are open and the branches, which are borne in 2s or 3s along the axis, tend to droop. The awn of the lemma is variable, reaching 5–20 mm long.

A common grass of open forests, reaching about 5000 ft, but seldom collected. Collections are known from the meadow at Longmire and from an unspecified place in the "upper valley of the Nisqually."

Festuca viridula Vasey
GREEN FESCUE, MOUNTAIN BUNCHGRASS

Generally about 60 cm tall, reaching 100 cm on favorable sites, this grass grows in conspicuous clumps that dot the landscape. Its lf blades are flat to slightly inrolled. The panicle is somewhat open and the branches,

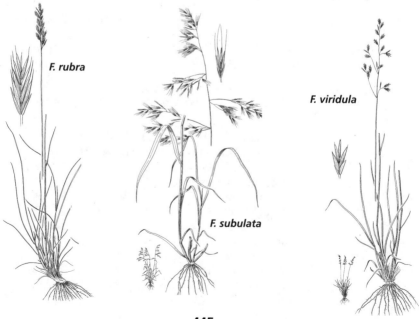

F. rubra

F. viridula

F. subulata

445

generally in pairs, are mostly erect. The lemma is not awned, or sometimes has a very short awnlike tip.

Described by C. V. Piper as "the finest grass on the slopes," this is probably the best-known grass in the Park, or at least the grass most frequently seen by visitors. This is the emerald-green bunchgrass so common across the meadow at Sunrise. It grows in meadows around the mountain, avoiding the wetter places. Besides around Sunrise, it can be found on moraines at Van Trump Park; in the upper reaches of Paradise Park to about 7000 ft; at Panhandle Gap at 6700 ft; in Moraine Park; and in Berkeley Park.

Glyceria R. Br.—MANNAGRASS

Perennial grasses spreading by rootstocks, almost exclusively of wet places, including streamsides, moist meadows, and the margins of ponds. Plants can sometimes be found growing in standing water. The shape of the spikelet is crucial in distinguishing the species: it is either long and cylindrical, at least 3 times as long as wide, or short and egg-shaped, not much more than twice as long as wide. The lemmas are prominently veined, an attractive feature under magnification.

The genus *Torreyochloa* has been segregated from *Glyceria* based on rather minor features: the lf sheath of the former is open to nearly its base and the upper glume is generally 3-veined. In *Glyceria*, the lf sheath is closed to near the top and the upper glume is 1-veined.

1	Spikelets cylindrical, more than 1 cm long, in a narrow panicle on strictly erect branches	2
1	Spikelets flattened and egg-shaped, less than 1 cm long, on spreading branches	3
2(1)	Spikelets 12–18 mm long; lemmas rounded at the tip and hairless or with hairs only on the veins	*G. leptostachya*
2	Spikelets 10–12 mm long; lemmas pointed at the tip and roughly hairy over the entire back	*G. borealis*
3(1)	Sheaths of the lf blades roughly hairy; lemmas greenish	*G. elata*
3	Sheaths of the lf blades smooth; lemmas purplish	*G. grandis* var. *grandis*

Glyceria borealis (Nash) Batch.
SMALL FLOATING MANNAGRASS

Usually about 100 cm tall, with flat lf blades 7–8 mm wide, this species bears its cylindrical spikelets in a narrow panicle on strictly erect branches. A close examination of the spikelet is necessary to distinguish this from *G. leptostachya*, as described in the key, above.

"Floating" mannagrass, because it frequently grows in standing water, at least early in the season. It can be found in wet ditches, and at the margins of ponds and lakes, below about 4000 ft.

Glyceria elata (Nash) M. E. Jones
TALL MANNAGRASS

Taller than the other mannagrasses, reaching 100–150 cm tall, the panicle of *G. elata* has widely spreading, sometimes drooping branches and small, egg-shaped spikelets. The plant is quite graceful.

Common along creeks and rivers below 4000 ft, generally growing in somewhat open riparian woods on the banks, frequently with cottonwood and alder. It can be found on the Carbon, Puyallup, Nisqually, Ohanapecosh, and White Rivers and their tributaries.

Glyceria grandis S. Watson var. *grandis*
AMERICAN MANNAGRASS

Best distinguished by the features in the key, this species is otherwise quite similar to *G. elata*.

Seldom found west of the Cascades, this species had not previously been known to occur in the Park. It was first verified for the Park by a collection at the University of Puget Sound made by Irene Creso at Sunshine Point.

Glyceria leptostachya Buckley
SLENDER-SPIKE MANNAGRASS

Besides the features mentioned in the key, this species is distinctive because its lf blades are often inrolled rather than flat. Like *G. borealis*, it has a narrow panicle, with relatively short, upright branches.

Another grass newly added to the Park flora, based on Creso collections housed at the University of Puget Sound and made along the Carbon R near the Park entrance and on Tahoma Cr on the way to Indian Henrys Hunting Ground, and an 1890s collection by O. D. Allen made in the lower Nisqually valley. The species typically grows on wet ground in the Puget lowlands.

Holcus L.—VELVET-GRASS

Holcus lanatus L.—VELVET-GRASS

Named for the abundance of short, soft hairs on the stems and lf blades, giving the plant the feel of velvet. The stems reach 100 cm and the panicle is congested, wider at the bottom and tapered to the top. The spikelet contains two fls: the lower bisexual and fertile, and the upper staminate or sterile. The lemma bears a curved awn, about 1 mm long.

Common weed on waste ground and along roads; also occasionally seen along trails in openings in woods.

Hordeum L.—BARLEY

The barleys are generally tall grasses and even the native species can appear weedy. The spikes are dense and cylindrical, with spikelets 3 per node on the axis: two spikelets, usually sterile, are below the central fertile one, and each, along with the glumes, bears an awn, giving the spikelet a brushy appearance.

1 Introduced, weedy grass; lf blades with well-developed, slender auricles at the base of the blade where it joins the stem .... *H. vulgare*
1 Native grasses (although *H. jubatum* is often found on disturbed ground); auricles absent 2

2(1) Awns of the lemmas 6–12 mm long, widely divergent .... *H. jubatum*
2 Awns of the lemmas 4–6 mm long, erect ....
.... *H. brachyantherum* ssp. *brachyantherum*

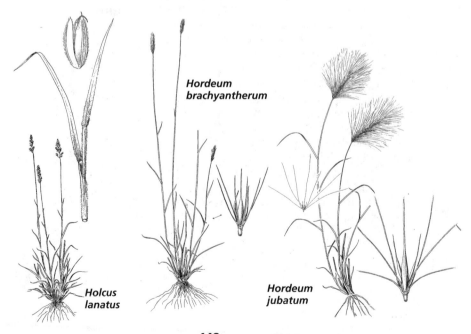

*Hordeum
brachyantherum*

*Holcus
lanatus*

*Hordeum
jubatum*

Hordeum brachyantherum Nevski—MEADOW BARLEY

Plants 30–90 cm tall and tufted, with a spike 5–10 cm long. The awns, around 1 cm long, are the shortest of the barleys in the Park, and the spikes, therefore, are the most compact-looking of the 3 species.

Uncommon in moist meadows, avoiding woods and dryer places, up to about 4500 ft, and most often seen on the south and west sides of the Mountain.

Hordeum jubatum L. ssp. *jubatum*
FOXTAIL-BARLEY

A common native perennial with gray-green lf blades, sometimes flowering its first year. The stems reach about 50 cm tall, with a brushy-looking spike, an effect due to the long, spreading awns of the glumes, which reach 2–6 cm in length (the lemma awns are shorter, 5–7 mm). The spike, overall, is usually a reddish purple color, more rarely pale green.

Widespread in the Park, in open woods, on disturbed ground in river bottoms, but most frequently seen at roadsides, reaching about 4000 ft.

Hordeum vulgare L.—COMMON BARLEY

Best distinguished from the other species, and especially *H. brachyantherum*, by the broad lf blades, usually more than 10 mm wide, with well-developed auricles. Once the mature spikelets drop, a crooked axis remains attached to the node.

An annual weed, of infrequent occurrence (and likely, therefore, to have been introduced repeatedly over the years and not persisting long). When it does appear, it is usually in the White R V, below about 3500 ft.

Lolium L.—RYEGRASS

Neither of these grasses is native and their occurrence in the Park tends to be sporadic; they are mostly confined to roadsides. Differing from wild-rye grasses, *Elymus*, in the way the spikelets are set on the axis: in *Lolium*, the flattened spikelet is turned so that an edge of the spikelet faces the axis; in *Elymus*, the flat side of the more or less cylindrical spikelet faces the axis.

1   Lemmas awned.. *L. perenne* ssp. *multiflorum*
1   Lemmas not awned .... *L. perenne* ssp. *perenne*

Lolium perenne L. ssp. *multiflorum* (Lam.) Husn.
ITALIAN RYEGRASS

An annual or biennial grass, but occasionally persisting longer than two years, with stems reaching 60–90 cm tall. The lf blades are a bright, glossy green color and inrolled when young. The lemmas bear awns 7–8 mm long.

Seemingly more common than ssp. *perenne*, Italian ryegrass is often found at roadsides and sometimes in grassy meadows at lower elevations, as at Longmire.

Lolium perenne L. ssp. *perenne*
PERENNIAL RYEGRASS

Perennial, and differing from ssp. *multiflorum* by the absence of awns on the lemmas. The young lf blades are folded rather than inrolled.

A weed of roadsides and around buildings, below about 3000 ft, throughout the Park.

Melica L.—MELIC

Melica subulata (Griseb.) Scribn.
var. *subulata*
ALASKA MELIC, ALASKA ONIONGRASS

Unique among the grasses of the Park: the bases of the lf blades are broadly expanded and thickened, resembling small bulbs. The stems are 60–120 cm tall and are tufted. The onionlike bulbs are borne on short, thick rootstocks, by which the plant spreads vegetatively. The lf blades are narrow, to 5 mm, flat and shiny. The panicle is rather narrow, with erect branches. The glumes are papery and translucent; the lemmas are not awned.

Uncommon, and found on the north and east sides of the Park along stream banks, in meadows, and on open slopes, above 4500 ft, reaching the summit at Chinook Pass.

Muhlenbergia Schreb.—MUHLY

Muhlenbergia filiformis (Thurb. ex S. Watson) Rydb.
PULL-UP MUHLY

A small, annual grass, with tufted stems to 25 cm tall, with a narrow, cylindrical panicle. The spikelets are 1-flowered; the lemma bears a rudimentary awn, less than 1 mm long.

Common in moist meadows and on stream banks. It has been collected at Mountain Meadows and on the Nisqually R near the Glacier View Bridge.

Pascopyrum A. Löve—WESTERN WHEATGRASS

Pascopyrum smithii (Rydb.) Barkworth & D. Dewey—BLUESTEM WHEATGRASS

A tall, rhizomatous perennial, to 100 cm, with flattened spikelets 2-ranked on the stems, but with short internodes so that the spikelets overlap. The stems and the inflorescence are markedly glaucous. The awn is less than 5 mm long.

Not common, and generally on sandy, loose soil on open slopes. It had been collected near Lk Allen.

Phalaris L.—CANARY-GRASS

Phalaris arundinacea L. REED CANARY-GRASS

A perennial, spreading strongly by rootstocks, with stems up to 150 cm tall, with flat lf blades 1–2 cm wide. The inflorescence is a tall, open panicle (although the branches become more erect as the seed matures). The spikelet tends to fall as a unit, leaving the glumes on the axis. The lemma is 3–4.5 mm long.

Listed by Park Botanist Regina Rochefort. There is doubt whether the species has ever been native in western Washington, although it is certain that the plants that are occasionally seen in the Park are introduced.

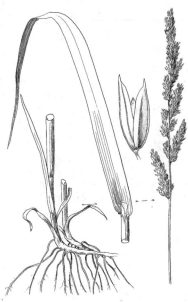

Phleum L.—TIMOTHY

Two very different timothy grasses are found in the Park. Both have 1-flowered spikelets in which both the glumes and lemmas are short-awned. The weedy common timothy can occur at high elevations, but the shape of the inflorescence easily differentiates it from mountain-timothy.

1   Inflorescence cylindrical, 5–10 cm long .... *P. pratense* ssp. *pratense*
1   Inflorescence elliptical, 2–3 cm long .... *P. alpinum*

Phleum alpinum L.—MOUNTAIN-TIMOTHY

The inflorescence is congested and elliptical to egg-shaped in outline, on a stem 20–60 cm tall. The awn of the lemma is 2–3 mm long.

Common above 6000 ft, sometimes lower, and reaching nearly 8000 ft, on slopes around the mountain. It has been found at Mountain Meadows; along the trail to Paradise Glacier and above Sluiskin Falls; on the trail between Tipsoo and Dewey Lks; and on Burroughs Mtn.

Phleum pratense L. ssp. *pratense*
COMMON TIMOTHY

A taller plant than the native mountain-timothy, reaching 100 cm tall, with a long, cylindrical, tightly congested inflorescence, less than 1 cm thick but 10 cm (or more) long.

An uncommon weed of roadsides and around buildings. Found at Longmire; in the developed area at Paradise; and around Ohanapecosh and intermittently along the highway north to the White R entrance.

Pleuropogon R. Br.—SEMAPHORE-GRASS

Pleuropogon refractus (Gray) Benth. var. *refractus*
NODDING SEMAPHORE-GRASS

A highly distinctive grass, with narrow spikelets, arranged like widely spaced pennants in a loose, 1-sided raceme. A perennial with stems about 100–150 cm tall and growing in loose tufts, with flat lf blades 5–7 mm wide. The awn of the lemma is 5–10 mm long.

Uncommon in the Park, found along streams and in moist places in open forests, below 3000 ft. It has been found in the Longmire meadow.

In 1994, Kartesz used *Lophochlaena* Nees for this genus. It's again *Pleuropogon* in the 1998 online version, as well as in *The Jepson Manual* (Hickman, 1993).

*Pleuropogon
refractus*

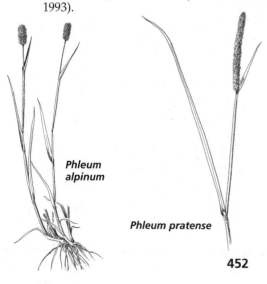

*Phleum
alpinum*

Phleum pratense

Poaceae

Poa L.—BLUEGRASS

With the most species of any genus of grasses in the Park, *Poa* species can be found in virtually all habitats. Several of the weedy species are common in lower-elevation meadows and in wet ditches. In most species, the lemmas have a distinctive purplish band, like a chevron, across the back near the tip. The presence and nature of the hairs on the lemma are critical in identifying some of the species.

1	Plant annual	*P. annua*
1	Plant perennial	2
2(1)	Most of the florets replaced by plantlets; base of the stems bulbous	*P. bulbosa*
2	Florets all normal; stems not bulbous	3
3(2)	Introduced grasses, more than 50 cm tall, usually of meadows and other moist places at lower elevations; plants with rootstocks or runners	4
3	Native grasses, mostly densely tufted and less than 50 cm tall, of meadows and slopes above 4500 ft (and frequently near or above timberline); rootstocks or runners absent	8
4(3)	Stems (including the nodes where the lf blades attach to the stem) strongly flattened, with 2 distinct edges	*P. compressa*
4	Stems round in cross section, not flattened or 2-edged	5
5(4)	Lemmas roughened over the back but not with a tangled web of hairs at the base	*P. wheeleri*
5	Lemmas roughened or variously hairy over the back, with a web of tangled hairs at the base	6
6(5)	Plant with strong rootstocks; sheaths of the lf blades open less than 3/4 of their length	*P. pratensis*
6	Plants with runners; sheaths open 3/4 of their length or more	7
7(6)	Veins on the back of the lemma prominent; lower glume 1-veined	*P. trivialis*
7	Veins of the lemma obscure; lower glume 3-veined	*P. palustris*
8(3)	Lemmas with a short to long web of tangled hairs at the base	9
8	Lemmas hairless or variously hairy, but not with a tangled web of hairs at the base	11
9(8)	Lemmas short-hairy on the back between the veins toward the base; stems 10–30 cm tall	*P. arctica* ssp. *grayana*
9	Lemmas hairless between the veins; stems of various heights	10
10(9)	Stems 10–30 cm tall; lf sheaths smooth	*P. leptocoma* ssp. *paucispicula*
10	Stems 30–60 cm tall; lf sheaths roughly hairy	*P. leptocoma* ssp. *leptocoma*
11(8)	Lemmas long-hairy on the keel and the margins	12
11	Lemmas hairless to short-hairy on the keel and margins	13

12(11)   Florets unisexual (each bearing stamens or a pistil but not both); lower lf blades flat or weakly inrolled, 2–5 mm wide ....
.... *P. fendleriana* ssp. *fendleriana*

12   Florets perfect (each with stamens and a pistil); lower lf blades inrolled, 1–2.5 mm wide .... *P. stenantha*

13(11)   Stems mostly about 50 cm tall (30–60 cm) .... *P. secunda*

13   Stems 10–30 cm tall 14

14(13)   Lemmas rounded on the back, about 5 mm long .... *P. suksdorfii*

14   Lemmas keeled on the back, about 3 mm long .... *P. lettermanii*

Poa annua L.—ANNUAL BLUEGRASS

The only annual bluegrass in the Park. The stems, which reach about 10 cm, tend to bend over and root at the nodes, forming small mats. The plant has an overall yellow-green color and the spikelets, in an open panicle, are pale green. The lemmas are hairy on the lower veins, but lack a tangle of hairs at their bases.

Low-growing and unobtrusive, yet spreading rapidly where it becomes established. Most common in lawns and planted areas around buildings, but also seen at roadsides below about 3000 ft. It is especially prevalent around Longmire.

Poa arctica R. Br. ssp. *grayana* (Vasey) A. Löve, D. Löve & B.M. Kapoor—ARCTIC BLUEGRASS

Tufted with creeping rootstocks, with stems 10–30 cm tall and a pyramidal panicle, the branches ascending to spreading. The lf blades are slender, about 3 mm wide. The lemmas are short-hairy on the lower part of the keel and have 5 veins, with a web of hairs at the base.

Fairly common in the Park, at high elevations on the north and east sides. It has been collected at Chinook Pass; at the Flett Glacier at an unspecified elevation; and at 10000 ft at an unspecified location, by O. D. Allen. It grows among rocky on slopes at and above timberline.

Poaceae

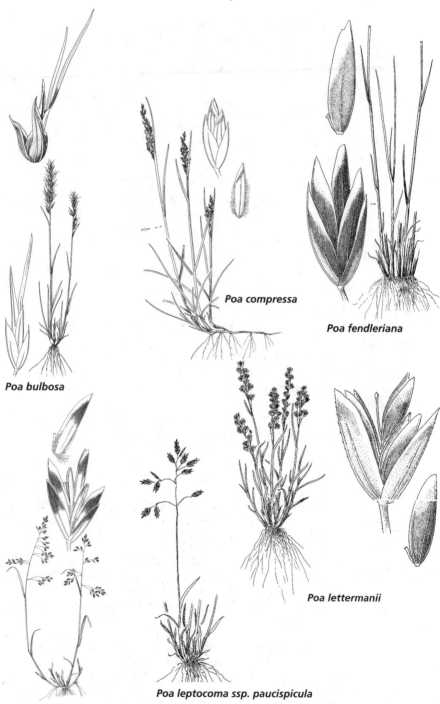

Poa compressa

Poa fendleriana

Poa bulbosa

Poa lettermanii

Poa leptocoma ssp. paucispicula

Poa leptocoma ssp. leptocoma

Poa bulbosa L.—BULBOUS BLUEGRASS

A unique species among the grasses of the Park, in which most of the florets are replaced by small, purplish plantlets, which resemble small bulbs. The panicle is cylindrical, the modified spikelets on short ascending branches, and rather crowded with the plantlets. The base of the stems is also somewhat bulbous, a feature shared with *Melica subulata*, though in that species all the florets are normal.

Not previously reported in the Park. Found among weeds along the road on the south side of Tumtum Peak and about 0.5 mi south of the White R entrance.

Poa compressa L.—COMPRESSED BLUEGRASS

A weedy perennial, rarely seen in the Park. The wiry stems, rising from a stout rootstock to 20–60 cm tall, are notably flattened, yielding 2 sharp edges. The lf blades are 3–4 cm wide and flat or folded. The panicle is dense, wider at the bottom, with short, ascending branches. The veins of the lemma are hairy and a web of tangled hairs may be present.

Known for the Park from a collection, in the Park herbarium, made in the Longmire meadow; it has not been previously listed in the Park.

Poa fendleriana (Steud.) Vasey ssp. *fendleriana* SKYLINE BLUEGRASS

Densely tufted, to 30 cm tall but often less. The lf blades are rather wide for an alpine grass, 2–5 mm. The panicle is narrow and compact. A close examination of the florets reveals that they are unisexual. The lemmas are weakly roughened, or sometimes smooth. The glumes may be blunt or pointed.

Common on high ridges and rocky slopes at and above timberline. Collections have been made at 6750 ft in Upper Van Trump Park; at 7400 ft on McClure Rock, at the top of Sluiskin Falls, and elsewhere in upper Paradise Park; at Sunrise and near Frozen Lk; and on Burroughs Mtn at 6800 ft.

Poa leptocoma Trin. ssp. *leptocoma*—MARSH BLUEGRASS

A taller version of ssp. *paucispicula*, with stems 30–60 cm tall. The lf blades are flat and 2–4 mm wide. The panicle is open, with spreading branches; the lemmas are blunt at the tip.

Herbarium vouchers were not located in the course of this study. Jones describes it as a plant of "boggy meadows" but gives no indication of its abundance. *Poa macroclada* Rydb. was once listed for the Park, probably on the basis of a misinterpretation of *P. leptocoma* ssp. *leptocoma*. *Poa macroclada* is native to the Rocky Mountains of Colorado.

Poa leptocoma Trin. ssp. *paucispicula* (Scribn. & Merr.) Tzvelev FEW-SPIKED BLUEGRASS

The common subspecies of *Poa leptocoma* in the Park, it is distinguished from ssp. *leptocoma* by its shorter stature, from 10 to 20 or 30 cm tall, narrower lf blades, about 2 mm wide, and lemmas that are pointed at the tip. Plants from very high elevations may be only 5 cm tall. The lemmas are hairy on the lower half of the veins, with a sparse web of tangled hairs.

Listed as "watch" in *The Endangered, Threatened, and Sensitive Vascular Plants of Washington*, where it appears as "*P. paucispicula*" (Washington Natural Heritage Program, 1997).

Found occasionally on rocky alpine slopes and ledges around the Mtn, 6500-11000 ft.

Poa lettermanii Vasey—LETTERMAN'S BLUEGRASS

A delicate-looking, much-dwarfed grass, growing in dense tufts to 10 cm tall, but often just 5 cm. The lf blades are up to 1.5 mm wide. The panicle is narrow and up to 3 cm high. The lemmas are smooth.

Common, but restricted in its distribution to high ridges and ledges, above timberline and reaching an elevation of 11000 ft. A collection from 8200 ft east of the Muir snowfield is in the Park herbarium and one from 6500 ft above Seattle Park is known. Although J. B. Flett's 1896 collection from 11000 ft bears no location information, it probably came from above Camp Muir.

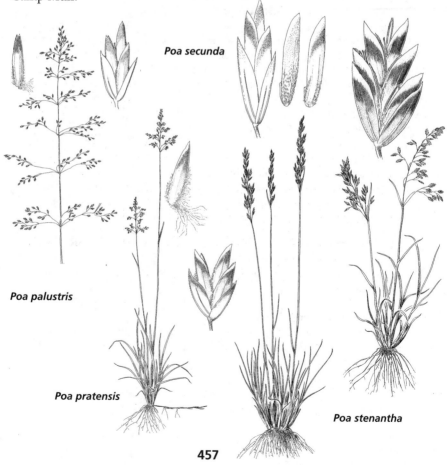

Poa secunda

Poa palustris

Poa pratensis

Poa stenantha

457

Poa palustris L.—FOWL BLUEGRASS
A tall, runner-bearing weedy grass about 100 cm high, with slender lf blades, 1.5–3 mm wide. The branches are slender, tending to droop, arranged in whorls, and bearing spikelets only near the tips. The veins of the lemmas are inconspicuous and hairy, and a web of tangled hairs is present at the base.

Rare in the Park, and apparently established only at Longmire.

Poa pratensis L.
KENTUCKY BLUEGRASS
Kentucky bluegrass, commonly used for lawns, has extensive rootstocks and often forms dense sod. The flowering stems can reach 120 cm tall, but are often shorter. The panicle is loose and open, with spreading branches arranged in whorls. The nerves of the lemma are prominent and long-hairy, with a tangle of hairs at the base of the lemma.

A common weed at roadsides, along streams, and occasionally in meadows up to about 5000 ft at Paradise.

Poa secunda J. Presl—ONE-SIDED BLUEGRASS,
SANDBERG'S BLUEGRASS
Some collections of this grass have been misidentified as *Poa arctica* ssp. *grayana*. While the lemmas of *P. secunda* do have long hairs on the lower half, they lack *P. arctica*'s characteristic tangled web of hairs at the base. Otherwise, the lf blades of *P. secunda* are more slender, just 1 mm wide, while the panicle is narrow, dense, and erect, with short, appressed branches.

Common on dry, rocky ground at and above timberline, most frequently seen on the west and south sides of the mountain. Collected along the trail between Klapatche and St. Andrews Parks; on the upper Van Trump mudflow at 6750 ft; at 6500 ft on the Skyline trail in Paradise Park; and at 7800 ft west of the Muir snowfield.

Kartesz does not recognize subspecies, while Robert Soreng, author of the treatment of *Poa* in *The Jepson Manual* (Hickman, 1993) uses ssp. *secunda* for plants of the main Sierra/Cascade axis.

Poa stenantha Trin.—NARROW-FLOWERED BLUEGRASS,
TRINIUS'S BLUEGRASS
Similar to *P. arctica* ssp. *grayana*, but without rootstocks. The lf blades are inrolled and 1–2 mm wide. The panicle is open, with spreading branches. The lemmas are hairy only on the keel and along the margins, not on the intervening veins.

A plant that only just enters the Park at Chinook Pass. Collections in the Park herbarium under this name, from Gobbler's Knob, proved to be *Calamagrostis sesquiflora*.

Poa suksdorfii (Beal) Vasey ex Piper
WESTERN BLUEGRASS

Tufted and very short, reaching just 20 cm tall at
most, with narrow, inrolled lf blades to 5 cm
long. The panicle is about 5 cm high and
narrow, with appressed branches, and
purplish in color. The lemmas are usually
smooth but occasionally roughened.

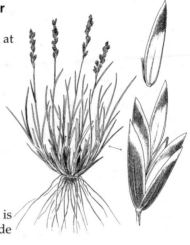

Listed as "watch" in Th*e Endangered,
Threatened, and Sensitive Vascular Plants of
Washington* (Washington Natural Heritage
Program, 1997).

Uncommon, and limited to rocky or sandy
alpine slopes, 8000-9000 ft; it reaches 10000 ft,
according to St. John and Warren. A collection is
known from Camp Curtis, on the northeast side
of the mountain alongside Emmons Glacier.

Poa trivialis L.—ROUGHSTALK BLUEGRASS

Forming clumps and spreading by runners, with stems
reaching 50–100 cm tall. The lf blades are flat and about 4
mm wide. The panicle is tall and pyramidal, with the
branches tending to be in whorls and spreading widely.
The lemma is smooth except for a long, tangled web of
hairs at the base.

A rare weedy grass, known only from a collection
made in the Longmire meadow.

Poa wheeleri Vasey
HOOKER'S BLUEGRASS

A native perennial spreading by rootstocks and
reaching 30–60 cm tall. The panicle is loose,
about 10 cm high, with relatively few spikelets.
The florets are somewhat large and almost
entirely female; florets with rudimentary
anthers are rare. The plant's former species
epithet, "*nervosa*," referred to the prominent
veins on the lemma. The lemmas have short
hairs across the back; a web of hairs is
absent.

Poa wheeleri should be looked for in
openings in forests below about 4000 ft. Known for the Park
only from two collections at the University of Washington,
made at an unrecorded location (or locations) by E. C. Smith
in the 1890s. The herbarium sheets were annotated in 1936 as
Poa pringlei and in 1964 as *P. nervosa* var. *wheeleri*. As
currently interpreted, *Poa nervosa* (Hook.) Vasey is a rare
plant of the lower Columbia Gorge.

459

Polypogon Desf.—BEARD GRASS

Polypogon monspeliensis (L.) Desf.
RABBITFOOT GRASS

Of medium height, to about 50 cm tall, with flat, roughened lf blades about 6 mm wide. The panicle is very dense, spikelike, with short appressed branches, and 5–10 cm tall. Both the glumes and the lemmas are awned, giving the panicle a bristly appearance.

A common weedy annual, found along roads, on disturbed ground, and occasionally at the edges of meadows below about 3000 ft. It also occurs in the meadow at Longmire.

Puccinellia Parl.—ALKALI-GRASS

Puccinellia distans (Jacq.) Parl.—SPREADING ALKALI-GRASS

A tufted perennial, growing to about 50 cm tall. The panicle is 10–15 cm tall, with wide-spreading branches forming an open pyramid. The spikelets are about 5 mm long, slender and more or less cylindrical, reminiscent of those of some *Glyceria* species. The lemma is rounded on the back, with the margins minutely serrate at the tip, and faintly 5-veined.

A rare weedy grass, reported to grow on wet ground, with a preference for alkaline soils, as at Longmire.

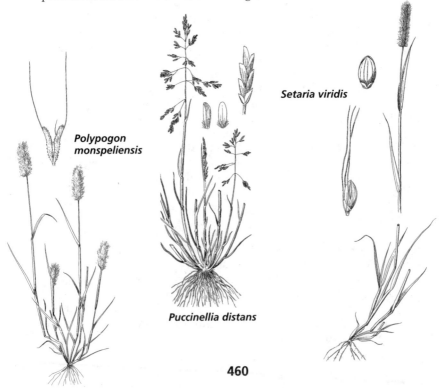

Setaria viridis

Polypogon monspeliensis

Puccinellia distans

Setaria Beauv.—BRISTLE-GRASS

Setaria viridis (L.) P. Beauv.—GREEN BRISTLE-GRASS

An annual weed, with stems about 50 cm tall. The lf blades are 5–10 mm wide. The panicles are remarkably dense and spikelike, cylindrical and about 5 cm long. Named for the bristlelike sterile branchlets that are set beneath each spikelet; the lemmas and glumes themselves are not awned.

An uncommon grass at roadsides; it is occasionally seen around Longmire.

Torreyochloa Church—FALSE MANNAGRASS

Torreyochloa pallida (Torr.) G.L. Church var. *pauciflora* (J. Presl) J. I. Davis
WEAK MANNAGRASS

Distinguished from *Glyceria* and *Puccinellia*, two very similar genera, by the lf sheaths, which are open their full length along the stem. The stems reach about 100 cm tall and the lf blades are very broad, to 15 mm wide. The panicle is loose, with slender, spreading branches, with purplish, oblong spikelets. The lemma is prominently veined.

Common along streams and in swamps, below about 4000 ft and most abundant in the Carbon R V and the lower Nisqually V.

Trisetum Pers.—TRISETUM

A group of common grasses, very different in appearance and occurring in a wide range of habitats. They have in common 2-flowered spikelets, in which the awned lemmas are shorter than the glumes. The awn rises from above the middle of the back of the lemma, which is toothed at its tip.

1 Panicle strongly contracted and spikelike; stems 10–40 cm tall, growing on alpine ridges and slopes .... *T. spicatum*
1 Panicle open to congested but not spikelike; stems 50–100 cm tall 2

2(1) Panicle open, the lower branches spreading or drooping ....
.... *T. cernuum* var. *cernuum*
2 Panicle congested, the lower branches erect .... *T. cernuum* var. *canescens*

Trisetum cernuum Trin. var. *canescens* (Buckley) Beal
TALL TRISETUM

Along with var. *cernuum*, this is a tall, tufted grass, reaching about 100 cm on favorable ground. It is distinguished by a narrow, congested panicle in which the ascending branches bear spikelets nearly to their bases. The flat lf blades are 4–7 mm wide.

Common along trails, in openings in forests, and on stream banks below 5000 ft, especially in the Nisqually V.

Trisetum cernuum Trin. var. *cernuum*
NODDING TRISETUM

Separated from var. *canescens* by its gracefully curved, drooping panicle, in which the spikelets are borne only near the ends of the spreading branches. The lf blades are 6–12 mm wide.

Found in similar habitats to the preceding variety.

Trisetum spicatum (L.) K. Richt.
SPIKE-TRISETUM

Densely tufted, with stems reaching 50 cm tall. The lf blades are stiff, usually folded, and 2–3 mm wide. The panicle is densely flowered and purplish, 10–15 cm long and spikelike.

Common at and above timberline, on open slopes and on rocky ridges. It has been collected at 5800 ft on Tolmie Ridge; in Paradise Park and at Panorama Point; at Sunrise and on Burroughs Mtn; and at 7200 ft on Fremont Peak.

Vahlodea Fries—ARCTIC HAIRGRASS

Vahlodea atropurpurea (Wahlenb.) Fr. ex Hartm.
MOUNTAIN HAIRGRASS

Formerly grouped with *Deschampsia*, *Vahlodea* differs in having wider, generally flat lf blades and lemmas that are awned at the middle of the back. *Vahlodea atropurpurea* is a loosely tufted, tall grass, reaching 100 cm. The panicle is open, with wide-spreading or drooping branches. The spikelets are about 5 mm long and purplish.

Common in meadows, 4500-7000 ft. Collections have been made at Mtn Meadows; in upper Van Trump Park at 6750 ft; through Paradise Park to 6000 ft; at the head of Stevens Canyon near the road; and at Berkeley Park.

Vulpia C. C. Gmelin—SIX-WEEKS GRASS

Vulpia microstachys (Nutt.) Munro var. *microstachys*—SMALL FESCUE

A fescue-like grass, with awned lemmas and glumes. A tufted annual growing to about 50 cm tall, *V. microstachys* differs from the true fescues in being an annual, with 1 stamen rather than the 3 of *Festuca*. The panicle is 10–20 cm high, with spreading branches.

An uncommon native grass that behaves, in the Park, in a weedy fashion, found at roadsides and around buildings.

Potamogetonaceae Dumort. PONDWEED FAMILY

(Naiadaceae in Jones)

Potamogeton L.—PONDWEED

Potamogeton natans L. FLOATING PONDWEED

A single pondweed has been observed in the Park. *Potamogeton natans* has slender stems 15–150 cm long, with two types of lvs: the submerged lf blades are linear and persist only a short time, followed by oval floating lvs 3–10 cm long and often tinged reddish. The inconspicuous fls are borne in whorls on spikes rising 2–5 cm above the water surface. Fl parts—pistils and stamens—are in 4s.

A collection was made in 1932 by Fred Warren at a small pond along upper Tahoma Cr.

Sparganiaceae Rudolphi—BUR-REED FAMILY

Sparganium L.—BUR-REED

Three species of this aquatic family are found in the Park, characterized by floating leaves and by round heads of inconspicuous flowers; the female flowers are followed by heads of conspicuous burlike fruits. They are fairly common in shallow ponds and swamps but are often only noted in mid-summer when the flowering stem rises above the water. [R. F. Thorne, in *The Jepson Manual* (Hickman, 1993), uses the cattail family, Typhaceae, for both *Sparganium* and *Typha*. J. L. Reveal's treatment, as separate families, is followed here.]

463

1   Male fls in 1 head; female heads (burs) 5–12 mm broad at maturity ....
.... S. natans

1   Male fls in 2 to several heads; female heads 10–30 mm broad at
maturity 2

2(1)   Lvs 7–15 mm wide; plant erect, with some lvs rising above the water
.... S. erectum

2   Lvs 2–4 mm wide; plants submersed, with lvs floating on the surface
.... S. angustifolium

Sparganium angustifolium Michx.
NARROW-LEAF BUR-REED

Characterized by slender stems and long, narrow
lvs that float on the water's surface. The 2–5
brownish male flowering heads top a stem that rises
above the water; below these are 2–4 green female
heads, 10–20 mm broad.

Common in ponds and lakes at low elevations;
sometimes not fruiting until late summer when
standing water has disappeared. Found in ponds at
the roadside on the south side of Tumtum Peak; at
Ohanapecosh in a small bog; and in the swamp at the
Stevens Canyon entrance station.

Sparganium erectum L.
SIMPLESTEM BUR-REED

More stout than *S. angustifolium* with
erect stems and lf blades. The lvs are
broader, averaging about 1 cm wide,
rather rigid, and keeled along their
length (and thus Y-shaped in cross
section). The 2–5 male heads are borne
above the 2–5 round female heads, which are 20–30 cm
broad at maturity.

Uncommon at lower and middle elevations, to about
4500 ft, in swamps and at the margins of ponds and
lakes.

Sparganium natans L.
SMALL BUR-REED

A slender aquatic, with stems 10–40
cm long and with long, slender
grasslike lvs floating on the water.

Small bur-reed has 1 somewhat oblong
male head at the top of the stem and 2 or 3 female
heads that are about 10 mm broad at maturity.

Common in a few places, growing in shallow
water at the edges of ponds and lakes at around

5000 ft. Collections have been made at Reflection Lks and Mystic Lk. The species name is spelled *"nutans"* by some authors.

Typhaceae Juss.—CATTAIL FAMILY

Typha L.—CATTAIL

Typha latifolia L.—BROAD-LEAF CATTAIL

A tall plant, with stems reaching 1–2 m tall and broad, flat lvs that are 2-ranked on the stem, sheathing the stem at their bases. The inflorescence is a long, cylindrical spike at the top of the stem, densely packed with innumerable small fls. The male fls are crowded at the top of the spike straw-colored when in fl; these soon fall away, leaving a bare stalk above the female part of the spike. The latter is dark brown, 10–20 cm long and 1–3 cm thick.

Quite common at the margins of lakes and ponds, and in marshy areas, below about 4000 ft, especially in the Nisqually R V and at Longmire.

Appendix

Plants Added to the Flora of Mount Rainier National Park

Listed here are 79 plants. About one-third were observed in the field by the author between 1996 and 1999. Fourteen were found in the herbaria at Pacific Lutheran University and at the University of Puget Sound, two institutions with much material collected by James R. Slater and Irene Creso. They did not place duplicates in herbaria at the University of Washington, Washington State University, or at the Park. These last three herbaria also hold species that had been overlooked by earlier students of the Park flora. Another 9 species represent newly made determinations by the author of collections in these herbaria. Regina Rochefort, Park Botanist, has noted the occurrence of a number of recently introduced weeds in the Park. Finally, some are "new" but are the result of changed taxonomic concepts rather than new discoveries.

Achlys californica Fukuda & Baker. *Achlys californica* was described in 1970. The distinctions between it and *A. triphylla* do not hold up terribly well in the Park.

Agoseris grandiflora (Nutt.) Greene. Observed by the author in 1999.

Agoseris heterophylla (Nutt.) Greene var. *heterophylla*. Observed by the author in 1999.

Amsinckia menziesii (Lehm.) A. Nelson & Macbr. Based on a collection made in the 1930s.

Antennaria umbrinella Rydb. Determination of collections from the 1930s, originally labeled "*A. alpina*."

Anthemis cotula L. Collected once in 1938 and not noted by earlier writers.

Brassica nigra (L.) W.D.J. Koch. Found by the author in the lower White River Valley in 1999.

Bromus sitchensis Trin. var. *sitchensis*. Two collections are known, made by Irene Creso around 1970 and by O. D. Allen in 1895.

Campanula persicifolia L. Observed by the author in 1998.

Carex nudata W. Boott. Based on a collection made by Irwin in 1951. The listing is tentative, for the exact location of the collection is not recorded—said to be Chinook Pass, but probably from some distance downhill.

Centaurea biebersteinii DC. Collected by Park Botanist Regina Rochefort.

Claytonia rubra (T. Howell) Tidestr. ssp. *rubra*. Observed by the author in 1996.

Corallorhiza trifida Chatel. Collected by J. R. Slater about 1960 and by Brockman in the 1940s. Observed by the author in 1996.

Cryptantha affinis (A. Gray) Greene. Determination of a collection made by F. A. Warren around 1935, originally called *C. muriculata*.

Cryptogramma cascadensis E. R. Alverson. First described by Ed R. Alverson in 1989; he cites several collections from the Park.

Draba verna L. Observed by the author in 1996.

Elymus canadensis L. Based on collections made between about 1950 and 1970, by Marcus Huntley and Irene Creso, that had gone unexamined in the herbaria of Pacific Lutheran University and the University of Puget Sound.

Appendix

Elymus x *hansenii* (Scribn.) J. G. Sm. Based on several early collections.

Epilobium densiflorum (Lindl.) Hoch & P.H. Raven. Said to occur in the Park, by J. B. Flett in a letter addressed to C. V. Piper, dated 1901.

Epilobium glaberrimum Barbey ssp. *glaberrimum*. Perhaps the subspecies was not recognized by earlier writers.

Epilobium halleanum Hausskn. Based on an H. W. Smith collection in the Park herbarium that had apparently been overlooked.

Equisetum variegatum Schleich. var. *variegatum*. Based on a collection made by J. R. Slater, in the herbarium at Pacific Lutheran University.

Erodium cicutarium (L.) L'Her. ex Aiton ssp. *cicutarium*. Observed by the author in 1997; perhaps a newly established weed.

Euphrasia stricta D. Wolff ex J. F. Lehm. Observed by the author in 1997; recently established in the lower White River Valley.

Geranium molle L. Observed by the author in 1996; perhaps recently established as a weed.

Geranium robertianum L. Reported by Park Botanist Regina Rochefort to have recently become established in the Carbon River Valley. Found by the author in the White River Valley in 1999.

Gilia capillaris Kellogg. Observed by the author in 1996; probably overlooked earlier.

Glyceria grandis S. Watson var. *grandis*. Based on a collection at the University of Puget Sound made by Irene Creso.

Glyceria leptostachya Buckley. Another plant found by Irene Creso in the 1970s, in the herbarium at the University of Puget Sound and overlooked by other researchers.

Hieracium atratum Fr. A weed that became established in the Park around 1975 and that has spread since then. Identification of plants from the Park by Geraldine Allen of the University of British Columbia.

Hieracium aurantiacum L. Evidently a recent arrival in the Park, it was noted in 1997 by Park Botanist Regina Rochefort.

Hieracium caespitosum Dumort. Found by the author in the lower White River Valley in 1999.

Hieracium scouleri Hook. var. *scouleri*. Determination of a collection made by F. A. Warren in the early 1930s.

Hordeum vulgare L. Based on another collection by Irene Creso in the early 1970s.

Huperzia chinensis (Christ) Czern. Segregated from *Lycopodium selago* L.

Huperzia haleakalae (Brackenridge) Holub. Segregated from *Lycopodium selago* L.

Lactuca serriola L. Known from two collections, one made in 1895 by O. D. Allen and the other in 1938 by C. L. Landes; evidently overlooked by earlier researchers.

Lemna minor L. Observed by the author in 1996.

Ligusticum canbyi J.M. Coult. & Rose. Determinations of several collections originally labeled *L. grayi* and *L. apiifolium*.

Linaria dalmatica (L.) Mill. Observed by the author in 1999.

Lotus corniculatus L. Observed by the author in 1998. Evidently a recent introduction thanks to erosion control plantings on Highway 410 north of the Park.

Lotus micranthus Benth. Observed by the author in 1996; perhaps overlooked earlier.

Lotus unifoliatus (Hook.) Benth. var. *unifoliatus*. Based on a J. M. Grant collection from about 1940, and observed in 1997.

Lupinus polycarpus Greene. Observed by the author in 1996.

Madia gracilis (Sm.) D.D. Keck & J. Clausen ex Applegate. Observed by the author in 1997.

Mimulus moschatus Douglas ex Lindl. var. *sessilifolius* Gray. Determination of a collection by Jenks in the 1960s, held in the Park herbarium.

Mimulus tilingii Regel var. *tilingii*. Based on an overlooked collection by J. R. Slater.

Montia fontana L. ssp. *fontana*. Based on a collection made by W. J. Eyerdam in the 1930s, held at the University of Washington and overlooked by earlier researchers.

Montia parvifolia (Moc. ex DC.) Greene ssp. *flagellaris* (Bong.) Ferris. Based on a collection made by Irene Creso in the early 1970s.

Mycelis muralis (L.) Dumort. Observed by the author in 1996. Just one overlooked collection is known, made in 1965 by Robert Wakefield.

Myosotis arvensis (L.) Hill. Based on an overlooked collection of Carole Olson, at Pacific Lutheran University.

Myosotis stricta (Hook.) Link ex Roem. & Schult. Observed by the author in 1996.

Platanthera dilatata (Pursh) Lindl. ex Beck var. *dilatata*. Determination of a collection, made by F. A. Warren, at the University of Washington.

Pleuricospora fimbriolata A. Gray. Observed by the author in 1996.

Poa bulbosa L. Observed by the author in 1996 and evidently a recent introduction.

Poa compressa L. Based on a collection by W. A. Silvenius, in the Park herbarium and dating from the 1940s.

Poa palustris L. Based on a collection by W. A. Silvenius, in the Park herbarium and dating from the 1940s.

Poa trivialis L. Based on a collection by W. A. Silvenius, in the Park herbarium and dating from the 1940s.

Polypodium amorphum Suksd. Microscopic examination of specimens at the University of Washington herbarium shows that most of what has passed as *Polypodium hesperium* is actually this species.

Prunella vulgaris L. ssp. *vulgaris*. Based on a collection made by Robert Wakefield in a 1965 exotic plants survey.

Pseudostellaria jamesiana (Torr.) W. A. Weber & R. L. Hartman. Determination of a collection in the Park herbarium which was originally called *Cerastium arvense*.

Rhinanthus minor L. ssp. *minor*. Observed by the author in 1998.

Sagina procumbens L. Observed by the author in 1996; a common weed, perhaps recently introduced.

Saxifraga lyallii Engl. ssp. *hultenii* (Calder & Savile) Calder & Savile. Observed by the author in 1996.

Selaginella scopulorum Maxon. Based on two collections made by Irene Creso.

Silene oregana S. Watson. Observed by the author in 1998. A possible collection was made by C. L. Landes in the 1930s, but the flowers are poorly preserved.

Sorbus sitchensis Roem. var. *sitchensis*. Determination from two Park herbarium collections.

Spergularia diandra (Guss.) Bois. A weed of roadsides and disturbed ground. Collected by C. L. Landes in the 1930s.

Stellaria graminea L. var. *graminea*. Observed by the author in 1998.

Teesdalia nudicaulis (L.) R. Br. Observed by the author in 1996.

Triteleia hyacinthina (Lindl.) Greene. Based on a Marcus Huntley collection made in 1951, at the University of Puget Sound. Also reported by J. B. Flett in a 1903 letter to C. V. Piper.

Vaccinium myrtillus L. Based on collections made by Irene Creso, in the herbarium at Pacific Lutheran University. Also widely observed by the author in the course of the present study.

Valeriana scouleri Rydb. Observed on the lower Carbon River by Mary Fries.

Veronica chamaedrys L. Observed by the author in 1997.

Veronica serpyllifolia L. ssp. *serpyllifolia*. Observed by the author in 1996.

Vicia sativa L. ssp. *nigra* (L.) Ehrh. Observed by the author in 1996.

Vicia tetrasperma (L.) Schreb. Observed by the author in 1996; a recently established weed.

Viola langsdorfii (Regel) Fisch. ex Ging. Based on a collection made by J. B. Flett, in the University of Washington herbarium. The place name used by Flett, "Simple Lake, Mount Tacoma," has not been identified.

Woodsia scopulina D. C. Eaton ssp. *scopulina*. Based on two collections, by J. R. Slater at Pacific Lutheran University and by Susan Cochrane in the Park herbarium.

Doubtful and Excluded Species

Agrostis rossiae Vasey—ROSS'S BENTGRASS

A name presently used for a grass native to thermal areas in Yellowstone National Park, *A. rossiae* (earlier spelled *"rossae"*) was used by C. V. Piper in his 1906 flora for plants now called *A. variabilis*. In his 1936 *Botanical Survey of the Olympic Peninsula, Washington*, G. N. Jones advanced a new name for these plants, *Agrostis exarata* Trin. var. *rossae* (Vasey) G. N. Jones. He went back to *"A. rossae"* in the 1938 Mount Rainier flora.

Arbutus menziesii Pursh—PACIFIC MADRONE

In his book, G. N. Jones wrote that this tree "is rare in Mount Rainier National Park; a few small trees [were] noted near Ohanapecosh in 1937 by Mr. C. Frank Brockman." However, Brockman himself wrote in 1938, in the Park's "Nature Notes" newsletter, that "a specimen of this plant was found, during the summer of 1937, just outside the park boundary in the southeast corner of the area near Ohanapecosh Hot Springs at an elevation of about 1600 feet." No herbarium material from the Park was found during this study and a reconnaissance of the Ohanapecosh area in 1996 did not reveal any trees or seedlings inside the Park.

Artemisia douglasiana Bess.—DOUGLAS'S SAGEBRUSH
Sometimes separable only with difficulty from *A. ludoviciana* ssp. *candicans*; *A. douglasiana* has broader leaves that are divided into fewer, less regular lobes, and glabrous to only slightly tomentose above. *Artemisia douglasiana* has a complicated synonymy, including *A. vulgaris* L. ssp. *heterophylla* (Bess.) H. & C. and *A. norvegica* Fr. ssp. *heterophylla* (Besser) H. & C. Both of these names have been used for plants from the Park, but in each case the plant referred to seems to be *A. ludoviciana* ssp. *candicans*.

Carex aperta Boott—COLUMBIA SEDGE
Irene Creso identified a plant collected at Paradise as this species. The specimens are not in the best condition, but the correct identity seems to be *Carex scopulorum*.

Carex halliana L. H. Bailey—HALL'S SEDGE
Listed for the Park by St. John and Warren (1937) as "*C. oregonensis* Olney," based on *Warren 1633*. *Warren 1633* is actually *C. lanuginosa*, from Longmire. Further south in the Cascade Range, Hall's sedge grows on pumice soils in dry forests at about 5000 ft. It is known to reach as far north as Mount Adams and it may one day be found in the Park.

Castilleja chrysantha Greenm.—WALLOWA PAINTBRUSH
St. John and Warren misapplied the name *Castilleja indecora* Piper to *C. cryptantha*. *Castilleja indecora* is a synonym for *C. chrysantha*, a plant of the Wallowa and Blue Mountains in northeastern Oregon.

Castilleja levisecta Greenm.—GOLDEN PAINTBRUSH
Long mistakenly cited for the Park, on the basis of St. John and Warren's misidentification of a J. M. Grant collection and of another made by H. W. Bailey. Both plants are *C. cryptantha*. *Castilleja levisecta* is a rare plant found on a few prairies in the Puget lowlands and Williamette Valley.

Castilleja parviflora Bong. var. *albida* (Pennell) Ownbey—SMALL-FLOWERED PAINTBRUSH
No herbarium collections were located during this study, but the plant has been reported to grow in Berkeley Park. However, reports of *C. parviflora* var. *albida* in the Park are almost certainly based on misidentifications of albino individuals of var. *oreopola*. According to Hitchcock and Cronquist, var. *albida* is found from the North Cascades of Washington into southern British Columbia.

Cistanthe tweedyi (A. Gray) Hershkovitz—TWEEDY'S LEWISIA
It's a great mystery that this plant [under the name *Lewisia tweedyi* (Gray) B. L. Rob.] has been included in the National Park Service flora for Mount Rainier, and it hasn't been mentioned by any earlier writer. It grows chiefly in the Wenatchee Mountains, although it is found occasionally northward in the Cascades to southern British Columbia. There would seem to be no suitable habitat in the Park: it prefers soils derived from metamorphic rocks, including serpentine, and is often found with ponderosa pine.

Appendix

Convallaria majalis L.—LILY-OF-THE-VALLEY
Lily-of-the-valley is listed in the National Park Service flora, and there
was a time, many years ago, when ornamental plantings may have
persisted around buildings, perhaps in the Longmire area. No herbarium
material was found in the course of this study, nor were plants found*
persisting in the field.

Dryopteris cristata (L.) A. Gray—CRESTED SHIELD-FERN
No basis is known for the listing of this species in the National Park
Service flora. It is not mentioned by the early writers, and Dunwiddie did
not list it in his 1983 paper. The occurrence of the species in the
Northwest is limited to the valleys of the Columbia and Kootenay Rivers
in southeastern British Columbia, just reaching into northern Idaho. It
seems unlikely that any *Dryopteris* in the Park would be mistaken for it: it
is quite distinctive in having fertile fronds that are much longer and
narrower than the sterile fronds.

Fragaria chiloensis (L.) Mill.—BEACH-STRAWBERRY
Not included in Jones's flora, it was listed by St. John and Warren, based
on *Warren 1758*, which was not found in the course of this study. It came
from "sandy soils, Stevens Crossing," possibly the point at about 4000 ft
where the Park road crosses Stevens Creek. Perhaps *Warren 1758*
approached *F. chiloensis* in characteristics of its leaves, but it is nearly
beyond belief that *F. chiloensis* would be found so far from its coastal
habitat. A specimen collected by C. L. Landes in the 1930s on the
"Ohanapecosh River trail" and labeled "*F. chiloensis*" is actually *F. crinata*.

Another possibility is suggested by a phrase in Piper's 1906 *Flora of
Washington*, where he says of *Fragaria cuneifolia*, "This is probably not
distinct from *F. chiloensis*." One collection of *F. cuneifolia* he cites was made
in 1883 at "Palace Camp," a popular stopping place in the Pierce County
foothills west of the present Park.

Lathyrus pauciflorus Fernald—FEW-FLOWERED PEA
This plant has long been listed for the Park, based on an error G. N. Jones
sought to rectify 60 years ago. In his flora, he wrote, "*L. pauciflorus*
Fernald, a species of the region east of the Cascade Mountains, has been
reported several times from Mount Rainier, apparently on the basis of a
specimen (*Allen 132*) "which belongs quite plainly to *L. polyphyllus*." The
Allen collection was made on Mount Wow and is in the Park herbarium.
(A duplicate, in the Ownbey Herbarium at Washington State University
was annotated *L. polyphyllus* by C. L. Hitchcock in 1949.) *Allen 132* is
indeed best regarded as an extreme example of *L. polyphyllus*; it differs in
many points from *L. nevadensis*.

Ligusticum tenuifolium S. Watson—SLENDER-LEAVED LOVAGE
St. John and Warren state that this species is "fairly common," in alpine
meadows between 5500 and 6500 feet." It has not been not verified by
herbarium collections, and it seems clear that the name was misapplied:
Jones himself wrote that "specimens [of *Ligusticum purpureum* (= *L. grayi*)]
with narrow leaflets have been reported as *L. tenuifolium*." The range of *L.*

tenuifolium is far to the east, just reaching the Northwest in the Wallowa and Blue Mountains.

Lithospermum californicum A. Gray—CALIFORNIA PUCCOON
Included in the National Park Service flora for the Park, but the basis for this is not known. No earlier writer on the Mount Rainier flora has mentioned it, and the species itself does not range north of southwestern Oregon, nor are any of the Park's borages likely to be confused with this species.

Lupinus burkei S. Watson ssp. *burkei*—BLUE-POD LUPINE
Previously listed for the Park as *L. polyphyllus* Lindl. and *L. polyphyllus* Lindl. var. *burkei* (S. Watson) C.L. Hitchc. There are two possible sources for reports of this plant in the Park. First, at Washington State University two collections can be found that were made at the turn of the century in the "Mount Rainier Forest Reserve," which at the time included land in Yakima County outside the present Park boundary. Both collections were made in the valley of Hell Roaring River: one from 7000 feet is *L. burkei* var. *burkei*, and one from 3500 feet appears to be *L. polyphyllus*. Second, some specimens of *L. latifolius* have leaves of a size suggestive of *L. polyphyllus*, but in these plants the strongly whorled inflorescence characteristic of *L. polyphyllus* (and *L. burkei*) is not seen.

Machaeranthera canescens (Pursh) Gray ssp. *canescens*—HOARY ASTER
An aster-like plant listed in the National Park Service flora but not by Jones, Brockman, or Dunwiddie. Perhaps said to occur in the Park based on a collection, at the University of Washington, made by G. N. Jones in Yakima County in the "Mount Rainier National Forest," which once included land east and south of the present Park boundaries. *Machaeranthera canescens* is common east of the Cascades, but grows at low elevations and is unlikely to climb high enough to cross into the Park.

Malva moschata L.—MUSK-MALLOW
Cited for the Park in the National Park Service flora list, but not listed by other writers and not verified in the course of this study, either in the field or by herbarium material. Of course, it would not be surprising to see it turn up as a weed from time to time.

Penstemon procerus Douglas ex Graham var. *procerus*—SMALL-FLOWERED PENSTEMON
The variety *procerus* is included in both Dunwiddie's 1983 list and the National Park Service list. It may be distinguished by its larger size, with flowering stems 15–30 cm tall, with 2–4 whorls of flowers (Strickler, 1997). No herbarium material verifying var. *procerus* was found during this study, nor were plants seen in the field that approached this size. Perhaps such an atypical specimen of var. *tolmiei* was once noted and given the name "*procerus*." Variety *procerus* is common across eastern Washington and approaches the eastern base of the Cascades.

Appendix

Penstemon rydbergii A. Nelson var. *oreocharis* (Greene) N. H. Holmgren—
MEADOW PENSTEMON
St. John and Warren listed "*P. hesperius*" for "gravelly soil near
Longmire," on the basis of *Warren 1674*, which is actually *P. procerus* var.
tolmiei, although "near Longmire," depending upon how near, is below
the usual elevations at which *P. procerus* var. *tolmiei* is found.

"*Penstemon hesperius*" is a synonym for *P. rydbergii* A. Nelson var. *varians*
(A. Nelson) Cronq., a common plant east of the Cascades but a species
not known to reach above middle elevations along the Cascade Divide. A
collection of *P. rydbergii* is in the Ownbey Herbarium at Washington State
University; it was made at "Bench Lake, Mount Adams," the closest
known location to Mount Rainier.

Phlox caespitosa Nutt.—CLUSTERED PHLOX
Listed by St. John and Warren (1937) as "common on grassy slopes 5000–
7000 feet," but regarded by Peter Dunwiddie (1983) as "dubious." It is
nevertheless included in the National Park Service flora. Not dissimilar to
Phlox diffusa Benth. ssp. *longistylis* Wherry, and quite possibly confused
with that common Park species by St. John and Warren.

Poa macroclada Rydb.
St. John and Warren include this species, saying it grows "on rocky soil,
Canyon Bridge," and basing the inclusion on *Warren 1763*. This specimen
has not been found, but the species is included nevertheless in the
National Park Service flora. Hitchcock (1959-67) notes that *P. macroclada*
has been reported for Idaho and Montana, but suggests a
misidentification of *P. reflexa* is involved. Otherwise, the species is known
from east of the Rocky Mountains. *Warren 1763* may have actually have
been *P. leptocoma*.

Populus tremuloides Michx.—QUAKING ASPEN
According to Jones, this tree grows "in or near bogs or swamps" and is
"not common." He placed it in the Humid Transition zone, by which he
probably meant the lower valley of the Nisqually River (or perhaps of the
Ohanapecosh River as well). Earlier, an unknown contributor to the
March 1, 1924, issue of the Park newsletter, "Nature Notes," stated that
"aspen . . . is not present in the Park." Neither herbarium vouchers nor
reliable reports have been found. It is still listed in the National Park
Service flora.

Quercus garryana Douglas ex Hook. var. *garryana*—OREGON WHITE OAK
G. N. Jones wrote that this tree is "known in Mount Rainier only from
near Ohanapecosh, where discovered by Mr. C. Frank Brockman in 1937."
Brockman, however, wrote in 1938 in the Park newsletter, "Nature
Notes," that "during the summer of 1937 one small tree of this species
was found just outside the southeastern corner of the Park, near where
the Ohanapecosh River crosses the south boundary." Brockman
speculated that it was not unlikely that eventually one or two of the trees
would be found growing within the Park boundaries, but this has not
happened to date.

Ranunculus occidentalis Nutt. var. *occidentalis*—WESTERN BUTTERCUP
A plant of the Puget lowlands similar to *R. uncinatus* but with petals much longer than the sepals, was collected by W. D. Eyerdam in 1937 "in a marsh, Mt Rainier district." Eyerdam made a number of collections in the "Mt Rainier district"; all are typically lowland plants. This collection, at the University of Washington, almost certainly came from outside the present Park boundaries. A collection made by O. D. Allen in the "upper valley of the Nisqually" was originally identified as *R. occidentalis* var. *lyallii* A. Gray; it is instead *R. uncinatus*.

Salix lasiolepis Benth. var. *lasiolepis*—ARROYO WILLOW
Listed by Brockman (1947), who claimed it was one of the willows "most apt to be noted by the average visitor." Brockman gave no location information and most likely confused *S. lasiolepis* with *S. lucida* ssp. *lasiandra*. Jones did not mention it, nor did Piper (1901) or St. John and Warren (1937). Arroyo willow is common along river courses east of the Cascades, and its occurrence in the Park would be remarkable.

Salix lutea Nutt.—YELLOW WILLOW
This one is complicated. St. John and Warren (1937) listed "*S. lutea*" from the Owyhigh Lakes. Peter Dunwiddie (1983) picked it up for his list, under the then-current name "*S. rigida* var. *watsonii*." Kartesz (1994) gives the name as, once again, *S. lutea*. According to Hitchcock and Cronquist, this (that is, var. *watsonii*) is a plant of the northern Rocky Mountains, barely reaching northeastern Oregon, and north-central Washington in forms transitional to var. *mackenzieana*.

The Owyhigh Lakes material was not located in the course of this study and it seems very likely that St. John and Warren had nothing more than a somewhat unusual specimen of what Jones called *Salix mackenzieana*, a plant called *Salix prolixa* in this book.

Salix monticola Bebb—MOUNTAIN WILLOW
Listed for the Park by St. John and Warren (1937) and said to be rare, growing in wet places between 2000 and 4000 feet. No herbarium specimens were found, and it seems likely the claim of *S. monticola* for the Park is based on a misidentification of *S. barclayi* (or just possibly *S. piperi*).

Saponaria officinalis L.—BOUNCING BET
Listed only on the National Park Service flora. Perhaps it once grew in flower gardens at Longmire or Ohanapecosh, but it is mentioned by neither Jones nor St. John and Warren. It was not observed in the course of this study and no collections were found in herbaria.

Senecio elmeri Piper—ELMER'S RAGWORT
A "monitor" status plant listed by the Washington Natural Heritage Program, Elmer's ragwort was listed for the Park by Peter Dunwiddie in 1983 and is included in the National Park Service flora for Mount Rainier. In her 1980 dissertation, Ola M. Edwards included the species in a list of rare and threatened taxa that, according to the Washington Natural Heritage Program, occurred in the Park. However, the latitude and

474

longitude given in the Program's database point to a location close to Hen Skin Lake, at about 5400 feet in the Crystal Mountain Ski Area, about 1 mi east of the Park boundary.

Senecio indecorus Greene—RAYLESS MOUNTAIN-BUTTERWEED
St. John and Warren (1937) list one collection (*Warren 776*) made at Longmire in "moist places." This specimen was not found during the present study and the Park is well west of the range of the species. Just what *Warren 776* might actually have been is not clear.

Vaccinium oxycoccos L.—SWAMP-CRANBERRY
In their 1937 list, St. John and Warren write: "Reported by Len Longmire (August, 1932) as growing in swamp at Longmire and near Bear Prairie Point, nearly forty years ago." Bear Prairie Point lies outside the southwest corner of the Park. If the plant did indeed once grow at Longmire, then it has not been reported for more than 100 years; no herbarium material has been found. Suitable habitat certainly exists within the Park and perhaps what Mr. Longmire reported was an introduction or short-lived invasion by the species.

Vaccinium uliginosum L.—WESTERN BLUEBERRY
The inclusion in earlier lists of the Park flora is based on two collections. A collection made by J. R. Slater in the 1950s at Spray Park (#E-20-a at the Creso Herbarium at Pacific Lutheran University) is labeled "*V. uliginosum*," but the fruit has been destroyed by insects and the identification cannot be confirmed. A collection at Washington State University (*Susan Field #88*, made in 1970, 0.5 mi east of Kautz Creek) is almost certainly *V. membranaceum*: the leaves are just beginning to unfurl, the branches are strongly angled, the flowers are about 4 mm long, and the calyx is shallowly lobed. (Dunwiddie listed this plant as *Vaccinium occidentale* Gray.)

Vicia americana Muhl. ex Willd. ssp. *minor* (Hook.) C. R. Gunn—AMERICAN VETCH
Peter Dunwiddie included this subspecies (he used "var. *minor* Hook.") in his 1983 list, citing C. Frank Brockman's 1947 popular flora. Brockman, however, does not clearly describe a plant distinguishable from ssp. *truncata*. In any case, in the course of this study no herbarium material or plants in the field were found which were clearly assignable to ssp. *minor*. Duane Isley, in his treatment of *Vicia* for *The Jepson Manual* (Hickman, 1993) writes, "Attempts to use leaflet form and hairs to define infraspecific taxa are untenable." Plants from the lower White River Valley are the most ambiguous in this regard.

Viola howellii A. Gray—HOWELL'S VIOLET
The only mention of this plant in the Park comes from Brockman's 1947 popular flora, where the violets are briefly discussed: *V. howellii* is mentioned with *V. adunca* and *V. palustris* as one of the blue-flowered violets in the Park. The plant was listed by Dunwiddie, citing Brockman, and is listed on the National Park Service flora. No herbarium material was found, and it seems almost certain that Brockman confused *V. howellii*, a plant of the Puget lowlands, with *V. adunca*.

Afterword

Mount Rainier has been called "an arctic island in a temperate sea." Today, 100 years following the establishment of the Park, Mount Rainier is an island in another, sadder sense: an ecological island, bordered by cut-over forests. The Park represents a reservoir of botanical diversity, a refuge, a home to species that have been eliminated from these lands or whose numbers there have been greatly reduced. It is doubly important, then, for the National Park Service to guard the integrity of the Park ecosystem, for it may well be that plants in the Park are the best, or only, hope, for restoring species to the surrounding forests as regrowth takes hold on the cut-over lands.

With this in mind, these words of Arthur R. Kruckeberg (1980), speaking of Mount St. Helens, will serve to close this attempt to catalog the Park's flora:

> *Impermanence, a key factor for most biotas, is accentuated in regions of active volcanism. The repeated eruptions of Mount Saint Helens should remind us of what may be in store for most of the Pacific Northwest volcanoes. As they too resume activity, their floras and faunas will suffer decimation only to rebound with rapid and tenacious resilience.*

Glossary

In the introductions of several families—especially the grasses, sedges, and sunflowers—will be found extended explanations of the unique anatomical features of those families.

achene—A dry, one-seeded fruit, not at all fleshy.

alpine—An elevational zone above timberline.

alternate—Describing leaves borne singly on a stem.

annual—A plant species that completes its life cycle within one growing season, typically with a taproot or fibrous roots, lacking a rootstock that survives more than one season.

anther—A sack in which pollen is produced.

ascending—Growing upward, usually curving, from a base or point of attachment.

awn—A needlelike or bristlelike tip of a leaf or flower part.

axil—The space in the angle where a leaf attaches to a stem.

basal—Positioned at the base; usually describing a cluster of leaves at the base of a stem.

beak—A (usually) short tip at the top of the seed (or at the end of the galea in some flowers).

biennial—A plant species that completes its life cycle within two growing seasons.

bilaterally symmetrical—Symmetric with respect to only one axis; typically an irregular flower.

bisexual—Describing a flower that has both stamens and pistils.

blade—The expanded portion of a leaf, typically attached to a leaf stalk. In grasses, the portion of the leaf that is free from the stem. Also used for the wider part of a petal that has a narrower, stalklike claw.

bloom—A white powdery or waxy coating on a surface.

bract—A small leaflike or scalelike structure that is placed beneath a flower, flower stalk, or branch.

calyx—The sepals of a flower, taken together, whether they are fused or not; usually green.

capsule—A dry, often thin-walled fruit, composed of more than one carpel, and containing (usually) numerous seeds.

catkin—A (usually) downward-hanging spike; the flowers are unisexual and inconspicuous, lacking petals.

claw—The narrowed, stalklike part of a petal that has a wider blade.

cordate—heart-shaped.

corolla—The petals of a flower, taken together, whether fused or not; usually colored.

cotyledon—Seed leaf; in the angiosperms, this is a modified leaf found within the seed.

cyme—A branched inflorescence in which the flower bud at the end of a branch opens first.

deciduous—A tree or shrub that loses all of its leaves at the end of the growing season and produces new leaves the following season.

deltate, deltoid—Shaped like an equilateral triangle, although usually with rounded corners.

dimorphic—Taking on two distinctive forms or shapes, as in some ferns where the fertile and sterile fronds are very different in appearance.

disk flower—In the sunflowers, an individual flower that lacks a ligule; usually making up the inner portion of the head. See ray flower.

dissected—Very deeply and irregularly divided, usually describing leaves.

divided—Describing a leaf that is deeply lobed or cut to the midvein.

drupe—A fruit like a cherry, with a stonelike seed at the center, surrounded by a pulpy flesh.

elliptical—Shaped like an ellipse; a flattened and stretched circle.

fibrous—Referring to a highly branched, spreading root system, each branch of about equal thickness. Contrast with taproot.

filament—The stalk of an anther.

follicle—A dry, thin-walled fruit, usually elongated and slender, with numerous seeds, opening along one side.

frond—The leaf of a fern.

glaucous—Having a whitish or silvery-gray waxy sheen, typically on leaves and stems.

globose—Nearly spherical, usually describing a fruit.

herbaceous—Not woody; a general term for plants that may be annual or perennial, in contrast to trees or shrubs.

indusium—A flap or veil of tissue that covers a sorus in some ferns.

inferior—Referring to an ovary that is placed below the petals and stamens.

inflorescence—The general term for an entire cluster of flowers, including the blossoms and stems.

involucre—The set of bracts beneath the flower head in the sunflowers. To be distinguished from a calyx.

irregular—A flower that is not radially symmetrical (that is, which does not follow the general petals-around-the-center model).

keel—A lengthwise ridge in some leaves or flower parts, usually seen on the lower surface. In pea flowers, a keel is formed by the two lowermost fused petals.

krummholz—German for "crooked wood." These are trees growing at timberline that are shrublike even when many decades old, with twisted branches and no leading trunk.

lanceolate—Shaped like the blade of a spear: widest at the base, elongated, and tapered to the tip.

lemma—In grasses, the lower of a pair of bracts at the base of the flower.

ligule—In some grasses, an extension of thin tissue partly or wholly surrounding the stem at the point where the sheath of a leaf joins the blade. In the sunflowers, the petal-like blade of flowers (in disk flowers, a ligule is absent).

linear—Narrow and elongated, with straight or nearly straight sides.

misapplied—A valid name for a species, but one that has been used in error for another species.

nutlet—A small, dry, fruit in which the single seed is surrounded by a hard shell or covering.

Glossary

oblanceolate—A shape the opposite of lanceolate, with the widest part of the above the middle.

obovate—A shape the opposite of ovate, with the widest part of the above the middle.

opposite—Placed directly across from one another, usually describing a pair of leaves on a stem.

ovary—The ovule-producing part of the female organ of a flower. See pistil.

ovate—Egg-shaped, usually describing leaves that are widest below the middle.

ovule—An immature seed, prior to fertilization.

palmate—Like the palm of a hand, with the parts radiating from a common point; used to describe a leaf shape or a pattern of veins.

panicle—A branched inflorescence in which the lowest flower on a branch opens first. A branched raceme.

pappus—In the sunflowers, a crown of awns, bristles, or scales at the top of the seed.

pectinate—Comb-like.

perennial—A plant species that lives for two years or more, typically with a heavy rootstock that survives more than one season. The stems may or may not survive the winter.

perianth—A collective term for the calyx and corolla; often used when the parts are difficult to distinguish.

petal—An individual part of the corolla, usually flattened and colored or white.

phyllary—A single bract in the involucre of a flower head in the sunflowers.

pinnate—Describing a compound leaf, in which the individual leaflets are placed on opposite sides of the midrib of the leaf. "Twice-pinnate" is used to describe the situation where the individual leaflets are again pinnately divided.

pistil—The female organ of a flower, comprising an ovary, (usually) a style, and a stigma.

pome—A fruit like an apple, in which the seeds are at the core of a fleshy body.

raceme—An unbranched inflorescence in which the flowers are stalked.

ray—An individual flower stalk in an umbel.

ray flower—In the sunflowers, an individual flower that has a ligule, usually forming an outer circle in the head.

receptacle—An enlargement of the tip of a flower stalk to which the parts of the flower are attached. In the sunflowers, the broad disk to which the individual flowers are attached.

reflexed—Bent or turned downward or backward.

rootstock—An underground, sometimes thickened, stem by which a plant may spread in a creeping fashion. Contrast with a runner.

rosette—A cluster of leaves, often flattened, at the base of a plant, arising directly from the crown of the roots of the plant.

runner—An above-ground, horizontal stem that roots as it creeps and from which new plants arise. Contrast with rootstock.

scree—An accumulation of rocks, similar to a talus, but the fragments relatively small, and not always located at the foot of a cliff.

sepal—An individual part of the calyx, usually greenish and seldom colored.

serrate—Sharply and rather coarsely toothed.

sheath—Describing a thin tissue at the base of a leaf that encircles ("sheaths") below the point at which the leaf is attached to the stem.

simple—Unbranched; not compound; not lobed.

sorus—In ferns, a cluster of sporangia, found on the underside of the frond. Plural: sori.

spike—An unbranched inflorescence in which the flowers are not stalked.

sporangium—In ferns, a tiny structure in which the spores develop. Plural: sporangia.

spore—The reproductive unit in ferns, grapeferns, horsetails, quillworts, and clubmosses.

sporophyll—In the clubmosses and grapeferns, a specialized leaf at the base of which, or on which, the spores are borne.

stamen—The male organ of a flower, with a stalklike filament and a pollen-producing anther; usually several stamens are present in a flower.

stigma—The expanded tip of a pistil, which receives the pollen; often on a short stalk called a style.

stomate—Pore or opening on a leaf surface, for gas exchange. Plural: stomata.

style—The slender stalk that elevates the stigma above the ovary in the pistil of a flower.

subalpine—An elevational zone just below timberline but above the reach of more or less continuous tree or shrub cover.

succulent—Thick and fleshy stems or leaves.

superior—Referring to an ovary that is placed above the petals and stamens.

talus—An accumulation of rock fragments on a slope at the foot of a cliff.

taproot—A main root of a plant, one that grows straight down.

tendril—An extension of the midvein of a leaf, which twines and helps support a stem.

terminal—Occurring at the end of a stem or branch.

umbel—An inflorescence in which the multiple flower stalks radiate from a common point, often flat-topped.

unisexual—Describing a flower that has either fertile stamens or pistils but not both.

whorl—Arranged in a ring, as in a stem where three or more leaves join the stem at the same level.

Bibliography

Abrams, L. 1940–1960. *Flora of the Pacific states: Washington, Oregon, and California.* Stanford: Stanford University Press. 4 vols.

Allen, G. F. 1922. *Forests of Mount Rainier National Park.* Washington, DC: U.S. Dept. of the Interior. National Park Service. 32 p.

Alverson, E. R. 1989. *Cryptogramma cascadensis*, a new parsley-fern from western North America. *American Fern Journal* 79: 95–102.

Barcott, B. 1997. *The measure of a mountain: beauty and terror on Mount Rainier.* Seattle: Sasquatch Press. 278 p. [The chapter titled "Meadow stomping" is recommended.]

Bilderback, D.E. 1987. *Mount St. Helens, 1980: botanical consequences of the explosive eruptions.* Berkeley: University of California Press.

Boeri, D. J. Studies on the pollination of sub-alpine plants at Mount Rainier National Park. M.S. thesis, University of Washington, Seattle. 156 p.

Brayshaw, T. C. 1996. *Trees and shrubs of British Columbia.* Vancouver: University of British Columbia Press. 374 p.

Brockman, C. F. 1947. *Flora of Mount Rainier National Park.* Washington, DC: U.S. Dept. of the Interior, National Park Service. 170 p.

———. 1949. *Trees of Mount Rainier National Park.* Seattle: University of Washington Press. 49 p.

Brummitt, R.K. and C.E. Powell. 1992. *Authors of plant names: a list of scientific names of plants, with recommended standard forms of their names, including abbreviations.* London: Royal Botanic Gardens, Kew. 732 p.

Buckingham, N. M., et al. 1995. *Flora of the Olympic Peninsula.* Seattle: Northwest Interpretive Association. 199 p.

Catton, T. 1997. The campaign to establish Mount Rainier National Park. *Pacific Northwest Quarterly* 88: 70–81.

Ceska, A. 1999. Rubus armeniacus - a correct name for Himalayan blackberries. *Botanical Electronic News* [electronic newsletter, August 25, 1999] <http://www.ou.edu/cas/botany-micro/ben230.html>

Chase, A. 1964. *First book of grasses: the structure of grasses explained for beginners.* Washington, DC: Smithsonian Institution Press. 127 p.

Coleman, R.A. 1995. *The wild orchids of California.* Ithaca: Cornell University Press. 201 p.

Cooke, S. S., editor. 1997. *A field guide to the common plants of western Washington and northwestern Oregon.* Seattle: Seattle Audubon Society & Washington Native Plant Society. 417 p.

Cox, B. J. 1974. A biosystematic revision of *Lupinus lyallii. Rhodora* 76: 422–445.

Crandell, D. R. 1968. *The geologic story of Mount Rainier.* Bulletin 1292. Washington, DC: U.S. Dept. of the Interior, Geological Survey. 43 p.

Cushman, M. J. 1981. The influence of recurrent snow avalanches on vegetation patterns in the Washington Cascades. Ph.D. dissertation, University of Washington, Seattle. 175 p.

Del Moral, R., and D. M. Wood. 1988. The high elevation flora of Mount St. Helens, Washington. *Madroño* 35: 309–319.

Detling, L. E. 1968. Historical background of the flora of the Pacific Northwest. Eugene: University of Oregon Museum of Natural History. 57 p.

———. 1953. Relict islands of xeric flora west of the Cascade Mountains in Oregon. *Madroño* 12: 39–47.

Dunwiddie, P. W. 1983. Holocene forest dynamics on Mount Rainier, Washington. Ph.D. dissertation, University of Washington, Seattle. 129 p.

———. 1987. A 6000-year record of forest history on Mount Rainier, Washington. *Ecology* 68: 58–68.

Edwards, O. M. 1979 [?]. Vegetation disturbance by natural factors and visitor impact in the alpine zone of Mount Rainier National Park: Implications for management. Pages 101–106 in *Recreational Impact on Wildlands Conference Proceedings*. Seattle: U.S. Dept. of Agriculture, Forest Service.

———. 1980. The alpine vegetation of Mount Rainier National Park: Structure, constraints, and development. Ph.D. dissertation, University of Washington, Seattle. 280 p.

Flett, J. B. 1903. The fern flora of Washington. *Fern Bulletin* 9: 79–85.

———. 1922. *Features of the flora of Mount Rainier National Park.* Washington, DC: U.S. Dept. of the Interior, National Park Service. 49 p.

Flora of North America Editorial Committee. *Flora of North America*, vols. 1–3. Internet URL: <http://www.fna.org/Libraries/plib/WWW/online.html>.

Franklin, J. F., and N. A. Bishop. 1969. *Notes on the natural history of Mount Rainier National Park.* Longmire, WA: Mount Rainier Natural History Association. 24 p.

Franklin, J. F., and C. T. Dyrness. 1988. Natural vegetation of Oregon and Washington. Corvallis, OR: Oregon State University Press. 452 p.

Franklin, J. F., et al. 1988. The forest communities of Mount Rainier National Park. Scientific Monograph Series No. 19. Washington, DC: U.S. Dept. of the Interior, National Park Service. 194 p.

Frehner, H. K. 1957. Development of soil and vegetation on the Kautz Creek flood deposit in Mount Rainier National Park. M.F. thesis, University of Washington, Seattle. 83 p.

Fries, M. A., B. Spring, and I. Spring. *Wildflowers of Mount Rainier and the Cascades.* Seattle: Mount Rainier Natural History Association and the Mountaineers. 208 p.

Frye, T. C. 1934. Ferns of the Northwest. Portland, OR: Binfords & Mort. 177 p.

Fukuda, I., and H. Baker. 1970. *Achlys californica* (Berberidaceae) — A new species. *Taxon* 19: 341–344.

Ganders, A. J. F., and F. R. Ganders. 1983. *Wildflower genetics: a field guide for British Columbia and the Pacific Northwest.* Vancouver: Flight Press. 215 p.

Grater, R. K. 1949. *Grater's guide to Mount Rainier National Park.* Portland, OR: Binfords & Mort. 134 p.

Haber, E. 1983. Morphological variability and flavonol chemistry of the *Pyrola asarifolia* complex (Ericaceae) in North America. *Systematic Botany* 8: 277–298.

Hamann, M. J. 1972. Vegetation of alpine and subalpine meadows of Mount Rainier National Park, Washington. M.S. thesis, Washington State University, Pullman. 120 p.

Harris, S. L. 1988. *Fire mountains of the West.* Missoula, MT: Mountain Press. 379 p.

Hemstrom, M. A. 1982. Fire in the forests of Mount Rainier National Park. Pages 121–126 in *Ecological research in national parks of the Pacific Northwest*. Corvallis: Oregon State University Forest Research Laboratory.

Hemstrom, M. A., and J. F. Franklin. 1982. Fire and other disturbances of the forests in Mount Rainier National Park. *Quaternary Research* 18: 32–51.

Henderson, J. A. 1972. *Flowers of the parks: Mount Rainier National Park, North Cascades National Park.* Longmire, WA: Mount Rainier Natural History Association.

Bibliography

———. 1973. Composition, distribution and succession of subalpine meadows in Mount Rainier National Park, Washington. Ph.D. dissertation, Oregon State University, Corvallis. 150 p.

Henderson, J. A., and D. Peter. 1981. *Primary plant associations and habitat types of the White River Ranger District, Mount Baker–Snoqualmie National Forest.* Portland, OR: U.S. Dept. of Agriculture, Forest Service, Pacific Northwest Region. 60 p.

Henderson, J. A., et al. 1992. *Field guide to the forested plant associations of the Mount Baker–Snoqualmie National Forest.* Seattle: U.S. Dept. of Agriculture, Forest Service. 196 p.

Hickman, J. C., editor. 1993. *The Jepson manual: Higher plants of California.* Berkeley: University of California Press. 1400 p.

Hitchcock, C. L., and A. Cronquist. 1976. *Flora of the Pacific Northwest.* Seattle: University of Washington Press. 730 p.

Hitchcock, C. L., et al. 1959–1967. *Vascular plants of the Pacific Northwest.* Seattle: University of Washington. 5 vols.

Hultén, E. 1968. *Flora of Alaska and neighboring territories: A manual of the vascular plants.* Stanford: Stanford University Press. 1008 p.

Jolley, R. 1988. *Wildflowers of the Columbia Gorge.* Portland: Oregon Historical Society Press. 331 p.

Jones, G.N. 1936. A botanical survey of the Olympic Peninsula, Washington. Seattle: University of Washington. 286 p.

Jones, G. N. 1938. *The flowering plants and ferns of Mount Rainier.* Seattle: University of Washington. 192 p.

Kartesz, J. T. 1994. *A synonymized checklist of the vascular flora of the United States, Canada, and Greenland.* 2nd ed. Portland, OR: Timber Press. 2 vols.

———. 1998. A synonymized checklist of the vascular flora of the United States, Canada, and Greenland. Internet URL: <http://www.csdl.tamu.edu/FLORA/newgate/cr1famzz.htm>.

Kartesz, J. T., and K. N. Gandhi. 1993. *Osmorhiza berteroi* (Apiaceae): The correct name for mountain sweet cicely. *Brittonia* 45: 181–182.

Kirk, R. 1973. *Exploring Mount Rainier.* Seattle: University of Washington Press. 104 p.

Kron, K. A., and W. S. Judd. 1990. Phylogenetic relationships within the Rhodorae (Ericaceae), with specific comments on the placement of *Ledum. Systematic Botany* 15: 57–68.

Kruckeberg, A. R. 1956. Notes on the *Phacelia magellanica* complex in the Pacific Northwest. *Madroño* 13: 209–221.

Lane, M. A., Z. Wang, and C. H. Haufler. 1993. *Rhododendron albiflorum* Hook. (Ericaceae): One taxon or two? *Rhodora* 95: 11-20.

Lellinger, D. B. 1985. *A field manual of ferns and fern-allies of the United States and Canada.* Washington, DC: Smithsonian Institution Press. 389 p.

Lescinsky, D. T., and T. W. Sisson. 1998. Ridge-forming, ice-bounded lava flows at Mount Rainier, Washington. *Geology* 26: 351–354. [Summarized in "When lava and ice clashed on Mount Rainier," *Science News* 153: 245.]

Little, R. L., D. L. Peterson, and L. L. Conquest. 1994. Regeneration of subalpine fir (*Abies lasiocarpa*) following fire: Effects of climate and other factors. *Canadian Journal of Forest Research* 24: 934–944.

Martinson, A. D. 1994. *Wilderness above the Sound: The story of Mount Rainier National Park.* Niwot, CO: Roberts Rinehart. 75 p.

McCune, B. and L. Geiser. 1997. *Macrolichens of the Pacific Northwest.* Corvallis, OR: Oregon State University Press. 386 p.

Miller, J. M. 1993. Nomenclatural changes and new taxa in *Claytonia* (Portulacaceae) in western North America. *Novon* 3: 268–273.

Moir, W. H. 1989. *Forests of Mount Rainier National Park: A natural history.* Seattle: Pacific Northwest National Parks and Forests Association. 111 p.

Moss, E. H. 1983. *Flora of Alberta.* 2nd edition, revised by J. G. Packer. Ontario: University of Toronto Press. 700 p.

Mount Rainier National Park. 1923–1927. *Nature Notes.* [newsletter.]

Mullineaux, D. R. 1974. *Pumice and other pyroclastic deposits in Mount Rainier National Park, Washington.* Bulletin 1326. Washington, DC: U.S. Dept. of the Interior, Geological Survey. 83 p.

Munz, P. A., and D. D. Keck. *A California flora with supplement.* Berkeley: University of California Press. 1905 p.

Ogilvie, R.T. 1991. *Vaccinium* in B.C. *Botanical Electronic News.* [electronic newsletter, Sept. 5, 1991] <gopher://gopher.freenet.victoria.bc:70/00/environment/ben/bengoph01>

Orr, E. L., and W. N. Orr. 1996. *Geology of the Pacific Northwest.* McGraw-Hill, New York. 408 p.

Piper, C. V. 1888–1926. Papers. Washington State University. Manuscripts, Archives, and Special Collections.

———. 1901. The Flora of Mount Rainier. *Mazama* 2: 93–117.

———. 1905. The Flora of Mount Rainier [addendum]. *Mazama* 2: 270–272.

———. 1906. *Flora of the State of Washington.* Contributions from the United States National Herbarium, vol. 11. Washington, DC: Government Printing Office. 637 p.

———. 1916. The Flora of Mount Rainier. Pages 254–286 in E. S. Meany, *Mount Rainier: A record of exploration.* New York: Macmillan.

Pojar, J., and A. MacKinnon. 1994. *Plants of the Pacific Northwest coast: Washington, Oregon, British Columbia and Alaska.* Vancouver: Lone Pine. 527 p.

Potash, L. L. 1991. *Sensitive plants and noxious weeds of the Mount Baker–Snoqualmie National Forest.* Seattle: U.S. Dept. of Agriculture, Forest Service. 112 p.

Reveal, J. L. 1998. USDA–APHIS concordance for family names. Internet URL: <http://www.inform.umd.edu/PBIO/usda/usdaindex.html>.

Rochefort, R. M., and S. T. Gibbons. 1992. Mending the meadow: High-altitude meadow restoration in Mount Rainier National Park. *Restoration and Management Notes* 10: 120–126.

———. 1993. Impact monitoring and restoration in Mount Rainier NP. *Park Science* 13: 29–30.

Rochefort, R. M., and D. L. Peterson. 1992. Effects of climate and other environmental factors on tree establishment in subalpine meadows of Mount Rainier National Park. *Bulletin of the Ecological Society of America* 73(2), supplement.

———. 1993. Genetic diversity and protection of alpine heather communities in Mount Rainier National Park. *Park Science* 13: 28–29.

———. 1996. Temporal and spatial distribution of trees in subalpine meadows of Mount Rainier National Park, Washington. *Arctic and Alpine Research* 28: 52–59.

Rochefort, R. M., R. L. Little, and A. Woodward. 1994. Changes in sub-alpine tree distribution in western North America: A review of climatic and other causal factors. *The Holocene* 4: 89–100.

Bibliography

St. John, H., and F. A. Warren. 1937. The plants of Mount Rainier National Park, Washington. *American Midland Naturalist* 18: 952–985.

Schmoe, F. C. 1925. Our greatest mountain: A handbook for Mount Rainier National Park. New York: G. P. Putnam's Sons. 366 p.

———. 1959. *A year in Paradise.* New York: Harper & Brothers. 235 p.

Schullery, P. 1987. Island in the sky: Pioneering accounts of Mount Rainier, 1833–1894. Seattle: Mountaineers. 197 p.

Sigafoos, R. S., and E. L. Hendricks. 1961. *Botanical evidence of the modern history of Nisqually Glacier, Washington.* Professional Paper 387-A. Washington, DC: U.S. Dept. of the Interior, Geological Survey. 20 p.

Standley, L. A. 1981. The systematics of *Carex* section *Carex* (Cyperaceae) in the Pacific Northwest. Ph.D. dissertation, University of Washington, Seattle. 247 p.

———. 1985. *Systematics of the* Acutae *group of Carex (Cyperaceae) in the Pacific Northwest.* Ann Arbor, MI: American Society of Plant Taxonomists. 106 p.

Starkey, E. E., J. F. Franklin, and J. W. Matthews. 1982. Fire in the forests of Mount Rainier National Park. Pages 121–126 in *Ecological research in national parks of the Pacific Northwest.* Corvallis: Oregon State University Forest Research Laboratory.

Strickler, D. 1997. *Northwest Penstemons: 80 species of Penstemon native to the Pacific Northwest.* Columbia Falls, MT: Flower Press. 191 p.

Taylor, R. J. 1971. Biosystematics of the genus *Tiarella* in the Washington Cascades. *Northwest Science* 45: 27–37.

Taylor, T. M. C. 1970. *Pacific Northwest ferns and their allies.* Toronto: University of Toronto Press. 247 p.

Tisch, E. 1991. Re: *Vaccinium. Botanical Electronic News.* [electronic newsletter, Dec. 31, 1991] <gopher://gopher.freenet.victoria.bc:70/00/environment/ben/bengoph01>

Turill, W. B., and J. R. Sealy. 1980. Studies in the genus *Fritillaria* (Liliaceae). Hooker's Icones Plantarum, XXXIX (I and II): 238–242.

U.S. Dept. of Agriculture, National Resources Conservation Service. 1988. PLANTS National Database. Internet URL: <http://plants.usda.gov/plants/>.

U.S. Geodynamics Committee. 1994. *Mount Rainier, active Cascade volcano: Research strategies for mitigating risk from a high, snow-clad volcano in a populous region.* Washington, DC: National Academy Press. 114 p.

Van Horn, J. C. 1979 [?]. Soil and vegetation restoration at the Sunrise developed area, Mount Rainier National Park. Pages 286–291 in *Recreational Impact on Wildlands Conference Proceedings.* Seattle: U.S. Dept. of Agriculture, Forest Service.

Van Pelt, R. 1996. *Champion trees of Washington State.* Seattle: University of Washington Press. 120 p.

Vitt. D.H., J.E. Marsh, and R.B. Bovey. 1993. *Mosses, lichens and ferns of Northwest North America.* 2nd ed. Edmonton: Lone Pine. 296 p.

Washington Natural Heritage Program. 1997. *Endangered, threatened, and sensitive vascular plants of Washington, with working lists of rare non-vascular species.* Olympia: Washington State Dept. of Natural Resources. [62] p.

Weather America: the latest detailed climatological data for over 4,000 places, with rankings. A.N. Garwood, editor. Milpitas, CA: Toucan Valley Publications, 1996. 1412 p.

Zwinger, A. H., and B. E. Willard. 1972. *Land above the trees: A guide to American alpine tundra.* New York: Harper & Row. 487 p.

Index

The index includes all accepted names, synonyms, names misapplied, and common names used in this book. Accepted names for families, genera, and species are shown in roman type; synonyms are in italics, followed by the accepted name within parentheses. Synonyms are taken from earlier writings on the flora of the Park, going back to Charles V. Piper in 1901. Names used in the standard flora for the region, *Flora of the Pacific Northwest* (Hitchcock and Cronquist, 1976), are also included as synonyms as appropriate. Occasionally, valid names have been incorrectly applied to plants in the Park; in these cases, a "misapplied" note is given.

The phrase "in Jepson 1993" indicates the name of a species as used in *The Jepson Manual: Higher Plants of California* (Hickman, 1993) and "in Kartesz 1998" indicates the name as used in the 1998 online version of *A Synonymized Checklist of the Vascular Flora of the United States, Canada, and Greenland.*

Index

487

Index

489

Index

491

Index

Index

495

Index

Index

Index

Index

Index